Lecture Notes
in Control and Information Sciences 271

Editors: M. Thoma · M. Morari

Springer-Verlag Berlin Heidelberg GmbH

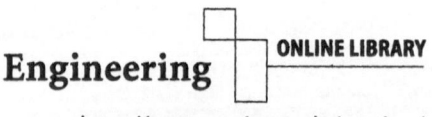

http://www.springer.de/engine/

Daniela Rus, Sanjiv Singh (Eds)

Experimental Robotics VII

Springer

Editors

Professor Daniela Rus
Dartmouth College
Department of Computer Science
Hanover, NH 03755
USA

Dr. Sanjiv Singh
Carnegie Mellon University
Robotics Institute
Pittsburgh, PA 15213
USA

Cataloging-in-Publication Data applied for
Die Deutsche Bibliothek – CIP-Einheitsaufnahme
Experimental robotics VII

(Lecture Notes in control and information sciences; 271)
ISBN 978-3-540-42104-7   ISBN 978-3-540-45118-1 (eBook)
DOI 10.1007/978-3-540-45118-1

http://www.springer.de

© Springer-Verlag Berlin Heidelberg 2001
Originally published by Springer-Verlag Berlin Heidelberg New York in 2001

Typesetting: Digital data supplied by author. Data-conversion by PTP-Berlin, Stefan Sossna
Cover-Design: design & production GmbH, Heidelberg
Printed on acid-free paper   SPIN 10796506   62/3020Rw - 5 4 3 2 1 0

Preface

Experimental Robotics is at the core of validating robotics research for both its systems science and theoretical foundations. Because robotics experiments are carried out on physical, sometimes complex, machines whose controllers are subject to uncertainty, devising meaningful experiments and collecting statistically significant results pose important and unique challenges in robotics. Robotics experiments serve as a unifying theme for robotics system science and algorithmic foundations. These observations have led to the creation of the International Symposia on Experimental Robotics in 1989. The meetings are bi-annual and focus on research where theories and principles have been validated by experiments.

The Seventh International Symposium on Experimental Robotics (ISER 2000) brought together a group of about 80 researchers to discuss recent results and relevant trends in experimental robotics. Held in Waikiki, Hawaii on December 11–13, the symposium was chaired by Prof. Daniela Rus (Dartmouth College) and Prof. Sanjiv Singh (Carnegie Mellon University). Prof. Song Choi from the University of Hawaii chaired the local arrangements committee. The meeting consisted of three invited talks, one invited panel, and fifty-seven contributed presentations in a single track. Each paper was refereed by the program chairs plus at least two members of the program committee. The program committee consisted of:Vincent Hayward (Canada), Oussama Khatib (USA), Herman Bruyninckx (Belgium), Alicia Casals (Spain), Raja Chatila (France), Peter Corke (Australia), Eve Coste-Maniere (France), John Craig (USA), Paolo Dario (Italy), Gerd Hirzinger (Germany), Jean-Pierre Merlet (France), Yoshihiko Nakamura (Japan), Daniela Rus (USA), Kenneth Salisbury (USA), Sanjiv Singh (USA), Tsuneo Yoshikawa (Japan), and Alex Zelinsky (Australia).

Topics reported at ISER 2000 included humanoids and human-robot interactions, perception systems, assembly and manipulation, medical and field applications, locomotion, multi-robot systems, modeling and motion planning, control, and navigation and localization. Several research projects presented at ISER 2000 are clearly breakthroughs in the field and will likely have a big impact in the future. Representatives include the three talks given by our invited speakers. Russell Taylor (Johns Hopkins University) presented an impressive medical robot system and argued a great case for how the impact of Computer-Integrated Surgery on medicine in the next 20 years will be as great as that of Computer-Integrated Manufacturing on industrial production over the past 20 years. Hirochika Inoue (University of Tokyo) described an amazing suite of humanoid robots and their tasks and presented a research agenda for a human-centered robotized society. Ralf Koeppe (DLR, Germany) described their progress with light-weight robotic manipulators and advocated an exciting range of applications to personal and service robotics.

We are very grateful to Dartmouth College and to Carnegie Mellon University for their generous financial support of ISER 2000. We would like to thank all the participants and their contributions, which made our meeting exciting and inspiring. We also thank David Bellows, Alan Guisewite, Monica Hopes, Catherine LaTouche, Dot Marsh and Alison Sartonov for their help with coordinating the meeting and producing the proceedings.

March 2001

Daniela Rus
Sanjiv Singh

Contents

1. Humanoids and Human-Robot Interaction

2. Perception

3. Assembly and Manipulation

4. Medical, Space, and Field Applications

5. Locomotion

6. Multi-robot Systems

7. Modeling and Motion Planning

8. Control

9. Navigation and Localization

Author Index

Haptically Augmented Teleoperation

Nicolas Turro

Inria/Stanford University Robotics Group

Nicolas.Turro@sophia.inria.fr

Oussama Khatib

Stanford University Robotics Group

ok@robotics.stanford.edu

Abstract: This article presents various experiments conducted at the Stanford Robotics Group to enhance teleoperation thru the use of movement constraints. Several types of constraints have been studied : Safety constraints, limiting the movements of the slave robot into safe areas and robot and guidance constraints, helping the operator to move according to a predefined geometrical path (along a line, on a plane). Those constraints have been implemented using force feedback on the master device and potential fields are used to compute the amplitude of the force feedback according to the constraints.

1. Introduction

Stanford Robotics Group has a long background of designing robotic assistants for the execution difficult tasks. As described in [1] and [2], the two platforms ROMEO and JULIET demonstrated how some basic behaviours can be performed by robots. The next logical step in this research is to include teleoperation capabilities to those platform, in order for a human to guide the robots remotely and use the various elementary behaviours when needed. Teleoperation and behaviour could also be combined in order, for example, to remotely wipe out a blackboard : the robotic behaviour automatically enforce a contact with the brush, with a predetermined force normal to the surface, while the operator guide the brush in the directions tangential to the surface.

The use of a haptic device as master for the teleoperation helps the operator to perform precise and complex tasks : the force feedback is computed using data from the slave robot (from force sensors, from ultrasound distance sensors), or from geometrical constraints computed on the master site. Using this feedback, the operator may guide the slave robot safely in narrow corridors without collisions, or write on a blackboard applying the rigth pressure with the pencil, or remotely interact with an human, or make perfect line movements.

In this paper, we present both the control framework we are using to achieve this haptically augmented teleoperation, and the first experimental results of our implementation.

D. Rus and S. Singh (Eds.): Experimental Robotics VII, LNCIS 271, pp. 1–10, 2001.

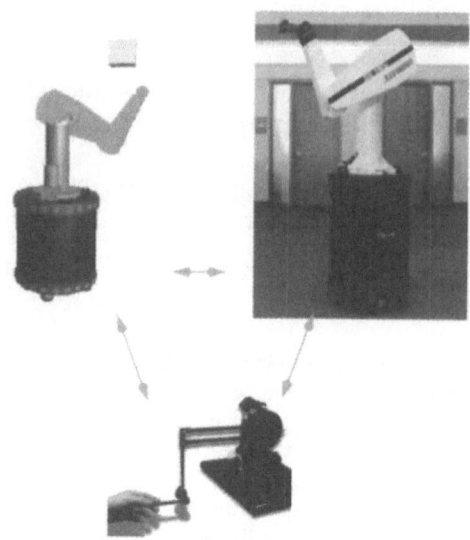

Figure 1. The teleoperation environment

2. Principles

As displayed in figure 1, our environment is built around three main components: a master robot, a slave robot and a virtual 3D representation of the robot and it's environment. The whole control scheme we evaluated is presented in figure (2). It comprises two servo loops to control the master device and the slave robot. The two servo loops communicate with each other by sending their respective positions and velocities.

2.1. The basic teleoperation control

The core control scheme of the teleoperation reproduces movements of the master device on the slave robot, and it also provides the operator with some feedback of the interaction between the robot and its environment. The master and the slave robots have dramatically different geometry and dynamics properties. Thus, the master device controls the position of the end effector of the slave using *operational space control* : the master device absolute position is used to control the position of the end effector of the slave robot. We compute the force F_s^* to apply to the end effector using a PD controller whose goal position is the master position x_m.

$$F_s^* = -K_p(x_s - Sx_m + x_0) - K_v(\dot{x}_s - \dot{x}_m). \qquad (1)$$

Then, the torque τ_s to send to the motors is computed using the dynamic (Mass Matrix Λ and gravity g) of the slave robot. Computing the dynamics of the robot is critical in our approach : A good position and velocity servoing could be achieved without it but would involve high gains K_p and K_v, making the robot stiff. Using the dynamics allows us to achieve the same performances, but with lower gains, making the robot more compliant and sensitive to external

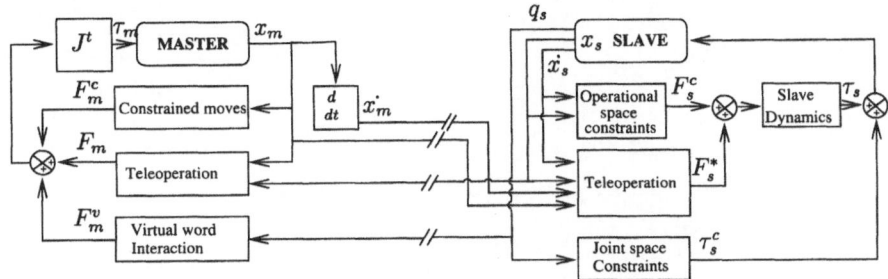

Figure 2. Complete control scheme

forces.

$$\tau_s = J^t \Lambda F_s^* + g. \tag{2}$$

2.2. Teleoperation feedback - Master control

On the master device, we provide a force feedback F_m computed using the offset between the master arm position and the position of the slave robot end effector. This way, when external forces (from environment) are applied to the slave robot, those forces will be felt on the master device because the error on the slave tracking will increase. This method doesn't require any force sensor. However, there is constantly an error between the master and the slave position, mostly due to friction on the slave and also to its inertia. Since we do not want to feel a constant drag on the master in free-space movements, we use a cubic function for the computation of F_m:

$$F_m = -K(x_m - \frac{1}{S}(x_s + x_0))^3$$

the gain K is chosen small to exploit the cubic function near zero, on its flat part. So, small tracking errors will be minimised whereas real contact forces will be amplified. We assume that the system is naturally damped (because the manipulator is handling the master device). Furthermore, since our master device is very light, and well balanced, its mass can generally be neglected and its dynamic does not have to be computed. Thus, the torques to send to the master device are computed directly from F_m using the master's Jacobian matrix.

2.3. Constraints

Our main contribution was to evaluate different ways to constrain our teleoperation system. Three alternative have been studied :

2.3.1. Slave-side constraints

Some constraints are critical for the safety of the slave robot, so it is preferable to plug them inside the *slave* controller, in case of a network or master device failure. For example, the master and the slave workspace are different most

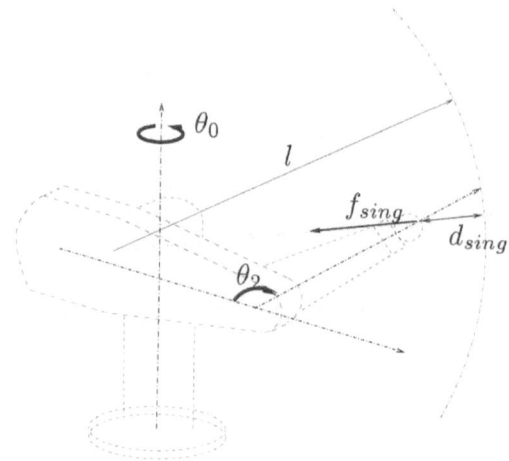

Figure 3. Singularity repulsive force

of the time: the joints limits and singularities are not the same. We propose
to enforce those limitations using *repulsion fields*. Those fields will prevent
the slave robot to move in given directions. Then, the basic control scheme
described above will haptically render those slave constraints on the master
device.

For example, our PUMA slave robot has a singularity on its second joint
(elbow singularity) when the angle θ_2 (figure 3) reach the value π, or the end
effector reaches a sphere whose radius is the length l of the arm fully extended.

For stability and safety reasons, we want to avoid this configuration, so we
add to the slave control force F^* in equation (1) an *operational space repulsion
force* f_{sing} whose direction is indicated on figure 3. The amplitude of f_{sing} is
proportional to the distance $\frac{1}{d_{sing}^3}$ between the end effector and a sphere whose
radius is the length l of the maximal extension of the robot (corresponding to
$\theta_2 = \pi$).

Since the amplitude of those repulsing forces grows faster than the PD
control force and torque, the slave robot will not reach the undesired configurations. Meanwhile, the operator will 'feel' fast-growing force feedback if he
tries to move the slave in those configurations. Thus, imposing constraints
on the slave robot, in its control scheme, both in operational and joint space,
actually constraints the operator's motions and makes them safer.

2.3.2. Virtual constraints

In this section, we will explain how potential fields can be used on the master's
control, in order to help him to carry out some complex tasks, like moving on
a perfect line, or help him stay out of some predefined zones.

Master servo loop constraints   We propose to help the operator move the
slave end effector on a predefined surface or curve by the following scheme.
We project the operators cartesian position on the desired trajectory. We call

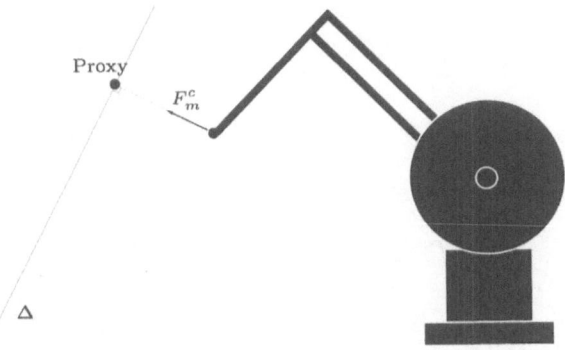

Figure 4. Attractive potential along a line

this point a *Proxy*. The curve, or the control surface to follow is surrounded by an *attractive potential field* whose amplitude increases with the distance between the master end effector and the proxy. Then we apply the corresponding attractive force F_m^c on the haptic device's end effector. By choosing the appropriate gains, the operator will easily move on the unconstrained directions, but will have to fight high torques on its master device to go away from it. Subsequently, the slave robot, following the master device will move according to the predefined constraints. Moreover, to further enforce the constraint, we can use the position of proxy to control the slave, instead of the real position of the master. Figure 4 displays the proxy and the corresponding force, when the constrained motion is along a line Δ.

This location of the constraints gives the better haptic feedback, since it runs at the speed of the master controller. However, despite ongoing work complex virtual interactions are very hard to computed in haptical real-time (in the thousand Hz range).

Master side virtual environment   Constraining the movements of the end effector on a predefined curve or surface is simple enough to be incorporated to the servo loop, and running at the same frequency as the control F_m. However, we would like to put constraints on the whole robot movements (not only its end effector) and interact with complex virtual scenes (represented by thousands of triangles, for example) leading to much more complex computations.

To fulfill this requirement, we propose to integrate a model of the real slave robot inside a virtual environment (figure 6). In this 3D environment, the robot will follow the moves of the real one, but will also interact with models of real objects as well as purely virtual obstacles. We compute this interaction asynchronously with respect to the master's control, usually at a much slower rate.

To model the interaction of the robot in its virtual environment, we define this environment with a set of n convex objects O_i (figure 5). Each object O_i is surrounded with a predefined repulsive potential field whose amplitude and range of action can be parameterized. Then, we compute the resulting forces of this potential field on each rigid moving part B_j of our robot. For each body

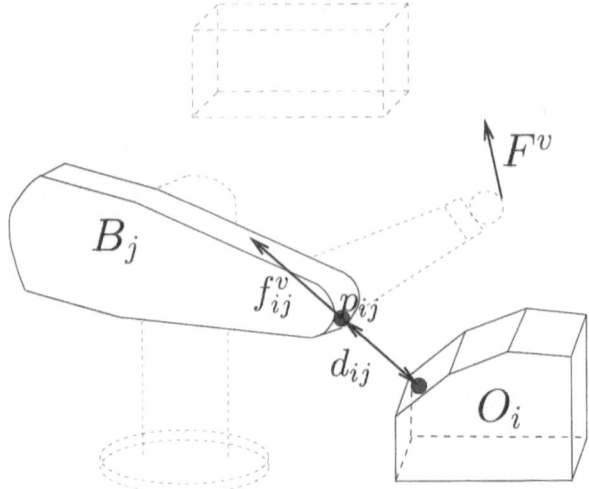

Figure 5. Interaction with the virtual environment: each Obstacle O_i produces a repulsion force f_{ij}^v on each part B_j of the robot, resulting in a force F_m^v at the end effector

B_j, we compute the shortest distance d_{ij} between the body and any obstacle. This distance computation will also provide the point of application p_{ij} and the direction of the partial virtual interaction force f_{ij}^{vp} between the part B_j of the robot and the object O_i. Once this is done for all couples of objects and robot bodies, the final virtual interaction force applied by the environment will be :

$$F_m^v = J^{t^{-1}} \sum_{i=1}^{n} \sum_{j=1}^{m} J_{p_{ij}}^t f_{ij}^{vp} \qquad (3)$$

where $J_{p_{ij}}$ is the intermediate Jacobian of the slave robot at the point p_{ij}. Once computed, this force F_m^v will be sent to the master servo loop, and be added to F_m. Due to the slow update frequency of this forces, the amplitude of f_{ij}^{vp} should not vary too fast in order to preserve the haptic feedback stability and is likely to need experimental tuning.

3. Experimental Setup and results

We applied this framework to two configuration : first a fixed puma 560, and then a mobile manipulator.

3.1. Hardware architecture and implementation

The master device is a Phantom with 6 degrees of freedom. On the used model, only the first three joints of this phantom can be read and controlled. Thus our experiments will deal only with positions of the end effector, not its orientation.

We use a linux PC quadri-processor Pentium Pro 200 MHz to control the Phantom, run the display and compute the virtual constraints on the master device. The slave controller runs on a PC Pentium II 333 MHz, using the QNX realtime OS. Both PCs are linked together thru a switch, using a dedicated

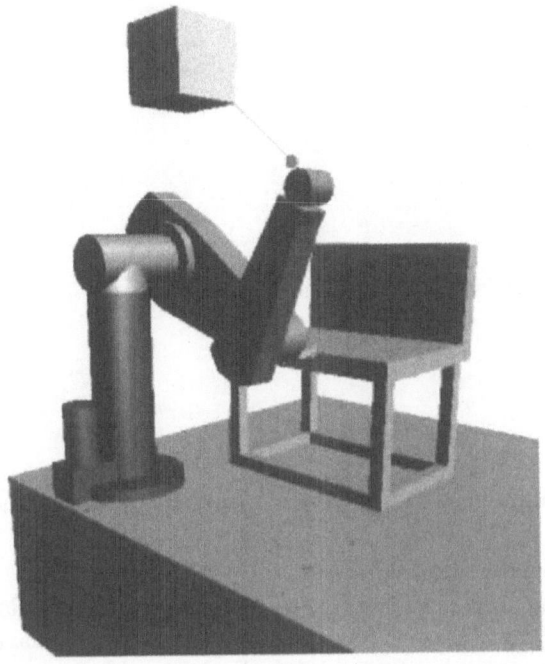

Figure 6. 3D graphical display of the scene, including the robot, real objects (the bench) and virtual ones (the floating cube).

100Mb/s ethernet line. On the Master PC, we use three separate processes to perform the different tasks, thus exploiting the physical processors. The three process communicate using UNIX UDP sockets. As a consequence, the distance computations described in section 2.3.2 is running asynchronously with the control, typically, much slower.

We implemented this distance computation using the Proximity Query Package (PQP, [3]), freely provided by the deparment of Computer Science of the university of North Carolina.

The graphical display of the scene is implemented using the client/server architecture described in [4], using *Mesa* implementation of openGL on a 3DFX graphical adapter (figure 6).

3.2. Performance

In the most complex case, moving on a line, with all virtual obstacles enabled, the master controller runs at 5000 Hz, which is appropriate to achieve a good feedback on the master device.

A servo rate of 600 Hz was proven sufficient to run smoothly the slave PUMA 560.

The interaction with the virtual environment was running at 200 Hz, beeing limited by the CPU power. A faster computation would greatly improve the haptic feedback.

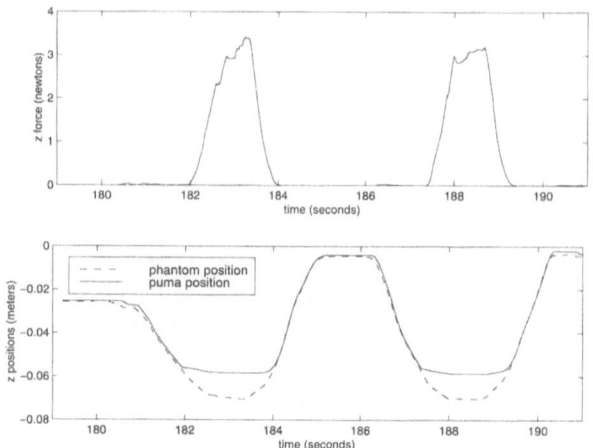

Figure 7. Force Feedback during a real contact on a plane $z = -0.06m$

3.3. Fixed Puma experiments

In this case, the slave and the master comunicate thru a dedicated 100Mb/s ethernet cable. The Master PC sends its position to the slave PC at the same rate as its servo loop, always providing the freshest value to the slave. In order to achieve optimal performance, this communication uses INET UDP sockets with fixed length 200 bytes packets. This size being smaller than the ethernet MTU (which is 1500 bytes), each position update is completely sent on the network in one shot, avoiding data fragmentation. The slave controller runs at a slower rate than the master controller, and sends its positions at the same rate as its control loop, which is slower than the masters one. In order to avoid accumulation of data coming from the master, we set the buffer size in the IP stack to the length of one of the packet we receive. Using this technique, each new update from the master erases any older data present in the socket buffer.

The following plots display the feedback (amplitude in the vertical direction) felt by the operator on the master device, during different interactions.

3.3.1. Real Contact

Figure 7 displays the force feedback on the master robot when the slave end effector makes two consecutive contacts with an horizontal plane (z= -0.06 m). When the robot moves in free space, the feedback force is almost zero. When the robot hits the plane, the feedback on the master device rapidly increases as the error between the slave and the master augment. Contact with any part of the robot (not necessarily its end effector) would give a similar feedback to the operator, in the direction of the applied force.

3.3.2. Virtual Obstacles: Repultion fields

Contact with a virtual obstacle results in a feedback profile displayed in figure 8. The best results for computing f_{ij}^{vp} in equation (3) were achieved using a cubic function of the distance to the obstacle. As in the previous section, the

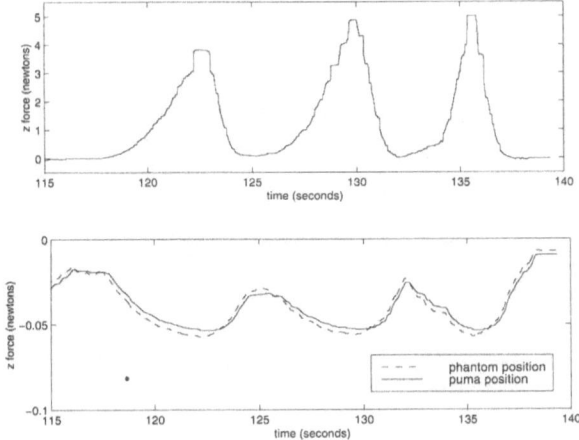

Figure 8. Force feedback during interaction with a virtual obstacle

obstacle is a z=-0.06m plane. The effects of the repulsion field are effectively felt at 1cm from the obstacle (about 2 Newtons), but the feedback profile is much less steep than the one associated with a real contact. Because of the relatively slow computation of the interaction with the virtual world (200 Hz), it was necessary to use low gains for the computations of f_{ij}^{vp} (eq. (3)) to ensure the master control stability. Small steps that can be observed on the plots are also due to this slow rate of computation.

3.3.3. Movement along a line: Attractive fields

Figure 9 display the force feedback during a constrained motion along an horizontal line (z = 0.018 m). Here, the feedback is much more stronger than in the two other exemples: when the user is moving along the constraint line (between start and $t = 48$ and $t = 50$ to the end), the feedback force is almost zero, but when the operator tries to go away from the line(between $t = 48$ and $t = 50$), the maximum saturated force of five newtons is quickly reached (for an error of 5 milimeters). This clearly shows the advantages of including the constraint forces computation inside the master servo loop whenever the computation time is compatible with 'haptic' real-time.

3.4. The mobile manipulator

The mobile manipulator consists in a puma arm mounted on an holonomic nomadics XR4000 base. The master and the slave are communicating using a RANGELAN2 radio ethernet network, which is the main difference from the fixed experimental platform.

The bandwidth of the radio link, which is supposed to be 1.2 Mbits per seconds was not sufficient to get more than a few tens of udpdates per seconds. This rate was experimentally proven sufficient to send goal position to the slave robot, but not to provide a realistic force feedback on the master device, as in the fixed plaform case. In consequence, information coming from the slave robot have been used only to update the 3D virtual representation of the

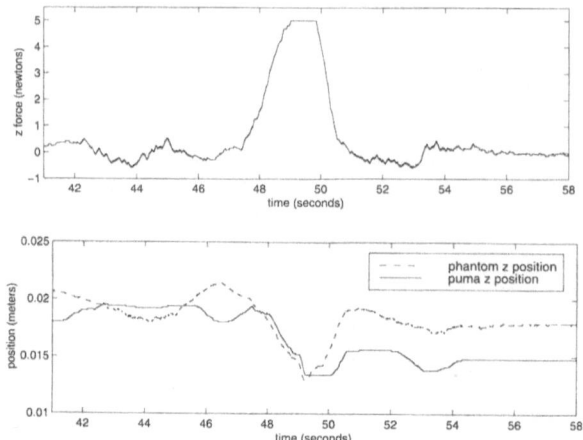

Figure 9. Force feedback during a constrained line movement: trying to go away from the constraint triggers a strong feedback

robot, and the force feedback on the master could only be computed using the local constraints in the master control loop (movements of the master device constrained along predefined geometrical primitives).

4. Conclusion

This framework was extensively experimented, providing a demonstration stable enough to be run tens of times for our visitors.

The experimental results in the fixed puma configuration are promising, and we plan to combine soon teleoperation and compliant motion with multi-contacts on the environment.

In the case of the mobile manipulator, our premiminary experiments show that our control framework is not adequate for low bandwidth. It could be complemented, for example, with a wave transmission scheme ([5]).

References

[1] Khatib O, 1999 Mobile Manipulation: The Robotic Assistant. *Journal of Robotics and Autonomous Systems*, vol. 26, 1999, pp. 175-183

[2] Khatib O, Yokoi K, Brock O, Chang K, Casal A Robots in Human Environments: Basic Autonomous Capabilities. *International Journal of Robotics Research*, vol. 18, no. 7, 1999, pp. 684-696

[3] Gottschalk S, Lin M, Manocha D 1996 OBB-Tree: A Hierarchical Structure for Rapid Interference Detection. *Proceedings of SIGGRAPH*, Annual Conference Series, pp 171-180, http://www.cs.unc.edu/ geom/SSV/

[4] Ruspini D C, Kolarov K, Khatib O, 1997 The Haptic Display of Complex Graphical Environments. *SIGGRAPH*, pp. 345-352

[5] Niemeyer G, Slotine J-J, 1998 Towards Force-Reflecting Teleoperation over the Internet. *Proc. of the 1998 IEEE Int. Conf. on Robotics and Automation*, pp. 1909-1915

Bilateral Teleoperation: Towards Fine Manipulation with Large Time Delay

Yasuyoshi Yokokohji, Takashi Imaida*, Yukihiro Iida
Department of Mechanical Engineering
Graduate School of Engineering
Kyoto University
Kyoto 606-8501, Japan
{yokokoji|iida24}@mech.kyoto-u.ac.jp
* currently with MITSUBISHI HEAVY INDUSTRIES, LTD.

Toshitsugu Doi**, Mitsushige Oda
National Space Development Agency of Japan
Tsukuba Space Center
2-1-1, Sengen, Tsukuba-shi
Ibaraki 305-8505, Japan
oda.mitsushige@nasda.go.jp
** currently with TOSHIBA CORPORATION

Tsuneo Yoshikawa
Department of Mechanical Engineering
Graduate School of Engineering
Kyoto University
Kyoto 606-8501, Japan
yoshi@mech.kyoto-u.ac.jp

Abstract: To conduct tasks that need high dexterity by teleoperation, a unified hand/arm master-slave system was developed. To measure the dexterity of individual teleoperation systems all over the world in equal condition, we propose toy block assembling (LEGO^{TM}) as the benchmark test for teleoperation systems. Meanwhile, a bilateral teleoperation experiment with ETS-VII (Engineering Test Satellite No.7) was conducted on November 22, 1999. Round-trip time for communication between the NASDA ground station and ETS-VII was approximately six seconds. These two results are the first step to improve the dexterity of teleoperation systems and to investigate the relationship between time delay and dexterity.

1. Introduction

In spite of great deal of research/development efforts in many years, application range of teleoperation has been limited mainly due to two factors: dexterity of teleoperators and communication time delay. Even bilateral teleoperation, which provides realistic kinesthetic coupling with a remote environment, cannot perform all tasks that human can do, since the dexterity of teleoperators is poorer than human dexterity as shown in Fig.1(a). It is one of our goals to improve the dexterity of teleoperator as close to the human dexterity as possible. To achieve this goal, we should be able to measure the dexterity of

D. Rus and S. Singh (Eds.): Experimental Robotics VII, LNCIS 271, pp. 11–20, 2001.

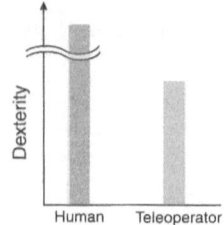   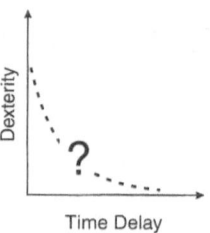

(a) Human vs. Teleoperator   (b) Dexterity degradation due to time delay

Figure 1. Dexterity and time delay.

teleoperation systems in quantitative manner. However, evaluating dexterity among individual teleoperation systems all over the world in equal conditions is a difficult problem. Therefore, we should establish a benchmark test to evaluate the dexterity of teleoperation systems.

Long distance between the operation site and the remote environment induces communication time delays, which make bilateral control difficult and sometimes unstable. As the time delay becomes longer, the dexterity degrades further as shown in Fig.1(b). It is not clear, however, how the dexterity actually degrades as the time delay becomes longer. Therefore, it is our second goal to investigate the relationship between dexterity and time delay. In other words, we would like to establish a guideline something like: *"To complete such kind of tasks by teleoperation, delay time should be less than..."*

In the first part of this paper, we present a unified hand/arm master-slave system that was developed for conducting tasks that require some dexterous operations. To evaluate the performance of individual teleoperation systems all over the world in equal condition, we propose toy block assembling (LEGO[1]) as the benchmark test. Using our hand/arm master-slave system, we measured completion time for assembling some block structures.

In the second part of this paper, the result of a ground-space bilateral tele-operation experiment using a robot arm mounted on the ETS-VII (Engineering Test Satellite No.7) is presented. The experiment was conducted on November 22, 1999, as the first step to investigate the relationship between time delay and dexterity. Round-trip time for data communication between the NASDA ground station and ETS-VII was approximately six seconds. Experimental results showed that force feedback to the operator is helpful even under such long time delay and improves the performance of the task.

2. Towards Fine Manipulation

2.1. Development of a unified hand/arm master-slave system

To build a teleoperation system with which humans can be really replaced, it is necessary to develop a system that has not only an arm but also a multifingered hand. However, it is surprising that there have been quite few teleoperation systems that have multifingered hands (e.g. [17]).

As the first step to improve the dexterity of master-slave systems, we have

---

[1]LEGO and LEGO duplo are trademarks of LEGO group.

 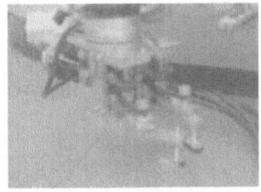 

(a) master hand   (b) slave hand   (c) overview

Figure 2. Developed master-slave hand and overview of the unified hand/arm system.

developed a two-fingered master-slave hand named "Kusshi", which means "one of the best" in Japanese. In order to configure a unified hand/arm system, this master-slave hand was designed so that it may be attached to the tip of an existing master-slave arm called "Ratsuwan", which means "superior ability" in Japanese.

The author proposed a design guideline of master arms considering operator dynamics[21]. According to this guideline, exoskeleton type is preferable and we applied this guideline when designing a multifingered master hand as well. The slave hand configuration should be also the same as the operator hand so that he/she can maneuver the slave hand in intuitive manners.

Since the existing arm is 3DOF SCARA type and moves in a horizontal plane, the developed hand has two fingers, each of which has two joints, mimicking thumb and index fingers, respectively. We have considered several design policies such as safety to the operators, compactness and low apparent inertia. For more detail of these design policies, refer to [22]. Figure 2(a) and Fig.2(b) shows the developed hand and Fig.2(c) shows an overview of the unified hand/arm system.

2.2. Benchmark test using LEGO™

Methods to evaluate teleoperation systems are classified into mainly two ways: theory-based[4][20] and experiment-based[6][2][13][14]. The theory-based approach enables us to evaluate the systems quantitatively, but we first need to identify the dynamics of the target system. In addition, it is hard to take an exact correlation between the obtained evaluation and the actual maneuverability in a specific task. In the experiment-based approach, one can set up a task as he/she likes and measure the performance of his/her system. This *local* test works well when comparing several control schemes or conditions *within* this specific system. However, it is very difficult to compare one system in one place with other systems in other places.

To evaluate the performance of individual teleoperation systems all over the world in equal condition, we must set up a standardized benchmark test. We propose toy block assembling (LEGO) as the benchmark test for teleoperation systems. Use of LEGO is advantageous in the following points: (i) easy to get, (ii) uniform quality, (iii) comparison to the skill level of human (in this case, children)[22].

Among the LEGO lineup, we chose LEGO duplo, blocks for children aged

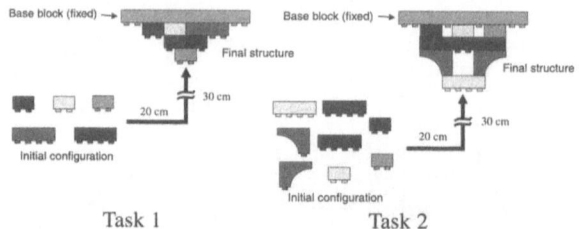

Task 1   Task 2

Figure 3. Tasks 1 and 2.

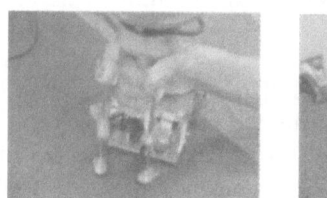 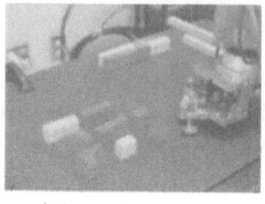 

(a) direct operation   (b) ready to go   (c) completed

Figure 4. Snapshots of direct operation and teleoperation before/after task 2.

1.5 through 6 years. Figure 3 shows two structures to be assembled in the evaluation test. Using our prototype system, we measured completion time to assemble these structures. In the experiment, we tried the following three methods. One is bilateral master-slave teleoperation. The other two are conducted in order to compare the performance of teleoperation. One of them is the operation by bare hand and the other one is direct operation through the slave hand as shown in Fig.4(a). In the bare hand case, the subjects are asked to use only two fingers (index and thumb) and their hand must always contact with the tabletop, restricting the motion in 2D.

Two male subjects participated the experiments. After the enough training phase (about 10 trials in 20 minutes), each subject performed tasks 1 and 2 by three methods seven times each and the completion times were measured. Figure 4(b) and Fig.4(c) shows snapshots of task 2 by teleoperation.

Figure 5 shows the experimental results, the averaged completion time and the standard deviation of seven trials. As expected, the bare hand operation was the fastest followed by the direct operation through the slave hand. Unfortunately, bilateral teleoperation was worst.

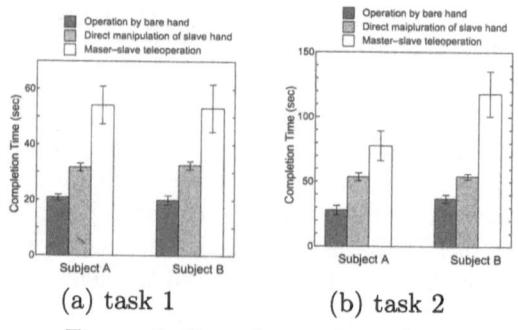

(a) task 1   (b) task 2

Figure 5. Experimental results.

Table 1. Amount of time delay in the previous works on teleoperation with force feedback

Author(s)	Model-Based?	Delay Time for Round Trip	Feature
[1] Anderson & Spong (1989)	No	80ms, 400ms, 4s	Scattering Theory
[15] Niemeyer & Slotine (1991)	No	1s	Wave Variables
[8] Kim et al. (1992)	No	1s	Shared Compliant Control
[11] Lawn & Hannaford (1993)	No	up to 1s	Comparison between Scattering Theory and Others
[9] Kosuge et al. (1996)	No	1.4s	Virtual Time Delay
[16] Obe & Fiorini (1998)	No	320ms	PD-type
[10] Kotoku (1992)	Yes	1s	Predictive Display with Force Feedback
[3] Funda et al. (1992)	Yes	3s	Teleprogramming
[19] Tsumaki et al (1996)	Yes	5s	Velocity/Damping Control
[18] Peñin et al. (2000)	Yes	5–7s	Truss Structure Experiment on ETS-7

2.3. Discussion

This benchmark test experiment has reminded us the following two points that should be considered when evaluating teleoperation systems.

Dependence on the body size of subjects: In most cases, it is not desirable to design a teleoperation system that suits *only* one specified operator, even if he shows an excellent performance. Therefore, to evaluate a system, one should prepare enough number of subjects among the variety of body size.

Easiness of learning: When evaluating teleoperation systems, we must consider not only the final performance but also the easiness to reach it. For example, one cannot definitely judge that a system which gives excellent performance but requires long time training is better than another system which gives a little worse performance with much less training efforts. Therefore, one should also state clearly how long the subjects have been trained before getting the shown performances.

The completion time by using our developed master-slave system were twice longer than that of direct manipulation and three times longer than that of bare-hand manipulation. This result clearly reflects the difference of dexterity between human and our current teleoperation system. It is our future work to improve the dexterity of our system as close to human dexterity as possible.

3. Towards a Distant Place—Space

3.1. Teleoperation with time delay

It is well known that even small communication delay may destabilize the system with conventional bilateral control methods, such as symmetric position servo and force reflecting servo. Anderson and Spong[1] proposed a bilateral control law that maintains stability under the communication delay using the scattering theory. Niemeyer and Slotine[15] studied further on this problem and introduced the notion of "wave variable".

It has been assumed, however, that bilateral control would not be effective under the time delay more than 1 sec[8][5][11]. Following what Peñin et al.[18] did, we also summarized previous works on teleoperation with force-feedback under the communication time delay as shown in Table 1. All of them conducted real experiments. These previous works can be divided into two groups:

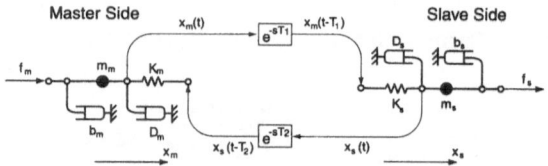

Figure 6. PD-type bilateral control

(i) direct bilateral teleoperation without any models of the remote site and (ii) model-based teleoperation with pseudo force feedback from the local model of the remote environment. From the table, it seems that when the delay time is longer than about 1 sec, the model-based approach would be the only solution. Instead of exactly drawing the limitation line at 1sec, however, our claim is, in a sense, quite natural as follows: "*Time delay limitation depends on the difficulty of the task. And even if the delay time becomes more than 1 sec, some tasks would be possible by direct bilateral teleoperation.*"

A ground-space bilateral teleoperation experiment using ETS-VII robot arm was conducted on November 22, 1999. Round-trip time for communication between the NASDA ground station and ETS-VII was approximately six to seven seconds. The purpose of this experiment was to investigate if the direct bilateral teleoperation is still valid under the time delay longer than 1 sec.

3.2. PD-based bilateral controller

One of the well-know approach to time delay is to use scattering transformation[1][15]. Besides the scattering-theory-based approach, there are several other approaches, which are less popular than the wave-variable approach. For example, Oboe and Fiorini[16] dealt with the time-varying delay problem over the Internet by using a simple PD-type controller.

We paid notice to this PD-type controller, which is shown in Fig.6. This PD-type controller is given by the following equations:

$$\tau_m = -K_m \left(x_m(t) - x_s(t - T_2) \right) - D_m \dot{x}_m, \tag{1}$$
$$\tau_s = K_s \left(x_m(t - T_1) - x_s(t) \right) - D_s \dot{x}_s, \tag{2}$$

where x_m and x_s denote positions of the master and slave arms, and τ_m and τ_s are actuator driving forces, respectively. K_m and K_s are position gains, and D_m and D_s are dumping gains. T_1 and T_2 denote delay times from master to slave and slave to master, respectively.

Oboe and Fiorini[16] analyzed the stability condition of this PD-type controller under time-varying delay conditions, but their analysis contains some errors and resultant condition is not true. We assumed constant time delays in both directions and derived the stability condition using Llewellyn's condition[12]. The derived condition is given by

$$(D_m + b_m)(D_s + b_s) \geq \frac{K_m K_s (T_1 + T_2)^2}{4}, \tag{3}$$

where b_m and b_s are the physical damping coefficients of the master and slave arms, respectively. As the delay time becomes longer, the dumping gains should

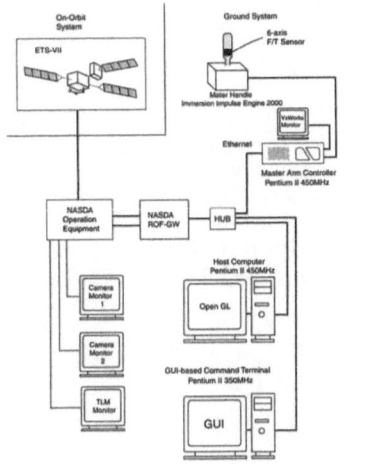

(b) Overview of control station

(a) System configuration

(c) Master handle

Figure 7. Experimental system.

be increased, resulting in sticky feeling. Unlike the scattering-theory-based controller, however, the apparent inertia keeps constant. Therefore, this PD-type controller is expected to be still useful even under a long time delay around the range of 5-7 seconds.

3.3. Experiment using ETS-VII robot arm

We planned to use the PD-type bilateral controller discussed in 3.2 for the ETS-VII experiment. Due to the limitation of the on-board arm controller specification of ETS-VII, however, we could not apply the original PD-type controller. Instead, we slightly modified the control scheme so as to match the on-board arm controller specification. Unfortunately the stability condition of the modified controller is not simple like eq.(3), but one can guarantee the system stability with appropriate gains[7].

Figure 7(a) shows the configuration of the experimental system. Figure 7(b) shows the overview of the control station. A 2-DOF force feedback joystick, which is shown in Fig.7(c), was used for the master handle. Figure 8 shows the robot experimental system on the ETS-VII and the task board used in the experiment.

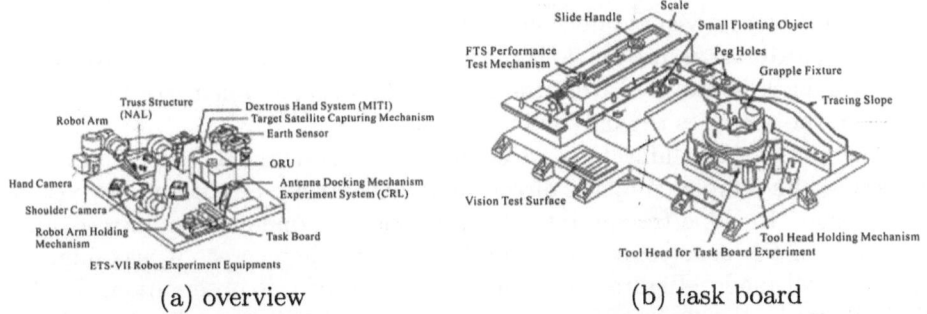

(a) overview

(b) task board

Figure 8. Robot system on ETS-VII.

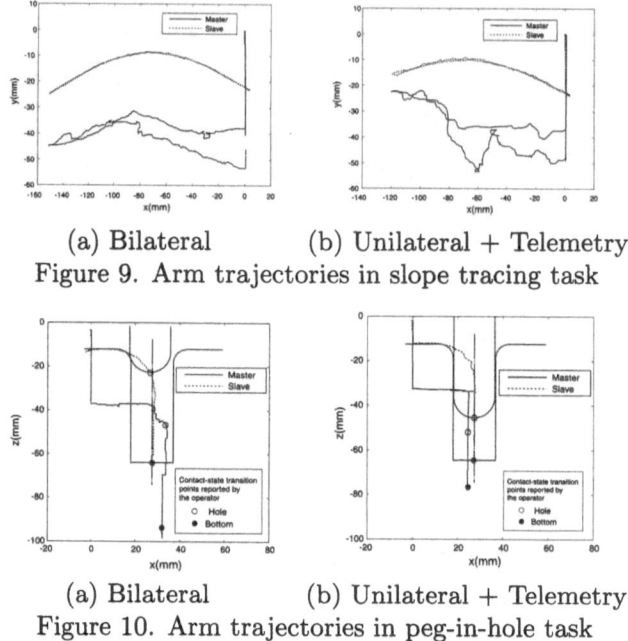

(a) Bilateral   (b) Unilateral + Telemetry
Figure 9. Arm trajectories in slope tracing task

(a) Bilateral   (b) Unilateral + Telemetry
Figure 10. Arm trajectories in peg-in-hole task

Unfortunately, we could not conduct the benchmark test using LEGO in space. Instead, several tasks, such as pushing task and slope tracing task, peg-in-hole task, and slide handle task, were carried out by using the experimental facilities on the task board shown in Fig.8(b). In pushing task, accuracy of force command was evaluated. In slope tracing task, we did not notify the operator the starting point on the slope and examined how accurately the operator can recognize the constraint surface shape. In peg-in-hole task, the accuracy of recognizing contact state transitions by the operator was checked. Finally, in slide handle task, we made the sliding direction unknown to the operator by inserting an arbitrary rotational coordinate transformation and evaluated the accuracy of recognizing this unknown constraint direction. In each task, the following three cases were tried:

Case 1 (bilateral mode + bar graph of force telemetry): The operator can get force feedback from the master handle. At the same time, he can monitor the telemetry force data displayed on the screen.

Case 2 (bilateral mode): The operator must operate with force feedback alone and no visual information is provided.

Case 3 (unilateral mode + bar graph of force telemetry): No force feedback is provided from the master handle. The telemetry force data on the screen is the only information.

Due to the space limitation, we cannot show the detailed results[7], but all tasks could be completed by the direct bilateral control even without any visual information. In slope tracing task, for example, the operator could recognize the shape of the constraint surface by bilateral mode whereas it was difficult by unilateral mode. Figure 9 explains the reason of this observation. In the bilateral mode, the trajectory of master handle is duplicating the slope shape while in unilateral mode it is difficult to estimate the slope shape from the master handle trajectory.

Figure 10 shows the arm trajectories in the peg-in-hole task. Actual peg position when the operator judged that the peg had reached the entrance of the hole is also drawn in the figure. In bilateral mode the operator could identify the transition of contact state accurately only from the force feedback, while the recognition by unilateral mode was inaccurate.

3.4. Discussion

From the experimental results including the questionnaire survey from the operators, we obtained several observations. First of all, all the operators paid most attention to the force feedback from the master handle even when the telemetry force data was displayed on the screen. We should also note that using the force feedback information from the master handle, the operators could recognize the shape of the tracing slope and the peg hole entrance with just a small local movement. Finally, we found that even a novice operator could complete the task, showing that the system requires no specific skill.

It was surprising even for us that the kinesthetic force feedback information was still useful and could improve the task performance even under such a long time delay. Of course the task should be performed slowly and the maneuverability is poor compare to the case without time delay. However, this experiment result has proven that it is still possible to complete some tasks by direct bilateral control under 6-7 sec time delay.

4. Conclusion

In the first part of this paper, a unified hand/arm master-slave system was introduced. To measure the dexterity of individual teleoperation systems all over the world in equal condition, we proposed toy block assembling (LEGO) as the benchmark test for teleoperation systems. Completion time to assemble LEGO blocks by our teleoperation system took three times longer than that by bare hand manipulation. We are expecting other research groups to try the same task in the future.

In the second part, the result of a bilateral teleoperation experiment with ETS-VII was shown. Form the experimental results, it was shown that force feedback is still helpful information to the operator and improves the maneuverability even with 6-7 seconds time delay. This experiment was probably the first ground-space teleoperation by direct bilateral control.

These two experimental results are the first step to improve the dexterity of teleoperation systems and to investigate the relationship between time delay and dexterity. The dexterity of our current system is still far from human dexterity as shown in Fig.1(a) and the motion is limited in 2D. We are planning to develop a 3D system as well as improving the current 2D system. The experiment with ETS-VII was a great experience for us. In this experiment, however, we could examine just a single delay time. Therefore, we need to investigate the system performance (using the benchmark test if possible) at various time delays and get the actual plot of Fig.1(b).

References

[1] Anderson R J, Spong M W 1989 Bilateral control of teleoperators with time delay. *IEEE Trans. on Automatic Control*, 34(5):494-501

[2] Das H *et al.* 1992 Operator performance with alternative manual control modes in teleoperation. *Presence*, 1(2):201-218

[3] Funda J *et al.* 1992 Teleprogramming: toward delay-invariant remote manipulation. *Presence*, 1(1):29-44

[4] Hannaford B 1989 A design framework for teleoperators with kinesthetic feedback. *IEEE Journal of Robotics and Automation*, 5(4):426–434

[5] Hirzinger G *et al.* 1993 Sensor-based space robotics –ROTEX and its telerobotic features. *IEEE Trans. on Robotics and Automation*, 9(5):649-663

[6] Hwang D Y, Hannaford B 1988 Teleoperation performance with a kinematically redundant slave robot. *Int. J. of Robotics Research*, 17(6):579-597

[7] Imaida T *et al.* 2001 Ground-space bilateral teleoperation experiment using ETS-VII robot arm with direct kinesthetic coupling. *Proc. IEEE ICRA 2001 (to appear)*

[8] Kim W S, Bejczy A K 1993 Demonstration of a high-fidelity predictive/preview display technique for telerobotic servicing in space. *IEEE Trans. on Robotics and Automation*, 9(5):698-702

[9] Kosuge K *et al.* 1996 Bilateral feedback control of telemanipulators via computer network. *Proc. IEEE/RSJ IROS'96*, 1380-1385

[10] Kotoku T 1992 A predictive display with force feedback and its application to remote manipulation system with transmission time delay. *Proc. IEEE/RSJ IROS'92*, 239-246

[11] Lawn C A, Hannaford B 1993 Performance testing of passive communication and control in teleoperation with time delay. *Proc. IEEE ICRA'93*, 3:776-783

[12] Llewellyn F B 1952 Some fundamental properties of transmission systems. *Proc. of the I.R.E.*, 40(5):271-283

[13] Matsuhira N *et al.* 1994 Maneuverability of master-slave manipulator with different configuration and its evaluation tests. *Advanced Robotics*, 8(2):185-202

[14] McLean G F *et al.* 1994 Teleoperated system performance evaluation. *IEEE Trans. on System, Man, and Cybernetics*, 24(5):796-804

[15] Niemeyer G, Slotine J J E 1991 Stable adaptive teleoperation. *IEEE J. of Oceanic Engineering*, 16(1):152-162

[16] Oboe R, Fiorini P 1998 A design and control environment for internet-based telerobotics. *Int. J. Robotics Res.*, 17(4):433-449

[17] Oomichi T *et al.* 1990 Development of working multifinger hand manipulator. *Proc. of the IEEE IROS'90*, 873-880

[18] Peñin *et al.* 2000 Force reflection for time-delayed teleoperation of space robots. *Proc. IEEE ICRA 2000*, 3120-3125

[19] Tsumaki Y *et al.* 1996 Virtual reality based teleoperation which tolerates geometrical modeling errors. *Proc. IEEE/RSJ IROS'96*, 1023-1030

[20] Yokokohji Y, Yoshikawa T 1989 Bilateral control of master-slave manipulators for ideal kinesthetic coupling -formulation and experiment. *IEEE Trans. on Robotics and Automation*, 10(5):605-620

[21] Yokokohji Y, Yoshikawa T 1993 Design guide of master arms considering operator dynamics. *ASME Journal of Dynamic Systems, Measurement and Control*, 115:253-260

[22] Yokokohji Y, Iida Y, Yoshikawa T 2000 "Toy problem" as the bechmark test for teleoperation systems. *Proc. IEEE/RSJ IROS 2000*, 996-1001

Virtual Exoskeleton for Telemanipulation

Josep Amat
IRI. Robotics Institute (UPC/CSIC).
Campus Nord UPC
08034 Barcelona, Spain

Manel Frigola and Alícia Casals
Dep. of Automatic Control and Computer Engineering
Universitat Politècnica de Catalunya.
Pau Gargallo, n° 5, 08028 Barcelona, Spain.
Email: {frigola, casals}@esaii.upc.es

Abstract: The growing number of robotics application fields, mainly in services, has led to the increase of new needs as well as the development of new facilities for teleoperation. Research in the design of more efficient and easy to use human-machine interfaces has propitiated the development of friendly communication systems such as those based on voice or gesture recognition. This work describes a vision based human-machine communication system that allows a computer or a control unit to "see and track" the position of the hands of a human. Thus, the vision system can be used as a virtual exoskeleton for simple telemanipulation tasks.

1. Introduction

Teleoperation as a means to operate a robot using the intelligence of a human requires the availability of adequate human-machine interfaces. The use of communication means such as natural language or gestures enables us to expand teleoperation to new application fields, making it possible for any kind of user to operate a robot in different work environments. This is possible because human-robot interaction becomes much more comfortable and easier.

The operation of a robot by means of a joystick is very common in areas such as civil engineering, in applications for parts manipulation in construction, or in the guidance, from a van, of mobile robots within sewers, among others. When the number of degrees of freedom to control is high or the operation to be performed requires certain ability, it is convenient to use more sophisticated devices. Different hand-held devices have been designed to facilitate this human-robot interaction. Other structures, such as the well known phantom devices, that introduce the concept of haptics, provide augmented reality in the interaction of the robot with the environment. These kinds of devices are extremely useful in application fields ranging from space to surgery, areas in which perception is essential to understand the evolution of a teleoperated task. In all these areas, such physical interfaces enable a human operator to interact with real or virtual

D. Rus and S. Singh (Eds.): Experimental Robotics VII, LNCIS 271, pp. 21–30, 2001.

environments, either for teleoperation works or for training applications, respectively.

The concept of exoskeleton, as the master device in teleoperation, started with the use of mechanical structures[1], like the Hardyman, a wearable articulated structure designed to amplify the human forces and movements in applications that require the manipulation of big or heavy loads. These devices have become lighter and they incorporate force feedback enabling us to perceive the effects of the performed actions by the robot, the slave, more effectively [2].

Electronic based devices aim to suppress the mechanical elements that, in some way, constrain the operator movements. Among such devices, data gloves are those that seem to be able to provide the best results.

With this same aim, to avoid the need of using mechanical structures, some efforts have been dedicated to designing computer vision systems to be used as the master, in a master-slave robotics configuration. Computer vision in industry and in robotic applications has progressed significantly, giving place to applications of detection of human movements and their gestures with the aim of interpreting signals or orders. Therefore, the human operator can avoid the need to wear a physical device over their body. Nevertheless, the detection and tracking of a human body in a natural environment presents certain difficulties if the background image is not homogeneous. Consequently, some of the developed systems use LED diodes or reflectors located in the body joints [3, 4, 5], or use colour information in applications where the user is forced to wear coloured clothes to facilitate the segmentation process [6, 7]. Other alternative systems are based either on magnetic position sensors or even on myoelectric sensors that convert the muscle's movements into signals, from which it is possible to detect the operator movements [8]. A mixture of them, magnetic sensors and cameras visualising some fiducial marks provide better performances since they combine the robustness of magnetic sensors with the precision of computer vision. Other researchers base their works on the analysis of the human movements, either through the use of optical flow [9, 10] or by means of the subtraction of successive images [11].

The present work is also based on the analysis of the human operator movements. In [12] the detection of movement is achieved working in highly contrasted environments. In our case, the system avoids the need to use specific *plateaus* with controlled lighting conditions or the need to wear special clothes or specific elements, or marks, on the body. This improvement on the working conditions is possible due to the fact that the system operates from the variations on the direction of the gradient between successive images. One such procedure notably improves the results obtained by the classical methods based on movement, enabling the availability of segmented images with very low noise level, even working in natural environments [13]. The computer vision system implements three basic functions: The detection of humans and segmentation of the hands of a person in a natural working environment, the tracking of the position of the upper limbs in 3D, and the control of the robotised arms, in accordance with the user's movements.

2. System structure

The aim of the vision system is the recognition and tracking of the arms and hands of a human operator to remotely control one or more robots. From such recognition, the postures and gestures of an operator have to be interpreted. This interpretation is based on the detection of an operator postures and the interpretation of the gestures derived from their movements are based on the location in a tridimensional space of the body more relevant parts for this application, using multiple views of the operating scenario. The location process is based on a first segmentation phase and a second one that validates, over a simplified model, the detection of the human figure.

Since the human detection and the image segmentation are based on the movement in the scene, it is necessary to distinguish a moving human body from other possible moving objects in the scene. Consequently, we use a 3D adaptable geometrical model of the human arms to make this detection more reliable, not only considering the target dimensions, but also its shape. The model is simple enough to be applicable in real time, but also complete enough to enable the description of the arms and hands position. The model is polycylindrical, articulated and tridimensional, and it is adaptable to changing shapes to fit in real time with the operator moving body.

The first step of the gesture human-machine communication interface is the detection and tracking of the arms and hands by estimating, over time, their position at every instant. The estimated position sequences will describe the body movement indicating the actions the person desires to express. The complete system, as shown in fig. 1, consists of the following tasks:

- Dedicated low level processing for movement detection
- Features extraction and detection of the arms singular points
- 3D position measurement, from stereo, and arms posture data validation using a simple geometrical model
- Filter and, operator-robotic arms, frames transformation

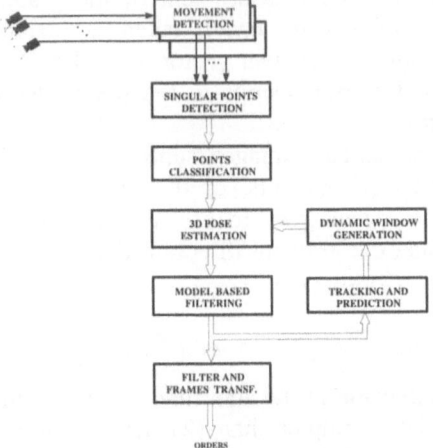

Fig. 1 Functional system structure.

3. Arms detection and tracking

The process of movement detection to interpret the human body orders to control the robot requires us to follow different steps. The vision process extracts the person silhouette from the image, for its analysis. Then from the human upper limbs silhouette the features that characterise their posture are obtained providing the data necessary to fit the human figure to a predefined adaptable human model. The process can thus be summarised in the three following steps: background extraction, features extraction and 3D model fitting.

3.1 Background extraction

Image segmentation is one of the main problems to face up to in computer vision. In complex scenes where it is not possible to find features discriminating enough to use satisfactorily any common segmentation process to extract the desired objects, it is possible to resort to the analysis of image sequences and to analyse the images' temporal variation, provided that the objects to be segmented are in movement.

Since the detection of a person's movements for the interpretation of her gestures requires the detection of the human figure and no more information from the scene is necessary, the level of image segmentation can be reduced to objects extraction from their background. The extraction of human figures in natural scenarios, either indoor or outdoor, should rely on segmentation techniques not dependent neither on the heterogeneity of the possible elements in the scene and its lighting conditions, nor on the person movement itself.

In this work, we use images subtraction, but to improve the system performances in complex scenarios the comparison pixel by pixel is carried out from the estimated gradient vector instead of using only its absolute value. In this case, images subtraction is performed as follows:

$$\mid \vec{G}_t(x,y) - \vec{G}_{t-1}(x,y) \mid \qquad \text{(Eq. 1)}$$

Where $\vec{G}_t(x,y)$ represents the measurement of the gradient vector computed at position x,y of the grey image $I(x,y)$, taken at instant t. The advantage of comparing images using the gradient vector instead of the gradient module is the increase from one to two dimensions in the pixels description, thus enhancing their characterisation.

In spite of this advantage, images subtraction is still sensitive to lighting variations. In natural environments, or in environments with fairly controlled lighting conditions it is necessary to use a more robust comparison. The new expression will include the gradient direction that does not depend on variations of lighting intensity, as follows:

$$\mid Atan2(\ \vec{G}_t(x,y)\ ) - Atan2(\vec{G}_{t-k}(x,y)\ )\mid \qquad \text{(Eq. 2)}$$

where Atan2 is the extension of the atan function to two dimensions.

Since equation (1) is simpler than (2) the former is normally used in this system when lighting conditions are fixed, and only when lighting variations could decrease reliability it is necessary to rely on (2). Fig. 2 shows the subtraction resulting from a sequence grey level images (a), using the gradient module (b), and using the gradient vector (c).

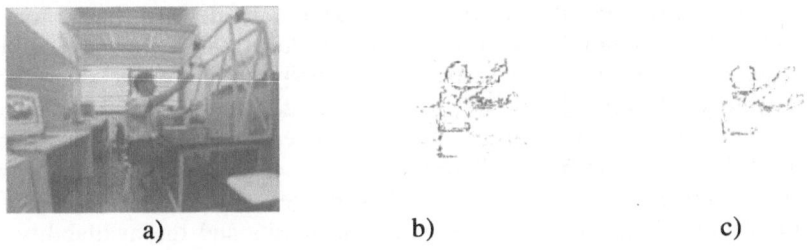

a)   b)   c)

Fig 2 Results of the segmentation operator, a) original image, b) segmented image using the gradient module, and c) using the gradient vector.

3.2 Features extraction for posture characterisation

From the human arms silhouette detected, the following step is the extraction of the features that characterise each operator's posture. The process followed to extract the required data, operating in different kind of environments and without imposing strict operating restrictions, is achieved by splitting the problem into three steps: features extraction, singular points detection and singular points classification.

The features selected to detect and to locate singular points are the clusters of pixels that verify some pre-established heuristic conditions. First, the clusters considered are the areas of the image whose distance among pixels are less than a maximum value, this value being chosen according to a compromise between efficiency and computing cost. A second parameter is the size of the clusters. The next step is to detect from these clusters the body singular points. The singular points considered are the most prominent ones of the silhouette. Fig. 3 shows some candidate clusters of pixels (a) and the set of points considered as arms singular points (b), corresponding to the scene in fig. 2.

a)   b)
Fig. 3 Features extraction. a) pixels clustering, b) singular points candidates

According to the geometry of the singular points distribution, they have to be identified, by means of a model, as the corresponding arm or hand parts, such as the elbow or a finger tip.

3.3 Tridimensional human model

With the aim of robustly detecting and recognizing a human upper limbs configuration or the arms posture and to avoid false detections, a model is defined to validate the extracted silhouettes. The model was defined based on a compromise between simplicity and speed on one hand and efficiency on the other. It was designed according to the human body structure, its shape and moving capability. The articulated body state will be defined by a set of variables that, at a given time instant, defines position, speed and acceleration of the different model constituting parts. Therefore, the model has been designed as an articulated structure composed of geometrical primitives.

The imposition of some anthropomorphic constraints and the availability of some dimensional measures make it possible to reject wrong detections without the need to apply the model, thus reducing the operation time. Therefore, it is possible to reject the shapes that do not fit to an adequate profile. The person model adopted is constituted by a set of cylinders that fit to the moving parts profile. The model consists of two coaxial cylinders that are adjusted to the head and body, and also a set of up to four cylindrical surfaces per arm, that are adjusted to the body overhanging elements, that can correspond to the arms and hands (Fig. 4).

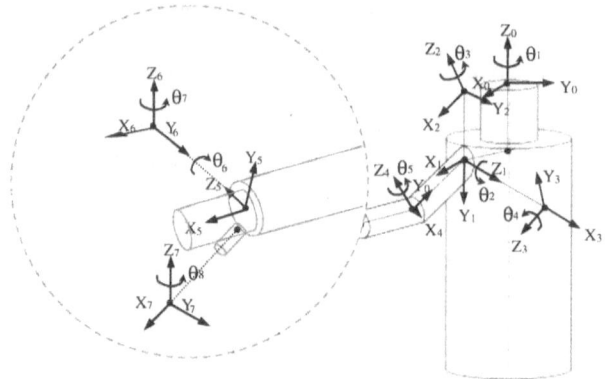

Fig. 4 The polycylindrical model with its joint reference frames.

From the pair of singular points in the two stereoscopic images, classified as belonging to the head, its corresponding 3D position is determined by triangulation. This head singular point defines the central axis of the two main cylinders of the body model (those corresponding to the head and to the trunk), and consequently the body position.

Based on this estimation of the body position, all the singular points that have also been detected, either those located at a distance too far or too close from the main axis, compared to a previously defined cylinder radius (R_c), are eliminated. In this way, we avoid the ambiguity derived from considering arm configurations that imply that the arm is too close or even in contact with the body.

Consequently, all the singular points located at a reasonable distance from the main axis form the set of points that will be used to generate the different hypothesis about the operator's gesture. Every point considered is again validated in the Cartesian world. Fig. 5 shows, part by part, the cylinders that fit with the relevant body parts for telemanipulation: the trunk, head, arms and hands.

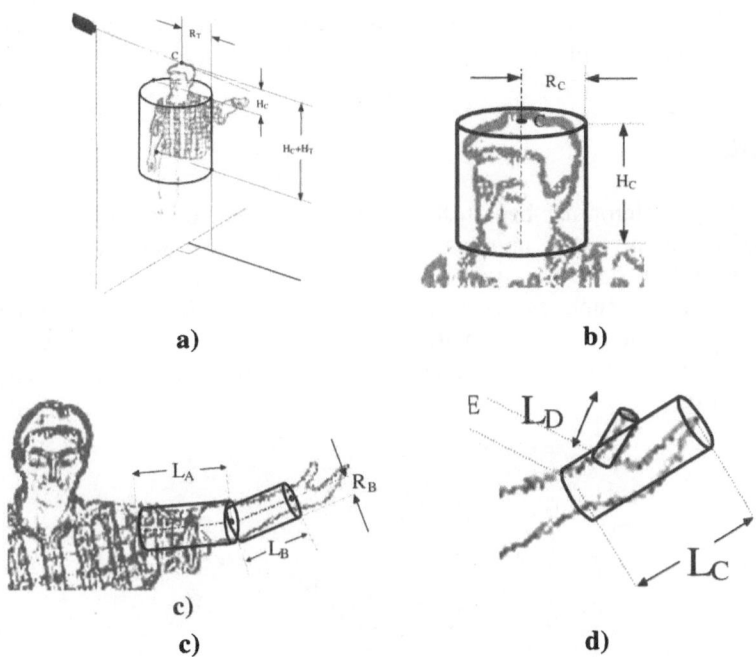

a)   b)

c)

c)   d)

Fig. 5 Cylinders parameters: a) and b) cylinders associated to main axis, trunk and head, c) arm cylinders and d) hand model.

4. Model fitting and frames transform

Once the position and orientation of the different model elements that fits with the operators posture has been obtained, it is necessary to transform the posture, or configuration of the operator's arms and hands, to the reference frames corresponding to the robot arms. This transform is necessary since the operator and the robot arms have different geometry, and it is set basically as an inverse kinematics problem.

With respect to these transforms there are two facts to consider. First, the changes of the head position indicate the potential user's displacements within the environment. Second, the elbows' orientation of the operator's model moves in such a way as to make them fit with the robot elbows' rotation, or that of the robotic platform supporting the teleoperated arms.

Later on, the inverse kinematics that provides the joint angles of the teleoperated arms is computed. This computation has to maintain the extreme points of the teleoperated arms at the same distance (with an adjustable scale factor) as the distances between the operator's arms extreme points and their shoulders. From the multiple possible configurations that enable us to achieve the same point, those that minimise the configuration changes are chosen.

We have not considered up to now other strategies, as for instance those that would avoid a collision between the robot elements, those that form the teleoperated structure, with the rigid objects in the scene. This is due mainly to the computing requirements, that would introduce an excessive delay that make the teleoperation difficult.

5. Results

The virtual exoskeleton has been tested on two different robots in our laboratory: a Cartesian robot and Garbí, an underwater vehicle provided with two arms.

Garbí is a low cost underwater robot, designed to carry out some simple manipulation tasks such as to collect some samples from the sea bed for applications in biology. To cope with such specifications its arms were designed to have uniquely three degrees of freedom each, plus the open-close movement of the gripper. Fig. 6

Fig. 6 Garbí, underwater robot.

Garbí has been the main test-bed for the experimentation of teleoperation tasks using the virtual exoskeleton. The Garbí exoskeleton itself, two simple articulated arms mounted on a chair and provided with potentiometers, in their joints, and push-buttons to order the open-close movements of the gripper, were replaced by the described vision based virtual exoskeleton.

The tasks more feasible to carry out, with the constraints derived from the simplicity of the structure of the arms that are not provided with any rotation in the wrists, can be classified into two types:

- Collecting samples of small parts, sea-weeds, stones, etc.
- Hoisting of objects using ropes and hooks steered from the assistant ship.

The system has proved to be efficient enough for sample collection, and the work carried out, fig. 7, has been always more comfortable than using the mechanical exoskeleton.

Other more complex tasks, such as the recovery of objects from the sea bed, have indeed been carried out. The experimentation environment has been a test laboratory scenario. The tasks experimented were oriented mainly to carry out operations for hoisting an object with ropes provided with a hook, fig. 8. The Garbí robot arms can be the means of transmitting the ability of the operator to the working place, but not the force required for hoisting the part. Consequently, the ropes arrive from the surface, from the support ship, and the forces required to proceed to the extraction of the hooked object are provided by the crane.

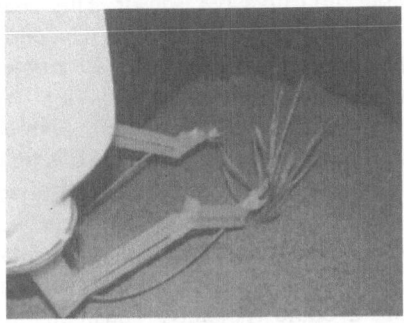 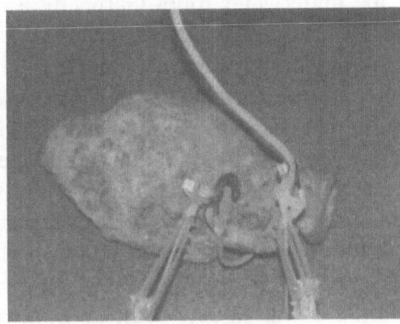

Fig. 7 Samples collection.   Fig. 8 Hoisting a part from the sea bottom.

In order to evaluate quantitatively the location and tracking precision some trials have been performed. The trials consisted of placing a finger tip over a reference stick, on the working table, and measurimg the error with which the robot the robot arms position themselves in the corresponding homologue points in their working space.

The robot positioning and repeatability observed over the vertical axis, can be seen in fig. 10. These errors, that in the worst case are within a radius of about 2 cm, are mainly due to the hysteresis of the robot arms movements. Nevertheless, these errors do not have a special effect in teleoperation since the users themselves continuously correct the operation visually.

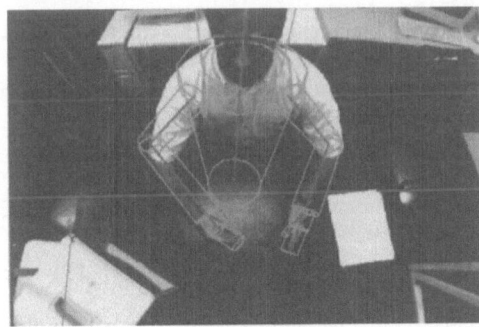 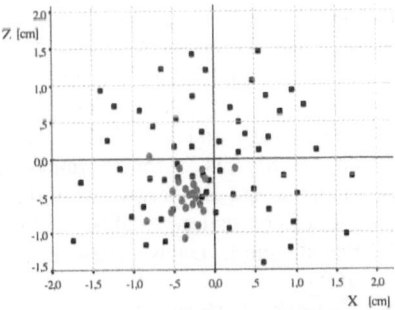

Fig. 9 The virtual exoskeleton.

Fig. 10 Dispersion of the arrival position of the hand, in the vertical plane, • from a fixed position, and ■ from random origin.

6. Conclusions

The described vision system, operating as an arms and hands position sensor, has permitted us to substitute the mechanical exoskeleton of the teleoperation working station of the underwater robot Garbí, by a remote, non invasive, sensor. This change makes the teleoperation tasks easier. The underwater robot, with its two teleoperated arms controlled from the surface, can grasp small samples of the sea bed, move objects, or manipulate a rope dedicated to hoisting objects and their retrieval.

Due to the simplicity of the upper-limbs human model used, the system developed can be used as a virtual exoskeleton in teleoperating tasks, in which the end-effector has up to 3 + 2 degrees of freedom, three for the wrist position and two for the end-effector orientation. The use of such friendly master-slave teleoperated structure opens the possibility of using teleoperation in environments or situations where it is not desirable or convenient to wear specific devices, or when a very intuitive way of user-robot communication is necessary, such as controlling a crane remotely. Thus, this work provides some advances towards "natural" human- machine interfaces.

References

[1] R. C. Goertz (1964) Manipulator systems development at ANL. In proc. Of the 12th Int. RSTD Conference.
[2] M. Bergamasco (1995). Force replication to the human operator: The development of arm and hand exoskeleton as haptic devices. In The seventh ISRR, Germany, pp. 173-182
[3] R. Azuma, G. Bishop (1994). Improved Static and Dynamic Registration in an Optical See-through HMD. In Proc. SIGGRAPH, Orlando.
[4] R. Rasid (1979). LIGHTS: A study in motion. In Proc. DARPA Image Understanding Workshop, pp. 57-68, Nov.
[5] M. Ward, R. Azuma, R. Bennett, S. Gottscalk, H. Fuchs (1992). A Demonstrated Optical Tracker with Scalable Work Area for Head-Mounted Display Systems. In Proc. of the Symposium on Interactive 3D Graphics, pp. 43-52.
[6] M. Yachida and Y. Iwai(1998). Looking at Human Gestures. Computer Vision for Human-Machine Interaction. Ed. by R. Cipolla and A. Pentland. Cambridge University Press.
[7] D.M. Gavrila and L.S. Davis (1996). 3-D model-based tracking of humans in action: a multi-view approach. IEEE Computer Vision and Pattern Recognition.
[8] O.A. Alsayegh and D.P. Brazakovic (1998). Guidance of Video Data Acquisition by Myoelectric Signals for Smart Human-Robot Interfaces. In Proc. of ICRA'98.
[9] A. Pentland and B. Horowitz (1991). Recovery of nonrigid motion and structure. IEEE Transactions on Pattern Analysis and Machine Intelligence, 13(7): 730-742.
[10] Y. Yacoob and L. Davis. Learned Temporal Models of Image Motion. ICCV'98, pp. 446-453, 1998
[11] H. Nugroho, J. Hwang and S. Ozawa (1994). Tracking Human Motion in a Complex Scene Using Textural Analysis. IECON 94, pp. 727-732.
[12] J.M. Buades, R. Mas & F.J. Perales (2000). Matching Human Walking Sequence with a VRML Synthetic Model. Works. Articulated Motion and Deformable Objects, pp. 145-158.
[13] J. Amat, A. Casals, M. Frigola. Stereoscopic System for Human Body Tracking in Natural Scenes. ICCV Workshop on Modelling People, MPEOPLE'99, 1999.

Design of Programmable Passive Compliance for Humanoid Shoulder
– Towards Skill of Compliance of Humanoid Robots –

Masafumi OKADA, Yoshihiko NAKAMURA and Shigeki BAN
Univ. of Tokyo
7-3-1 Hongo Bunkyo-ku Tokyo 113-8656, JAPAN
okada@ynl.t.u-tokyo.ac.jp

Abstract: We have developed the 3-DOF humanoid's shoulder mechanism 'Cybernetic Shoulder'. Whose advantages are human-like motion, introduction of the passive compliance, large mobile area and singularity free. In this paper, we develop the second prototype of the cybernetic shoulder which has the programmable passive compliance mechanism using a redundant actuator, which is an essential function for humanoid robots to realize the human skill. The programmability of this mechanism is evaluated by an experiment.

1. Introduction

Humanoid robots that share the space and environments with human should have compliance for human friendliness, safety issue and relief of impacts. There are two strategies to develop the robot compliance. One is active compliance on which many researches have been reported [1]~[6], the other is passive compliance. The active compliance is realized by actuators. The compliance of robot joints is developed using control theories such as impedance matching method. It has high programmability however cannot cope with fast responses because of the low resolution of sensors, a long sampling time of control and noises of sensors. The passive compliance means mechanical compliance of members of robot arm. It works effectively in all frequency (both fast and slow responses) but its programmability is low.

Our research focuses on the ' Skill of Compliance ', which means the tuning of passive compliance, planning of swing pattern and design of the control law. In the casting of fishing, the potential energy is accumulated in the rod by taking the swing and the large kinetic energy is obtained by discharging the potential energy in the instant to throw the prickle farer. In this motion, the passive compliance of the rod is tuned, the swing pattern of the rod and the force control of our arm are well designed. In the sports, the faster motion needs the higher compliance and the harder hit needs the lower compliance. The implementation of the skill of the compliance to the humanoid robot approaches to the development of the new friendly and safe humanoid robots.

So far, we have developed the cybernetic shoulder[7] that is the three degree-of-freedom mechanism for humanoid robots. It has human-like motion

D. Rus and S. Singh (Eds.): Experimental Robotics VII, LNCIS 271, pp. 31–40, 2001.

and passive compliance using the closed kinematic chain. And we have proposed Active / Passive hybrid compliance design method using the cybernetic
shoulder and H_∞ control method in the frequency domain[10]. In this paper,
we design the programmable passive compliance mechanism for the cybernetic
shoulder (Programmable Passive Compliance Cybernetic Shoulder) and obtain
the compliance ellipsoid[4] of this mechanism. The programmability of the
designed mechanism is evaluated by experiments.

2. Passive compliance

2.1. Compliance, control law and swing pattern

In this section, we show the skill of the passive compliance. Consider the two

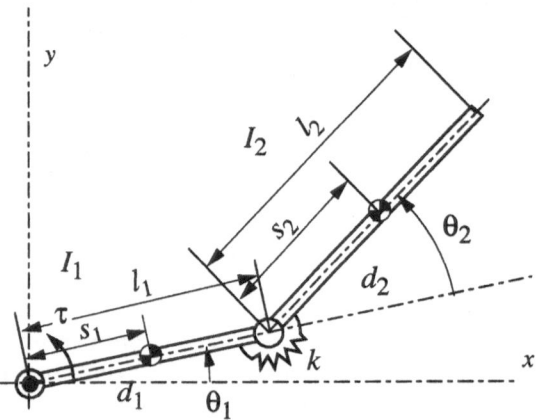

Figure 1. Two links manipulator in the horizontal plane

links manipulator in the horizontal plane shown in Fig.1. One joint is actuated
and another is free joint that has passive compliance. ℓ_i are the length of links
(we set $\ell_1 = 0.3$ [m], $\ell_2 = 0.5$ [m]), s_i are the positions of the center of gravity of
links ($= \ell_i/2$), I_i are the inertias of links, d_i are the coefficients of the viscosity
of joints ($d_1 = 0.3$ [Nms/rad], $d_2 = 1.0$ [Nms/rad]), θ_i are the rotation angles
of the links, k is the spring constant of the passive joint and τ is the torque of
the motor. θ_1 is controlled by PD controller K as shown in Fig.2. P is the two

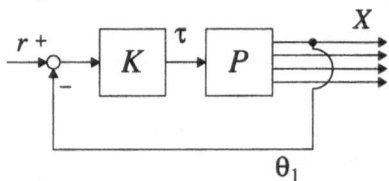

Figure 2. Control system of the two links manipulator

links manipulator, r is the reference signal for θ_1 and X is as follows.

$$X = \begin{bmatrix} \theta_1 & \dot\theta_1 & \theta_2 & \dot\theta_2 \end{bmatrix}^T \tag{1}$$

The dynamics of the two links manipulator is as fallows.

$$M(\theta_2)\ddot{\Theta} + C(\Theta, \dot{\Theta}) = U \tag{2}$$

$$\Theta = \begin{bmatrix} \theta_1 & \theta_2 \end{bmatrix}^T \tag{3}$$

$$M = \begin{bmatrix} a + 2b\cos\theta_2 + c & b\cos\theta_2 + c \\ b\cos\theta_2 + c & c \end{bmatrix} \tag{4}$$

$$C = \begin{bmatrix} -b\cos\theta_2(2\dot{\theta}_1^2 + \dot{\theta}_2^2)\dot{\theta}_2 & b\sin\theta_2 \cdot \dot{\theta}_1^2 \end{bmatrix}^T \tag{5}$$

$$U = \begin{bmatrix} \tau - d_1\dot{\theta}_1 & -k\theta_2 - d_2\dot{\theta}_2 \end{bmatrix}^T \tag{6}$$

$$a = m_1 s_1^2 + m_2 \ell_1^2 + I_1 \tag{7}$$

$$b = m_2 s_2 \ell_1 \tag{8}$$

$$c = m_2 s_2^2 + I_2 \tag{9}$$

Setting the reference signal as

$$r(t) = -\sin(2\pi t), \quad 0 \le t \le 1 \tag{10}$$

we get the optimal spring constant k_{opt} which minimizes the following cost function J.

$$J = \sum_{i=1}^{2} w_i J_i \tag{11}$$

$$J_1 = \max_t (\dot{\theta}_1(t)\tau(t)) \tag{12}$$

$$J_2 = \frac{1}{\max_t (v_y(t))} \tag{13}$$

$$v_y(t) = \dot{\theta}_1 \ell_1 \cos\theta_1 + (\dot{\theta}_1 + \dot{\theta}_2)\ell_2 \cos(\theta_1 + \theta_2) \tag{14}$$

$$w_1 = 1, \quad w_2 = 500 \tag{15}$$

J_1 aims at reduction of the actuator power. J_2 aims at maximizing the velocity of the end of the arm along with y axis. Maximization of the velocity means that the two links manipulator can throw fastball. Though the optimized spring constant depends on the motor controller (control law) and reference signal in equation (10) (swing pattern), we optimize only the spring constant (compliance) in one situation that fixes control law and swing pattern. The values of J and J_1, J_2 due to the spring constant k are shown in Fig.3, 4 respectively, which are given from the numerical simulations. These figures show that the optimal spring constant k_{opt} is given as

$$k_{opt} = 2.15 \tag{16}$$

and the maximum velocity is 6.19 [m/s]. These results show that by using the passive compliance, the two links manipulator can throw the faster ball by small consumption of the motor energy.

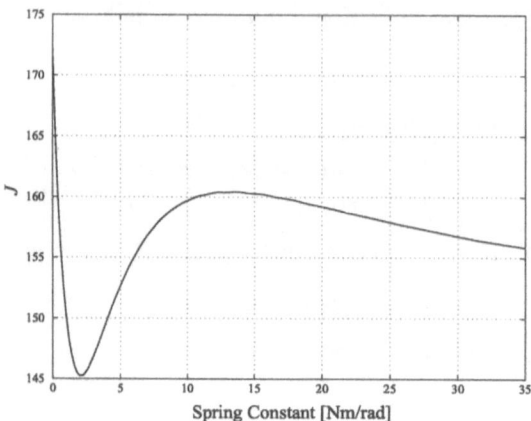

Figure 3. Value of J versus spring constant k

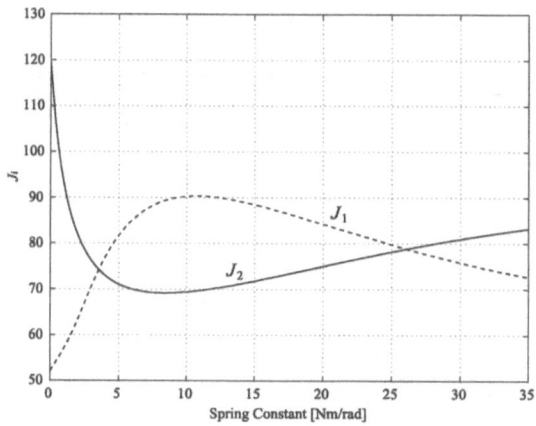

Figure 4. Value of J_i versus spring constant k

2.2. Compliance ellipsoid

2.2.1. Cybernetic shoulder

In this section, we briefly introduce the cybernetic shoulder. The cybernetic shoulder is the three degree-of-freedom mechanism for the humanoid shoulder mechanisms [7] that has human like motion and passive compliance. Figure 5 shows the model of the cybernetic shoulder. β and δ are two degree-of-freedom gimbal mechanisms, d is a ball joint, b is a two degree-of-freedom universal joint, a is a four degree-of-freedom spherical and prismatic joint and e is a prismatic joint. Moving point A within vertical plane alters the pointing direction of the main shaft G, which determines, along with the constraints due to the free curved links E between points b and d, the direction of the normal vector of D. Because the length of free curved links decide the orientation of the disk

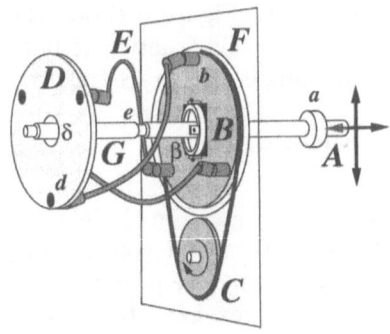

Figure 5. The cybernetic shoulder

D, the compliance of the free curved links decides that of the endplate D.

2.2.2. Compliance ellipsoid of the cybernetic shoulder

The compliance ellipsoid [4] is helpful for the foundation of the swing pattern and motor control law.

Consider the compliance matrix C defined as

$$C = JK^{-1}J^T \tag{17}$$

Here, J is the jacobian matrix and K is the spring constant matrix. Using the singular value decomposition of C,

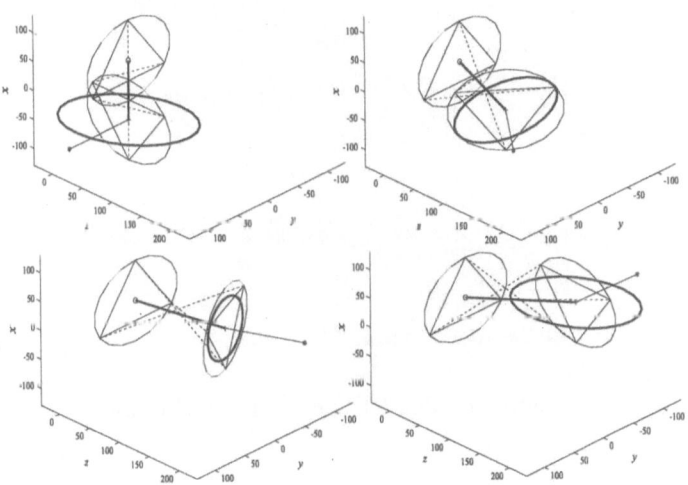

Figure 6. Compliance ellipsoid of the cybernetic shoulder

$$
\begin{aligned}
C &= USV^T &\qquad(18)\\
&= [\, U_1 \quad U_2 \quad \cdots \quad U_n \,]\, \mathrm{diag}\{\, s_1 \quad s_2 \quad \cdots \quad s_n \,\} V^T &\qquad(19)
\end{aligned}
$$

the compliance ellipsoid is defined in the n dimensional space whose axes are $s_i U_i (i = 1, 2, \cdots n)$. In this paper, we consider the two-dimensional compliance ellipsoid of the cybernetic shoulder. Figure 6 shows the compliance ellipsoid in accordance with the motion of the cybernetic shoulder. These ellipsoids are calculated in each orientation using equation (SVD).

3. PPC Cybernetic Shoulder

3.1. PPC Mechanism

Because the optimal spring constant given in the previous section depends on the weight of links and swing pattern and control low, the spring constant should be tunable, which is achieved by the programmable passive compliance (PPC). So far some PPC mechanisms have been developed [6, 8, 9]. The difficulties of these mechanisms are as follows.

Development of the multi DOF mechanism Development of the multi degree-of-freedom mechanism assembling the single degree-of-freedom mechanism, it gets heavy weight and large volume.

Control of redundant actuators The programmable passive compliance is realized by two redundant actuators whose outputs should be well controlled. Otherwise the joint may rotate or has an oscillation.

To overcome these problems, we develop the PPC mechanism using a closed kinematic chain shown in Fig.7. The advantages of this mechanism are as

Figure 7. The PPC cybernetic shoulder

follows.

PPC mechanism We replace the prismatic joint e in Fig.5 with a linear actuator (4.5[W] DC motor and ball screw) as shown in Fig.8. By changing the length of L to $L + \Delta L$ by this linear actuator, the internal force is applied to members E, which causes the programmability of the passive compliance when E has nonlinear relationship between strain and stress.

Compactness and small backlash The universal joints on the point b and d are replaced with elastic universal joints as shown in Fig.9. It has the

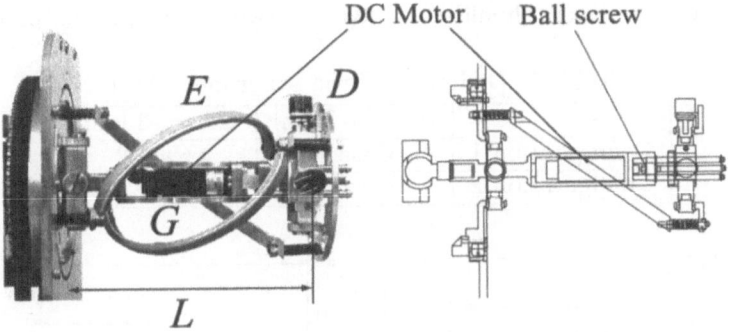

Figure 8. PPC mechanism

Figure 9. Elastic universal joint

same structure as a flexible coupling. This is for the compactness and for the small backlash.

Multi DOF compliance Because the end disk D has a gimbal mechanism in its center, the PPC cybernetic shoulder has two degree-of-freedom compliance around the rotation axis of the gimbal mechanism. Because the center rod G is rigid, the PPC cybernetic shoulder has high stiffness for any other degree-of-freedom of compliance.

3.2. Evaluation of the programmability

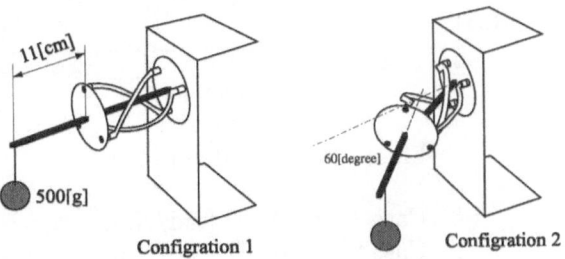

Figure 10. Configurations of the PPC cybernetic shoulder

In this section, we evaluate the programmability of the passive compliance of the PPC cybernetic shoulder.   We set two configurations of the PPC

Table 1. Definition of the experimental set

	$\Delta L = 0$ [mm]	$\Delta L = -3$ [mm]
Configuration 1	Case 1	Case 2
Configuration 2	Case 3	Case 4

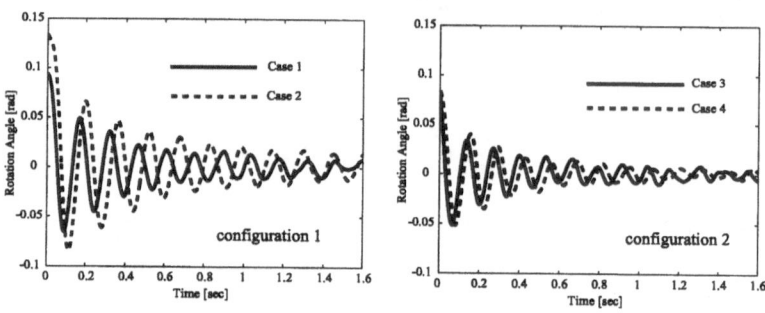

Figure 11. Responses on the cybernetic shoulder

cybernetic shoulder as shown in Fig.10. By cutting the 500[g] weight hung from the end of the arm, the external force is applied. The torque of the external force becomes 0.539 [Nm]. Two cases are adopted on each configuration, in one case $\Delta L = 0$ [mm], in another case $\Delta L = -3$ [mm]. Each occasion is defined as Table 1. The responses of each case are shown in Fig.11. In this prototype, the members E are rigid but joints (elastic joints) have compliance. The passive compliance of this mechanism is realized by the joint compliance. The compliance on each case is as follows which is calculated from the rotation angle in time zero.

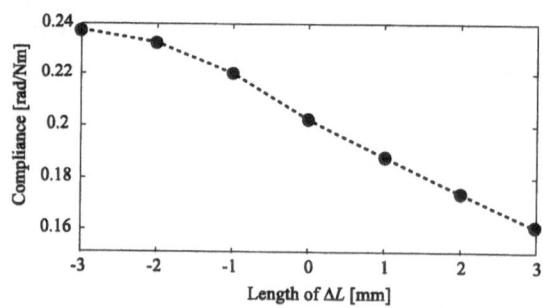

Figure 12. PPC due to ΔL

Case 1 : 0.202 [rad/Nm]
Case 2 : 0.237 [rad/Nm]
Case 3 : 0.156 [rad/Nm]
Case 4 : 0.170 [rad/Nm]

In configuration 2, the compliance cannot be changed so much. In configuration 1, we measure the passive compliance by small resolution of changing ΔL. Figure 12 shows the compliance due to ΔL in the configuration 1. The shorter L yields the higher compliance. The elastic universal joints have high compliance for yaw and pitch direction but have low compliance along thrust direction, that yield the passive compliance of the PPC cybernetic shoulder. The more dominant the compliance along thrust direction decides the compliance of the end plate, the lower the passive compliance of the PPC cybernetic shoulder becomes.

Consider a humanoid robot with the PPC cybernetic shoulder shown in Fig.13. Suppose that a 200 [g] weight falls from 1 [m] height and collide with an

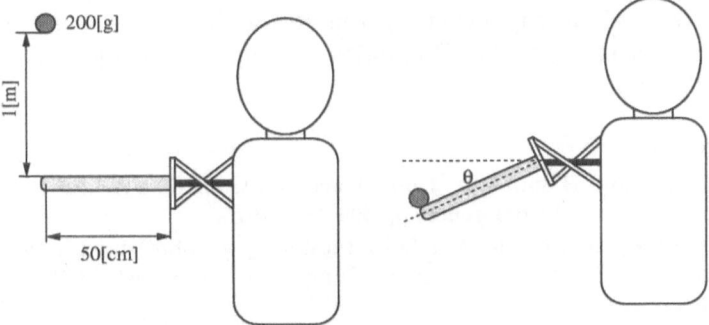

Figure 13. Configuration of the ball hit

arm. The rotation angles θ are shown in Fig.14 on each ΔL. This result shows

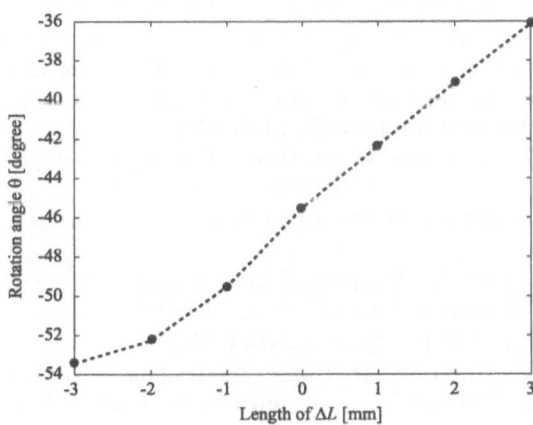

Figure 14. Rotation angle of θ due to ΔL

that by changing L, we can get large change of the passive compliance, that means the PPC cybernetic shoulder has high programmability of the passive compliance.

4. Conclusions

In this paper, we focus on the skill of compliance whose elements are tunable passive compliance, planed swing pattern and designed control low. The results of this paper are as follows.

1. We simulate the relationship between the compliance, control law and swing pattern.
2. The geometry of the shoulder compliance is obtained using the compliance ellipsoid, which is for the foundation to design swing pattern and motor control law.
3. We fabricate the programmable passive compliance cybernetic shoulder that is the shoulder mechanism for humanoid robots.
4. The PPC cybernetic shoulder has high tunablity of the passive compliance by the redundant actuator.

This research is supported by the Research for the Future Program, the Japan Society for the Promotion of Science (Project No. JSPS-RFTF96P00801).

References

[1] R.P.C.Paul and B.Shimano : Compliance and Control, Proc. of the 1976 Joint Automatic Control Conference, pp.694–699, 1976.

[2] H.Hanafusa and H.Asada : Stable Pretension by a Robot Hand with Elastic Fingers, Proc. of the 7th International Symposium on Industrial Robots, pp.361–368, 1977.

[3] N.Hogan : Mechanical Impedance Control in Assistive Devices and Manipulators, Proc. of the 1980 Joint Automatic Control Conference, pp.TA10-B, 1980.

[4] J.K.Salisbury : Active Stiffness Control of a Manipulator in Cartesian Coordinates, Proc. of the IEEE Conference on Decision and Control, 1980.

[5] N.Hogan : Impedance Control: An Approach to Manipulation: Part 1~3, ASME Journal of Dynamic Systems, Measurement and Control, Vol.107, pp.1–24, 1985.

[6] K.F.L-Kovitz, J.E.Colgate and S.D.R.Carnes: Design of Components for Programmable Passive Impedance, Proc. of IEEE International Conference on Robotics and Automation, pp.1476–1481, 1991.

[7] M.Okada and Y.Nakamura: Development of the Cybernetic Shoulder – A Three DOF Mechanism that Imitates Biological Shoulder-Motion –, Proc. of IEEE/RSJ International Conference on Intelligent Robots and Systems , Vol.2, pp.543-548 (1999)

[8] T.Morita and S.Sugano: Design and Development of a new Robot Joint using a Mechanical Impedance Adjuster, Proc. of IEEE International Conference on Robotics and Automation, pp.2469-2475 (1995)

[9] H.Kobayashi, K.Hyodo and D.Ogane: On Tendon-Driven Robotic Mechanism with Redundant Tendons, Int. J. of Robotics Research, Vol.17, No.5, pp.561-571 (1998)

[10] M. Okada, Y. Nakamura and S. Hoshino: Design of Active/Passive Hybrid Compliance in the Frequency Domain — Shaping Dynamic Compliance of Humanoid Shoulder Mechanism —, Proc. of IEEE International Conference on Robotics and Automation, pp.2250-2257 (2000)

Design, Implementation, and Remote Operation of the Humanoid H6

Satoshi Kagami   Koichi Nishiwaki   James J. Kuffner Jr.
Tomomichi Sugihara   Masayuki Inaba   Hirochika Inoue
Dept. of Mechano-Informatics, Univ. of Tokyo.
7-3-1, Hongo, Bunkyo-ku, Tokyo, 113-8656, Japan.
{kagami,nishi,kuffner,zhidao,inaba,inoue}@jsk.t.u-tokyo.ac.jp

Abstract: The paper describes the humanoid robot "H6", which was designed to serve as a platform for experimental research on the development of advanced humanoid-type robots. The key features of the design of H6 include: 1) complete, 35-DOF human-shaped body with joints having sufficient torque to support full-body motion, 2) high-performance PC/AT compatible on-board computer running RT-Linux, a real-time OS that facilitates simultaneous low-level and high-level control, 3) fully self-contained on-board power supply and wireless network connection, that allows remote operation via radio ethernet, 4) software for dynamic walking trajectory generation, motion planning, and 3D stereo vision. We give an overview of both the hardware and software design components of H6, and discuss some experimental results involving remote operation.

1. Introduction

Recently, research on humanoid-type robots has become increasingly active, and many fundamental issues are under investigation. In particular, techniques for bipedal dynamic walking, soft tactile sensors, motion planning, and 3D vision continue to progress. However, in order to achieve a humanoid robot which can safely operate in a human environment together with human beings, not only the fundamental components themselves, but also the successful integration of these components will be required. At present, almost all humanoid robots that have been developed have been designed for bipedal locomotion experiments. In order to satisfy the functional demands of locomotion as well as high-level behaviors, humanoid robots require good mechanical design, hardware, and software which can support the integration of tactile sensing, visual perception, and motor control.

The child-sized, full-body humanoid "H5" (127cm height, 33kg mass) was previously developed for conducting research on dynamic bipedal locomotion, and techniques for dynamically-stable trajectory generation have been proposed[1, 2, 3]. However, a humanoid robot which can operate safely in a human environment requires fully self-contained on-board processing, sensing systems, power supply, and sophisticated software. The humanoid robot H6 was developed with the aim of satisfying these requirements.

In this paper, the functional requirements, design and implementation,

D. Rus and S. Singh (Eds.): Experimental Robotics VII, LNCIS 271, pp. 41–50, 2001.

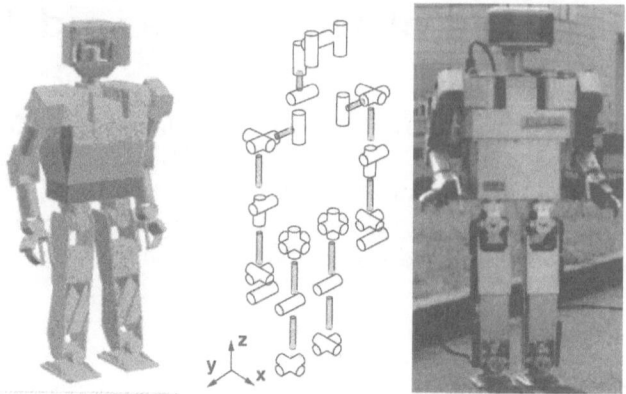

Figure 1. Geometric Model in Euslisp, DOF arrangement, Photo of H6

and remote-operation experiments using the humanoid robot H6 as a research platform for perception-action integration are described.

2. Conceptual Design and Specifications of H6

2.1. Design Criteria

The following criteria were considered in selecting the design for H6:

- Compact and light weight body.

- Modular structure for ease of maintenance and enhancement.

- Fully self-contained, including batteries, CPU, and network connection to a LAN via wireless ethernet.

- Sufficient DOF, range of movement, and maximal joint torque and speed, that enables dynamic walking as well as the ability to stand up should the robot fall down.

- Head-mounted dual cameras capable of looking straight down at the feet, as well as having vergence control motors for 3D vision.

- Smooth body surface, suitable for mounting tactile sensor skin made of pressure sensors and air-chambers.

- On-board high performance PC/AT computer running RT-Linux as the primary system controller.

- Software libraries for dynamically-stable walking trajectory generation, motion planning, 3D vision, and voice recognition and synthesis.

2.2. Specifications

The humanoid robot H6 was designed and developed according to the above requirements. The robot weighs a total of 55.0[kg] (including batteries), and measures $285(l) \times 598(w) \times 1361(h)$ $[mm]$ when standing at rest. H6 has a total of 35 degrees of freedom (DOF): 6 for each leg, 1 for each foot (toe joint), 7 for each arm, 1 for each gripper, 2 for the neck, and 3 for the head-mounted

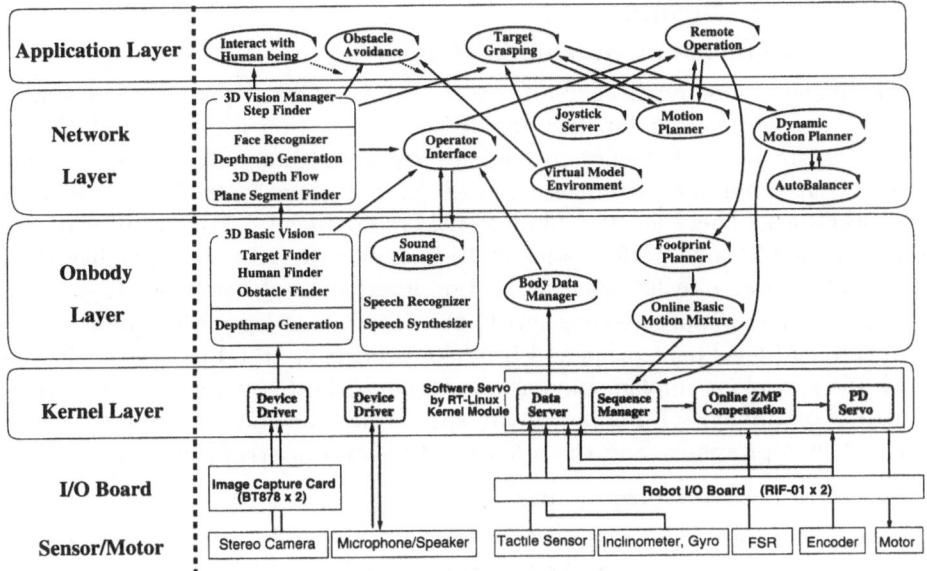

Figure 2. H6 Software Components

cameras (including vergence control). All major joints are driven by DC motors and Harmonic drive gears. An onboard PC/AT computer equipped with dual PentiumIII-750MHz processors running RT-Linux is used for real-time servo and balance compensation, as well as coordinating high-level 3D vision and motion planning component software modules. The system is connected to the network via wireless ethernet. Thus, the robot is fully self-contained (it can be operated without any external cables).

A 24V DC power supply is used for H6, thus two (12V, 5.0A, 2kg) lead-acid batteries are stored inside the torso for both controllers and motors. This configuration can supply power for about 10 to 15 minutes of normal operation on average (5 minutes for continuous walking). When squatting down, the maximum current drawn by the motors is about 29A, with the average current at approximately 4A(96W). The power consumption for the computer is 4.3A(102W) on average.

Twelve force sensing resistor (FSR) sheets are attached to the soles of each foot, so that the total force along the vertical axis and the ZMP position can be measured. An inclinometer and accelerometer are also mounted inside the center of the torso. Data from these sensors are measured using RIF-01 A/D boards.

A tactile skin sensor suit consisting of air chambers with pressure sensors was designed to provide both tactile feedback and shock absorption.

3. Software Design

3.1. Requirements of Humanoid Robots

A humanoid robot research platform should satisfy many aspects of experimental research, from low-level quick/smooth motion control to high-level
vision/sensor-based behavior in complex environments. Therefore, a transparent software-layer system is adopted for real-time control and high-level
computations. There are two fundamental requirements: 1) efficient software
servoing, and 2) high-performance multi-tasking with network capability, such
as remote resource utilization, application interface to developers, etc.

In order to simultaneously satisfy the demands of legged locomotion control and high-level perception and behavior processing, RT-Linux[4] is adopted
with the servo loop implemented as a kernel module. Since Linux is not a real-
time OS originally, RT-Linux has two special mechanisms: one is a scheduler
for real-time processes, and the other is a two-level interrupt handler.

3.2. H6 Software Components

There are six software components in H6(Fig.2). i) realtime servo-loop, online
ZMP compensation mechanisms for servoing and walking, ii) online footprint
planning mechanisms, iii) onbody low-level 3D vision processing, voice processing functions, iv) motion planning and obstacle-avoidance functions, v) a
vision, sound, and other sensor data server, in order to achieve network through
data processing, vi) high-level 3D vision functions, voice recognition, and other
high-level recognition functions are distributed on other network computers.

3.3. Joint Servo Unit

All 28 joints except those in the head are controlled by one RT loop which runs
at a 1msec cycle (Motor servo in Fig.2). It is essentially PD control.

3.4. Online ZMP Compensation

For humanoid-type robots, it is difficult to "replay" dynamically-stable walking
trajectories correctly in the real world, even if they satisfy the ZMP constraints.
Therefore, various local compliance control methods have been proposed [5, 6,
7, 8]. Currently, we have adopted a torso position compliance method to track a
given ZMP trajectory. This method attempts to track a given ZMP trajectory
by adjusting the horizontal motion of the torso. It consists of two parts: one
is a ZMP tracking mechanism, and the other is inverse pendulum control used
to maintain dynamic balance (Online ZMP compensation in Fig.2)[9].

3.5. Walk Trajectory Generation

For walking, it is difficult to generate controls for every DOF interactively.
We have proposed an offline dynamically-stable trajectory generation method
for humanoid robots[9]. From a given input motion and the desired ZMP
trajectory, the algorithm generates a dynamically-stable trajectory using the
relationship between the robot's center of gravity and the ZMP. A simplified
robot model is introduced that represents the relationship between its center
of gravity and ZMP. It can then be shown that a horizontal shift of the torso
can satisfy the given desired ZMP trajectory.

Let the z axis be the vertical axis, and the x and y axes be the other components of that sagittal and lateral plane respectively. First, we introduce a model of humanoid-type robot by representing the motion and rotation of the center of gravity (COG). Let the total mass of the robot be m_{total}, the position of the center of gravity be $r_{cog} = (r_{cog_x}, r_{cog_y}, r_{cog_z})$, and the total reaction force that the robot feels be $f = (f_x, f_y, f_z)$. The ZMP $p_{cog} = (p_{cog_x}, p_{cog_y})$ around the point $p = (p_x, p_y, h)$ on the horizontal place $z = h$ is defined as the point where the total moment around point p is $T = (0, 0, Tz)$. Then following differential equation is obtained.

$$p_{cog}^{err}(t) = r_{cog}^{err}(t) - \frac{m_{total} r_{cog_z}(t) \ddot{r}_{cog}^{err}(t)}{f_z^o(t)} \tag{1}$$

Here, p_{cog}^{err} is the error between the ideal ZMP p_{cog}^* and the current measured ZMP p_{cog}, and r_{cog}^{err} is the error between the ideal center of gravity trajectory r_{cog}^* and the current measured trajectory r_{cog}.

Finally, an iterative numerical method is adopted to eliminate approximation errors arising from the simplified model.

3.6. Online Mixture and Connection of Pre-designed Motions

In order to implement interactive and adaptive behaviors, an online walking pattern generation function is developed. Enhancing "Dynamically Stable Mixture of Pre-designed Motions[2]", an appropriate body position, posture and velocity can be generated by mixing pre-calculated candidate motions online (Footprint planner in Fig.2). The step cycle time is fixed, and arbitrary footprint positions on the plane can be achieved.

Utilizing the properties of the ZMP, a dynamically-stable mixture of pre-designed motion is carried out to generate a desired walking motion. This mixture consists of three stages: 1) offline typical stepping pattern generation, which calculates 21 pre-designed basis motions using the previously-described offline trajectory generation method for translational motions, 2) mixing the pre-designed basis patterns independently along the X&Y axes, and then mixing the resulting patterns, 3) connecting subsequent stepping motions by selecting patterns in which the torso velocities at the boundaries are smooth enough.

The 2nd and 3rd stages of this mixture method incur a relatively low computational cost, so that desired walking patterns can be generated in real-time. The user need only to designate the direction and speed of the motion using a pointing device. Fig.3(Top row) shows a joystick control experiment using the humanoid H6.

3.7. Autobalancer

The "AutoBalancer" software reactively generates dynamically-stable motions of a standing humanoid robot on-line, given an input motion ([10, 3]). The system consists of two parts: 1) a planner for considering state transitions defined by the nature of the contacts between the legs and the ground, and 2) a dynamic balance compensator which maintains balance by formulating and solving a second order nonlinear constrained optimization problem. The latter can compensate for the centroid position and the tri-axial moments of any standing motion, using all joints of body in real-time. The complexity of AutoBalancer is $O((p+c)^3)$, where p is the number of DOFs and c is the number of constraint equations (Autobalancer in Fig.2).

3.8. 3D Vision Processing

Real-time 3D Vision functions are fundamentally important for a robot that behaves in the real world. Recently, several real-time 3D depth map generation systems have been proposed in the computer vision literature (e.g. [11, 12]) and some commercial products are also currently available (e.g. [13]). However, these solutions typically require special hardware. Since an onboard real-time system is needed for mobile robotics (or other camera moving) applications, it is difficult to build such a system given the extra hardware requirements.

In order to solve this problem, we proposed a real-time depth map generation system using only standard PC hardware and a simple image capture card [14]. Four key issues were considered in order to achieve real-time performance and to obtain accurate range data: 1) use of a recursive (normalized) correlation technique, 2) cache optimization, 3) an online consistency checking method, 4) utilizing the MMX/SSE(R) multimedia instruction set.

So far we have developed real-time 3D vision functions that include: 1) depth map generation[14], 2) 3D depth flow generation[15], and 3) a plane segment finder[16]. Th real-time depthmap generation system, along with an application for finding and tracking human targets can run on the onboard PC. Other high-level vision functions such as the plane segment finder, face recognition software, and so on, consume much more computational resources. Thus, they are run on other processors distributed over the network.

3.9. Dynamically-stable Motion Planning

Since humanoid robot has many DOFs, it is difficult to calculate a full-body trajectory, such as reaching towards a target object, without colliding with itself or an obstacle in the environment.

We have developed an approach to path planning for humanoid robots that computes dynamically-stable, collision-free trajectories from full-body posture goals. Given a geometric model of the environment and a statically-stable desired posture, we search the configuration space of the robot for a collision-free path that simultaneously satisfies dynamic balance constraints [17].

Our approach is to adapt a variation of the randomized planner described in [18] to compute full-body motions for humanoid robots that are both dynamically-stable and collision-free. This planner (RRT-Connect) and its variants utilize Rapidly-exploring Random Trees (RRTs) [19] to connect two search trees, one from the initial configuration and the other from the goal. These methods have been shown to be efficient in practice and converge towards a uniform exploration of the search space.

3.10. Sound Processing

Since a humanoid robot has many motors, a significant amount of sound is generated during its operation. Therefore, voice recognition software should be able to filter internal noises. We adopted voice recognition software developed by Dr.Hayamizu at ETL. This software has the advantage that it can run on the onboard processor (it runs on Linux), and a programmer can very easily manage its dictionary. Leveraging this advantage, task-based dictionaries which contain only a limited set of words specific to a given task are prepared, which makes the recognition function robust in terms of noise. The speech synthesis is done using commercial software (Fujitsu), which also runs on Linux. Fig.3 2nd row shows a voice-command based walking experiment.

4. Remote Operation of a Humanoid Robot

The software components described in the previous section are still rather primitive considering the kinds of sophisticated autonomous behaviors likely to be expected of future humanoid-type robots. Thus, a graphical user interface for network remote operation with limited autonomy was adopted. It requires low-level autonomy for stability and task execution, but can accept high-level commands. Currently the Humanoid Robotics Project(HRP:MITI Japan) contains a sub-project for network-based, remote tele-operation of a humanoid robot [20]. However, instead of developing low-level autonomy, it is strongly dependent on virtual reality technology and hardware, and the availability and training of a human operator to carefully control the robot.

Increasingly sophisticated low-level autonomy will be required to enable higher-level autonomous behavior research on humanoid-type robots. Therefore, not only tele-operated applications will benefit, but also this autonomy will be useful for building other direct applications which require high-level autonomous behavior and adaptability to environments with uncertainty.

4.1. Network Operation Interface for Humanoid Robot

There exists three major requirements for a network-operated humanoid robot interface: 1) control interface, 2) environment interface, and 3) interaction interface.

The first requirement is the interface for controlling the joints of the robot body. Due to the number of DOFs, it is tedious and nearly impossible to control a robot by adjusting each joint position manually. Even using a master-slave control interface, it is very difficult to control the robot so as to maintain dynamic stability. Thus, we have designed a "virtual-puppet" graphical interface that combines the Autobalancer, and walking controls. An operator can issue high-level commands to the humanoid robot, such as to manipulate a particular object, or walk in a desired direction.

The second requirement is an interface for recognizing the environment. A remotely-operated robot working in a real environment typically has cameras with a relatively narrow field of view, thus it is difficult to obtain an overall 3D picture of the environment surroundings. Using a graphic display, the internal robot environment model, state display, and 3D vision results can be visualized and compared to local internal models. With this interface, an operator can simply designate a target object to grasp or a direction to walk with relative ease.

The third requirement consists of an interface between human beings who are a) working in the same environment as the remotely-controlled humanoid robot, and b) the user in front of the remote operation interface. Voice recognitionspeech and human findingface recognition functions are denoted. An operator can communicate with other human beings in the robot's environment without having to spend a lot of effort to do so. Thus, he or she can better concentrate on the robot's given task at hand.

5. Conclusion

The paper describes the humanoid robot "H6", which was designed to serve as a platform for experimental research on the development of advanced humanoid-type robots. The key features of the design of H6 include: 1) complete, 35-DOF human-shaped body with joints having sufficient torque to support full-body motion, 2) high-performance PC/AT compatible on-board computer running RT-Linux, a real-time OS that facilitates simultaneous low-level and high-level control, 3) fully self-contained on-board power supply and wireless network connection, that allows remote operation via radio ethernet, 4) software for dynamic walking trajectory generation, motion planning, and 3D stereo vision. We hope that H6 can become a common test-bed for experimental research aimed at developing humanoid-type robots with higher levels of both performance and autonomy.

Acknowledgements

This research has been supported by Grant-in-Aid for Research for the Future Program of the Japan Society for the Promotion of Science, "Research on Micro and Soft-Mechanics Integration for Bio-mimetic Machines (JSPS-RFTF96P00801)" project.

References

[1] K. Nagasaka, M. Inaba, and H. Inoue. Walking Pattern Generation for a Humanoid Robot based on Optimal Gradient Method. In *Proc. of 1999 IEEE Int. Conf. on Systems, Man, and Cybernetics No. VI*, 1999.

[2] K. Nishiwaki, K. Nagasaka, M. Inaba, and H. Inoue. Generation of reactive stepping motion for a humanoid by dynamically stable mixture of pre-designed motions. In *Proc. of 1999 IEEE Int. Conf. on Systems, Man, and Cybernetics No. VI*, pp. 702–707, 1999.

[3] S. KAGAMI, F. KANEHIRO, Y. TAMIYA, M. INABA, and H. INOUE. Autobalancer: An online dynamic balance compensation scheme for humanoid robots. In *Proc. of Fourth Intl. Workshop on Algorithmic Foundations on Robotics (WAFR'00)*, pp. SA–79–SA–89, 2000.

[4] V. Yodaiken and M.Barabanov. *RT-Linux*. http://www.rtlinux.org.

[5] Ken'ichirou NAGASAKA, Masayuki INABA, and Hirochika INOUE. Stabilization of Dynamic Walk on a Humanoid Using Torso Position Compliance Control. In *Proceedings of 17th Annual Conference on Robotics Society of Japan*, pp. 1193–1194, 1999.

[6] Honda Co. Ltd. *Walking Control System for Legged Robot*. Japan Patent Office (A) 5-305583, 1993.

[7] Honda Co. Ltd. *Walking Control System for Legged Robot.* Japan Patent Office (A) 5-200682, 1993.

[8] Honda Co. Ltd. *Walking Pattern Generation System for Legged Robot.* Japan Patent Office (A) 10-86080, 1998.

[9] S. KAGAMI, K. NISHIWAKI, T. KITAGAWA, T. SUGIHARA, M. INABA, and H. INOUE. A fast generation method of a dynamically stable humanoid robot trajectory with enhanced zmp constraint. In *Proc. of IEEE International Conference on Humanoid Robotics (Humanoid2000)*, 2000.

[10] Y. Tamiya, M. Inaba, and Hirochika Inoue. Realtime balance compensation for dynamic motion of full-body humanoid standing on one leg. *Journal of the Robotics Society of Japan*, Vol. 17, No. 2, pp. 268–274, 1999.

[11] K. Konolige. Small Vision Systems: Hardware and Implementation. In Y. Shirai and S. Hirose, editors, *Robotics Research: The Eighth International Symposium*, pp. 203–212. Springer, 1997.

[12] T. Kanade, A. Yoshida, K. Oda, H. Kano, and M. Tanaka. A Stereo Machine for Video-rate Dense Depth Mapping and Its New Applications. In *Proc. of the 1996 International Conference on Computer Vision and Pattern Recognition*, pp. 196–202, Jun 1996.

[13] Point Grey Research Inc. *Triclops Stereo Vision System.* http://www.ptgrey.com.

[14] S. KAGAMI, K. OKADA, M. INABA, and H. INOUE. Design and implementation of onbody real-time depthmap generation system. In *Proc. of International Conference on Robotics and Automation (ICRA'00)*, pp. 1441–1446, 2000.

[15] S. Kagami, K. Okada, M. Inaba, and H. Inoue. Real-time 3d optical flow generation system. In *Proc. of International Conference on Multisensor Fusion and Integration for Intelligent Systems (MFI'99)*, pp. 237–242, 1999.

[16] S. Kagami, K. Okada, M. Inaba, and H. Inoue. Plane segment finder. In *5th Robotics Symposia*, pp. 381–386, 2000.

[17] J. J. Kuffner, S. KAGAMI, M. INABA, and H. INOUE. Dynamically-stable motion planning for humanoid robots. In *Proc. of IEEE International Conference on Humanoid Robotics (Humanoid2000)*, 2000.

[18] J.J. Kuffner and S.M. LaValle. RRT-Connect: An efficient approach to single-query path planning. In *Proc. IEEE Int'l Conf. on Robotics and Automation (ICRA'2000)*, San Francisco, CA, April 2000.

[19] Steven M. LaValle and Jr James J. Kuffner. Rapidly-exploring random trees: Progress and prospects. In *Proc. of Fourth Intl. Workshop on Algorithmic Foundations on Robotics (WAFR'00)*, 2000.

[20] H. Inoue and S. Tachi and K. Tanie and K. Yokoi and S. Hirai and H. Hirukawa and K. Hirai and S. Nakayama and K. Sawada and T. Nishiyama and O. Miki and T. Itoko and H. Inaba and M. Sudo. HRP: Humanoid Robotics Project of MITI. In *Proc. of IEEE International Conference on Humanoid Robotics (Humanoid2000)*, 2000.

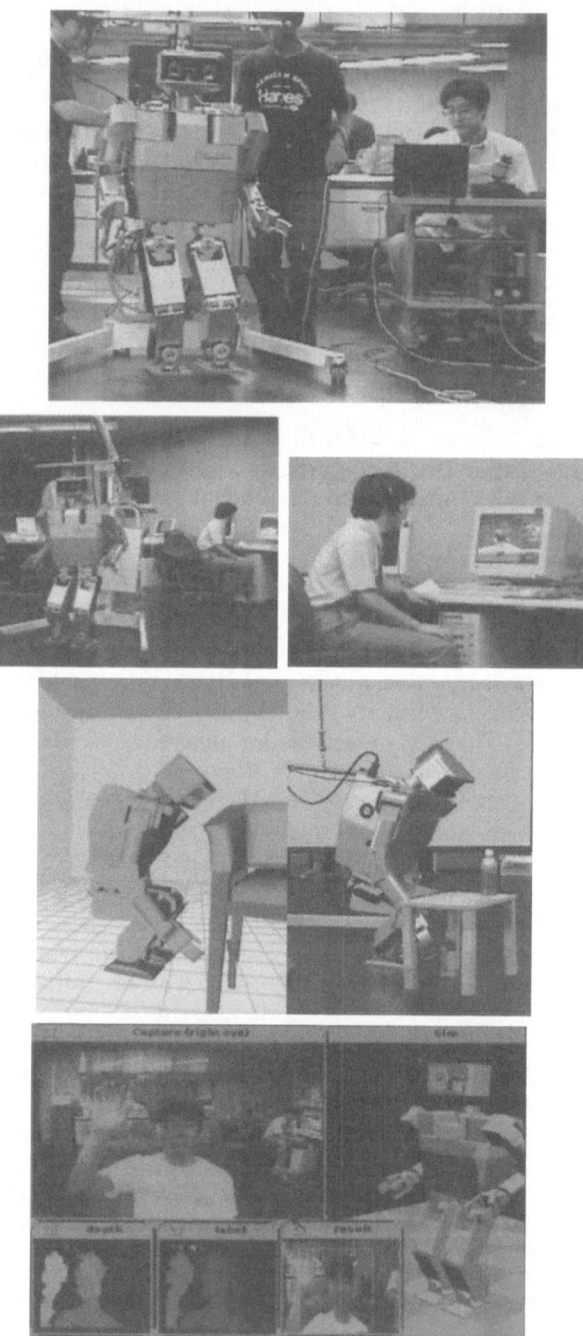

Figure 3. H6 Experiments : Top row: H6 Controlled by joystick (man on the right has joystick), 2nd row: Voice control, 3rd row: Dynamically-stable motion planning, 4th row: 3D vision system which finds and tracks human beings.

Cooperative Human and Machine Perception in Teleoperated Assembly

Thomas Debus and Jeffrey Stoll
Boston University
Boston. MA
tdebus@bu.edu and jstoll@bu.edu

Robert D. Howe
Harvard University
Cambridge, MA
howe@deas.harvard.edu

Pierre Dupont
Boston University
Boston. MA
pierre@bu.edu

Abstract: This paper presents results on a teleoperator expert assistant – a system that in cooperation with a human operator estimates properties of remote environment objects in order to improve task performance. Specifically, an undersea connector-mating task is investigated in the laboratory using a PHANToM master and WAM remote manipulator. Estimates of socket orientation are obtained during task performance and conveyed to the operator through a graphical display. Task performance, measured by completion time and peak insertion force, is compared for operators using combinations of video images, the graphical display and a shared control mode in which the connector automatically rotates to the estimated socket orientation. The graphical display and automatic orientation controller reduce task completion times and contact forces by over one-third for inclined sockets when the video signal is noisy, e.g., due to water turbidity.

1 Introduction

At present, teleoperation is the only way that robots can perform sophisticated manipulation tasks in unstructured environments. In this control mode, the human operator performs all required sensing and planning, and generates all motion commands based on feedback from the remote environment. In practical teleoperation systems (e.g. undersea operations [1] and surgery [2]), the sensory feedback is often limited to video images without force feedback, which greatly restricts dexterity and productivity. We have been working to alleviate this situation by using information from the remote robot arm's sensors to assist in teleoperated manipulation tasks [3][4][5]. We have derived algorithms that identify essential properties of objects in the remote environment including geometry, mass, compliance, and friction.

D. Rus and S. Singh (Eds.): Experimental Robotics VII, LNCIS 271, pp. 51–60, 2001.

In this paper, we focus on a specific practical application, undersea connector mating, where a cylindrical connector is inserted into a socket [1]. This peg-in-hole task is commonly performed in the offshore oil industry to provide hydraulic or electrical power to equipment on the sea floor. There are a number of factors that make this task troublesome. First, the clearance between the connector and receptacle is small, so a few degrees of angular misalignment can cause the connector to become jammed [6]. Second, sensory feedback is limited: because of stringent cost and reliability requirements, the master control device does not provide force feedback (Figure 1). Visual information is also restricted to monocular cameras that may be obstructed by sediment from nearby drilling operations. Third, manipulator arms for this application are hydraulically driven, with little compliance that would facilitate insertions. All these limitations make timely completion of this task difficult; in some cases it can take over an hour to insert a single connector, resulting in significant costs for the oil platform operator.

For a solution to find acceptance with the robot vendor and offshore operators, it must involve a minimum of modification to the robot, its controller and its software, as well as meeting rigorous reliability constraints. Thus, traditional robotic solutions to the insertion problem, such as mechanical or programmed remote center of compliance, are precluded. The ideal solution is one that can be implemented on an existing robot installation. Our proposed solution is to use the information from the joint angle sensors on the remote manipulator arm to determine the key geometric parameters of the remote environment. Specifically, we have developed an algorithm that automatically determines the relative orientation of the connector and the socket during the insertion task.

The process begins as the operator brings the connector into contact with the planar surface surrounding the opening of the socket, and slides the connector over the surface near the opening. The recorded joint sensor values are then combined with a model of the contact constraint between the cylindrical connector and the planar surface. Solution of the constraint equations yields a value for the orientation of the planar surface. This new information is presented to the operator as a graphical display to assist in orienting the connector. To explore the performance limitations imposed by the "minimum modification" rule, we have also implemented a shared control mode. In this approach, the connector is automatically oriented according to the estimated socket axis, while the operator controls translation to complete the insertion. This approach does not involve modification of the teleoperator system, but only access to joint encoder values. In this paper, we present experimental results confirming the benefits of the approach.

2   Methods

The laboratory teleoperator testbed used in these experiments consists of a PHANToM haptic interface as the master controller and a Barrett Whole Arm Manipulator (WAM) as the remote robot (Figure 2). To emulate the undersea application, the PHANToM (Model 1.5, Sensable Technologies, Cambridge, Mass., USA) is used as a passive 6 degree of freedom input device, and the motors are not activated. The WAM (Barrett Technologies, Cambridge, Mass., USA) is a redundant arm with 7 degrees of freedom, but only 5 axes are required for this insertion task, so the upper arm roll and final wrist roll axes are locked. Optical encoders measure the joint position on both robots, and velocities are computed using filtered backward differences. The workspace is roughly 0.2 m in diameter for the master robot and 1.0 m in diameter for the remote robot. The WAM robot is controlled by a dedicated RISC processor (Model DS1103, dSpace GmbH, Paderborn, Germany) running at a 10 kHz servo rate. The PHANToM joint data is read by a PC at a rate of 1 kHz and written into memory shared by the PC and the RISC processor.

 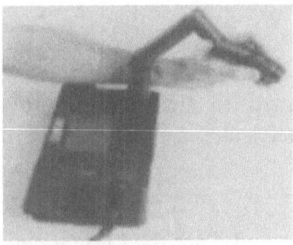

Figure 1. Schilling Robotic Systems Titan II manipulator, one of the leading commercial robots for undersea applications. (a) Remote manipulator arm is hydraulically powered. (b) The passive master arm provides no force feedback.

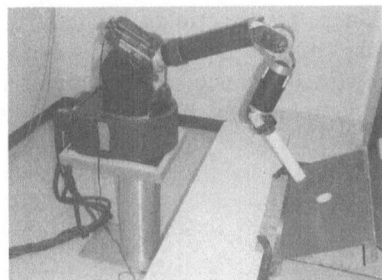 

Figure 2. (a) WAM remote robot arm with connector mating apparatus. (b) PHANToM master arm.

Teleoperation is accomplished with a simple proportional-derivative controller with feedforward gravity and motor torque ripple compensation on the arm and integral feedback on the wrist. In this control method, incremental Cartesian position, velocity and orientation of the master robot are mapped to the remote workspace, converted to remote robot joint positions and velocity using inverse kinematics, and then to torque commands by the following control laws.

$$\tau_i^{\text{remote}} = K_{P_i}(\theta_i^{\text{master}} - \theta_i^{\text{remote}}) + K_{V_i}(\dot{\theta}_i^{\text{master}} - \dot{\theta}_i^{\text{remote}}) + \tau_i^{\text{gravity}} + \tau_i^{\text{ripple}}, \qquad \text{arm}$$

$$\tau_i^{\text{remote}} = K_{P_i}(\theta_i^{\text{master}} - \theta_i^{\text{remote}}) + K_{V_i}(\dot{\theta}_i^{\text{master}} - \dot{\theta}_i^{\text{remote}}) + K_{I_i}\int(\theta_i^{\text{master}} - \theta_i^{\text{remote}})dt, \quad \text{wrist}$$

(1)

Here $\theta_i, \dot{\theta}_i, \tau_i, \tau_i^{\text{gravity}}$ and τ_i^{ripple} are the ith components of joint position, velocity, torque, gravity compensation and motor torque ripple compensation.

2.1 Environment modeling and operator assistance.

To find the orientation of the socket axis, we use techniques developed in our previous work [4],[5]. This approach estimates the remote object parameters by combining constraint equations describing the geometry of the contacting surfaces, closed loop kinematic relations, and kinematic data from the remote robot's joint sensors. In the present case, we are concerned with describing the contact point as the operator slides the cylindrical connector over the planar surface surrounding the socket. We assume that the connector is rigidly gripped by the remote robot in a known configuration. The contact point $p^{tool} = [x, y, z]^T$ on the connector is constrained to lie along its outer circular rim, which may be written in tool coordinates as

$$x^2 + y^2 - r^2 = 0, \quad z - l = 0 \tag{2}$$

where r and l are the known peg length and radius. The planar surface contact point $p^{world} = [u, v, w]^T$ may be written in world coordinates as

$$au + bv + cw + d = 0 \tag{3}$$

where a, b, c, and d are the desired parameters describing the orientation of the socket surface; we assume that the socket axis is normal to this surface, so that estimation of these parameters yields the correct connector orientation for insertion. These descriptions of the contact point p are related by the kinematic transformation between tool and world frames

$$T_{world}^{tool}(\theta) \, p^{tool} = p^{world} \tag{4}$$

As the connector slides over the planar surface, the joint positions θ are sampled, from which the transformation $T_{world}^{tool}(\theta)$ is calculated. Simultaneous solution of the above equations provides an estimate of the plane parameters a, b, c, and d and thus the correct connector orientation. This typically took 15 seconds to accomplish and 25 points were selected for the least squares estimation process using master-remote position error to infer contact.

The output is an estimate of the socket orientation for the insertion task. We are investigating a number of methods for conveying this information to the operator. The first is a 3D graphical model showing the manipulator, connector, and estimated plane location and orientation (Figure 3a), which is displayed on the monitor of the PHANToM–WAM interface computer. The model moves in real time to reflect the WAM's motion, and the operator can select the optimum rendering angle and distance. The second display shows a pair of targeting circles (Figure 3b). The larger circle represents the proximal end of the connector and the smaller circle the distal end, as projected onto the socket plane. The operator rotates the connector until the circles are concentric, which corresponds to the correct orientation, then proceeds with the insertion.

In addition to these displays, we have implemented a shared control mode. Following the estimation procedure, the orientation of the connector is automatically driven to the estimated socket angle, and orientation changes at the PHANToM stylus are ignored. The operator retains control of translational motion of the remote robot to perform the connector insertion. Additional display and shared control methods under development are described in the Discussion section below.

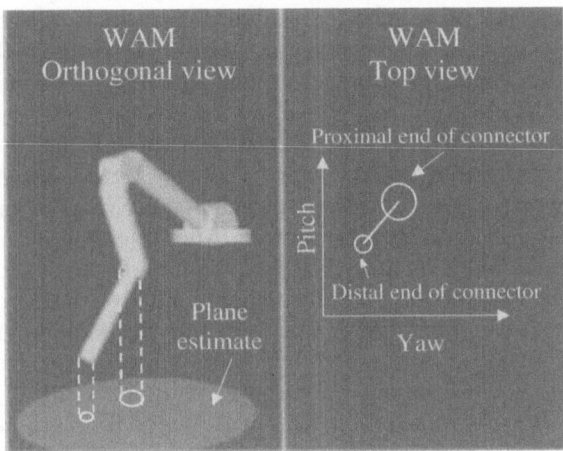

Figure 3. Graphical display (a) WAM model and (b) target circles, which are peg ends projected on the estimated socket plane.

2.2 Experimental protocol

The connector-socket apparatus was simulated by a pair of PVC plastic tubes (Figure 2a). The connector was 50 mm in diameter and 315 mm long. The socket had an inside diameter of 53 mm, and was mounted perpendicular to a planar surface that could be pivoted to a range of inclination angles between 0 and 57 degrees from horizontal. The operators used a monitor to view visual feedback from the remote robot via a video camera mounted adjacent to the shoulder joint of the WAM (Figure 4).

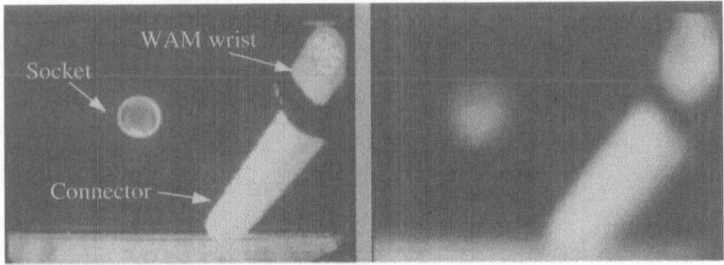

Figure 4. Typical visual feedback to operators from remote robot, showing the WAM wrist, connector, and socket opening. (a) Clear video signal. (b) Noisy video signal produced by defocusing video lens.

Two operators experienced in use of the system performed the insertion task under three different control modes and two visibility conditions, for a range of socket orientations. The first control mode provided only visual feedback with no estimation of the socket orientation, as in the current undersea connector mating task. The second mode added the two graphical displays of the estimated socket orientation described above, and the third used the shared control mode to automatically set the connector orientation in addition to the video and graphical displays. To simulate poor undersea visibility conditions, in some trials the video signal was degraded by defocusing, as shown in Figure 4(b). Each operator performed the task five times in each combination of control mode, visibility, and socket orientation.

In each trial that involved the estimation algorithm, the operator brought the end of the connector into contact with the planar surface surrounding the socket opening. The operator then depressed a switch that activated recording of joint angle data, and slid the connector over the surface. After releasing the switch, the operator proceeded with the insertion task. The time required for the estimation process and for the rest of the insertion task was recorded. In addition, the forces produced in the insertion process were approximated by recording the position errors at the remote robot and multiplying by the controller gain. This ignores inertial, impact, and frictional forces, but because of the clean drive train of the WAM robot and the quasi-static nature of the insertion task, it provides a reasonable estimate of the contact forces generated in the task.

3   Experiment results

In initial tests, we assessed the estimation algorithm's accuracy in determining the orientation of the planar surface around the socket opening. Five trials were conducted at five values of the orientation, at 0, 11, 30, 41 and 57 degrees. Figure 5 compares the estimated and actual orientations. These results show good agreement between the estimated and actual angles: the largest error is 1.3 degrees, and the largest standard deviation of the estimates at each angle is 0.6 degrees.

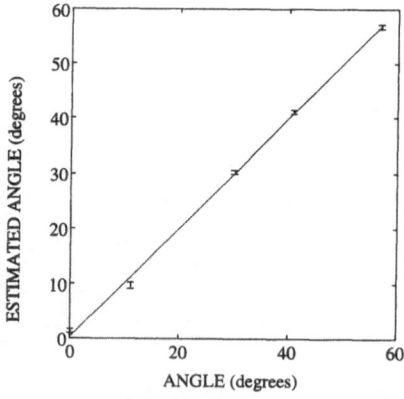

Figure 5. Estimates of the orientation of the planar surface around the socket opening.

Figure 6 and Figure 7 show the time-to-completion results for the connector-mating task for the three control modes and two visibility conditions at two surface orientations. For the two modes that used the estimation algorithm (i.e., graphic display and auto-orientation) separate results are indicated for the time to perform the estimate, the time to perform the insertion, and the sum of these, which is the total time to perform the task.

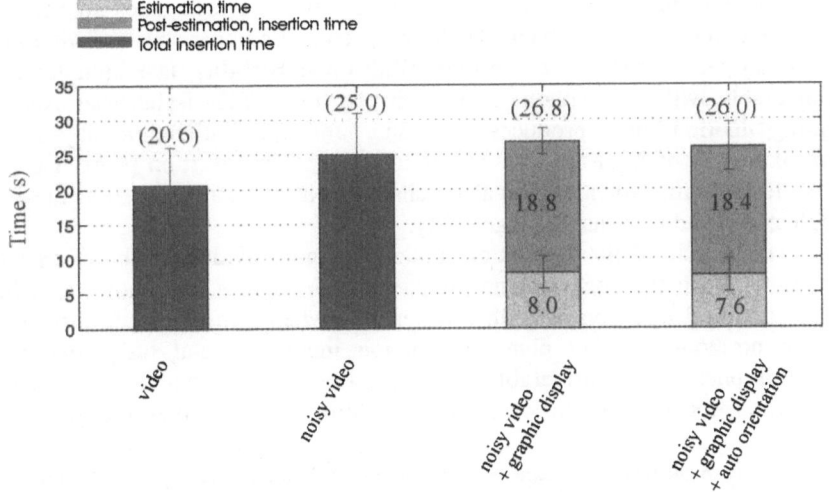

Figure 6. Task completion time as a function of control modes and visibility conditions for a horizontal surface. Symbols indicate mean; bars indicate standard deviations.

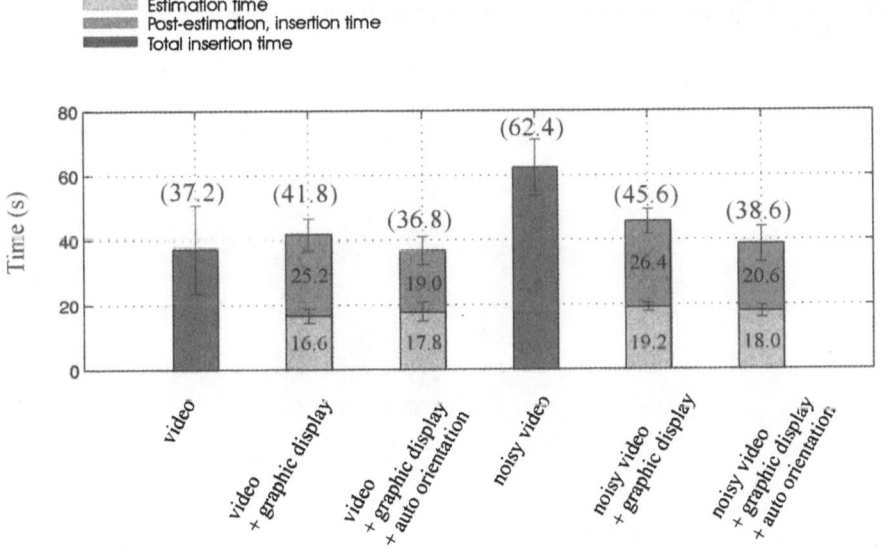

Figure 7. Task completion time as a function of control modes and visibility conditions for a surface inclined at 57 degrees.

As the angle of inclination was decreased from 57 degrees to horizontal, task completion time decreased uniformly. This is due to the assistance that gravity provides in correctly orienting the relatively compliant WAM wrist for a non-contact surface (Wrist compliance is due to tendon drive elasticity as well as deliberately reduced feedback gains, the latter in order to avoid damage-inducing contact forces). In addition, the steeper surface was less intuitive for the operators, especially during the estimation phase

These results indicate that, not surprisingly, the task takes significantly less time with clear video feedback. The more pertinent comparisons are within the clear and noisy video feedback conditions. With good visibility, task completion time is comparable with and without estimation algorithm. This is because executing the sliding motion that produces the data for the estimation algorithm takes approximately 40% of the task completion time. The insertion phase itself was 30-50% faster with either graphical or shared control assistance from the estimated angle compared to visual feedback only.

The effects of the estimation modes are more evident in the cases with poor visibility. With the socket plane inclined at 57 degrees, the entire task, including estimation and insertion, was 38% faster than when using visual information alone. For a horizontal socket plane, estimation increased total task time, while the insertion phase was considerably faster. For the non-zero inclination cases that used estimation, the shared control mode was uniformly faster than the graphic display mode.

Figure 8 shows estimates of the normalized peak force levels generated during the task for a socket plane inclined at 57 degrees. Normalization was based on video-only insertion force. As with the task completion time measure, the main difference of interest is among modes with poor visibility; the good visibility case produces the lowest forces. Both of the estimation-based modes produced peak forces over a third lower than the noisy video only mode.

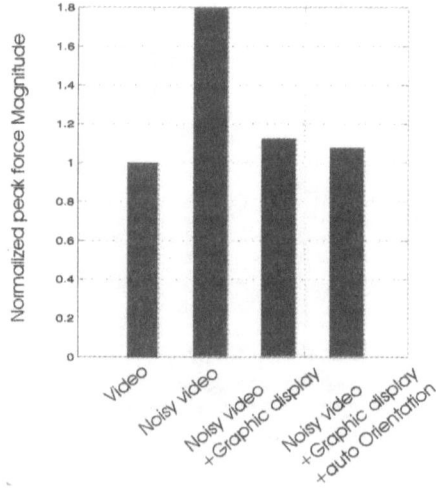

Figure 8. Normalized peak force estimates as a function of control mode for a socket plane inclined at 57 degrees.

4   Discussion

These results demonstrate that parameter estimation based on remote robot kinematic sensors can produce useful information for teleoperated manipulation tasks. Accuracy of surface orientation estimates was within 1.6 degrees, with a standard deviation of less than 0.6 degrees. Operators were able to accomplish the insertion phase of the task far more quickly with the estimation-based modes in all cases. These results indicate the viability of the technique and suggest a clear benefit for the intended application.

The approach has a number of features that make it well suited to the undersea application. The estimation algorithm is based on joint angle sensors only, which are standard components of the manipulators used in offshore applications. The algorithm does not require relatively expensive and fragile force sensing (although force information may improve the estimates and enable identification of additional properties [3]). This means that it is possible to configure a system that "looks over the shoulder" of the operator, without interfering with normal task execution and without significant modifications to the existing manipulator. If a graphical display is used to communicate with the operator, the computer that executes the estimation algorithm simply monitors the remote robot's joint sensors. If the estimation system fails, it has no impact on the manipulator's capabilities, so operations can continue as with the original manipulator system. This is an essential consideration for the offshore oil industry, where manipulator downtime can incur large costs.

Several important issues must be addressed prior to undersea implementation. One important issue is the difference in compliance between the WAM and undersea teleoperated robots, which are hydraulically powered, resulting in high endpoint impedance. High impedance makes insertions particularly difficult because the connector does not conform to contact forces and torques generated by misalignments. The relatively high compliance of the WAM resulted in much faster insertions than commonly observed in the undersea application. The low compliance of the hydraulic undersea manipulators makes the proposed estimation-based approach particularly beneficial, but it also makes it more difficult for the operator to slide the connector smoothly over the surface surrounding the socket. Further work will be directed at increasing the effective stiffness of the WAM arm to permit better investigation of these competing issues.

This application uses only a portion of the techniques we have developed for identifying the properties of remote environments during teleoperation [3],[4],[5]. In addition to estimating a variety of properties, these techniques can segment the data stream from the robot sensors and determine which contact states are active at each time. This enables a much greater range of applications. For example, the assumptions in the model, such as known length and radius of the connector, and perpendicular orientation between the socket axis and the surrounding surface, can be relaxed. A similar approach can be used to estimate other useful geometric parameters, such as the location of the socket opening. Such parameters can be presented to the operator using a variety of display modalities. Alternatively, they can be used to effect sophisticated shared controlled strategies. This capability promises to not only simplify teleoperation, but may represent a new level of perceptual capability for autonomous manipulation as well.

Acknowledgements

The support of the National Science Foundation under grant IIS-9988575 is gratefully acknowledged.

References

[1] Dennerlein J.T., Millman P., and Howe R.D. (1997). Vibrotactile Feedback for Industrial Telemanipulators. *Sixth Annual Symposium on Haptic Interfaces for Virtual Environment and Teleoperator Systems*, ASME International Mechanical Engineering Congress and Exposition, Dallas, Nov. 15-21, 1997, DSC-Vol. 61, pp. 189-195.

[2] Hill J.W, Jensen J.F. (1999). Telepresence technology in medicine: Principles and applications. *Proc. IEEE*, 86(3):369-380.

[3] Dupont P., Schulteis T., Millman P., and Howe R.D. (1999). Automatic Identification of Environment Haptic Properties. *Presence* 8(4):392-409.

[4] Debus T., Dupont P., and Howe R.D. (1999). Automatic Property Identification via Parameterized Constraints. *Proc. 1999 IEEE Conference on Robotics and Automation*, Detroit, MI, May, p. 1876-81.

[5] Debus T., Dupont P., and Howe R.D. (2000). Automatic Identification of Local Geometric Properties During Teleoperation. *Proc. 2000 IEEE Conference on Robotics and Automation*, San Francisco, CA, April, p. 3428-34.

[6] Whitney DE (1982). Quasi-Static Assembly of Compliantly Supported Rigid Parts, *ASME Journal of Dynamic Systems, Measurements, and Control* 104:65.

Regulation and Entrainment in Human-Robot Interaction

Dr. Cynthia Breazeal
MIT Artificial Intelligence Lab
Cambridge, MA 02139 USA
cynthia@ai.mit.edu

Abstract:

Newly emerging robotics applications for domestic or entertainment purposes are slowly introducing autonomous robots into society at large. A
critical capability of such robots is their ability to interact with humans,
and in particular, untrained users. This paper explores the hypothesis that
people will intuitively interact with robots in a natural social manner provided the robot can perceive, interpret, and appropriately respond with
familiar human social cues. Two experiments are presented where naive
human subjects interact with an anthropomorphic robot. Evidence for mutual regulation and entrainment of the interaction is presented, and how
this benefits the interaction as a whole is discussed.

1. Introduction

New applications for domestic, health care related, or entertainment based
robots motivate the development of robots that can socially interact with, learn
from, and cooperate with people. One could argue that because humanoid
robots share a similar morphology with humans, they are well suited for these
purposes – capable of receiving, interpreting, and reciprocating familiar social
cues in the natural communication modalities of humans.

However, is this the case? Although we can design robots capable of
interacting with people through facial expression, body posture, gesture, gaze
direction, and voice, the robotic analogs of these human capabilities are a crude
approximation at best given limitations in sensory, motor, and computational
resources. Will humans readily read, interpret, and respond to these cues in
an intuitive and beneficial way?

Research in related fields suggests that this is the case for computers [1]
and animated conversation agents [2]. The purpose of this paper is to explore
this hypothesis in a robotic media. Several expressive face robots have been
implemented in Japan, where the focus has been on mechanical engineering
design, visual perception, and control. For instance, the robot in the upper left
corner of figure 1 resembles a young Japanese woman (complete with silicone
gel skin, teeth, and hair [3]. The robot's degrees of freedom mirror those of
a human face, and novel actuators have been designed to accomplish this in
the desired form factor. It can recognize six human facial expressions and can

D. Rus and S. Singh (Eds.): Experimental Robotics VII, LNCIS 271, pp. 61–70, 2001.

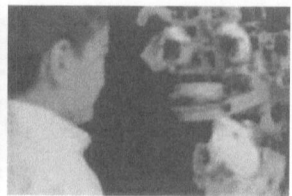

Figure 1. A sampling of robots designed to interact with people. The far left picture shows a realistic face robot designed at the Science University of Tokyo. The middle left picture shows *WE-3RII*, an expressive face robot developed at Waseda University. The middle right picture shows *Robita*, an upper-torso robot also developed at Waseda University to track speaking turns. The far right picture shows our expressive robot, *Kismet*, developed at MIT. The two leftmost photos are courtesy of Peter Menzel [6].

mimic them back to the person who displays them. In contrast, the robot shown in the upper right of corner of figure 1 resembles a mechanical cartoon [4]. The robot gives expressive responses to the proximity and intensity of a light source (such as withdrawing and narrowing its eyelids when the light is too bright). It also responds expressively to a limited number of scents (such as looking drunk when smelling alcohol, and looking annoyed when smoke is blown in its face). The lower right picture of figure 1, shows an upper-torso humanoid robot (with an expressionless face) that can direct its gaze to look at the appropriate person during a conversation by using sound localization and head pose of the speaker [5].

In contrast, the focus of our research has been to explore dynamic, expressive, pre-linguistic, and relatively unconstrained face to face social interaction between a human and an anthropomorphic robot called Kismet (see lower right of figure 1). For the past few years, we have been investigating this question in a variety domains through an assortment of experiments where naive human subjects interact with the robot. This paper summarizes our results with respect to two areas of study: the communication of affective intent and the dynamics of proto-dialog between human and robot. In each case we have adapted the theory underlying these human competencies to Kismet, and have experimentally studied how people consequently interact with the robot. Our data suggests that naive subjects naturally and intuitively read the robot's social cues and readily incorporate them into the exchange in interesting and beneficial ways. We discuss evidence of communicative efficacy and entrainment that results in an overall improved quality of interaction.

2. Communication of Affective Intent

Human speech provides a natural and intuitive interface for both communicating with humanoid robots as well as for teaching them. Towards this goal, we have explored the question of recognizing affective communicative intent in robot-directed speech. Developmental psycholinguists can tell us quite a lot about how preverbal infants achieve this, and how caregivers exploit it to regulate the infant's behavior. Infant-directed speech is typically quite exaggerated in the pitch and intensity (often called *motherese*). Moreover, mother's intuitively use selective prosodic contours to express different communicative intentions. Based on a series of cross-linguistic analyses, there appear to be at least four different pitch contours (approval, prohibition, comfort, and attentional bids), each associated with a different emotional state [7]. Figure 2 illustrates these four prosodic contours.

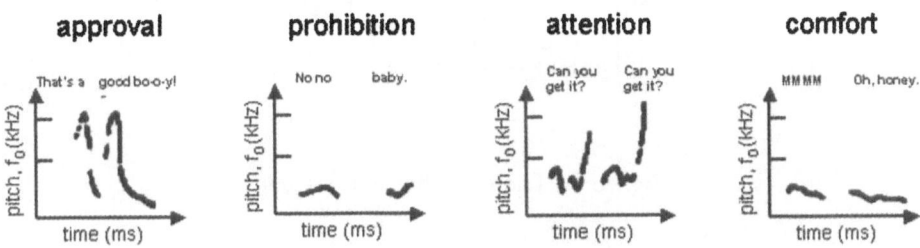

Figure 2. Fernald's prototypical prosodic contours for approval, attentional bid, prohibition, and soothing.

Mothers are more likely to use falling pitch contours than rising pitch contours when soothing a distressed infant [8], to use rising contours to elicit attention and to encourage a response [9], and to use bell shaped contours to maintain attention once it has been established [10]. Expressions of approval or praise, such as "Good girl!" are often spoken with an exaggerated rise-fall pitch contour with sustained intensity at the contour's peak. Expressions of prohibitions or warnings such as "Don't do that!" are spoken with low pitch and high intensity in staccato pitch contours. Fernald suggests that the pitch contours observed have been designed to directly influence the infant's emotive state, causing the child to relax or become more vigilant in certain situations, and to either avoid or approach objects that may be unfamiliar [7].

Inspired by these theories, we have implemented a recognizer for distinguishing the four distinct prosodic patterns that communicate praise, prohibition, attention, and comfort to preverbal infants from neutral speech. We have integrated this perceptual ability into our robot's *emotion system*, thereby allowing a human to directly manipulate the robot's affective state which is in turn reflected in the robot's expression.

2.1. The Classifier Implementation

We made recordings of two female adults who frequently interact with Kismet as caregivers. The speakers were asked to express all five communicative intents (approval, attentional bid, prohibition, soothing, and, neutral) during the interaction. Recordings were made using a wireless microphone whose output was sent to the speech processing system running on Linux. For each utterance, this phase produced a 16-bit single channel, 8 kHz signal (in a .wav format) as well as its corresponding pitch, percent periodicity, energy, and phoneme values. All recordings were performed in Kismet's usual environment to minimize variability in noise due to the environment.

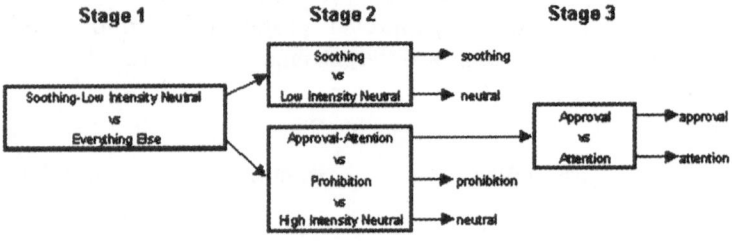

Figure 3. The classification stages.

The implemented classifier consists of several mini classifiers executing in stages (as shown in figure 3). In all training phases we modeled each class of data using the Gaussian mixture model, updated with the EM algorithm and a Kurtosis-based approach for dynamically deciding the appropriate number of kernels [11]. In the beginning stages, the classifier uses global pitch and energy features to separate the classes based on arousal measures (see fig 4). The remaining clustered classes were then passed to later classification stages that used features that carefully encoded the shape of the contours (as suggested by Fernald). These findings are consistent with Fernald's work and proved useful in separating the *difficult* classes. The classifier's structure follows logically from these observations.

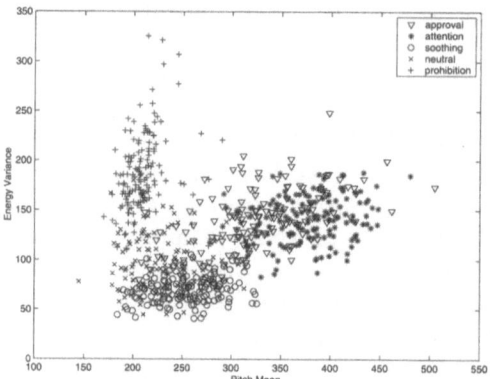

Figure 4. Feature space of all five classes.

The output of the recognizer is integrated into the rest of Kismet's synthetic nervous system as shown in figure 5. Due to space limitations, we leave the details to the interested reader as described in [12]. For our purposes here, the result of the classifier is passed to the robot's higher level perceptual system where it is combined with other contextual information. The result of the classifier can bias the robot's affective state by modulating the arousal and valence parameters of the robot's *emotion system*. The emotive responses are designed such that praise induces positive affect (a happy expression), prohibition induces negative affect (a sad expression), attentional bits enhance arousal (an alert expression), and soothing lowers arousal (a relaxed expression). The net affective/arousal state of the robot is displayed on its face and expressed through body posture [13], which serves as a critical feedback cue to the person who is trying to communicate with the robot. This expressive feedback serves to close the loop of the human-robot system.

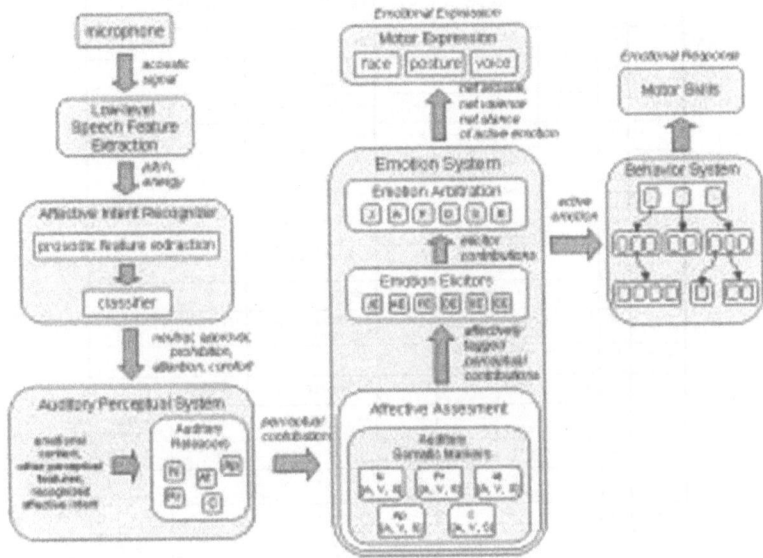

Figure 5. The output of the affective intent classifier is passed to the robot's *emotion system*, where it can influence the robot's affective state, its facial expression, and its behavior. The classifier output is first combined with other contextual information in the higher level perceptual system. These perceptions are then assessed for affective impact with respect to how they contribute to the robot's arousal, valence and stance parameters. This information is used to elicit the most relevant emotional response, that subsequently modulates the robot's expressive and behavioral response.

2.2. Affective Intent Experiment

Communicative efficacy has been tested with people very familiar with the robot as well as with naive subjects in multiple languages (French, German, English, Russian, and Indonesian). Female subjects ranging in age from 22 to

54 were asked to praise, scold, soothe, and to get the robot's attention. They were also asked to signal when they felt the robot "understood" them. All exchanges were video recorded for later analysis.

Intent	Tr	# phrase	Robot's Cues	Correct?	Subject's response	Change in prosody	Subject's comments
Praise	1	1	Ears perk up	No	Smile and acknowl		
	2	1	Ears perk up, little grin	no	Smile and acknowl		
	3	2	Look down	no	Lean forward	Higher pitch	
	4	2	Look up	no	Smile and acknowl	Higher pitch	
	5	1	Ears perk up, smile	yes	Lean forward, smile, acknowledge		"That's it"
	6		Lean forward, smile	yes	smile		
	7	2	smile	yes	Lean forward, smila, acknowledge	Higher pitch	
	8	3	smile	yes	Lean forward, smile, acknowledge	Higher pitch	
	9	4	attending	no	ignore		
	10		smile	yes	Lean forward, smile, acknowledge		
Alert	11	3	Make eye contact	no	Smile, acknowledge	Higher pitch	
	12	1	attending	yes	acknowledge		
	13	1	attending	yes	acknowledge		
	14	1	attending	yes	acknowledge		
	15	2	Lean forward, eye contact	yes	Lean forward, ack		
	16	2	Lean further, eye contact	no	Lean furhter, ack		
	17		Look down, frown		ignore		
	18	4	Look up	no	Lean forward, smile, acknowledge	Higher pitch	
Scold	19	4	look down	no	Lean forward, talk		
	20	4	frown	yes	acknowledge	Lower pitch	
	21	6	Look down, small grin	no	Lean forward, talk	giggle	"Volume would help"
	22	2	frown	yes	Pause, acknowledge	louder	
Soothe	23	4	Look up, eye contact	yes	Pause, acknowledge		
Scold	24	6	frown	yes	Pause, acknowledge		

Figure 6. Sample experiment session of a naive speaker, S3.

Figure 6 illustrates a sample event sequences that occurred during experiment sessions of a naive speaker. Each row represents a trial in which the subject attempts to communicate an affective intent to Kismet. For each trial, we recorded the number of utterances spoken, Kismet's cues, subject's responses and comments, as well as changes in prosody, if any.

2.3. Discussion

Recorded events show that subjects in the study made ready use of Kismet's expressive feedback to assess when the robot "understood" them. The robot's expressive repertoire is quite rich, including both facial expressions and shifts in body posture. The subjects varied in their sensitivity to the robot's expressive feedback, but all used facial expression, body posture, or a combination of both to determine when the utterance had been properly communicated to the robot. All subjects would reiterate their vocalizations with variations about a theme until they observed the appropriate change in facial expression. If the wrong facial expression appeared, they often used strongly exaggerated prosody to "correct" the "misunderstanding". In trial 20–22 of subject S3's experiment session, she giggled when kismet smiled despite her scolding, commented that volume would help, and thus spoke louder in the next trial. In general, the subjects used Kismet's expressive feedback to regulate their own behavior.

Kismet's expression through face and body posture becomes more intense as the activation level of the corresponding emotion process increases. For instance, small smiles verses large grins were often used to discern how "happy" the robot appeared. Small ear perks verses widened eyes with elevated ears and craning the neck forward were often used to discern growing levels of "interest" and "attention". The subjects could discern these intensity differences and several modulated their own speech to influence them. For example, in trials 1 and 2, Kismet responded to subject S3's praise by perking its ears and showing a small grin. In the next two trials the subject raised her pitch while praising Kismet to coax a stronger response. In trials 6–8 Kismet smiles broadly. We found that subjects often use Kismet's expressions to regulate their affective impact on the robot.

During course of the interaction, several interesting dynamic social phenomena arose. Often these occurred in the context of prohibiting the robot. For instance, several of the subjects reported experiencing a very strong emotional response immediately after "successfully" prohibiting the robot. In these cases, the robot's saddened face and body posture was enough to arouse a strong sense of empathy. The subject would often immediately stop and look to the experimenter with an anguished expression on her face, claiming to feel "terrible" or "guilty". In this emotional feedback cycle, the robot's own affective response to the subject's vocalizations evoked a strong and similar emotional response in the subject as well. This empathic response can be considered to be a form of entrainment.

Another interesting social dynamic we observed involved *affective mirroring* between robot and human. For instance, for another female subject (S2), she issued a medium strength prohibition to the robot, which caused it to dip its head. She responded by lowering her own head and reiterating the prohibition, this time a bit more foreboding. This caused the robot to dip its head even further and look more dejected. The cycle continues to increase in intensity until it bottoms out with both subject and robot having dramatic body postures and facial expressions that mirror the other. We see a similar pattern for subject S3 while issuing attentional bids. During trials 14–16 the subject mirrors the same alert posture as the robot. This technique was often employed to modulate the degree to which the strength of the message was "communicated" to the robot. This dynamic between robot and human is further evidence of entrainment.

3. Proto-Dialog

Achievement of adult-level conversation with a robot is a long term research goal. This involves overcoming challenges both with respect to the content of the exchange as well as to the delivery. The dynamics of turn-taking in adult conversation are flexible and robust. Well studied by discourse theorists, humans employ a variety of para-linguistic social cues, called *envelope displays*, to regulate the exchange of speaking turns [2]. Given that a robotic implementation is limited by perceptual, motor, and computational resources, could such cues be useful to regulate the turn-taking of humans and robots?

Kismet's turn-taking skills are supplemented with envelope displays as posited by discourse theorists. These paralinguistic social cues (such as raising of the brows at the end of a turn, or averting gaze at the start of a turn) are particularly important for Kismet because processing limitations force the robot to take-turns at a slower rate than is typical for human adults. However, humans seem to intuitively read Kismet's cues and use them to regulate the rate of exchange at a pace where both partners perform well.

3.1. Envelope Display Experiment

To investigate Kismet's turn-taking performance during proto-dialogs, we invited three naive subjects to interact with Kismet. Subjects ranged in age from 12 to 28 years old. Both male and female subjects participated. In each case, each subject was simply asked to carry a "play" conversation with the robot. The exchanges were video recorded for later analysis. The subjects were told that the robot did not speak or understand English, but would babble to them something like an infant.

Often the subjects begin the session by speaking longer phrases and only using the robot's vocal behavior to gauge their speaking turn. They also expect the robot to respond immediately after they finish talking. Within the first couple of exchanges, they may notice that the robot interrupts them, and they begin to adapt to Kismet's rate. They start to use shorter phrases, wait longer for the robot to respond, and more carefully watch the robot's turn taking cues. The robot prompts the other for their turn by craning its neck forward, raising its brows, and looking at the person's face when it's ready for them to speak. It will hold this posture for a few seconds until the person responds. Often, within a second of this display, the subject does so. The robot then leans back to a neutral posture, assumes a neutral expression, and tends to shift its gaze away from the person. This cue indicates that the robot is about to speak. The robot typically issues one utterance, but it may issue several. Nonetheless, as the exchange proceeds, the subjects tends to wait until prompted.

Before the subjects adapt their behavior to the robot's capabilities, the robot is more likely to interrupt them. There tend to be more frequent delays in the flow of "conversation" where the human prompts the robot again for a response. Often these "hiccups" in the flow appear in short clusters of mutual interruptions and pauses (often over 2 to 4 speaking turns) before the turns become coordinated and the flow smoothes out. However, by analyzing the video of these human-robot "conversations", there is evidence that people entrain

		time stamp (min:sec)	time between disturbances (sec)
subject 1	start @ 15:20	15:20 – 15:33	13
		15:37 – 15:54	21
		15:56 – 16:15	19
		16:20 – 17:25	70
	end @ 18:07	17:30 – 18:07	37+
subject 2	start @ 6:43	6:43 – 6:50	7
		6:54 – 7:15	21
		7:18 – 8:02	44
	end @ 8:43	8:06 – 8:43	37+
subject 3	start @ 4:52 min	4:52 – 4:58	10
		5:08 – 5:23	15
		5:30 – 5:54	24
		6:00 – 6:53	53
		6:58 – 7:16	18
		7:18 – 8:16	58
		8:25 – 9:10	45
	end @ 10:40 min	9:20 – 10:40	80+

	subject 1		subject 2		subject 3		avg
	data	%	data	%	data	%	
clean turns	35	83%	45	85%	83	78%	82%
interrupts	4	10%	4	7.5%	16	15%	11%
prompts	3	7%	4	7.5%	7	7%	7%
significant flow disturbances	3	7%	3	5.7%	7	7%	6.5%
total speaking turns	42		53		106		

Figure 7. The left table shows data illustrating evidence for entrainment of human to robot. The right table summarizes Kismet's turn taking performance during proto-dialog with three naive subjects. Significant disturbances are small clusters of pauses and interruptions between Kismet and the subject until turn-taking become coordinated again

to the robot (see the table to the left in figure 7). These "hiccups" become less frequent. The human and robot are able to carry on longer sequences of clean turn transitions. At this point the rate of vocal exchange is well matched to the robot's perceptual limitations. The vocal exchange is reasonably fluid. The table to the right in figure 7 shows that the robot is engaged in a smooth proto-dialog with the human partner the majority of the time (about 82%).

4. Conclusions

Experimental data from two distinct studies suggests that people do use the expressive cues of an anthropomorphic robot to improve the quality of inter-action between them. Whether the subjects were communicating an affective intent to the robot, or engaging it in a play dialog, evidence for using the robot's expressive cues to regulate the interaction and to entrain to the robot were observed. This has the effect of improving the quality of the interaction as a whole. In the case of communicating affective intent, people used the robot's expressive displays to ensure the correct intent was understood to the appropriate intensity. In the case of proto-conversation, the subjects quickly used the robot's cues to regulate when they should exchange turns. As the result, the interaction becomes smoother over time with fewer interruptions or awkward pauses. These results signify that for social interactions with humans, expressive robotic faces are a benefit to both the robot and to the human who interacts with it.

5. Acknowledgements

Support for this research was provided by ONR and DARPA under MURI N00014–95–1–0600, by DARPA under contract DABT 63-99-1-0012, and by NTT.

References

[1] B. Reeves and C. Nass 1996, *The Media Equation.* CSLI Publications. Stanford, CA.

[2] J. Cassell 2000, "Nudge Nudge Wink Wink: Elements of face-to-face conversation for embodied conversational agents". In: J. Cassell, J. Sullivan, S. Prevost & E. Churchill (eds.) *Embodied Conversational Agents*, MIT Press, Cambridge, MA.

[3] F. Hara 1998, "Personality characterization of animate face robot through interactive communication with human". In: *Proceedings of IARP98*. Tsukuba, Japan. pp IV-1.

[4] H. Takanobu, A. Takanishi, S. Hirano, I. Kato, K. Sato, and T. Umetsu 1998, "Development of humanoid robot heads for natural human-robot communication". In: *Proceedings of HURO98*. Tokyo, Japan. pp 21–28.

[5] Y. Matsusaka and T. Kobayashi 1999, "Human interface of humanoid robot realizing group communication in real space". In: *Proceedings of HURO99*. Tokyo, Japan. pp. 188-193.

[6] P. Menzel and F. D'Alusio 2000, *Robosapiens*. MIT Press.

[7] A. Fernald 1985, "Four-month-old Infants Prefer to Listen to Motherese". In *Infant Behavior and Development, vol 8*. pp 181-195.

[8] Papousek, M., Papousek, H., Bornstein, M.H. 1985, The Naturalistic Vocal Environment of Young Infants: On the Significance of Homogeneity and Variability in Parental Speech. In: Field,T., Fox, N. (eds.) *Social Perception in Infants*. Ablex, Norwood NJ. 269–297.

[9] Ferrier, L.J. 1987, Intonation in Discourse: Talk Between 12-month-olds and Their Mothers. In: K. Nelson(Ed.) *Children's language, vol.5*. Erlbaum, Hillsdale NJ. 35–60.

[10] Stern, D.N., Spieker, S., MacKain, K. 1982, Intonation Contours as Signals in Maternal Speech to Prelinguistic Infants. *Developmental Psychology*, 18: 727-735.

[11] Vlassis, N., Likas, A. 1999, A Kurtosis-Based Dynamic Approach to Gaussian Mixture Modeling. In: *IEEE Trans. on Systems, Man, and Cybernetics. Part A: Systems and Humans*, Vol. 29: No.4.

[12] C. Breazeal & L. Aryananda 2000, "Recognition of Affective Communicative Intent in Robot-Directed Speech". In: *Proceedings of the 1st International Conference on Humanoid Robots (Humanoids 2000)*. Cambridge, MA.

[13] C. Breazeal 2000, "Believability and Readability of Robot Faces". In: *Proceedings of the 8th International Symposium on Intelligent Robotic Systems (SIRS 2000)*. Reading, UK, 247–256.

Advancing Active Vision Systems by Improved Design and Control

Orson Sutherland, Harley Truong, Sebastien Rougeaux,
& Alexander Zelinsky

Robotic Systems Laboratory
Department of Systems Engineering, RSISE
Australian National University
Canberra, ACT 0200 Australia
http://syseng.anu.edu.au/rsl/

Abstract: This paper presents the mechanical hardware and control software of a novel high-performance active vision system. It is the latest in an ongoing research effort to develop real-world vision systems based on cable-drive transmissions. The head presented in this paper is the laboratory's first fully cable-driven binocular rig, and builds on several successful aspects of previous monocular prototypes. Namely, an increased payload capacity, a more compact transmission, and a design optimised for rigidity. In addition, we have developed a simple and compact controller for real-time tracking applications. It consists of two behavioural subgroups, saccade and smooth pursuit. By using a single trapezoidal profile motion (TPM) algorithm, we show that saccade time and motion smoothness can be optimised.

1. Introduction

A brief overview of previously built active vision devices reveals a trend towards smaller, more agile systems. In the past the goals were to experiment with different configurations using large systems with many DOFs, like the KTH active head [6] with its 13 DOFs and Yorick 11C [8] with a 55cm baseline and reconfigurable joints. More recently, smaller active heads such as the palm-sized Yorick 55C [7] and ESCHeR [3] with an 18cm baseline have been designed to mount on mobile robots for active navigation and for telepresence applications.

The trend towards smaller active vision systems comparable in size to the human head is pushing the limit of motor, gearbox and camera design. In most systems, the size of the motors and cameras limit the compactness of the active head and the motors themselves add to the inertia of moving components. A notable exception to this is the Agile Eye [2] where no motor carries the mass of any other motor. Such a parallel mechanical architecture was the inspiration for the drive system in our active head (Figure 1).

D. Rus and S. Singh (Eds.): Experimental Robotics VII, LNCIS 271, pp. 71–80, 2001.
© Springer-Verlag Berlin Heidelberg 2001

Figure 1. Fully assembled active head.

Another issue in the pursuit of faster and more accurate active heads is the choice of transmission system. The need for backlash-free speed reduction is critical for high-speed applications and the most common way this is solved is with harmonic-drive gearboxes. All three versions of Yorick as well as ESCIIeR use harmonic-drive technologies. A disadvantage of the technology is an unavoidably large speed-reduction ratio that limits the output speed to less than 100rpm [9]. This limitation is seldom a problem in applications like smooth pursuit where joint velocities rarely saturate. But during high speed movements like saccades, where the motors are driven at maximum acceleration to travel from one extreme position to the other, velocity saturation is of concern. Cable drive technology is an alternative to harmonic drive gearboxes that does not have speed limitations. The advantages of cable drive are discussed in later sections.

Our earlier prototype built at the ANU Robotic Systems Laboratory proved the usefulness of cable-drive transmissions and parallel mechanical architectures in a 2 degree-of-freedom active 'eye' system [11]. The prototype was fast, responsive and accurate. CeDAR applied the knowledge learnt from the earlier design, but in a stereo configuration.

This paper documents the design of the CeDAR system from initial performance specifications through to the choice of kinematics, transmission system and mechanical architecture as well as the hardware components used and the results of performance testing. In addition a control system that makes use of TPM to optimise saccade time and smoothness of pursuit in tracking applications is presented. Finally, a brief synopsis of future developments is given.

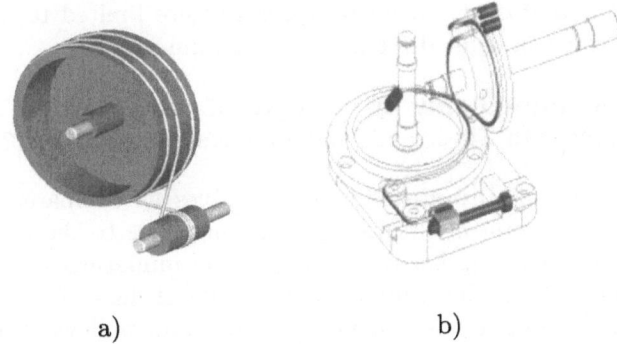

a)   b)

Figure 2. a) Cable transmission system b) Cable drive equivalent of bevel gear.

2. Mechanical Design

2.1. Kinematics

There are two widely used configurations for stereo active platforms, the Helmholtz and the Fick configuration. A description of the merits of each design is given in [5]. CeDAR is arranged in the more popular Helmholtz configuration with three axes: left vergence, right vergence and a common tilt (elevation) axis.

An important kinematic property of the design is that the axes intersect at the optical center of each camera. For vision processing this reduces translational effects and the number of unknown parameters that need calibration.

2.2. Transmission System

The transmission system used in CeDAR is the same as the one used in our first prototype [11]. It was inspired by a cable driven manipulator [10]. A cable drive transmission consists of a pulley, a smaller diameter pinion and a cable that wraps around both the pulley and the pinion (Figure 2.a).
The principle is the same as in gear transmissions except force is transmitted by tension in the cables and not by contact between gear teeth. Speed reduction, similar to gear transmissions, is proportional to the ratio of pulley and pinion diameters. There are many advantages in using cables:

- **No backlash**: force is transmitted by tension in the cables rather than contact forces between gear teeth.

- **No slippage**: unlike belt drive, the cables are terminated at each end and torque is transmitted to the pinion by several turns of cable to prevent slippage.

- **No lubrication**: the cables do not experience wear or friction like gearboxes and therefore do not require lubrication.

- **High efficiency**: typically 96% [10] compared to 80% for planetary gearboxes [4].

- **No speed limits**: harmonic gearboxes are limited to less than 100rpm [HD Systems], cable drive has no speed limitations.

- **Torque limited only by strength of cables**: We use a 1.12mm diameter cable with 343 strands and a breaking strength of 77kg.

There are some disadvantages in using cables as compared to conventional gear trains. The first is a finite angular range due to the cables not forming a continuous loop. Another is the difficulty in miniaturizing the transmission. The limiting factor is the minimum bend radius of the stainless steel cables that prevents the use of smaller diameter pinions and pulleys. Future prototypes may use other types of cables like synthetic fibres that have better strength to thickness ratios and more flexibility.

However, in well designed active heads, the disadvantages just mentioned are not relevant because (i) the angular ranges of the joints are limited to 90° (Table 1), and (ii) if the pulleys are integrated into structural members, then the size of the transmission is no longer an issue. For example, in our active head, the final stage bevel is part of the camera mounting bracket (Figure 3).

Figure 3. Rear view showing cable circuits.

An interesting part of the cable system is the bevel transmission that transmits torque across orthogonal shafts (Figure 2.b). The key part of the design is the use of two cables: one for forward motion and one for backward motion. Each bevel has two cable-wrapping surfaces with different diameters so that there are two points of intersection between the bevels for the cables to jump across. If there were only one wrapping surface per bevel, then both cables would have to cross over at the exact same point, which is physically impossible.

2.3. Mechanical Architecture

Inspired by devices such as the Agile Eye [2], the active head has a parallel mechanical architecture. Figure 3 shows how all the motors are fixed to the base so that they do not contribute mass to any of the joints. The advantage in doing so as opposed to locating the motors on the tilt joint itself is that the load placed on the tilt motor is lessened. Another advantage is that cable management is easier: the motor and encoder wires do not have to pass through awkward joints to reach the base.

The penalty of having a parallel architecture is that it makes the device more complex. Indeed, adding a fourth degree of freedom, a global pan (neck) joint, and still keeping to the parallel drive architecture would be challenging.

Finally, the rig has been optimised for maximum stiffness and minimum weight. This was necessary not only to increase the speed of the head but also its accuracy.

3. Hardware Overview

Figure 1 shows the fully assembled active head. It weighs 3.5kg with a moving mass of 1.7kg including the 700g payload.

The video and control hardware consists of Sony digital cameras (DFW-VL500), a Motion Engineering Inc. motion card (PCX/DSP), Maxon DC motors (RE25 and RE36) and a Pentium III computer.

4. Performance Specifications and Testing

Table 1 lists the performance specifications for the active head. The maximum range, payload and baseline specifications were based on the potential use of larger motorised-zoom cameras. The saccade rate and pointing accuracy were chosen based on the desired performance of the device in its intended application. Real-time tracking is the desired task and there is a direct relationship between our task-oriented specifications and the minimum requirements for effective tracking [1].

A software routine was written to test the speed and accuracy of CeDAR. The results are summarised in Table 1. To test speed, the joints were driven to perform repeated saccades. To test accuracy laser pointers were mounted on the sides of the cameras (Figure 1). By programming the head to follow geometric patterns on a wall 5 meters away, we were able to prove **repeatability**, **angular resolution** and **coordinated motion**. Table 1 shows that all specifications were met convincingly. CeDAR's performance compares extremely

Specification	Test Tilt	Test Vergence	Spec Tilt	Spec Vergence
Max Velocity	$600°.s^{-1}$	$800°.s^{-1}$	$600°.s^{-1}$	$600°.s^{-1}$
Max Acceleration	$18000°.s^{-2}$	$20000°.s^{-2}$	$10000°.s^{-2}$	$10000°.s^{-2}$
Saccade Rate	$5Hz$	$6Hz$	$5Hz$	$5Hz$
Ang Repeatability	$0.01°$	$0.01°$	$0.01°$	$0.01°$
Ang Resolution	$0.01°$	$0.01°$	$0.01°$	$0.01°$
Max Range	$90°$	$90°$	$90°$	$90°$

Table 1. Performance specifications and test results.

well to existing heads in addition to its ability to carry a wide range of payloads. Table 2 compares CeDAR's peak vergence velocity and acceleration to two key designs.

Specification	CeDAR	ESCHeR	Agile Eye
Max Velocity	$800°.s^{-}1$	$400°.s^{-}1$	$1000°.s^{-}1$
Max Acceleration	$20000°.s^{-}2$	$16000°.s^{-}2$	$20000°.s^{-}2$

Table 2. Comparison with two leading designs.

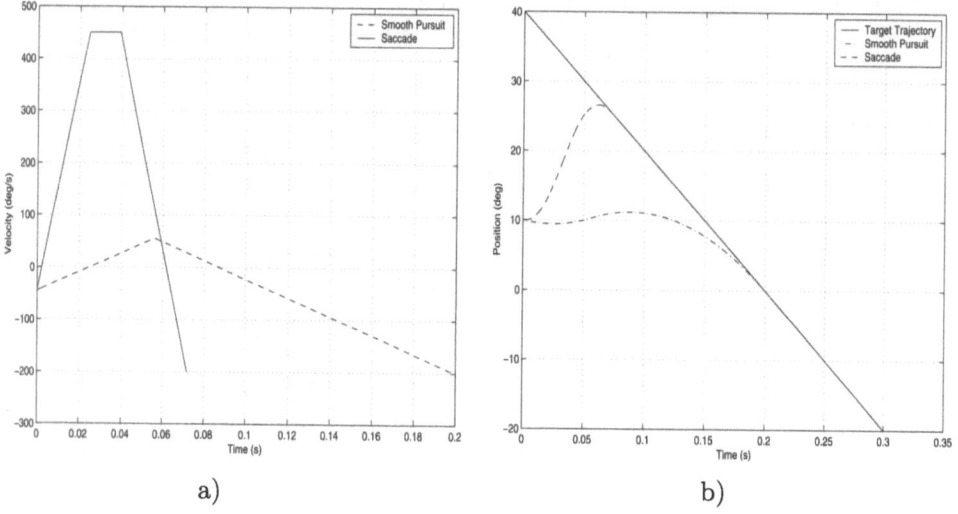

a)   b)

Figure 4. a) Velocity profiles b) Position profiles, during trapezoidal motion for saccade and smooth pursuit.

5. Control

CeDAR's controller is an extension of preliminary work undertaken by [5] on TPM. In particular our approach allows for the implementation of a single algorithm for both saccade and smooth pursuit, enhancing the simplicity and compactness of the controller's design.

5.1. Trapezoidal Profile Motion

The essence of the TPM problem is to catch a target initially a distance x_0 from the image center either in the shortest time possible, in the case of saccade, or in the smoothest way possible, in the case of smooth pursuit. Both the joints' and the target's starting velocities are potentially non-zero and disparate. Specifically it causes an axis to accelerate at a constant acceleration to a pre-calculated ceiling velocity[1], coast at this velocity for a given period and then decelerate at the same constant rate as the initial acceleration until the target velocity is reached (see figure 4.a). From a mathematical perspective it is a 4 dimensional problem, where the bang acceleration a, ceiling velocity v, move time T and total distance travelled x are the unknowns. The initial joint velocity v_1, the target velocity v_2 and the target's initial distance from the image center x_0 are the givens.

---

[1]The maximum absolute velocity of the TPM trajectory

The Algorithm

Since the acceleration a is assumed to be constant, the time taken by the head to accelerate from its initial velocity v_1 to the ceiling velocity v is

$$T_a = \frac{s \cdot v - v_1}{s \cdot a},$$ (1)

where s is positive if the head accelerates from v_1 to v and negative if it decelerates.

Similarly the time taken to decelerate from the ceiling velocity to the target velocity v_2 is

$$T_d = \frac{s \cdot v - v_2}{s \cdot a}.$$ (2)

Note that the rate of deceleration is equal to the rate of acceleration.

If T_c is the time spent coasting at the ceiling velocity, the total time of the trapezoidal profile motion is

$$T = T_a + T_c + T_d.$$ (3)

The distance traveled by the head during the action is

$$x = \frac{s \cdot v + v_1}{2} T_a + s \cdot v T_c + \frac{s \cdot v + v_2}{2} T_d,$$ (4)

but can also be considered as the sum of the initial distance of the target from the foveal center x_0 and the distance travelled by the target during the move

$$x = x_o + T v_2.$$ (5)

With these general TPM equations the specifics of saccade and smooth pursuit can now be developed.

Saccade

Saccades involve changing the head's current position and velocity state to that of the target, as inferred by its previous states, in the shortest time possible (see Figure 4.b). Motion smoothness is not a concern and hence acceleration is set to its maximum possible magnitude. Two cases can arise:

- The ceiling velocity required for the action is less than the maximum allowed velocity and hence no time is spent coasting.

- The theoretical ceiling velocity required for the action is greater than the maximum allowed velocity and hence some time must be spent coasting (see Figure 4.a)

It is useful to assume T_c as initially being zero so that (1), (2), (4) and (5) yield

$$s \cdot v = v_2 \pm \frac{1}{2} \sqrt{4sx_0 a - 2(v_1^2 + v_2^2) - 4v_1 v_2},$$ (6)

where the smaller of the two values is taken if v_2 is greater than v_1 and vice-versa. But if both of these values are in excess of the maximum allowed velocity, acceleration and velocity in (1), (2), (4) and (5) are replaced by their respective maxima from which

$$T_c = \frac{1}{v_2 - s \cdot v}((\frac{s \cdot v + v_1}{2} - v_1)T_a + (\frac{s \cdot v + v_2}{2} - v_2)T_a - x_0), \qquad (7)$$

is calculated and hence T is deduced.

Equation (6) also defines the value of s. In particular it must be such that the operand of the radical is greater or equal to zero, hence:

$$\begin{cases} s = +1 & \text{if } (v_1 - v_2)^2 + 4x_0a \geq 0 \\ s = \text{-1} & \text{otherwise} \end{cases}$$

Smooth Pursuit

In smooth pursuit we wish to move from one position and velocity state to the next in a given amount of time with the optimal smoothness (see figure 4.b). To achieve this the acceleration in moving to and from the ceiling velocity must be as small as possible. Again both the coasting and no-coasting cases mentioned above are relevant and again we start by assuming that the coasting time is zero initially, from which (1), (2), (4) and (5) yield:

$$v = \frac{x}{T} \pm \frac{1}{2T}\sqrt{4x^2 - 4Tx(v_1 + v_2) + 2T^2(v_1^2 + v_2^2)}. \qquad (8)$$

If these values are in excess of the maximum allowable velocity of the head, the time constraint is unrealiseable. In this eventuality CeDAR's controller has been implemented to initiate a saccade.

6. Future Work

6.1. Applications

As already mentioned, CeDAR's mechanical and control architectures were designed for real-time tracking. Zero-Disparity filtering and Optic Flow algorithms are in the process of being integrated into the system. Coupled with the TPM controller this should allow for robust tracking. In particular, we intend to demonstrate CeDAR's ability to locate and track a tennis ball during a tennis match.

6.2. Hardware Improvements

Most applications in active vision, like tracking and especially mobile navigation require devices with a global pan joint (neck). Further improvements on the active head would implement this feature using a harmonic drive motor. Since the neck joint does not need to move rapidly, there is no need to implement the joint in parallel with the other joints. A simple serial design where the fourth motor would sit beneath the existing head is a straightforward way to do this.

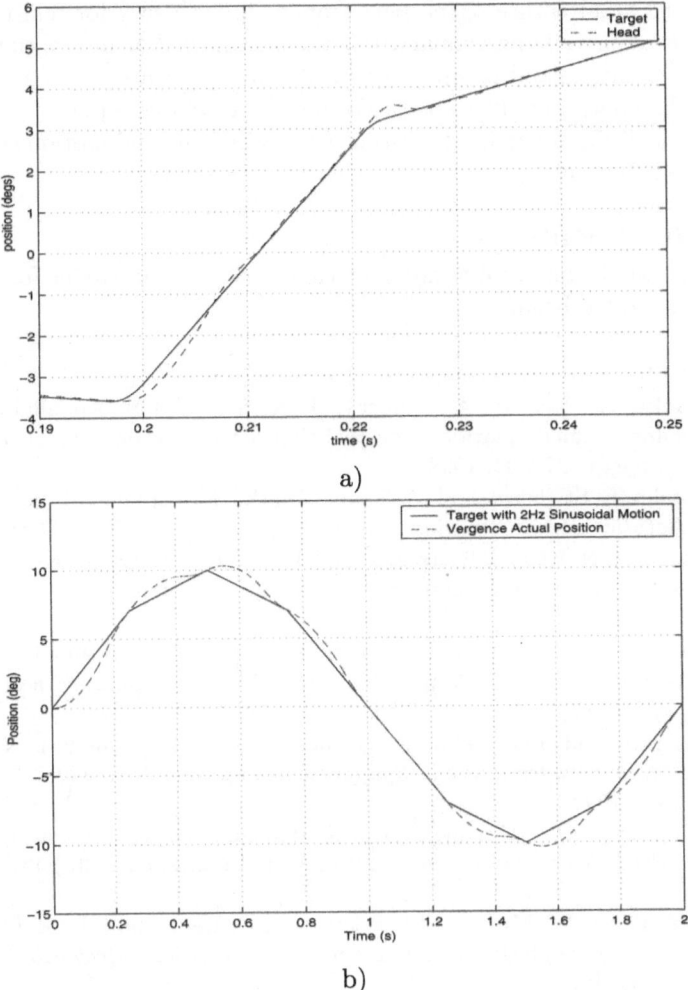

Figure 5. a) Two successive real-time saccades b) Smooth pursuit of a 0.5Hz sinusoid, sampled at 4Hz.

6.3. Future Prototypes

The use of plastics and other polymers in a cable-drive system should see a significant reduction in size and weight. It would also be a low cost way to manufacture moderate to high numbers of active heads. Another idea would be to use the Fick configuration to build a head with two independent 'eyes' similar to the pan-tilt device. The advantage in doing so would be to reduce the inertia to essentially only the cameras.

7. Conclusion

This paper has outlined a novel approach to the design of a fast and accurate 3 DOF stereo active head. Performance was achieved using cable transmissions and a parallel architecture. Such performance is necessary for real-time applications such as surveillance, navigation and human/robot interaction.

In addition, a simple and compact controller was presented that allowed both saccade and smooth pursuit to be performed for tracking applications. It made use of a single TPM algorithm to optimise saccade time and motion smoothness during smooth pursuit.

Acknowledgments

The authors would like to extend their thanks to Dr Jon Keifer for his initial design of the active head.

References

[1] A. Brooks, G. Dickens, A. Zelinsky, J. Kieffer, and S. Abdallah. A high-performance camera platform for real-time active vision. *Field and Service Robotics*, pages 527–532, 1998.

[2] C. Gosselin, E. St-Pierre, and M. Gagne. On the development of the agile eye. In *IEEE Robotics and Automation Magazine.*, pages 29–37. IEEE, December 1996.

[3] Y. Kuniyoshi, N. Kita, S. Rougeaux, and T. Suehiro. Active stereo vision system with foveated wide angel lenses. In *Asian Conf. on Computer Vision*, Singapore, 1995.

[4] Maxon Motors. High precision drives and systems. www.maxonmotor.com, 1999.

[5] D.W. Murray, F. Du, P.F. McLauchlan, I.D. Reid, P.M. Sharkey, and M. Brady. *Active Vision*, pages 155–172. MIT Press, 1992.

[6] K. Pahlavan and J.O. Eklundh. A head-eye system - analysis and design. *CVGIP: Image Understanding: Special issue on purposive, qualitative and active vision*, July 1992.

[7] P.M. Sharkey, D.W. Murray, and J. J. Heuring. On the kinematics of robot heads. *IEEE Transactions on Robotics and Automation*, 13(3):437–442, June 1997.

[8] P.M. Sharkey, D.W. Murray, S. Vandevelde, I.D. Reid, and P.F. McLauchlan. A modular head/eye platform for real-time reactive vision. *Mechatronics Journal*, 3(4):517–535, 1993.

[9] HD Systems. Harmonic drive systems. www.hdsystemsinc.com.

[10] W. T. Townsend. *The effect of transmission design on force-controlled manipulator performance*. PhD thesis, MIT Artificial Intelligence Laboratory, April 1988.

[11] S. Truong, J. Kieffer, and A. Zelinsky. A cable-driven pan-tilt mechanism for active vision. In *Proceedings Australian Conference on Robotics and Automation*, pages 172–177, Brisbane 1999.

S-NETS: Smart Sensor Networks

Yu Chen
University of Utah
Salt Lake City, UT 84112 USA
yuchen@cs.utah.edu

Thomas C. Henderson
University of Utah
Salt Lake City, UT 84112 USA
tch@cs.utah.edu

Abstract: The utilization of nonmobile, distributed sensor and communication devices by a team of mobile robots offers performance advantages in terms of speed, energy, robustness and communication requirements. Models of mobile robots with on-board sensors, a communication protocol and the *S-Net* system are established. Algorithms are defined for the *S-Net* which perform cooperative computation and provide information about the environment. Behaviors include robots going to or surrounding a temperature source. The simulation experiments show that the *S-Net* performs well, and is particularly robust with respect to noise in the environment. System cost versus performance is studied, and guidelines are formulated for which the *S-Net* system out-performs the non-*S-Net* system.

1. Introduction

At one extreme, mobile robots can be provided with a wealth of on-board sensing, communication and computational resources [1, 2]; at the other extreme, robots with fewer on-board resources can perform their tasks in the context of a large number of stationary devices distributed throughout the task environment [3]. We call the latter approach the *Smart Sensor Network*, or the *S-Net*. In this study, all the results are from simulation experiments using software (C and Matlab), and the performance of robot tasks with and without the presence of an *S-Net* (i.e., a set of distributed sensor devices) is evaluated in terms of various measures. See [4] for a more detailed account.

This approach can be exploited widely and across several scales of application; e.g., fire fighting robots. If mobile robots are used to fight forest fires, there may be several hot spots to extinguish or control. If sensor devices can be distributed in the environment, then their values and gradients can be used to direct the behavior of fire fighting robots and to transport fire extinguishing materials from a depot to the closest fire source. During this movement to and from the fire, collision avoidance algorithms can be employed. Sometimes coordinated activities are necessary and communication models are also important.

D. Rus and S. Singh (Eds.): Experimental Robotics VII, LNCIS 271, pp. 81–90, 2001.

This study provides models for various components of study: (1) mobile robots with on-board sensors (2) communication, (3) the *S-Net* (includes computation, sensing and communication), and (4) the simulation environment. We have developed algorithms for the *S-Net* which perform cooperative computation and provide global information about the environment. Local and global frames are defined and created. A method for the production of global patterns using reaction-diffusion equations is described and its relation to multi-robot cooperation is demonstrated.

We provide the results of a set of simulation experiments designed to help us better understand the benefits and drawbacks of the *S-Net*. For behaviors of one mobile robot going to a temperature source, and multiple mobile robots surrounding a temperature source, in the ideal situation (which means no noise), the *S-Net* takes more time and distance. But when noise is added in, which is more realistic, the *S-Net* system is more robust than the non-*S-Net* system. For the task of multiple mobile robots going back and forth to a temperature source, there are thresholds above which the *S-Net* system out-performs the non-*S-Net* system.

2. Models

We have developed a mobile robot model, sensor models, an *S-Net* model, a communication model, and a model of the environment. The simulation provides a computational framework for the interaction of these models in terms of mobile robots performing useful tasks in the environment, and we define our simulation model as well. In order to act, a robot must receive current environmental information and calculate its movement based on the information received. On-board sensors (e.g., temperature, range, etc.) provide information about the environment and inform the robot's behaviors. In addition, the mobile robot may be able to communicate with other robots or the *S-Net*. The robot achieves movement by rotating or translating based on turning and motion primitives with given rotational and linear speeds.

The high-level behavior of the robot is specified by a program which maps the robot state and environmental information to primitive behavior sequences. For example, the behavior for the mobile robot to go to the closest temperature source includes: mobile robot sensing to get environmental temperature and gradients, turning as well as going forward to the source, and finally stopping when it reaches a certain distance from the temperature source.

Functions define source distribution of energy, material, etc. (e.g., heat, chemicals, etc.). The formula for distribution of temperature is:

$$T(x,y) = \frac{C}{\sqrt{(x - x_s)^2 + (y - y_s)^2 + 1}} \tag{1}$$

where C is a constant related to the temperature source, (x_s, y_s) is the location of the temperature source, and (x, y) is the location at which we want to know the temperature.

Sensor models for temperature and range are both of the form:

$$\hat{T}(x,y) = T(x,y) + N(\mu, \sigma) \tag{2}$$

where $T(x, y)$ is the actual value at location (x, y) in the environment and $N(\mu, \sigma)$ is a normal distribution function with mean μ and variance σ.

The *communication model* consists of a protocol, a message layout, error model and performance characteristics. The protocol specifies the meaning of the bits in a message, as well as a set of commands for communication between robots and *S-Net* elements (*S-elements*). A group of *S-elements* sharing a common local frame is called an *S-clique*.

S-Net devices consist of three essential components: computation, sensing and communication. The *computation element* is described by the speed of the processor, its storage capacity, power requirements and cost. Sensors used by the *S-Net* devices are modeled as described above, but also include bandwidth, latency, power requirements and cost. The *communication model* is like that given for mobile robots, but includes power requirement and cost as well.

We use discrete event simulation with a fixed time step. In that we must model and simulate continuous events (e.g., during robot motion) as well as discrete events, we allow for an *every-time-step* event which can be put at the head of the event queue and must be handled every time step. Any number of these may be added to the event queue. The event list is a table recording all events that will happen in each time step. At the beginning of each time step, we copy it to a temporary list, and new events will be generated and added to the event list during the movement of the robots. During each time step, all scheduled events are handled and new events will be generated and added to the event list according to the different robot behaviors. At the end of each time step, the resulting state is evaluated to determine its feasibility. Once a possible state is achieved, the status of each robot such as position and local direction is updated. This procedure is repeated until the simulation terminates.

3. Algorithms

The *S-Net* is a collection of individual devices (*S-elements*) which have sensors, communication and computation abilities and are distributed either in a specified pattern or randomly to provide environmental information. The *S-Net* provides the framework for information gathering, analysis and presentation — it is an "information field" for the mobile robot. We propose three major algorithmic infrastructures for *S-Nets*:

- *Coordinate frames:* both local and global frames can be determined and exploited for robot tasks.

- *Pattern formation:* global patterns in the *S-Net* can be calculated and used to provide information for the robots [5, 6]; we use a stripe pattern to coordinate robot motion.

- *Level sets:* the ability to model and compute moving boundaries in the *S-Net* adds significant capabilities for robot tasks, including constrained shortest path information (see [7]).

4. Behaviors

A mobile robot's moving and turning behaviors are based on information provided by either its on-board sensors or the *S-Net*. A mobile robot may have four on-board temperature sensors located in different positions, thus providing four different spatial samples from which to compute the temperature gradient. The temperature gradient is used to control the heading of the robot. At the other extreme, with the *S-Net*, the robot will obtain the gradient information from the scattered *S-elements*.

The behaviors studied include:

- T_1: A single robot goes to a temperature source.

- T_2: Multiple mobile robots move to the temperature source and then cooperate and communicate to maintain a certain distance from the temperature source and to surround it.

- T_3: Multiple mobile robots going back and forth between the temperature source and a *Home* location (with the *S-Net*, *Home* is chosen as the origin of the *S-clique* that sensed the lowest temperature). Stripe patterns are formed along the gradient of the temperature source to *Home*. With no *S-Net*, an arbitrary location is selected.

5. Performance

We have compared the performance of mobile robots with and without the *S-Net* while solving the tasks: T_1, T_2, and T_3. For the robot behaviors that do not exploit the *S-Net*, the mobile robot obtains the information about the source (e.g., the temperature gradient) by itself. The mobile robot moves along the gradient towards the source, until the detected value (e.g., temperature) is a local maximum or is above some limit.

This set of tasks represents typical mobile robot tasks and can be configured to exploit many of the constraints described earlier. For example, a robot's path may be required to be the shortest, the gradient may be followed, or patterns in the *S-Net* may be used as road markers. Moreover, the last two tasks provide a setting to use multiple robots, ranging from few to many robots. In addition, robot interactions are necessary, at least as far as avoiding collisions. For each of these tasks, we propose a relevant set of performance measures, as well as a discussion of parameters and their possible values. Finally, we give the performance results and compare the two approaches.

Our goal is to find out under what conditions the *S-Net* system can outperform a non-*S-Net* system, and to study robustness and cost; cost is measured by time taken, distance traveled or total system cost. To summarize our results, we found that, for the first two behaviors (T_1 and T_2), the *S-Net* system does not perform better under ideal conditions (which means no noise at all). But when noise is added to the sensor data, we found that the *S-Net* system is more robust, especially in very noisy situations. For the third behavior (T_3), the *S-Net* system not only performs much better under realistic conditions, but also under ideal conditions. For certain round trip distance requirements, the

S-Net system can support more robots on the route while preventing collisions between robots. On the other hand, if there are too many robots on the route, in the system without the *S-Net*, some robots cannot move properly to prevent collision.

In these simulation experiments we test performance time and distance traveled with respect to sensor noise or variance (0 to 25), number of *S-elements* (100 to 300), and broadcast distance (1m to 2.5m) for the *S-elements*. According to [8, 9, 10], noise of a sensor includes inherent noise, transmitted noise, mechanical noise and so on. The temperature sensor model we choose here has a range of $[0, 1000]$, and we believe that 0.05% is a reasonable tolerance for the temperature sensors. That is why we choose σ^2 ranges from 0 to 25. Using standard statistical techniques, we compute 90% confidence intervals.

One Robot Goes to a Temperature Source and Multiple Robots Surround a Temperature Source: Our results in these two cases show that when the noise variance is above 10, for the mobile robot that utilizes the *S-Net*, the successful result does not change much. But for the mobile robot that uses on-board sensors, it so happens that the robot fails to locate the temperature source correctly – it takes the maximum time allowed for the task and does not locate the source. We believe that this is because the four on-board sensors are located too close to each other and cannot overcome the affect of noise. When noise is added to each sensor, the gradient computed from their values can have large error, which will further change the direction the mobile robot moves. One proposed solution to this problem is to have the mobile robot move to four widely spaced locations and get samples across a greater spatial scale to compute the correct gradient. This will certainly cost much more in time and energy. In fact, it also reduces the accuracy with which the robot can locate the source.

From all these measurement and comparison of the two systems, we can see that in the ideal situation, which means no noise, the *S-Net* takes more time and distance. Compared to the system without the *S-Net* (time used $=$ 3.22 sec, distance traveled $=$ 3.21m), our cost of time ranges from 3.6 sec to 5.5 sec, and distance traveled from 3.6m to 5.2m. But when noise is added, which is more realistic, the *S-Net* system basically does not change much, but the system without the *S-Net* gets much worse. We conclude that in real situations, especially a tough situation with lots of noise, the *S-Net* system will be more robust than the system without the *S-Net*.

Multiple Robots Go Back and Forth to the Temperature Source: This experiment is designed to explore the benefits of using the *S-Net* with regard to multiple cooperating mobile robots. The same behavior is used in each robot, so that by satisfying the same set of constraints, the robots can achieve the desired final result.

In the case that mobile robots use the *S-Net* (500 *S-elements*), *Home* is chosen as the origin of the *S-clique* that sensed the lowest temperature; this provides the longest path to the maximum temperature *S-element*. Then stripe patterns are formed along the gradient of the temperature source to *Home* (using the reaction-diffusion method). The straight line from *Home* to the temperature source (located at [10, 20]) is in the middle of the white stripe (pattern value is 1), the width of each stripe is a constant (5m in our case); black stripes (pattern value is 0) alternate spatially with white stripes. Different stripe patterns are formed for different random streams. The robots will move along the white stripe toward the temperature maximum and follow the black stripe *Home*. When a robot detects that a collision is about to happen, it will slow down to prevent the collision.

In the case that mobile robots do not use the *S-Net*, *Home* is arbitrarily located at the origin of environment (0, 0), and the temperature source is located according to the average distance from *Home* to the temperature source in the *S-Net* experiments. The purpose of this is to make sure that the distances of the round trips are basically the same for both setups. We believe that the gradient of temperature source to *Home* does not affect our experiment, so we choose the temperature source located in (40, 46). When robots detect an environment collision, they make a right turn and then try to get back on track again.

For the *S-Net system*, when the number of robots and trips increases, the average time used and distance traveled by each robot increase linearly, and there are no major deviations from linear. When we take a close look at the data collected, we find that on occasion, due to the particular random number streams, the result is not ideal, which means the robots cannot exactly follow the stripe, but get lost looking for the correct stripe. Under detailed analysis, we found that it is caused by some particular distributions of the *S-elements*. Since we use the origin with lowest temperature as the *Home*, it is sometimes possible that it is on the border of the stripe. There may not be enough *S-elements* on the border for black stripes, and then when the robots try to follow the stripe to go *Home*, they may not get enough information to keep on the black stripe, and thus move away. This can be solved by making more *S-elements* on the border or making *Home* far away from borders.

For the non-*S-Net* system, when the number of robots and trips increases, the average time used and distance traveled by each robot increase linearly. After a detailed analysis of the data collected, we found that when there are more than eight robots on the same path, several robots may lose control. This is related to the robot behavior chosen. While there are lots of other behaviors, we believe that this is a rather standard collision avoidance algorithm and representative of many implementations in physical systems. Using methods described in [11], we find that the confidence interval for the mean difference in these two experimental setups is (0.2647, 9.5464). Since it does not include zero, we can say with 90% confidence that there is no evidence to suggest that there is not a statistically significant difference. Figures 1 and 2 show how noise affects the performance in both cases. From these figures, we found that the one using the *S-Net* can handle noise very well, when the noise variance is about 10, the performance and trace of robots generally stay the same. But for the case that does not use the *S-Net*, noise variance has a huge effect on the performance of the robots, where even a tiny variance as little as 0.5 can cause the robots to lose control (the robots wander erratically). We conclude

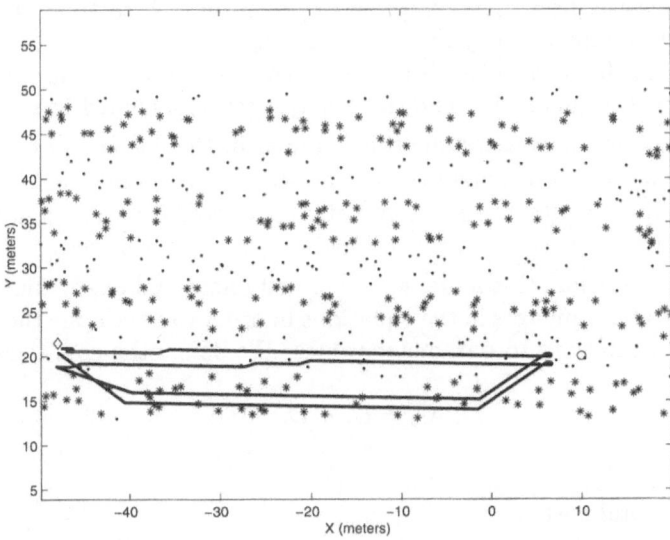

Figure 1. Trace of robots going back and forth with *S-Net* (2 robots, 2 round trips, noise = 10)

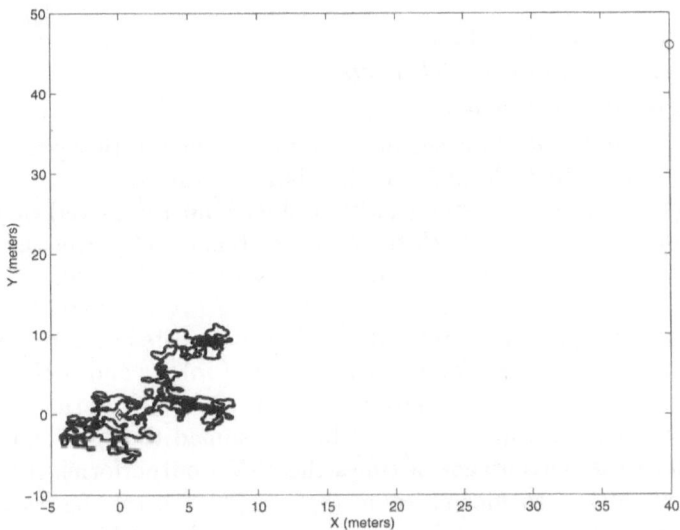

Figure 2. Trace of robots going back and forth without *S-Net* (2 robots, 2 round trips, noise = 0.5)

that, in terms of noise, the *S-Net* performs much better.

The performance measures used to this point have looked at success/failure, time to goal and distance traveled. Another crucial aspect is the more qualitative users' defined cost which is, in general, a function of the physical performance measures. For example, it may be that timeliness is extremely important, and the user may assign an exponential cost to time. Even if the cost is linearly related, it may have a steep slope.

To explore this aspect of performance cost, we have set up two models: (1) linear, and (2) quadratic. The three major terms included are:

- robot cost: we always assume this is linear in the number of robots.
- *S-Net* cost: we always assume this is linear in the number of *S-elements*.
- physical quantity (e.g., time and distance determined from simulation experiments): we apply a linear or quadratic form to this term.

In order to explore this issue, we examined linear and quadratic cost functions in terms of parameters in the equations in order to determine the existence of various cost relations to parameter values. We define the cost relation as:

$$C_l = C_s + C_p$$

where:

- C_l is the total cost
- C_s is the cost of the system infrastructure
- C_p is the cost of performance
- $C_s = N_r * C_r + N_{s-el} * C_{s-el}$
- $C_p = a_t * t^k + a_d * d^k$
- N_r is the number of robots
- N_{s-el} is the number of *S-elements*
- a_t and a_d are coefficients.

in which $k = 1$ in the linear case, and $k = 2$ in the quadratic case, t is the time taken to complete the task, and d is the distance traveled.

We compare the two systems (with and without the *S-Net*) by computing the percentage of cases for which the *S-Net* system outperforms the non-*S-Net* system (over the 100 cases of experiment – 1 to 10 robots making 1 to 10 round trips).

To establish C_s, we investigated mobile robot costs and a reasonable projection for *S-element* costs. For a given number of robots and *S-elements*, these costs are fixed and the cost variation comes from the C_p term. Rather than look at particular fixed a_t and a_d, we have assumed they are equal. Figures 3 and 4 show the percentages of times the *S-Net* outperforms the non-*S-Net* as a function of the coefficient value ($a_t = a_d$). As can be seen, for both the quadratic and linear cost function, there are thresholds below which the non-*S-Net* out-performs the *S-Net*. This indicates that for any particular implementation, a specific detailed analysis should be done to determine which is preferred.

In the quadratic distribution, we found that when a_t and a_d are chosen greater than 2, the percentage of times that the *S-Net* costs less is above 50%, which means the *S-Net* system is a better choice. In the linear distribution, when a_t and a_d are chosen greater than 2200, the percentage of times that the *S-net* costs less is above 50%. These graphs show that it is very likely that even in the ideal conditions, the *S-Net* is the better choice for a system with a

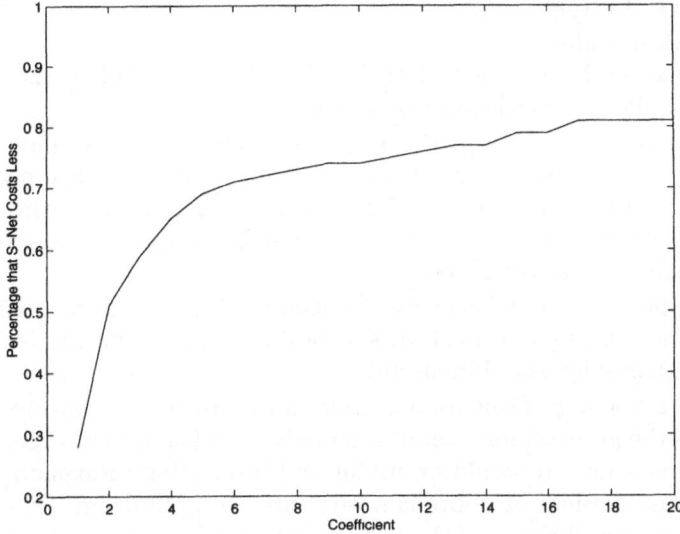

Figure 3. System cost comparison vs. coefficient in quadratic distribution

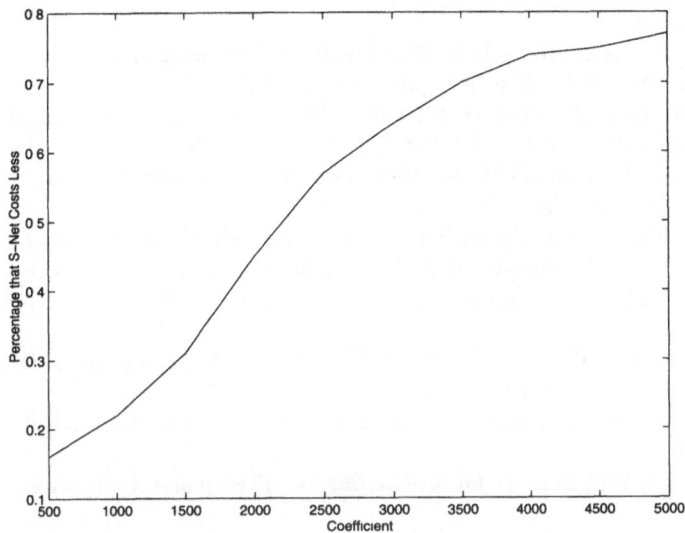

Figure 4. System cost comparison vs. coefficient in linear distribution

quadratic cost function. As an example, the measure of the circumference of a fire burning outward in a circular pattern grows quadratically with time, which could mean quadratic cost. In the linear cost case, it seems that the *S-Net* is a less performant option. However, it should not be overlooked that when noise is present, the *S-Net* dominates in performance.

6. Future Work

Future work includes:

- Physical implementation of the *S-Net* system, including the *S-elements* and mobile robots which exploit them.

- Optimization of the communication message code or layout, and study of communication errors, such as bad bits or lost message problems. In realistic situations, communication is never perfect, and communication error is an important aspect that needs to be handled in order to maintain the robustness of the *S-Net*.

- Development of a wider array of patterns to help multiple mobile robots cooperate. Computation of the shortest path with respect to realistic maps and topography would be useful.

- Exploration of gradient computation in the *S-Net*. The optimal computation of the gradient for a set of randomly sampled data has been solved for one dimension. It would be useful and interesting to expand the theory to two dimensions or above is a very interesting problem. By solving this problem, the efficiency of the *S-Net* can be further improved, and it's likely that the robustness of the system will be improved.

References

[1] Bares J E, Wettergreen D S 1999 Dante II: Technical description, results, and lessons learned. *Int J Rob Res.* 18(7):621-649 July

[2] Smith R, Frost A, Probert 1999 P A Sensor System for the navigation of an underwater vehicle. *Int J Rob Res.* 18(7):697-710 July

[3] Henderson, T C, Dekhil M, Morris S, Chen Y, Thompson W B 1998 Smart Sensor Snow. *IEEE Conf IROS.* Oct, pp 1377-1382

[4] Chen Y 2000 S-Nets: Smart Sensor Networks. MS Thesis, University of Utah

[5] Murray J 1993 *Mathematical Biology.* Springer-Verlag, New York

[6] Turing A 1952 The chemical basis of morphogenesis. *Phil Trans Roy Soc London.* B237:37-72

[7] Sethian J A 1999 *Level Set Methods and Fast Marching Methods.* Cambridge University Press, Cambridge UK

[8] Fraden J 1993 *AIP Handbook of Modern Sensors.* Americal Institute of Physics, New York

[9] Everett H R 1995 *Sensors for Mobile Robots: Theory and Applications.* A K Peters Ltd, MA

[10] Prasad L, Iyengar S S, Min H 1995 *Advances in Distributed Sensor Integration.* Prentice-Hall Inc, New Jersey

[11] Lilja D J 2000 *Measuring Computer Performance, A Practitioner's Guide.* Cambridge University Press, Cambridge UK

Six Degree of Freedom Sensing For Docking Using IR LED Emitters and Receivers

Kimon Roufas, Ying Zhang, Dave Duff, Mark Yim
Systems and Practices Lab, Xerox Palo Alto Research Center
Palo Alto, CA 94304
{kroufas, yzhang, dduff, yim}@parc.xerox.com

Abstract: Six DOF offset sensing between two plates is important for automatic docking mechanisms. This paper presents an easy and inexpensive implementation of such a system using four commercial-off-the-shelf (COTS) infrared (IR) light emitting diode (LED) emitters and two COTS IR receivers on each of two docking plates. The angular intensity distribution of an emitter and the sensitivity distribution of a receiver allow for estimation of the angle and distance between them. Simple experiments have been conducted indicating that such a setup is able to give positional offset in any of 6 degrees of error (x, y, z, pitch, roll, and yaw) within a range. A theoretical framework is also established using least squares minimization. The theoretical framework is general and applies to other configurations of emitter and receiver parts and positioning.

1. Introduction and Motivation

Six degree of freedom (DOF) offset sensing between two plates is critical for automatic active docking of self-reconfigurable robot systems such as PolyBot[1] (see Figure 1). Automated active docking requires that the offset errors are measured and then corrected by an automated control system.

Figure 1. PolyBot in spider configuration.

The PolyBot system uses repeated modules all with identical docking mechanisms, or *interface plates*. Since two interface plates that may dock with each other are identical, they need to have hermaphroditic connection mechanisms. Other systems such as [2][3] may have male and female connection mechanisms, however, the sensing method described in this paper is general and extends to these systems as well.

This paper presents an easy and cheap implementation of 6 DOF sensing system, using four commercial-off-the-shelf (COTS) infrared (IR) light emitting

D. Rus and S. Singh (Eds.): Experimental Robotics VII, LNCIS 271, pp. 91–100, 2001.
© Springer-Verlag Berlin Heidelberg 2001

diode (LED) emitters and two COTS IR receivers on each of the opposing plates. Each of the eight emitters is lit in sequence and an analog reading is taken from both opposing receivers. The angular intensity distribution of an emitter and the sensitivity distribution of a receiver allow for estimation of the angle and distance between them. Simple experiments have been conducted indicating that such a setup is able to give positional offset in any of 6 degrees of error (x, y, z, pitch, roll, and yaw) within a range. A theoretical framework is also established using the least squares minimization. The theoretical framework is general and applies to other configurations of emitter and receiver parts and positioning.

For PolyBot, a first order analysis of the open loop errors indicated that the system can place the interface plates within 30mm of each other. The mechanical features of the plates are designed to passively mate with up to 3mm of positional error. The 6 DOF docking sensor system described in this paper is the system developed to close the loop and bring those errors from 30mm down to 3mm. IR LEDs and sensors were chosen for this system for their low cost, small size, minimal interface requirements and low processing overhead.

There are a variety of other means for determining the relative position of two objects with 6 DOF, although most are expensive or not suitable for docking. In the Virtual Reality (VR) hardware domain, the use of 6 DOF trackers is a staple. Such systems include linkage based systems, electro-magnetic field based systems[4][5], ultrasonic ranging[6], inertial tracking and vision based methods[7][8]. The ultrasonic and inertial tracking methods have not been extended to 6 DOF in a robust fashion. The vision based methods tend to be computationally intensive and expensive. The electro-magnetic based methods are the most popular for VR however do not work well for self-reconfigurable systems since they are prone to interference by metallic objects and are expensive. Other non-light based positional measurement methods include eddy-current sensing, hall-effect sensor or capacitance based[9] methods. Both of these methods may work however the intimate presence of electric motors may cause too much noise to make the sensing feasible.

Low cost measurement and actuation components may enable applications beyond robotic docking such as: cars that park themselves, jacks on the back of the computer or stereo that move into position as you fumble to plug them in, a gas nozzle that finds the car's tank opening, or robot appliances that automatically dock to recharge, fluid or supply interfaces in your home.

The paper is organized as follows. Section 2 describes the mechanical and electronic design; Section 3 focuses on obtaining the IR intensity model; Section 4 presents the methods of computing 6D offsets; Section 5 discusses experimental results and Section 6 concludes the paper and gives some directions for future work.

2. Mechanical and Electronic Design

Figure 2 shows the mechanical design of the plate, where the four small squares at the corners are IR emitters, and two hemispheres along the middle line are IR receivers.

The electrical design ensures that each receiver detects and samples the intensity from each emitter on the opposite plate at a distinct time. In order to do that, each of the eight emitters is lit in sequence and an analog reading is taken from both opposing receivers. Readings are also taken in between the times when the IR LEDs are emitting to measure the ambient IR. Figure 3 shows the control signals for the first four emitters to be lit. The algorithm decides which side will emit in 'time slot 1' and which in 'time slot 2.' The two sides are synchronized so that the opposing plates measure at the correct time.

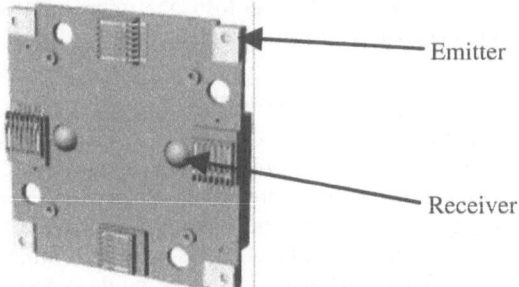

Figure 2. Mechanical design of the IR 6D sensing device on a PolyBot faceplate.

At the end of a time period, each of the receivers (total four receivers, two on each plate) will have four readings from their opposing emitters, and four ambient readings totaling 32 measurements. The ambient IR readings are subtracted from the preceding sample to make the system more robust. Hence we end up with 16 pieces of data.

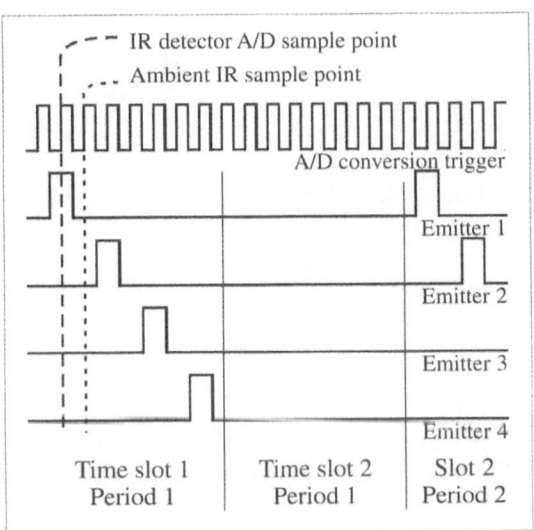

Figure 3. Emitting and receiving sequence.

The current design uses two synchronized Motorola MPC555 PowerPC embedded-controllers to collect the data. In each processor a TPU3 (Time Processing Unit) generates the trigger and emitter control signals. The trigger is fed back into the QADC64 (Queued Analog to Digital Converter) external trigger input line to obtain the readings from IR detectors. A list of conversions is initially programmed into the A/D queue and a single interrupt is generated at the end of each complete period. The interrupt service routine is responsible for subtracting the ambient measurements and sending the data to the master computation thread. One or both of the MPC555s must send this data over the CANbus network to the master, which is also an MPC555 micro-controller where the algorithm for finding the 6D position offset is implemented.

3. IR Intensity Model

The theory behind this design is based on the fact that the intensity detected by a receiver is a function of the distance and/or angle between the emitter and receiver, i.e., $I = f(e, r, d)$ where I is the intensity reading, e and r are emitter and receiver angles, respectively, and d is the distance between the emitter and the receiver. An accurate model can be obtained by model fitting for given emitters and detectors.

Our model was constructed by decomposing f into three functions, $f_e(e)$, $f_r(r)$, $f_d(d)$ and let $I = Af_e(e)f_r(r)f_d(d)$ where A is a scale factor. We did two separate data collections, one was to fix r to 0 degrees and change e from 0 to 90 in 5 degree increments and d from 0.5 inches to 5 inches in 0.5 inches increments (see Figure 4(a)); and the other was to fix e to 0 and change r from 0 to 90 and d from 0.5 to 5 inches (see Figure 4(b)). The results of this data collection are plotted in Figure 5 (a) and (b), respectively.

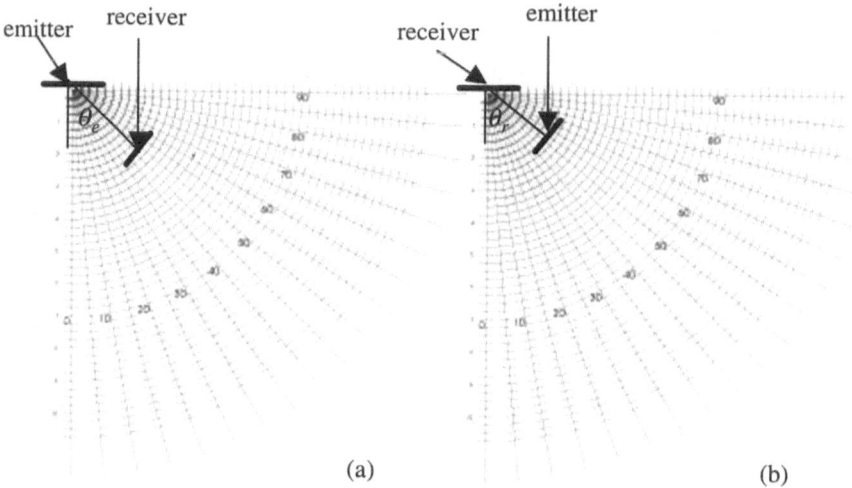

(a)   (b)

Figure 4. (a) Fix receiver angle to 0 degree and change emitter angles and distance. (b) Fix emitter angle to 0 degree and change receiver angle and distance.

We use function $f_e = (A - O)e^{-(\theta_e - \theta_e^0)^2 / \sigma_e^2} + O$, where A is 1000, O is 150 and θ_e^0 is 20 degrees, to fit the data in a least squares fashion, and obtain the parameter σ_e as 0.2660. Similarly, we use function $f_r = (A - O)e^{-(\theta_r - \theta_r^0)^4 / \sigma_r^4} + O$ with θ_r^0 as 40 degrees to fit the receiver data, and obtain the parameter σ_r as 0.3694. As a result, we obtain the IR intensity model

$I = (A-O)e^{-(\theta_e - \theta_e^0)^2/\sigma_e^2}e^{-(\theta_r - \theta_r^0)^4/\sigma_r^4} + O$. Using this model, we plot the corresponding model data in Figure 6(a) and (b). Compared with Figure 5, the model fits the data relatively well.

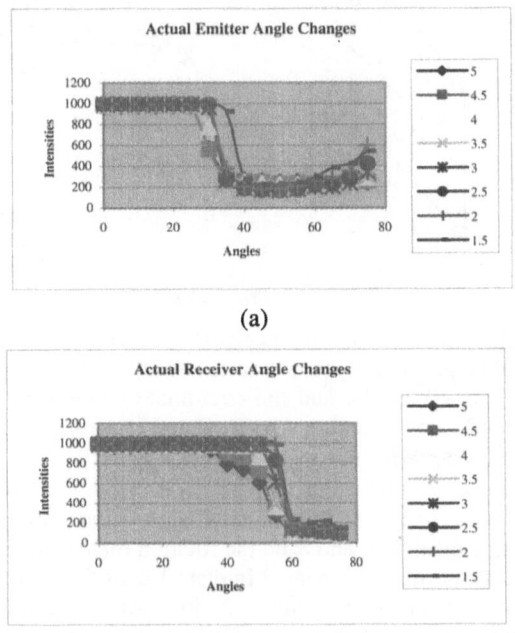

(a)

(b)

Figure 5. Actual intensity changes w.r.t. emitter and receiver angles.

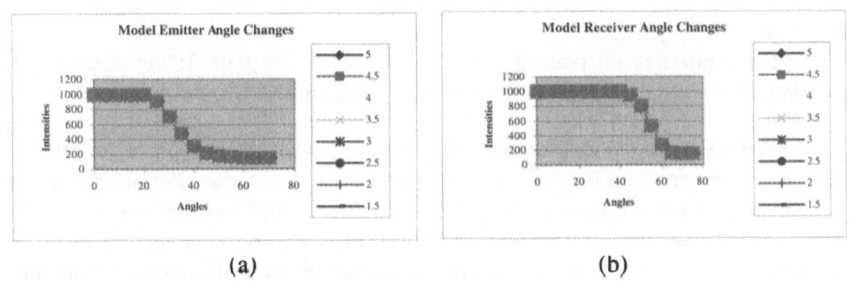

(a)   (b)

Figure 5. Intensity changes computed by the model w.r.t. emitter and receiver angles.

4. Six Dimensional Offset Estimation Methods

For each plate, we attach a frame as shown in Figure 7 (in this case Plate 1 and Plate 2 are facing each other). Given an offset between the two plates, the spatial relationship between each pair of emitter and receiver is determined.

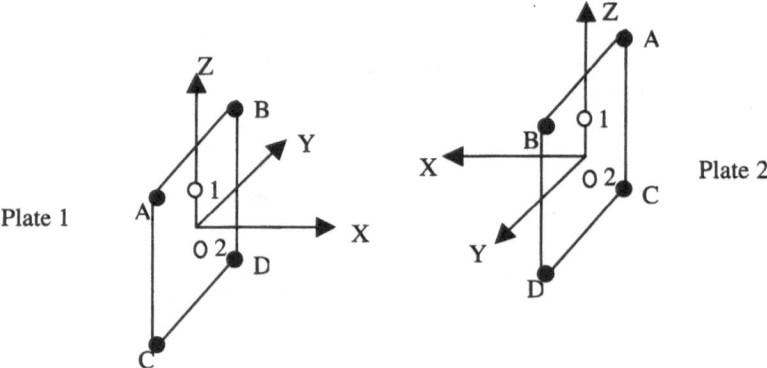

Figure 7. Frames for plates.

Let d be the distance from the receiver to the center of the plate, and w and h be the width and height of the position of the emitters. The coordinate of receiver 1 in its own frame is $<0, 0, d>$, and the coordinate of the receiver 2 is $<0, 0, -d>$; similarly, the coordinates of emitters A, B, C and D are $<0, -w, h>$, $<0, w, h>$, $<0, -w, -h>$ and $<0, w, -h>$, respectively. Let $<x, y, z, \alpha, \beta, \gamma>$ be the offset of the frame of plate 2 with respect to the frame of plate 1 (in the case of two plates facing each other, the offset is $<x, 0, 0, \pi, 0, 0>$) and let T be the transform matrix from plate 1 to plate 2 obtained by the offset, and R be the rotation matrix of T. The norm of plate 1 is $<1, 0, 0>$ and the norm of the plate 2 in plate 1 coordinates is $R<1, 0, 0>$. Let $<x_e, y_e, z_e>$ be the coordinate of the emitter in its own frame and $<x_r, y_r, z_r>$ be the coordinate of the receiver of the opposing plate in its own frame. There are two cases:

The emitter is on plate 1 and the receiver is on plate 2: the position of the emitter is $o = <x_e, y_e, z_e>$ and the position of the receiver is $q = Tp$ where $p = <x_r, y_r, z_r>$ and $q = <x_r', y_r', z_r'>$.

The emitter is on plate 2 and the receiver is on plate 1: the position of the receiver is $o = <x_r, y_r, z_r>$, and the position of the emitter is $q = Tp$ where $p = <x_e, y_e, z_e>$ and $q = <x_e', y_e', z_e'>$.

Given two points in space, o and q, and the norms of their plates, n_o and n_q, the distance between them is $|q-o|$, the angle at o is $arccos(n_o \bullet (q-o)/|q-o|)$ and the angle at q is $arccos(n_q \bullet (o-q)/|q-o|)$. Therefore, the emitter and receiver angles as well as the distance between the receiver and the emitter can be obtained for each of the sixteen pairs of emitters and receivers. Given the IR intensity model we obtained in the previous section, we get a model from each 6D offset between two plates to 16 readings of intensities, i.e., $I_i = f_i(x, y, z, \alpha, \beta, \gamma)$ for I=1 to 16.

4.1. Absolute 6 DOF position sensing

Theoretically, the problem of 6D offset estimation becomes a problem of data fitting, i.e., solving $<x, y, z, \alpha, \beta, \gamma>$ given sixteen data readings. In particular, let R_i, $i =1..16$, be the sixteen readings and let E be an energy function to be minimized,

$$E = \frac{1}{2} \sum_{1}^{16} (R_i - f_i(x, y, z, \alpha, \beta, \gamma))^2 \quad \text{which} \quad \text{transforms} \quad \text{to} \quad \text{six} \quad \text{equations:}$$

$$\frac{\partial E}{\partial p} = \sum_{1}^{16} (R_i - f_i(x, y, z, \alpha, \beta, \gamma)) \frac{\partial f_i}{\partial p} = 0 \text{ where } p \text{ is } x, y, z, \alpha, \beta, \text{ and } \gamma.$$

This set of equations can then be solved using Newton's method. We used Singular Value Decomposition (SVD) for solving linear equations at each Newton step. The use of SVD greatly reduces the risk of reaching a singularity that is very common in problems involving the inverse of matrices. It also achieves a better result in both under- (minimum change) and over-constrained (minimum error) situations.

We applied this method to a set of known offset positions of two plates. To our surprise, the result was not as good as we expected. We examined the problem further and found that the following maybe the major causes:

The particular emitter and receiver pairs we chose were not ideal. First, they have a small range before saturating, and second, the slope in the valid range is too steep, which makes the data extremely sensitive.

The emitters and receivers are sensitive to the mounting position alignment. The resulting variation is very difficult to capture by a simple model. IR ranging using combinations of many plates compounds this problem.

If we fix the two hardware problems in the future, we should be able to get a good absolute position estimation. (This is a good example of good theory that does not necessarily end up with good results in practice).

4.2. Relative 6D offset sensing

In the close loop control of the docking process, it is not necessary to have absolute 6D position sensing, as long as (1) it can tell the direction of the offset and (2) it can tell if the offset between the plates are small enough; (1) is used to guide the motion of the plates and (2) is to trigger the latches in the plates to open and close at the right time.

We developed a centering method based on the idea of signal "balancing" when the plates are centered and facing each other. Let Xij represent a reading where X is the emitter ID (A, B, C, or D), i is the receiver ID (1 or 2) and j is the plate ID (1 or 2), see Figure 7, and let _ represent the case that holds for both plate 1 and 2. When two plates are centered and facing each other, we have a set of equations, e.g., A1_=B1_, A2_=B2_, C1_=D1_, C2_=D2_. In practice, even when the two plates are exactly centered, the equations may not hold because of noise and slight variations when mechanically assembling the plates. The difference, however, can be used as a guideline for a relative offset. For example, (A1_-B1_)+(A2_-B2_)+(C1_ D1_)+(C2_-D2_) gives offset in Y direction, while (A1_-C2_)+(A2_-C1_)+(B1_-D2_)+(B2_-D1_) gives relative offset in Z direction. This method has been used to successfully dock two plates in a plane, i.e., a special case with 3D offset.

To follow this path further, we discovered six groups of "balancing" equations, each of which corresponds to an invariant with respect to a subset of 6D offset:

1. Horizontal Group (x, z, β invariant): eight equations, four for each plate: A1_=B1_, A2_=B2_, C1_=D1_, C2_=D2_.

2. Vertical Group (x, y, α invariant): eight equations, four for each plate: A1_=C2_, A2_=C1_, B1_=D2_, B2_=D1_.
3. Diagonal Group (x, γ invariant): eight equations, four for each plate: A1_=D2_,A2_=D1_,B1_=C2_,B2_=C1_.
4. Horizontal Cross Group (x, y, γ invariant): eight equations between two plates in horizontal direction: A11=B12, A21=B22, B11=A12, B21=A22, C11=D12, C21=D22, D11=C12, D21=C22.
5. Vertical Cross Group (x, y, z invariant):   eight equations between two plates in vertical direction: A11=D22, A21=D12, B11=C22, B21=C12, C11=B22, C21=B12, D11=A22, D21=A12.
6. Diagonal Cross Group (x, z, γ invariant): eight equations between two plates in diagonal direction: A11=C22, A21=C12, B11=D22, B21=D12, C11=A22, C21=A12, D11=B22, D21=B12.

We developed a minimization method that can be used for one or more equations. For example, for equation $A11=B11$, we define an energy function

$E = \dfrac{1}{2}(A11-B11)^2$. Note that this energy function does not have the explicit IR

model as the one used for the absolute position sensing. The goal of centering is to move to the direction where the energy function can be minimized. In order to

minimize E, we calculate $J = \left\langle \dfrac{\partial E}{\partial x}, \dfrac{\partial E}{\partial y}, \dfrac{\partial E}{\partial z}, \dfrac{\partial E}{\partial \alpha}, \dfrac{\partial E}{\partial \beta}, \dfrac{\partial E}{\partial \gamma} \right\rangle$ and H where

$\dfrac{\partial E}{\partial p} = (A11-B11)(\dfrac{\partial A11}{\partial p} - \dfrac{\partial B11}{\partial p})$, p is $x, y, z, \alpha, \beta, \gamma$ and H is a 6x6 matrix with

$H_{pq} = \dfrac{\partial^2 E}{\partial p \partial q} \approx (\dfrac{\partial A11}{\partial p} - \dfrac{\partial B11}{\partial p})(\dfrac{\partial A11}{\partial p} - \dfrac{\partial B11}{\partial q})$, in which, p, q are x, y, z, α, β or γ. By

using SVD to solve the linear equation $H\Delta p+J=0$ where $\Delta p=<\Delta x, \Delta y, \Delta z, \Delta \alpha, \Delta \beta, \Delta \gamma>$, we obtain the direction of the offset movement to minimize the energy function defined by the equation.

Given a set of equations 1..k, we define the energy function as the sum of the

energy functions of each equation $E = \sum_1^k E_i$. Therefore $\dfrac{\partial E}{\partial p} = \sum_1^k \dfrac{\partial E_i}{\partial p}$ and

$H_{pq} = \dfrac{\partial^2 E}{\partial p \partial q} = \sum_1^k \dfrac{\partial^2 E_i}{\partial p \partial q}$ in which, p, q are x, y, z, α, β or γ. By solving the linear

equation $H\Delta p+J=0$ where $\Delta p=<\Delta x, \Delta y, \Delta z, \Delta \alpha, \Delta \beta, \Delta \gamma>$, we obtain the direction of the offset movement to minimize the energy function defined by the set of equations.

The groups of equations we defined can be used to calculate the subset of offsets that are not invariant of the equations. For example, Group 1 equations can be used to calculate y, α and γ, Group 2 equations can be used to calculate z, β and γ, Group 5 equations can be used to calculate α, β and γ. Also we can combine all the groups and calculate y, z, α, β and γ. To calculate x, we use energy function

$E = \frac{1}{2} \sum_{\substack{i=1,2 \\ j=1,2}} (Aij^2 + Bij^2 + Cij^2 + Dij^2)$, based on the fact that all the readings go to

minimum when x approaches 0 in centered position. For simplicity, assuming the plates are centered, we have

$$\Delta x = -\frac{\partial E}{\partial x} / \frac{\partial^2 E}{\partial^2 x}, \quad \text{where} \ \frac{\partial E}{\partial x} = \sum_{\substack{i=1,2 \\ j=1,2}} (Aij \frac{\partial Aij}{\partial x} + Bij \frac{\partial Bij}{\partial x} + Cij \frac{\partial Cij}{\partial x} + Dij \frac{\partial Dij}{\partial x})$$

and $\dfrac{\partial^2 E}{\partial^2 x} \approx \sum\limits_{\substack{i=1,2 \\ j=1,2}} ((\frac{\partial Aij}{\partial x})^2 + (\frac{\partial Bij}{\partial x})^2 + (\frac{\partial Cij}{\partial x})^2 + (\frac{\partial Dij}{\partial x})^2)$.

5. Experimental Results

We developed an experimental setup for measuring 6D offset. The setup includes two PolyBot modules each of which has an IR plate and one module for calculating the offset. The three modules communicate via CANbus and each contains an MPC555. The outputs of the 6D offset are sent from the computing MPC555 to CANalyzer, which is a CAN interface program running on a PC. The experiments are done basically by fixing one IR module and moving the other IR module in space (see Figure 9).

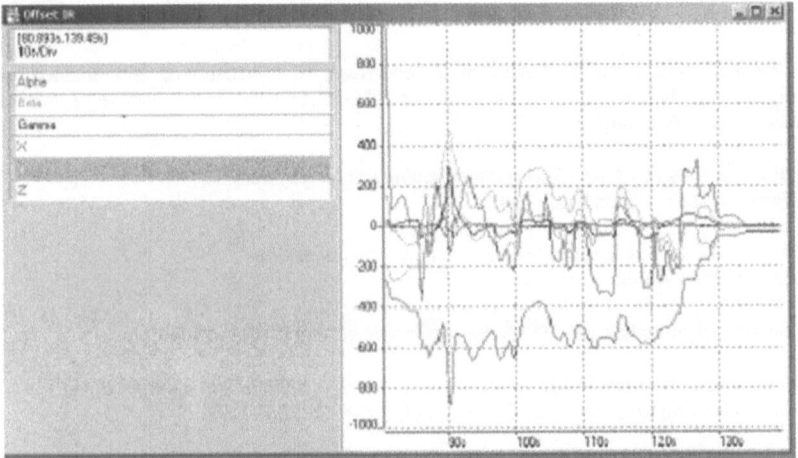

Figure 9. CANalyzer trace graphics window.

We first experimented with each set of equations individually to find the sensitivity of each of the six dimensions offset with respect to the set of equations. We found that using individual groups for calculating individual offsets, in this case Group 1 for y, Group 2 for γ, Group 4 for z, Group 5 for α and Group 6 for γ works better than using all the equations to calculate all the offsets at once. Offset x will only be calculated if other offsets are small.

When the plates are very close, using the current emitter-receiver placement, all the readings tend to approach zero that results in loss of sensitivity. This is an implementation limitation and not a limitation of the method.

6. Conclusions and Future Work

We have presented an integrated system, with mechanical-electrical design and embedded software for obtaining a six degrees of freedom offset between two opposing plates for the purpose of docking. The system is simple and cheap, using eight IR emitters and four IR receivers. The software is general and robust using minimization techniques. The same algorithm used for 6 DOF offset estimation is also used for inverse kinematics, similar to [10].

For future work, we plan to improve the IR curve to reduce the saturation range and extend sensitivity in the unsaturated range, so that 6D absolute positioning may be obtained. We also plan to rearrange the positions of IR emitters and receivers so that the receivers still measure signals when the two plates are docked without loss of sensitivity.

References

[1]Yim, M., Duff, D.G., Roufas, K., "PolyBot: a Modular Reconfigurable Robot", in *Proc. of the 2000 IEEE Intl. Conf. on Robotics and Automation,* 2000.

[2] Will, P., Castano, A. and Shen W-M., "Robot modularity for self-reconfiguration," in *Proc.of SPIE Sensor Fusion and Decentralized Control in Robotic Systems II*, Vol. 3839, pp. 236-245, 1999.

[3] Fukuda, T., Nakagawa, S., Kawauchi, Y., and Buss, M., "Structure decision method for self organising robots based on cell structures-CEBOT'", in *Proc. of the 1989 IEEE Intl. Conf. on Robotics and Automation*, 1989.

[4] Polhemus FASTRACK® 6D electromagnetic tracking system
http://www.polhemus.com/ftrakds.htm.

[5] Ascension Flock of Birds 6D tracker, http://www.ascension-tech.com/products/flockofbirds/flockofbirds.com.

[6] Logitech Headtracker and 3D Mouse (no longer commercially available).

[7] Wang, C.C., "Extrinsic Calibration of a vision sensor mounted on a robot" in *IEEE Transactions on Robotics and Automation*, Vol. 8, No. 2, April 1992.

[8] Northern Digital Inc. OPTOTRAK optical tracking system,
http://www.ndigital.com/optotrak.html.

[9] Kizhner S, "On capaciflector sensor imaging and imaging applications for robot control", in *SPIE* Vol. 2057, pp. 443-453, 1993.

[10] Markus P. J. Fromherz, Maia Hoeberechts, Warren B. Jackson, "Towards Constraint-based Actuation Allocation for Hyper-redundant Manipulators", in *CP'99 Workshop on Constraints in Control (CC'99)*, Alexandria, VA, October 16, 1999.

Height Estimation for an Autonomous Helicopter

Peter Corke, Pavan Sikka and Jonathan Roberts
CSIRO Manufacturing Science & Technology
Pullenvale, AUSTRALIA.
pic,p.sikka,jmr@cat.csiro.au

Abstract: Height is a critical variable for helicopter hover control. In this paper we discuss, and present experimental results for, two different height sensing techniques: ultrasonic and stereo imaging, which have complementary characteristics. Feature-based stereo is used which provides a basis for visual odometry and attitude estimation in the future.

1. Introduction

We have recently started development toward an autonomous scale-model helicopter, see Figure 1. Our first target is to achieve stable autonomous hovering which involves the subtasks of: height control, yaw control, and roll and pitch control. For initial purposes we consider these as independent control problems, though in reality the dynamics of various degrees of freedom have complex coupling.

This paper is concerned only with the problem of estimating height of the vehicle above the ground which is an essential input to any height regulation loop. We are currently investigating two approaches: ultrasonic for very close range ($< 2\,\text{m}$) and stereo imaging for medium range ($< 20\,\text{m}$). Stereo vision methods are discussed in

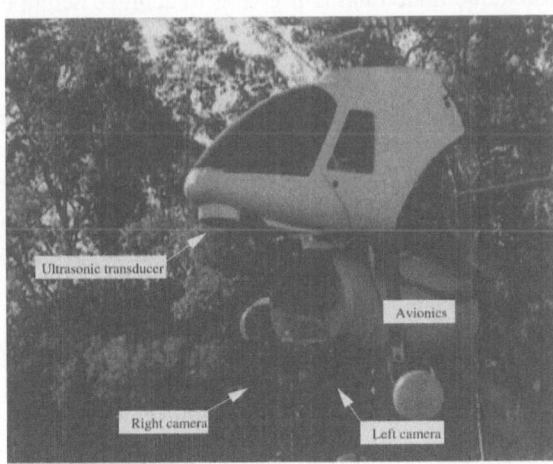

Figure 1. The CMST scale-model helicopter in flight.

D. Rus and S. Singh (Eds.): Experimental Robotics VII, LNCIS 271, pp. 101–110, 2001.
© Springer-Verlag Berlin Heidelberg 2001

Figure 2. The small stereo camera head used on the helicopter in a downward looking configuration.

Section 2, ultrasonics in Section 3, and combined results in Section 4. Conclusions and future work are the subject of Section 5.

2. Stereo-based height estimation

Our previous work has concentrated on dense area-based stereo vision[1], but for this work we have focussed on feature-based techniques. The disparity between corresponding image features in the left and right cameras can be used to provide sparse range information which is sufficient for height control purposes. In addition the features can be tracked in time[5] to provide information about the vehicle's change in attitude and motion across the ground.

Stereo imaging has many desirable characteristics for this application. It can achieve a high sample rate of 60Hz (NTSC field rate), is totally passive, and gives height above the surface rather than height with respect to some arbitrary datum (as does GPS). Its disadvantages are computational complexity, and failure at night[1]. In our experiments the helicopter is flying over a grassy area that provides sufficient irregular texture for stereo matching. Our goal is to achieve height estimation without artificial landmarks or visual targets.

2.1. The camera system

The ST1 stereo head from Videre Design[2], shown in Figure 2, comprises two 1/4 inch CMOS camera modules mounted on a printed circuit board that allows various baselines — we are using the maximum of 160 mm. The two cameras are tightly synchronized and line multiplexed into an NTSC format composite video signal. Thus each video field contains half-vertical resolution images, 320 pixels wide and 120 pixels high, from each camera. The cameras have barely adequate exposure control and no infra-red filters.

The lenses are 12 mm diameter screw fit type with a nominal focal length of 6.3 mm. The lenses are made of plastic and have fairly poor optical quality — barrel distortion and poor modulation transfer function are clearly evident in the images we obtain.

The essential camera parameters are given in Table 1. Since the two cameras are multiplexed on a line-by-line basis the effective pixel height is twice that of the

---

[1]Infra-red cameras could be used but size and cost preclude their use with this vehicle.
[2]http://www.videredesign.com

Parameter	Value
Baseline	160mm
Pixel width	14.4um
Pixel height	13.8um
Focal length (f)	6.3mm
$\alpha_x f$	438
$\alpha_y f$	457

Table 1. Camera and stereo head parameters. α is pixel pitch (pixels/m).

actual photosite. Mechanically the stereo head is manufactured so that scan lines are parallel and vertically aligned to reasonable precision. The mounting configuration also ensures that the optical axes are approximately parallel but in practice we find that the cameras are slightly convergent. Another PCB, parallel to the main one and connected via nylon spacers stiffens the structure but again in practice we find that the boards flex so the geometry varies with time.

2.2. Calibration

For parallel camera geometry, which this stereo head approximates, the horizontal disparity is given by

$$d = \frac{(\alpha_x f)b}{r} = \frac{70.1}{r} \text{pixels} \tag{1}$$

where α_x is the pixel pitch (pixels per metre), f is the focal length of the lens, b is the baseline and r is the range of the point of interest.

The derived parameters $\alpha_x f$ and $\alpha_y f$ given in Table 1 are the lumped scale parameters that appear in the projection equations. The computed values in the table agree well with simple calibration tests which give values of $\alpha_x f = 445$ and $\alpha_y f = 480$ — consistent with a slightly higher focal length of approximately 6.5 mm. Figure 3 shows results of another calibration experiment which gives the empirical relationship

$$d = \frac{67.7}{r} - 6.65 \text{pixels} \tag{2}$$

This indicates that the cameras are slightly verged with the horopter at 10.2 m — beyond which disparity is positive. For our target working range of 2 m to 20 m the disparity would lie in the range -4 to 27 pixels, or a 32 pixel disparity search range. The estimated numerator constant is quite similar to that given in (1) based on camera data.

2.3. Architecture for visual processing

An overall view of the proposed system is given in Figure 4. The helicopter control computer is based on the PC104/PC104+ bus, which is a stackable architecture based on the ISA/PCI specifications. The framegrabber is a PC104+ module which writes images directly over the PC104+ (PCI) bus into system memory. The vision processing software runs under the on-board real-time operating system, LynxOS, and uses a custom streaming video driver for the framegrabber. The device driver supports three modes of operation: *demand*, *streaming* and *field-streaming*. In demand mode, frame capture is triggered by the application program. In streaming mode, the framegrabber

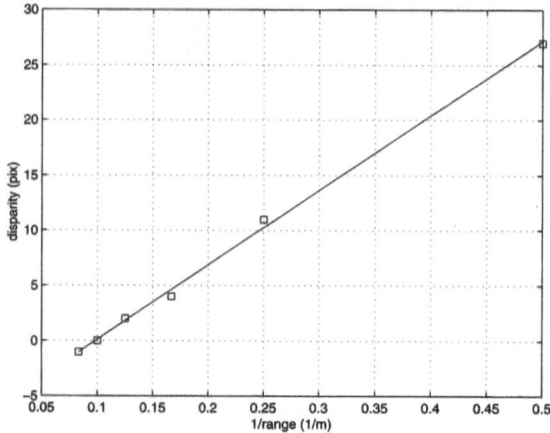

Figure 3. Experimental disparity versus inverse range.

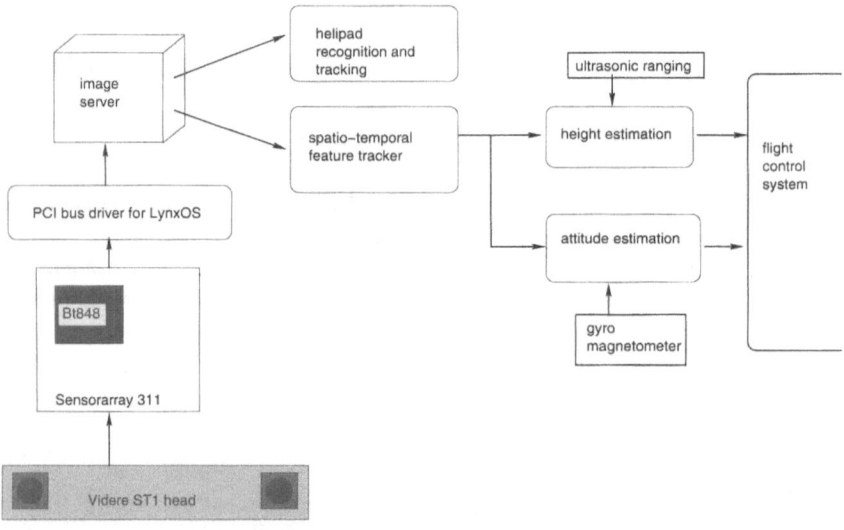

Figure 4. Overall system architecture.

is setup to capture images continually into a user-provided ring of image buffers. This makes it possible to capture images at frame rate. Field-streaming mode is a variant of streaming mode in which each field is considered an image. This allows for image capture at two times the frame rate.

The PCI bus has a theoretical maximum bandwidth of 132 Mbyte/s and full-size PAL color video at frame rate consumes only 44 Mbyte/s. However, there is other activity on the system and depending on the number of devices using the system bus, this can be a significant issue. The disk bandwidth becomes an issue when video has to be written to disk at frame rate. The disk subsystem requires almost the same bandwidth as the framegrabber and that could prove to be too much for the system. For the hardware we use, we have found that it is possible to capture monochrome

192x144 sized images to the 'raw' disk device at frame rate. A bigger limitation was the hard-disk shutting down due to vibration — we now use solid state disks.

2.4. Stereo vision

Our previous work with stereo was concerned with dense matching[1]. However for the helicopter application we chose to use feature-based matching for the following reasons. Firstly the non-parallel imaging geometry meant that since the epipolar lines were not parallel, and given the large range over which we are matching, the vertical disparity becomes significant. Projective rectification was investigated but is computationally too expensive for this application and assumes the geometry is time invariant. Secondly, for height control we do not need dense range data. A few points, sufficient to establish consensus, will be adequate. Thirdly, features can be temporally tracked to provide odometry and attitude information.

The main processing stages are as follows:

for i=L, R **do**
    for each pixel in I_i **do**
        $C_i = \text{plessey}(I_i)$
    end for
    X_i = coordinates of N strongest local maxima in C_i
end for
for each feature $X_{L_i} \in X_L$ **do**
    for each candidate feature $X_{R_j} \in X_R$ **do**
        compute similarity $s_{ij} = \text{ZNCC}(X_{L_i}, X_{R_j})$
    end for
    best match $s_i^* = \max_j s_{ij}$
    if $s_i^* > \tilde{s}$ **then**
        if X_{R_j} matches back to X_{L_i} **then**
            $d_i = \text{disparity}(X_{L_i}, X_{R_j})$
        end if
    end if
end for
$d = \text{median}(d_i)$

2.5. Corner operators

Corner features are those which have high curvature in orthogonal directions. Many corner detectors have been proposed in the literature[3] but we considered only two well known detectors: Plessey[2] and SUSAN[6]. Roberts[5] compared the temporal stability for outdoor applications and found that the Plessey operator was superior.

The Plessey operator requires computation, for each pixel, of

$$M = \begin{bmatrix} \overline{I_x^2} & \overline{I_x I_y} \\ \overline{I_x I_y} & \overline{I_y^2} \end{bmatrix}$$

where $I_x = I \otimes (-1, 0, 1) \approx \partial I / \partial x$, $I_y = I \otimes (-1, 0, 1)^T \approx \partial I / \partial y$, and $\overline{X} = X \otimes w$ where w is a Gaussian smoothing function. The Plessey corner function is $\text{trace} M / \text{det} M$ which is invariant to illumination scale. This is very important when looking for features in images from two different cameras.

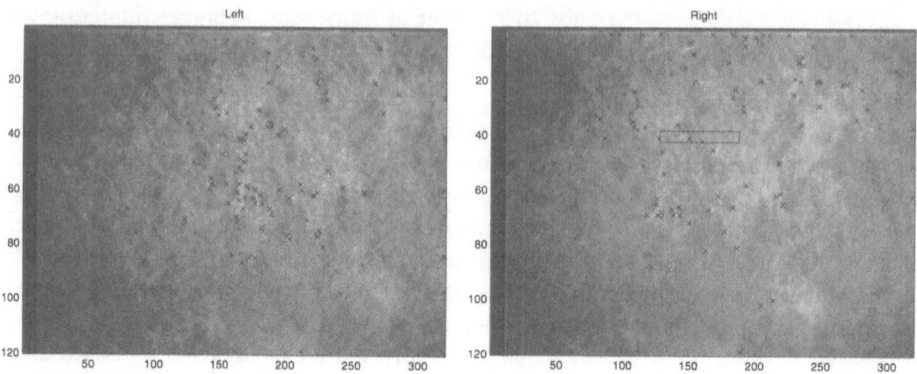

Figure 5. Left and right hand camera image from helicopter image sequence with corner features indicated. Also shown is a feature search box corresponding to the left image feature marked by a circle.

The features used for the matching stage are the coordinates of non-local corner maxima that have been sorted by strength. The strongest 100 features are used. Figure 5 shows a typical flight image and detected corner features.

2.6. Geometric constraints

The corner detection stage provides 100 features in each image for which we must establish correspondence. Naively applying a similarity measure would result in a 100×100 matching problem. Applying epipolar constraints can greatly reduce the order of this problem since we know that the corresponding point is constrained to lie along the epipolar line.

In practice we observe, see Figure 6, considerable variation in slope for the epipolar lines over time. The epipolar lines were computed using the method of Zhang[7] for each frame in the sequence. We believe two effects are at work here: flexing of the stereo head results in changes to the camera geometry, and correlation with vertical scene motion. This latter may be due to timing issues in the camera's line multiplexing circuitry or CMOS readout timing.

For each feature in the left image we know an approximate region of the right hand image in which the corresponding feature must lie. The vertical extent of this region is a function of uncertainty in the vertical disparity, and the horizontal extent is a function of the disparity which may or may not be approximately known, see Figure 5. We assume here no apriori knowledge of disparity. By this means we reduce the number of region comparisons from 10,000 to typically less than 600.

2.7. Similarity measure

The first approach was based on earlier work by Roberts[5] in temporal feature tracking. Corners are defined in terms of a 3-element feature vector containing the intensity and the orthogonal gradients and similarity is defined by Euclidean distance between features. This is computationally inexpensive but in practice exhibited poor discrimination in matching.

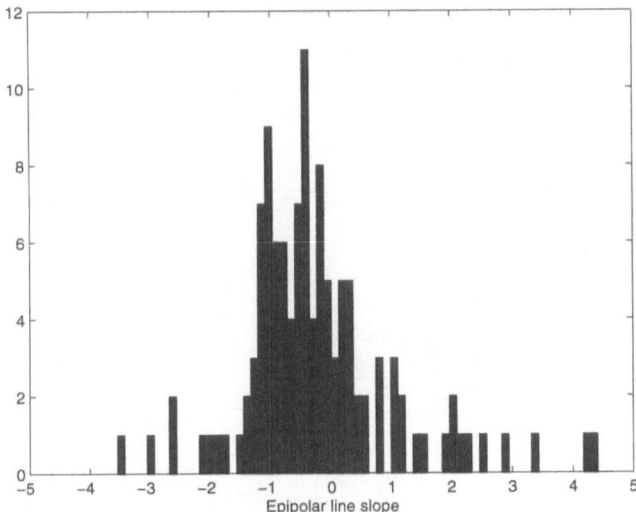

Figure 6. Histogram of epipolar line gradient.

Instead we used a standard ZNCC measure

$$\text{ZNCC}(L, R) = \frac{\sum_{ij}(L_{ij} - \overline{L})(R_{ij} - \overline{R})}{\sum_{ij}(L_{ij} - \overline{L})\sum_{ij}(R_{ij} - \overline{R})}$$

which is invariant to illumination offset and scale. The best match within the right image search region that exceeds a ZNCC score of 0.8 is selected. The role of the two images is then reversed and only features that consistently match left to right, and right to left are taken. Typically less than 50 features pass this battery of tests.

2.8. Establishing disparity consensus

The disparities resulting from the above procedure still contains outliers, typically very close to zero or maximum disparity. We use a median statistic to eliminate these outliers and provide an estimate of the consensus disparity from which we estimate height using (2). Figure 7 shows disparity and the number of robustly detected features for successive frames from a short flight sequence.

2.9. Time performance

The code runs on a 300 MHz K6 and is written carefully in C but makes no use, yet, of MMX instruction as does[4]. Corner strength, maxima detection and sorting takes 78 ms for both images. Matching takes 6.5 ms ($20\,\mu$s for an 11×11 window) with a further 1.3 ms to verify left-right consistency. With other onboard computational processes running we are able to obtain height estimates at a rate of 5 Hz.

3. Ultrasonic height estimation

When close to the ground the stereo camera system is unable to focus and the blurred images are useless for stereo matching. Ultrasonics are an alternative sensing modality which are theoretically suited to this height regime, but we were not very optimistic

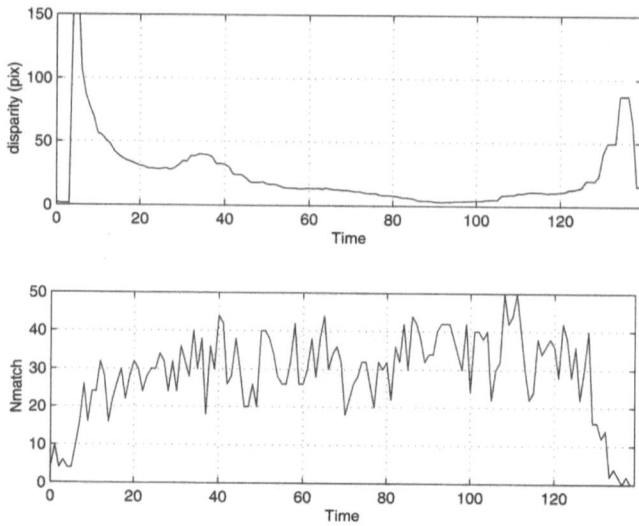

Figure 7. Disparity and number of robustly detected corner features for short flight sequence.

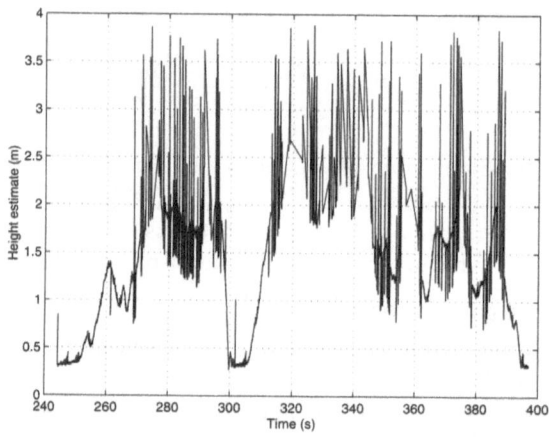

Figure 8. Raw height estimate from the ultrasonic transducer.

that it would work in this application. The helicopter provides an acoustically noisy environment, both in terms of mechanically coupled vibration and "dirty" air for the ultrasonic pulses to travel in. We attempted to shield the sensor from the main rotor downwash by positioning it beneath the nose of the vehicle.

The ultrasonic sensor we chose is a Musto 1a unit which has a maximum range of 4 m. The sensor has been adjusted so that the working distance range spans the output voltage range of 0 to 5 V. The sensor emits 10 pings per second, and when it receives no return it outputs the maximum voltage.

Figure 8 shows raw data from the sensor during a short flight. The minimum height estimate is 0.3 m which is the height of the sensor above ground when the

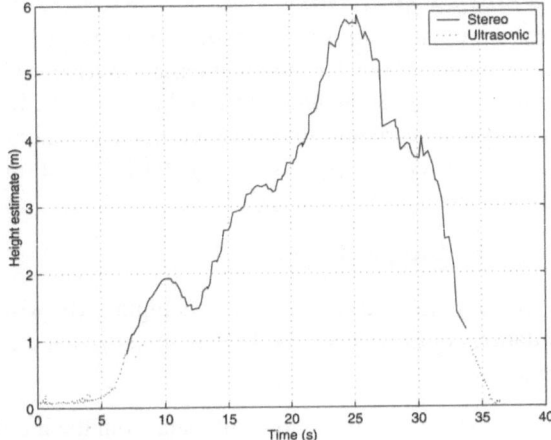

Figure 9. Combined height estimate from ultrasonic and stereo sensors with respect to the stereo camera.

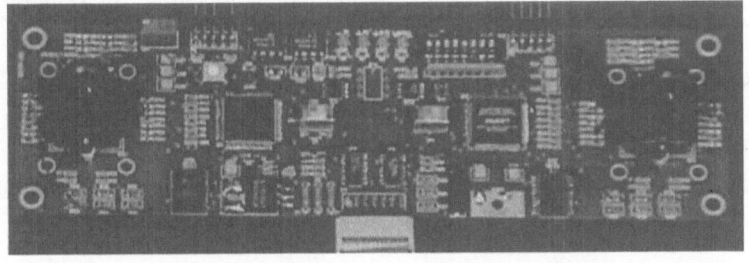

Figure 10. New integrated stereo camera sensor.

helicopter is on the ground. Good estimates are obtained for heights below 1.2 m. Above 1.5m the signal is extremely noisy which we believe is due to poor acoustic signal to noise ratio. Whether this is improved by isolating the sensor from the body of the helicopter has not yet been tested. Nevertheless there is just sufficient overlap between the operating range of the stereo and ultrasonic sensors to provide some cross-checking.

4. Combined results

Figure 9 shows the results of combining raw ultrasonic data with stereo disparity data. We have plotted all ultrasonic range values below 1.2m and added the height differential between the ultrasonic and stereo sensor systems. The stereo disparity data is converted to range using (2).

On takeoff there is a very small amount of overlap and we have good stereo data to below 1m (disparity of 75 pixels). The landing occurred in a patch of long grass which led to very blurry images in which no reliable features could be found.

5. Conclusions and further work

We have shown that stereo and ultrasonic sensors can be used for height estimation for an autonomous helicopter. They are complementary in terms of the height regimes in which they operate. Both are small and lightweight sensors which are critical characteristics for this application. Feature-based stereo is computationally tractable with the onboard computing power, and provides sufficient information for the task. The next steps in our project are:

1. "Close the loop" on helicopter height.

2. Fly our new custom stereo-camera head, see Figure 10, which will overcome the many limitations of the current ST-1 head, in particular exposure control and readout timing.

3. Fuse stereo-derived height information with data from the inertial sensor and use the estimated height to control the disparity search range and gate disparity values at the consensus stage.

4. Implement temporal feature tracking and use feature flow for visual gyroscope and odometry.

5. Visual-servo control of helicopter with respect to a landmark.

Acknowledgements

The authors would like to thank the rest of the CMST autonomous helicopter team: Leslie Overs, Graeme Winstanley, Stuart Wolfe and Reece McCasker.

References

[1] J. Banks. *Reliability Analysis of Transform-Based Stereo Matching Techniques, and a New Matching Constraint.* PhD thesis, Queensland University of Technology, 1999.

[2] D. Charnley, C. G. Harris, M. Pike, E. Sparks, and M. Stephens. The DROID 3D Vision System - algorithms for geometric integration. Technical Report 72/88/N488U, Plessey Research Roke Manor, December 1988.

[3] R. Deriche and G. Giraudon. A computational approach for corner and vertex detection. *Int. J. Computer Vision*, 10(2):101–124, 1993.

[4] S. Kagami, K. Okada, M. Inaba, and H. Inoue. Design and implementation of onbody real-time depthmap generation system. In *Proc. IEEE Int. Conf. Robotics and Automation*, pages 1441–1446, April 2000.

[5] J. M. Roberts. *Attentive Visual Tracking and Trajectory Estimation for Dynamic Scene Segmentation.* PhD thesis, University of Southampton, UK, dec 1994.

[6] S. M. Smith. A New Class of Corner Finder. In *Proceedings of the British Machine Vision Conference, Leeds*, pages 139–148, 1992.

[7] Z. Zhang, R. Deriche, O. Faugeras, and Q.-T. Luong. A robust technique for matching two uncalibrated images through the recovery of the unknown epipolar geometry. Technical Report 2273, INRIA, Sophia-Antipolis, May 1994.

Ladar-based Discrimination of Grass from Obstacles for Autonomous Navigation

Jose Macedo
University of San Diego
Industrial and Systems Engineering
San Diego, CA 92110
jmacedo@acusd.edu

Roberto Manduchi, Larry Matthies
Jet Propulsion Laboratory
California Institute of Technology
Pasadena, CA 91109
{manduchi,lhm}@telerobotics.jpl.nasa.gov

Abstract: Autonomous navigation in vegetated terrain requires the ability to discriminate obstacles from grass, a non-trivial problem when the sensorial world of the robot is based only on range information as provided, for example, by a laser rangefinder (ladar). We present a statistical analysis of the range data produced by a single-axis ladar in different situations, including the case of an obstacle partially occluded by grass. Such analysis inspired a simple classification algorithm, which has been tested on real range data acquired by JPL's urban robot.

1. Introduction

Autonomous vehicles have great promise for applications in the military, agriculture, space exploration, and other domains. Moreover, rapid progress in miniaturization and improved cost-effectiveness of navigation sensors, cameras, and computers is accelerating the maturation of robotic vehicles. However, a key limitation remains for domains in which robots must navigate in tall grass, small bushes, or forested areas, because existing perception systems cannot do effective obstacle detection in these situations. Most obstacle detection systems to date rely exclusively on range data from ladar, stereo vision, radar, or ultrasonic sensors to perceive scene geometry and assume implicitly that the scene consists of relatively large, solid surfaces [7]. When driving in vegetated terrain, the notion of "obstacle" needs to be revisited. For example, a small bush can be considered an obstacle based solely on geometric speculation, although it probably can be driven over without damaging the vehicle. Thus, for efficient navigation in vegetated terrain, a higher level of reasoning must intervene, based on both the geometric description of the scene and the composition of the terrain cover.

In this paper, we are interested in determining whether an "obstacle" is a rock (non-traversable) or a patch of grass (traversable). Terrain cover classification can be based on color features [1], but such an approach won't work at night. Visual texture is another promising approach, but it is computationally expensive and the technology is not mature yet. In this work, we discuss a simple approach based on the statistical analysis of the range data as provided

D. Rus and S. Singh (Eds.): Experimental Robotics VII, LNCIS 271, pp. 111–120, 2001.

by a laser rangefinder (ladar). Intuitively, range data on grass and bushes will be spatially scattered, while range data on bare soil or rock will tend to be more "regular" and lie on a relatively smooth surface.

In the following section we derive theoretical probability distributions for range data from a single-axis ladar in vegetated terrain. We consider a number of "canonical" situations, including the case of a field of randomly distributed grass containing a partially occluded rock. The theoretical results of Section 2 are validated in Section 3, where we show histograms of real range data. We also introduce a simple and fast algorithm for the classification of grass versus rocks based on statistical measures over moving windows. The range data used in the experiments was collected by the urban robot developed by JPL as part of the Tactical Mobile Robotics (TMR) Program funded by DARPA (see Figures 4,8). This robot is equipped with stereo cameras, an omnidirectional camera, an uncooled thermal infrared camera, and a 2-axis scanning ladar (although the data for this work has been acquired by rotation around the vertical axis only) [2, 6]. Autonomous navigation capabilities to date include obstacle avoidance, visual servoing to goals, and autonomous stair climbing.

2. Statistics of Range Data on Grass

The statistics of range measurements can provide us with precious information about the terrain cover. We introduce here a model for the range distribution which can be used to design a classifier of grass versus other obstacles, as discussed in Section 3. It is assumed here that the laser rotates around a vertical axis, and that the laser beam width is infinitesimal. We also assume that the ladar always receives a return when the laser ray hits a surface, and that the measurements are noiseless. These assumptions are discussed in more detail in Section 2.4

In the next subsections, we will consider three "canonical" situations: when the robot is in the middle of an homogeneous field of grass, when the robot is placed at a certain distance from a patch of grass, and when a rock is partly visible through the grass.

2.1. Case 1: Homogeneous grass field

To model an homogeneous field of grass, we will assume that the blades of grass have constant circular section of diameter d, and that the centers of the grass blades are distributed according to an isotropic Poisson point process in space with intensity λ [4]. This means that within a unit area of soil we expect to find λ blades[1]. In the case of an infinitesimal laser ray, one easily proves that the probability density function (pdf) of the range r is

$$p(r) = \lambda d e^{-\lambda d r} U(r) \tag{1}$$

where $U(\cdot)$ is the Heavyside function. In other words, we expect the range data on grass to behave as an exponential distribution with mean $\mu = 1/\lambda d$. A similar exponential behavior was predicted and observed in the case of range

---

[1] This model is not entirely correct because it assumes that grass blades can intersect [5].

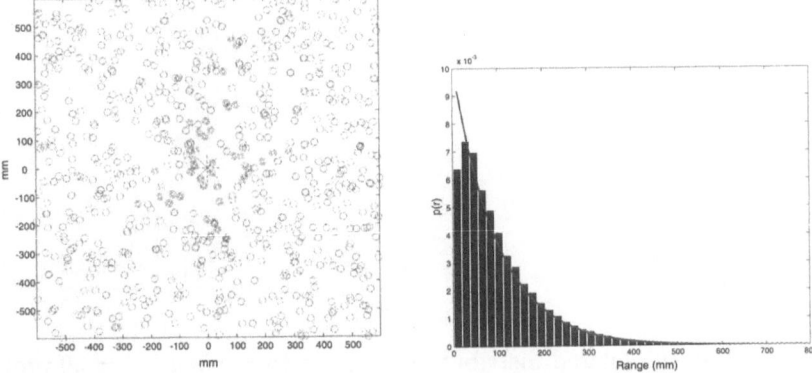

Figure 1. Left: Simulated distribution of grass ($\lambda = 500\text{m}^2, d = 20\text{mm}$). The laser is placed in the middle; the acquisition angular period is $0.5°$. The circles filled in red represents blades of grass hit by a laser ray. Right: Normalized histogram of range over 500 trials.

data measured in a forest [3], as a consequence of the random distribution of tree trunks.

A synthetic example of grass distribution is shown in Figure 1. Each grass blade has diameter d equal to 20 mm, and there are on average λ=500 blades per square meter. It is assumed that the ladar acquires data with an angular period of $0.5°$. The blades of grass that are hit by the laser ray are represented with a red kernel. Figure 1 also shows the normalized histogram of the range measured over 500 trials. An exponential density curve has been fitted to the data and superimposed to the histogram in the figure. Such best-fitting exponential has mean μ equal to 0.098/m, which is very close to the expected value of 0.1/m.

From the pdf of the range (1) we may compute its second-order moment (variance, σ^2) and its third-order moment (skewness[2], sk), which will be used in Section 3:

$$\sigma^2 = \mu^2 = 1/(\lambda d)^2 \, , \, sk = 2\mu^3 = 2/(\lambda d)^3 \qquad (2)$$

The skewness measures the degree of symmetry of a distribution around its mean. Positive (negative) values of the skewness indicate that the distribution extends towards the right (left) tail. The case $sk = 0$ corresponds to a symmetric distribution, such as the normal. In the case under exam the skewness is positive, meaning that the distribution is skewed to the right.

2.2. Case 2: Grassy patch seen from the distance

An instance of this case is shown in Figure 2 ($\lambda = 85/\text{m}^2$, $d = 30$ mm). The robot is at a distance $D = 200$ mm from the rectilinear edge of a patch of

---

[2]By identifying the skewness sk with the third-order moment of the distribution, we adhere to the definition given in [8]. Note that other authors (e.g. [4]) define the skewness as the third-order moment divided by $\sigma^{3/2}$.

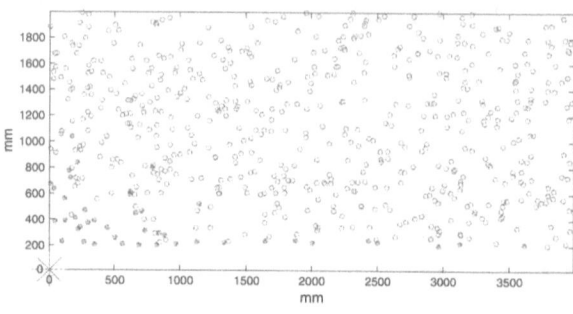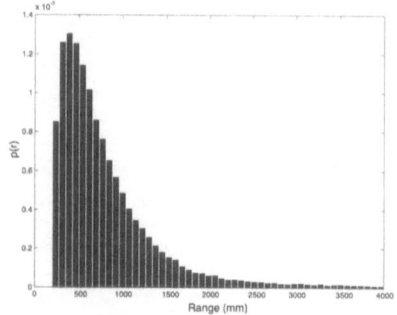

Figure 2. Left: Simulated distribution of grass ($\lambda = 85/m^2, d = 30$mm). The laser is placed in the lower left corner, at a distance of 200 mm from the patch of grass. Right: Normalized histogram of range over 500 trials.

grass. The distribution of the range data along a fixed line oriented at an angle α with respect to the normal to the patch edge is now an exponential shifted by a value of $D/\cos(\alpha)$. However, the pdf of the range computed over the whole angular span (equal to 90° in this example) does not have an exponential shape. Indeed, the normalized histogram shown in Figure 2 (computed over 500 trials) shows a heavier tail than in the case of an exponential density.

2.3. Case 3: Rock behind the grass

Suppose the robot is looking at a patch of grass which contains a rock (or any non-traversable object). Clearly, if the rock is right in front of the robot, the distribution of the range concentrates around the actual distance of the rock. If the rock is very far away from the robot, all the rays from the ladar are likely to be stopped by the grass before reaching the rock surface. However, if the rock is at an intermediate distance, it is quite possible that some rays from the ladar reach the rock surface, while the other rays hit some blades of grass on the way. This situation is shown in Figure 3, where the parameters of the grass distribution are $\lambda = 300/m^2$, $d = 10$ mm (the rock surface is represented by the blue circles.) To simplify the computation of the range distribution, we will assume that all points of the rock surface are at the same distance T from the ladar (this hypothesis is clearly not verified in Figure 3, where the rock surface is at a slanted angle). In this case, the pdf of the range r is simply an exponential truncated at $r = T$, followed by a peak at $r = T$ of area $e^{-\lambda dT}$. The larger the distance T to the rock, the smaller the area of the peak, and the further to the right its position. This expected behavior can be noted in the histogram of Figure 3 (computed over 3000 trials within the angular sector subtended by the rock surface). The histogram is clearly bimodal, although the second peak spreads out due to the slant of the rock surface. The mean μ, variance σ^2 and skewness sk of the range are easily computed from our model distribution:

$$\mu = \frac{1}{\lambda d} \left(1 - e^{-\lambda dT}\right) \tag{3}$$

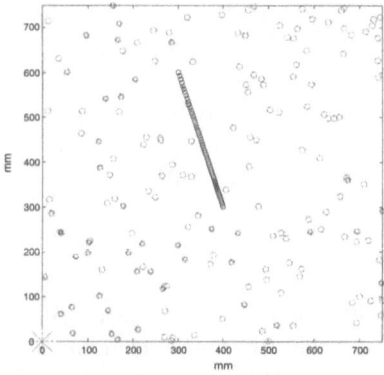 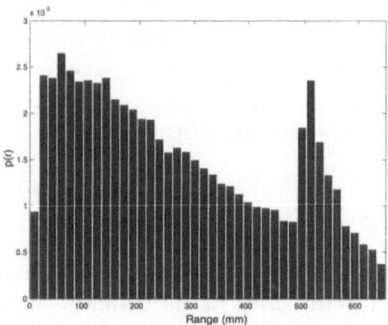

Figure 3. Left: Simulated distribution of grass ($\lambda = 300/\mathrm{m}^2, d = 10\mathrm{mm}$). The blue circles represent an impenetrable surface. The laser is placed in the lower left corner. Right: Normalized histogram of range over 300 trials.

$$\sigma^2 = \frac{1}{(\lambda d)^2} - e^{-\lambda dT}\frac{2T}{\lambda d} + \frac{e^{-2\lambda dT}}{(\lambda d)^2} \tag{4}$$

$$sk = \frac{2}{(\lambda d)^3} - e^{-\lambda dT}\frac{3T^2}{\lambda d} - e^{-2\lambda dT}\frac{6T}{(\lambda d)^2} - e^{-3\lambda dT}\frac{2}{(\lambda d)^3} \tag{5}$$

By comparing (5) with (2), one maintains that the skewness in this case is always smaller than in the case of homogeneous grass, and indeed it takes on negative values for sufficiently small T. In other words, the presence of the peak centered at $r = T$ makes the distribution more skewed to the left.

2.4. Validity of our theoretical assumptions

While the theoretical results derived in the previous subsections are useful to understand the behavior of the range data, and indeed have inspired the simple classification algorithm of Section 3, we should comment on the shortcomings of our analysis. Firstly, the assumption that the ladar always receives a return when the laser ray hits a surface is not very realistic. For example, it is quite possible that the ladar does not read a return when the ray hits the outer edge of a grass blade. Secondly, the return is integrated over a non-infinitesimal time window while the ladar is revolving. This means that the measured range actually represents the average value of the actual profile within a small angular sector. The measured range is thus a "smoothed" version of the actual range profile. Finally, a more realistic analysis would take into account the fact that laser ray has a non-null divergence γ (e.g., the laser used in the experiments has $\gamma = 1$ mr). It can be shown that in this case the distribution of the range along a fixed direction can be modeled by the following form:

$$p(r) = \lambda d(1 + ar)e^{-\lambda d(r-D)(1+a(r+D)^2/2)}U(r - D) \tag{6}$$

where D is the distance to the edge of the grassy patch along the ray and $a = 2\tan(\gamma/2)/d$. For small values of γ and D, the distribution of the range data does not differ substantially from the infinitesimal ray case. This is because the ray will normally hit a blade of grass before "thickening". However, if the patch of grass is far from the ladar (i.e., for large enough values of aD^2), it is seen from (6) that the effect of finite ray size cannot be neglected. In this case, the ray will have thickened noticeably before hitting a blade of grass. Intuitively, this corresponds to using an infinitesimal ray in a field of grass where the density of the blades increases proportionally with the distance from the ladar.

3. Experimental Results

In this section we present some experiments on real range data, collected by our urban robot at La Canada, near JPL, in environments that include grass with some rocks and trees. The rotation rate of the ladar was set to 5Hz and the angular sampling period to 0.7°.

Figure 4 shows the spatial distribution of range measurements with the robot placed in front of a patch of grass. The grass is visible in the upper half-plane; the dots in the lower plane correspond to soil, which was visible because the rotation axis was not perfectly vertical. Two histograms of range are presented in Figure 5, covering respectively a narrow and a large angular sector of the grassy patch. As expected, the first histogram has a shape very similar to the exponential curve of (1). The best-fitting exponential (super-imposed on the histogram) has mean $\mu = 1/\lambda d$ with $\lambda d=3.7$/m. The second histogram has a much heavier tail, as predicted by the discussion in Section 2.2. Note in passing that the histograms have been computed over a number of revolutions while the robot stood still. A light breeze is sufficient to keep the grass blades in constant shaking motion, enabling a good statistical sampling. Another reason for scan-to-scan differences is that in each revolution the measurements may not be taken at identical angular positions.

Figure 6 shows a situation with some grass and two rocks (one of which partially occluded by grass). The histograms, computed within the angular sector corresponding to three different situations (grass, rock, rock behind the grass), match the densities predicted in Section 2 rather well. In particular, the histogram corresponding to the rock behind the grass is clearly bimodal (see also Figure 8).

A classifier of grass versus non-grass may be based on the statistical properties of the range discussed above. We implemented a very simple and fast algorithm, based the local estimation of the variance and of the skewness of the range distribution. A suitable threshold set on the variance may allow us to classify rather robustly grass from an obstacle with an exposed smooth surface. However, this technique will fail when the obstacle is partially occluded by grass, because the range measurements will still have a high variability. To deal with this situation, we may exploit the fact that the skewness is lower than in the case of homogeneous grass (see Section 2.3). Thus, our improved statistical test classifies a point as grass if the local variance is above a threshold t_1 and the local skewness is above a threshold t_2.

An important issue here is the choice of the sample size for the computation of the variance. A small sample size causes the variance to vary widely

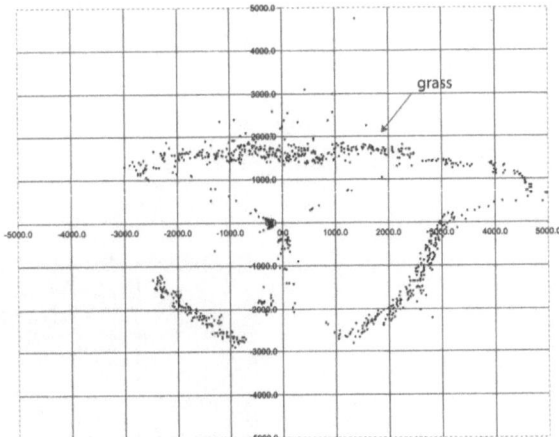

Figure 4. Spatial distribution of range measurements around the robot. In this figure as well as in the following ones, the axis ticks represent millimiters. The cluster of points at the immediate left of the center is actually a part of the robot within the field of view of the ladar.

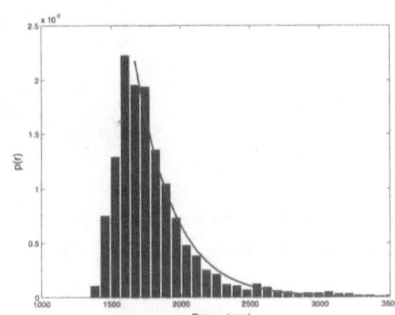
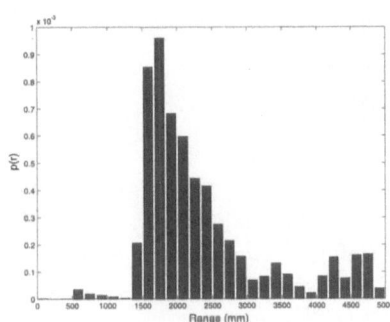

Figure 5. Normalized histogram of range measured over 30 revolution at angles between 60° and 120° (left) and between 5° and 140° (right) for the case of Figure 4.

within the same terrain cover class. A large sample size covers a broad area, and therefore reduces the spatial resolution of the estimation. It was found by extensive testing that a sample size of 9 for the variance and of 29 for the skewness represents a good compromise between resolution and stability. Figure 9 shows the classification results using our algorithm in two different situations.

4. Conclusions

We presented an analysis of the statistics of range measurements in a vegetated environment, and showed its use in the design of a classifier of grass versus obstacles. Our technique has given good results in real-world experiments,

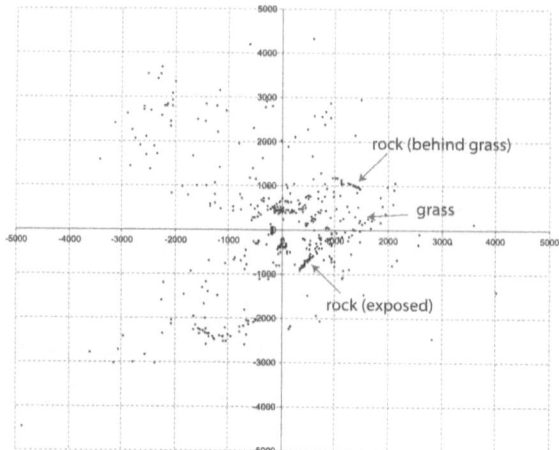

Figure 6. A situation with two rocks and grass.

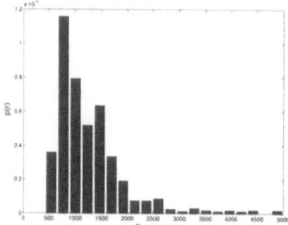
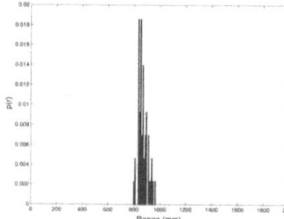
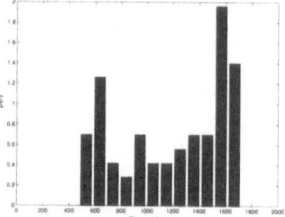

Figure 7. Normalized histogram of range measured at angles between $-30°$ and $30°$ (left), between $-40°$ and $-60°$ (center) and between $35°$ and $50°$ (right) for the case of Figure 6.

even when obstacles were partially occluded by grass. Future work will extend our analysis to more complex situations (involving, for example, discriminating patches of visible soil), as well as to the case of two-dimensional range data.

Acknowledgments

The research described in this paper was carried out at the Jet Propulsion Laboratory, California Institute of Technology, and was sponsored by the Defense Advanced Research Projects Agency under the Mobile Autonomous Robot Software program and under the Tactical Mobile Robotics program. Reference herein to any specific commercial product, process, or service by trade name, trademark, manufacturer, or otherwise, does not constitute or imply its endorsement by the United States Government or the Jet Propulsion Laboratory, California Institute of Technology.

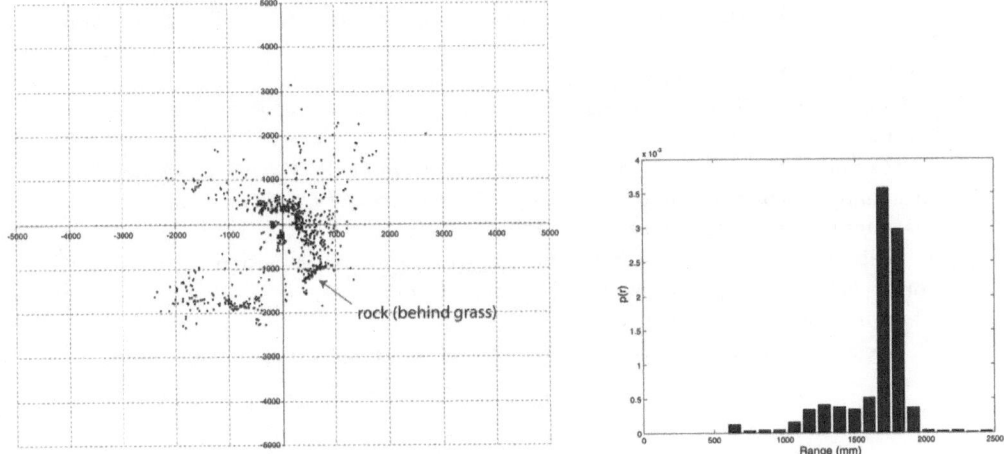

Figure 8. A situation with a rock behind the grass (left) and the normalized histogram of range measured over 3 revolution at angles between −45° and −70° (right).

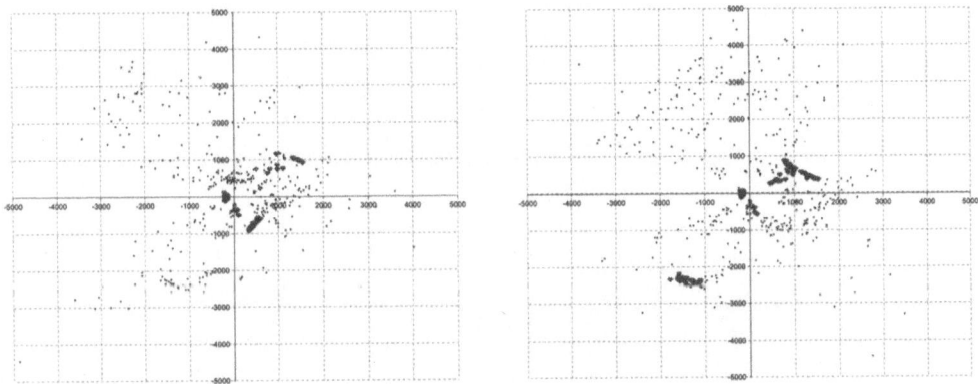

Figure 9. Classification results using local measurements of range variance and skewness for two different instances. The red enlarged dots have been identified by the algorithms as non-grass.

References

[1] P. Bellutta, R. Manduchi, L. Matthies, K. Owens, A. Rankin, "Terrain Perception for DEMO III", *IEEE Intelligent Vehicles Symposium 2000*, Dearborn, MI, October 2000

[2] C. Bergh, B. Kennedy, L. Matthies, A. Johnson, "A compact and low power two-axis scanning laser rangefinder for mobile robots", *Seventh Mechatronics Forum International Conference*, Atlanta, Georgia, September 2000.

[3] J. Huang, A.B. Lee, D. Mumford, "Statistics of range images", *IEEE Conference on Computer Vision and Pattern Recognition*, Hilton Head, June 2000.

[4] M. Kendall, A. Stuart, *The advanced theory of statistics*, Macmillan Publishing, 1977.

[5] T. Leung, J. Malik, "On perpendicular texture or: Why do we see more flowers in the distance?", *IEEE Conference on Computer Vision and Pattern Recognition*, San Juan, Puerto Rico, June 1997.

[6] L. Matthies, Y. Xiong, R. Hogg, D. Zhu, A. Rankin, B. Kennedy, "A portable, autonomous, urban reconnaissance robot", *International Conference on Intelligent Autonomous Systems*, Venice, Italy, July 2000.

[7] L. Matthies, A. Kelly, T. Litwin, G. Tharp, "Obstacle detection for unmanned ground vehicles: a progress report", *Robotics Research: Proceedings of the 7th International Symposium*, 1996.

[8] C. Nikias, A. Petropulu, *Higher-order spectral analysis*, Prentice-Hall, 1993.

Reality-based Modeling with ACME: *
A Progress Report

Dinesh K. Pai, Jochen Lang, John E. Lloyd, and Joshua L. Richmond
Department of Computer Science
University of British Columbia
Vancouver, Canada
{pai|jlang|lloyd|jlrichmo}@cs.ubc.ca

Abstract: We describe the current state of ACME, the UBC Active Measurement facility, a telerobotic facility designed for building computational models of everyday physical objects. We show how ACME is being used to acquire models of contact texture (including friction and roughness), contact sounds, and contact deformation.

1. Introduction

The UBC Active Measurement Facility (ACME) is a robotic measurement facility designed for building reality-based models (i.e., models of real objects constructed from measurements). These reality-based models are intended for applications in virtual environments, computer animation, computer games, and e-commerce; the focus is on easily creating rich multi-modal models of objects suitable for human interaction, rather than on highly accurate models of specific attributes.

At ISER 99, we described the design of the system and initial results [9]. In this paper, we describe the current state of the ACME facility (§2), and how ACME has been used to build new types of models, including contact texture models (§3), contact sound models (§4), and contact deformation models (§5).

2. ACME Facility

ACME consists of a variety of sensors and actuators (Fig. 1), all of which can be controlled using small Java programs called **Experiments**. More details are available in [9]. Briefly, the subsystems include: a 3 DOF Test Station used for precise planar positioning of the test object; a Field Measurement System (FMS), consisting of a Triclops trinocular stereo vision system, a high resolution RGB camera, and a microphone, all mounted on a 5 DOF positioning gantry; and a Contact Measurement System (CMS) consisting of a Puma 260 robot equipped with a force/torque sensor, mounted on a linear stage. The entire facility can be teleprogrammed in Java from any location on the Internet.

The current ACME system has been enhanced in a number of ways. A special **Sensor** interface (Sect. 2.1) has been added to facilitate the real-time

* Supported in part by grants from NSERC and IRIS NCE.

D. Rus and S. Singh (Eds.): Experimental Robotics VII, LNCIS 271, pp. 121–130, 2001.

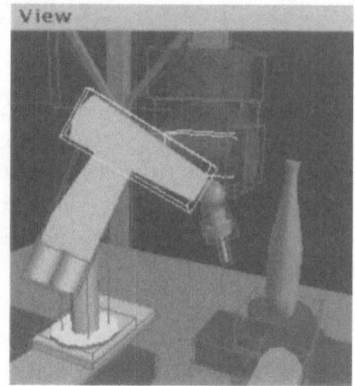

Figure 1. ACME facility being used to map surface friction and roughness of a bottle.

Figure 2. ACME simulation detail showing the bounding boxes. Boxes involved in a collision are highlighted.

collection of data from the system's various sensors, and the ACME simulator has been redesigned to facilitate the easy addition of classes for simulating sensor input. The entire ACME facility is now modeled using an enhanced Java3D scene graph, containing all the geometric, positional, and appearance information used in simulation, run-time checking, and graphic display. One of the features of this scene graph is a collision detector, which performs rapid on-line collision detection using oriented bounding boxes [4] placed around principal system components (Fig. 2).

2.1. Sensor interface

An ACME experiment collects data from the system's sensors (e.g., cameras, actuator position sensors, force/torque sensors, and microphones) through implementations of a **Sensor** interface. This provides an abstract interface to the underlying sensing framework which has several important features. Most importantly, it enables asynchronous real-time data acquisition from a Java program. The user's Java **Experiment** controls data acquisition by specifying trigger conditions (which can be sophisticated methods written in Java) to decide when to start and stop data acquisition based on the sensed data. The interface provides a number of utility methods that permit the experiment to adjust the frequency at which data is returned from the sensor, enable/disable the time stamping of data, and adjust the buffering within the **SensorInputStream**. Finally, the framework allows the **Experiment** to connect either to a remote **SensorServer** object running on the main ACME server, or (when an experiment is run in simulation) to a **SensorServer** object running in the simulator on the ACME client.

3. Contact Texture Modeling

Friction and roughness are important components of haptic texture and are required for contact simulation. Real objects usually have multiple materials and surface finishes that vary over the surface of the object. We can measure these in ACME using the CMS and force sensor, by analyzing the force profile produced by stroking the surface with a round-tipped probe (Fig. 1). The task is non-trivial for several reasons. First, friction estimates must account for variation in the surface normal due to curvature and uncertain geometry. Second, we need to be able to extract from the noisy force profile a simple model of surface roughness. Third, exploring the entire surface can be very time consuming if done in an exhaustive fashion, and so a more efficient approach is needed.

3.1. Friction Estimation

The problems in estimating friction can be handled using a differential measurement approach. The probe strokes the surface along a short path using a compliant motion that applies pressure to the surface. This will result in a normal force component $\mathbf{f}_n$ and a tangential component of magnitude $\|\mu\mathbf{f}_n\|$ (where μ is the coefficient of friction) in a direction opposite to that of the motion. If $\mathbf{f}_n$ were known, then μ could be computed directly. Unfortunately, $\mathbf{f}_n$ is not accurately known because the surface normal is itself not known accurately and may also vary along the path. To compensate for this, we stroke the surface again, along the same path but in the opposite direction. At any point along the path, we then have a force value $\mathbf{f}$ which was measured during the forward motion, and another value $\mathbf{f}^*$ which was measured during the reverse motion.

Now referring to Fig. 3, $\mathbf{f}$ and $\mathbf{f}^*$ each have components parallel to the surface normal $\mathbf{n}$, and friction components $\mathbf{f}_f$ and $\mathbf{f}_f^*$ which lie opposite to the directions of motion and are perpendicular to $\mathbf{n}$. Now even if $\mathbf{n}$ is unknown, and the magnitudes of $\mathbf{f}$ and $\mathbf{f}^*$ differ, we can still estimate μ by calculating the angle between $\mathbf{f}$ and $\mathbf{f}^*$:

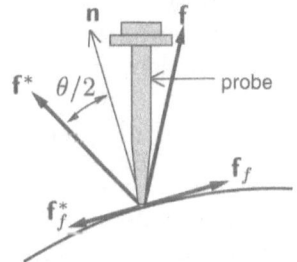

$$\mu = \tan(\theta/2). \qquad (1)$$

This calculation is independent of the speed of travel, the force of the probe against the surface (in either direction), the orientation of the probe, and of course the surface normal itself. By averaging the values μ obtained at various points along the path, a reasonable estimate of the friction coefficient may be obtained.

Figure 3. Forces associated with motion in two different directions along an object surface.

3.2. Roughness characterization

The force profiles obtained by stroking the surface can also be used to identify the surface roughness (Fig. 4). We wish to characterize this roughness using

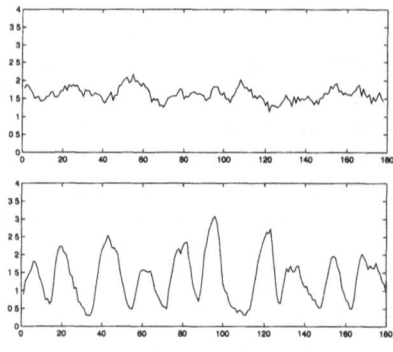

Figure 4. Force profiles for two different types of material: rubber gasket sheeting (top), and masonite (bottom).

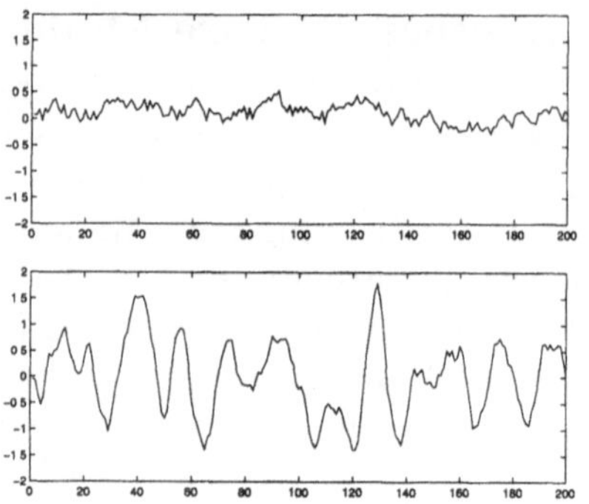

Figure 5. Zero-meaned AR(2) simulations based on models extracted from the profiles in Fig. 4. Top: rubber gasket sheeting, with $a_1 = 0.60$, $a_2 = 0.25$, and $\sigma = 0.11$. Bottom: masonite, with $a_1 = 1.61$, $a_2 = -0.73$, and $\sigma = 0.16$.

a simple model that is suitable for easy simulation, particularly in haptic displays. Intuitively, we can see that the roughness is largely indicated by the "noisiness" of the profile, as characterized by the noise amplitude (variance), and spatial density. In addition, the profile will often contain a non-random periodic component, particularly in man-made materials, due to manufacturing processes, or functional or aesthetic requirements (as evidenced by the masonite in Fig. 4).

These qualities can be captured by treating the force profile $f(x)$ as a discrete random series described by a second-order autoregressive model, denoted AR(2):

$$f(kL) \equiv f_{(k)} \approx a_1 f_{(k-1)} + a_2 f_{(k-2)} + \sigma \epsilon_{(k)}, \qquad (2)$$

where k is the sample number, L is the length of the spatial discretization, a_i are the model coefficients, and $\sigma \epsilon_{(k)}$ is a sample from a white noise sequence with standard deviation σ.

The model of Eq. 2 is obviously very efficient to simulate, and is completely defined by the parameters a_1, a_2, and σ. These can be readily determined, for instance by autoregressive parameter estimation via the covariance method (e.g., the arcov function in Matlab). The parameters and associated simulated signals for the profiles in Fig. 4 are shown in Fig. 5. Higher order AR(p) models or ARMA models could also be used if necessary.

3.3. Efficient surface exploration

To avoid having to sample contact texture exhaustively over the entire surface, we employ a hierarchical exploration approach, similar to that used in the acoustical modeling described in Sect. 4 of this paper. The surface geometry is described using a Loop subdivision mesh [8], and the robot initially samples friction and/or roughness at locations corresponding to the vertices of coarsest resolution mesh. If the contact texture properties at adjacent vertices differ significantly, we then sample between them, at a vertex of the next higher resolution mesh.

This method was used to map the surface friction for the bottle shown in Fig. 1, to which sandpaper patches had been applied to provide regions of high friction. The resulting map is shown in Fig. 6; the sample locations are shown by small line segments. The sampling decisions were made entirely by the planner: notice that the hierarchical exploration correctly clusters the samples near the boundaries between areas of constant friction.

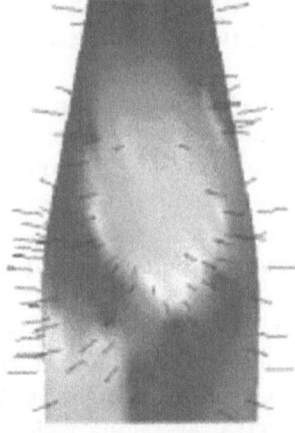

Figure 6. Measured friction map of bottle. Dark regions represent low friction ($\mu \approx 0.13$ for the glass), and light regions represent high friction ($\mu \approx 0.5$ for the sandpaper patches shown).

Figure 7. Contact sound measurement

4. Contact Sound Modeling

Contact sounds provide important cues for the shape and material properties of objects [6], and can be efficiently simulated [13]. By "contact sounds", we mean the sounds produced by the interaction of two objects' surfaces (e.g., scraping, knocking, rolling, etc.). Such sounds are useful in many applications, including interactive virtual environments, physical simulations, tele-operative controls and robotic perception of material [7].

A sound model may be viewed as the acoustic analog to a texture map in graphics. The parameters of contact sound models vary over an object's surface as they are affected by shape, material, composition and surface texture. For most everyday objects, their complex geometry and material composition render analytical solutions to the sound model difficult, if not impossible. An empirical solution is more tractable and practical, particularly when constructing sound models for many arbitrary objects.

Many techniques for acquiring the measurements needed to create empirical sound models have been tried, especially in the realm of musical instrument modeling [1, 3, 14]. Many of these techniques, though appropriate for their application, are inaccurate in either location or impact profile. Also, acquiring densely sampled measurements by these manual techniques is often tedious and subject to human error. ACME was recently equipped to perform the first automatic measurements and estimations of contact sound models [10].

We measure the sound response of an object in ACME using the "sound effector" device [10]. This push-type solenoid mounted on the end-effector of the CMS robot is used to apply light near-impulsive impacts at specified locations on the surface, and the resulting sound is recorded using a microphone (see Fig. 7).

As in our contact texture modeling experiments (§3), we employ an adaptive algorithm to select sample locations automatically. The vertices of the object's surface are selected as an initial set of sample points. As sound models are constructed at adjoining vertices, the sounds are compared, and new sample locations added if the difference between the two sounds models is perceptually relevant. Models are compared by a sound metric derived from perceptual studies of contact sounds [6].

An additional advantage of using ACME to perform sound model acquisition is repeatability; multiple measurements can be recorded at the same surface location. Spectral averaging is performed across samples at the same location to produce a prototypical sound. Spectral averaging yields a 10 to 20% reduction in parameter estimation error [11].

The parameters of the sound model are estimated from the prototypical sound at each location. Complete sound models have been created for a variety of objects using ACME. Figure 8 demonstrates the fidelity of the model at one location on the surface of the glass bottle of Fig. 1. Note that the amplitude, decay and frequency of the most dominant frequency modes have been correctly identified in the model.

5. Deformation Modeling

Modeling the deformation behavior of real objects is a major challenge. There has been recent progress in real-time simulation of linear elastic objects [5, 2]. To construct such models requires measuring the deformation of an object in response to applied forces.

In particular, a linear elastic map Ξ relates known (prescribed) tractions and displacements x at the boundary to the unknown complementary set of tractions and displacements b.

$$\Xi x = b$$

The matrix Ξ represents the discrete Green's functions of the boundary value problem associated with the object's elastostatic behavior. These Green's functions can be used to quickly solve boundary value problems encountered in haptic simulation, e.g. produced by touching the object and moving the point

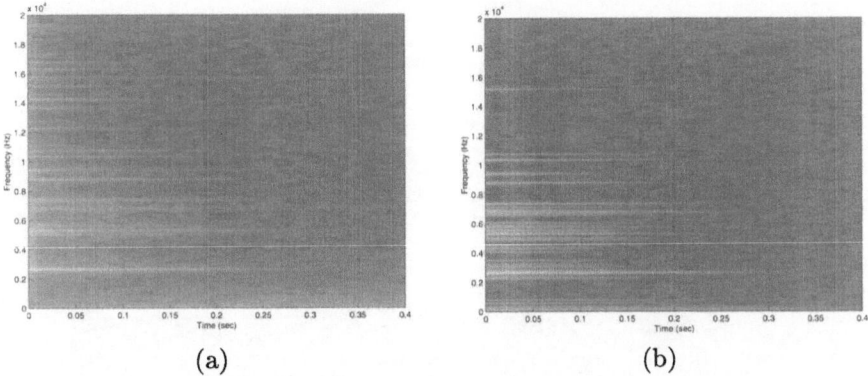

(a)   (b)

Figure 8. A spectral comparison of a recorded acoustic impulse (a) and an acoustic impulse synthesized from a prototypical model (b). The prototypical model was created by spectral averaging over three recordings of the bottle at the same surface location. The initial amplitude, decay rate and frequency of the forty most-dominant modes were estimated to create the model. These modes are illustrated by the horizontal peaks in the spectrogram.

of contact over the surface [5]. The Green's functions for an object with known geometry and known homogeneous, isotropic linear-elastic material properties can be calculated in various ways. One approach is to apply the boundary element method and obtain the map directly (for details and further references, see [5]); another approach is to discretize the object's volume and apply a finite element method followed by a condensation step [2].

In practice, for most everyday objects, the geometry and material properties are not known. We would like to directly estimate the Green's functions based on active deformations performed using ACME.

The Green's functions correspond to a particular reference boundary configuration. In the reference configuration, the object's fixed support surface has a displacement boundary condition, and the free surface has a traction boundary condition. The contact measurement system (CMS) in ACME applies a sequence of point-like forces to the free surface. The CMS's sensors provide a local measurement of the displacement of the contact point and the applied force. The global object deformation is sensed using the Triclops stereo vision system[1] mounted on the field measurement system (FMS).

We have developed a novel surface deformation measurement technique utilizing the Triclops trinocular stereo-vision system. The measurement of surface deformation is based on three-dimensional scene flow. Optical flow (see Fig. 9) is calculated simultaneously through a sequence of calibrated stereo images. The stereo images are also matched utilizing the Triclops Stereo Vision library. Combining the optical flow in the sequence of stereo triplets with the depth information from stereo provides three-dimensional scene flow (see Fig. 10). A geometric model of the undeformed surface of the object under test

---

[1] Point Grey Research, Vancouver, Canada, *http://www.ptgrey.com/*.

Figure 9. Segmented image flow during deformation. (Zoom shows every 4th flow vector.)

is either assumed known or acquired beforehand. This geometric model and the tracking capabilities of the system allow the three-dimensional flow to be segmented into surface flow and non-surface flow.

Fig. 9 shows the segmented image flow for deforming a test object; here a stuffed toy tiger. The initial geometry is represented as a triangular mesh (Fig. 10). The mesh is estimated from Triclops data with the aid of a volumetric surface reconstruction software from the National Research Council of Canada [12]. The resulting displacement at mesh nodes is shown in Fig. 10. The applied force and local displacement is simultaneously measured using the CMS.

In finding an object's Green's functions, an estimation step based on the measured local and global deformations is next. We are currently exploring ways to robustly perform this estimation. Questions which ACME will enable us to explore include estimation robustness, the adequacy of linear elastic maps for non-homogeneous bodies and a comparison of simulation and measurement.

6. Conclusions

The ACME facility has matured and is demonstrating the promise of reality-based modeling. Several improvements make the system more robust and easier to program. The abstract sensor framework allows the acquisition of large amounts of data in a timely fashion. The improved simulation protects the facility from many operator errors. ACME can now be used to acquire models of a wide variety of everyday objects from metallic vases to stuffed toys. The object's response in multiple modalities can be excited and measured, enabling the modeling of contact texture, sound, and deformation.

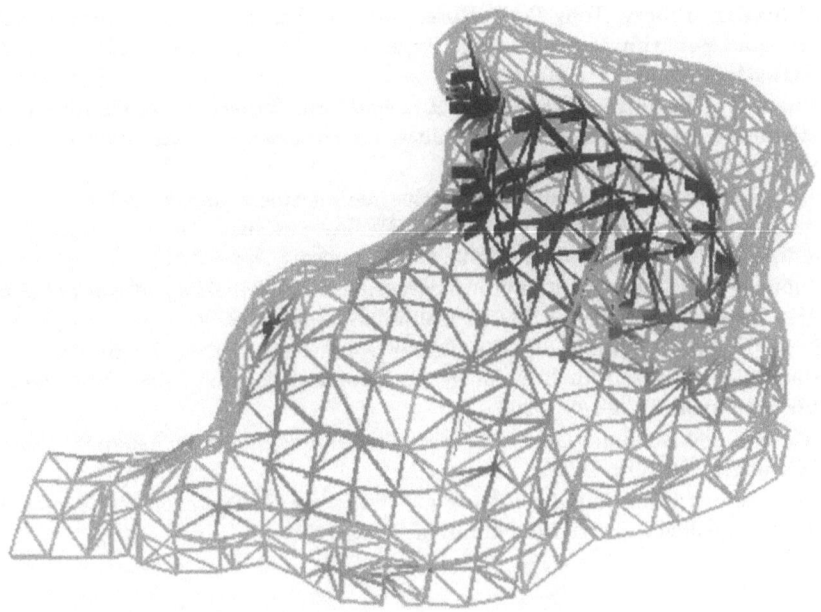

Figure 10. Mesh node displacements. (Darker nodes indicate larger displacements.)

References

[1] Perry R. Cook and Dan Trueman A database of measured musical instrument body radiation impulse responses, and computer applications for exploring and utilizing the measured filter functions. International Symposium on Musical Acoustics, Acoustical Society of America, Woodbury, NY, 1998.

[2] S. Cotin, H. Delingette, J-M Clement, V. Tassetti, J. Marescaux, and N. Ayache. Geometric and physical representations for a simulator of hepatic surgery. In *Proceedings of Medicine Meets Virtual Reality IV*, pages 139–151. IOS Press, January 1996.

[3] Robert S. Durst and Eric P. Krotkov Object classification from analysis of impact acoustics. In *Proceedings of the IEEE/RSJ International Conference on Intelligent Robots and Systems*, 1:90–95, 1995.

[4] Stefan Gottschalk, Ming Lin, and Dinesh Manocha. Obb-tree: A hierarchical structure for rapid interference detection. In *Computer Graphics (ACM SIG-GRAPH 96 Conference Proceedings)*, pages 171–180, August 1996.

[5] Doug L. James and Dinesh K. Pai. ARTDEFO, accurate real time deformable objects. In *Computer Graphics (ACM SIGGRAPH 99 Conference Proceedings)*, pages 65–72, August 1999.

[6] R. L. Klatzky, D. K. Pai, and E. P. Krotkov. Hearing material: Perception of material from contact sounds. *PRESENCE: Teleoperators and Virtual Environments*, 9:4, pages 399–410, The MIT Press, August 2000.

[7] Eric Krotkov. Robotic perception of material. In *Proceedings of the Fourteenth International Joint Conference on Artificial Intelligence*, pages 88–94, 1995.

[8] Michael Lounsbery, Tony D. DeRose, and Joe Warren. Multiresolution analysis for surfaces of arbitrary topological type. *ACM Transactions on Graphics*, 16(1), pages 34–73, January 1997.

[9] Dinesh K. Pai, Jochen Lang, John E. Lloyd, and Robert J. Woodham. Acme, a telerobotic active measurement facility. In *Proceedings of the Sixth Intl. Symp. on Experimental Robotics*, 1999.

[10] J. L. Richmond and D. K. Pai. Active measurement and modeling of contact sounds. In *Proceedings of the 2000 IEEE International Conference on Robotics and Automation*, pages 2146–2152, San Francisco, April 2000.

[11] Joshua L. Richmond *Automatic measurement and modeling of contact sounds*. MSc. thesis, University of British Columbia, August, 2000.

[12] G. Roth and E. Wibowoo, An efficient volumetric method for building closed triangular meshes from 3-D image and point data. In *Proceedings Graphics Interface*, pages 173–180, 1997.

[13] Kees van den Doel and Dinesh K. Pai. The sounds of physical shapes. *Presence*, 7(4), pages 382–395, 1998.

[14] Kees van den Doel *Sound synthesis for virtual reality and computer games*. PhD thesis, University of British Columbia, May, 1999.

Grasp Strategy Simplified by Detaching Assist Motion (DAM)

Makoto Kaneko, Tatsuya Shirai, Kensuke Harada and Toshio Tsuji

Department of Industrial and Systems Engineering

Hiroshima University.

Higashi–Hiroshima, Japan

{kaneko, kharada, tsuji}@huis.hiroshima-u.ac.jp

shirai@hfl.hiroshima-u.ac.jp

Abstract:

For a small object placed on a table, human easily captures it within the hand by changing the finger posture from upright to curved ones after each finger makes contact with the object. A series of this motion is called as Detaching Assist Motion (DAM). This paper discusses a generalized grasp strategy where a multi–fingered robot hand can approach an object and finally envelop it, irrespective of the size of object.

1. Introduction

For considering the grasp strategy of robot hand, human motion often provides us with good hints. Due to this reason, many researchers have discussed the classification of either final grasp patterns or grasp postures[1]–[3] learnt by human motion. On the other hand, we are particularly interesting to consider the whole grasping procedure where the hand first approaches an object placed on a table and finally achieves an enveloping grasp. Through the observation of human grasping, we learnt that human changes his (her) grasping strategy according to the size of objects, even though they have similar geometry. We called the grasp planning *Scale Dependent Grasp*[4],[5]. Through these works, we found that human roughly switches the grasp pattern three times depending upon the size of objects. The most complicated pattern is observed for an object whose representative size is smaller than that of our fingertip, while a simple grasp pattern can work for a relatively large object. For such small objects, two characteristic patterns are observed. One is that human first picks up the object from the table, and then finally achieves the target grasp through a grasp transition from the fingertip to the enveloping grasps, as shown in Fig.1(a). The other one which is observed just by chance is that human first approaches the object until fingertips make contact with the object, and then the finger posture is changed from upright to curved ones gradually, as shown in Fig.1(b). This is what we call *Detaching Assist Motion (DAM)*. From the viewpoint of robot application, the most attractive feature of *DAM* is its extremely simple finger motion, while the grasp pattern in Fig.1(a) is so complicated that the robot hand may often fail especially in changing the phase from fingertip to enveloping

D. Rus and S. Singh (Eds.): Experimental Robotics VII, LNCIS 271, pp. 131–140, 2001.

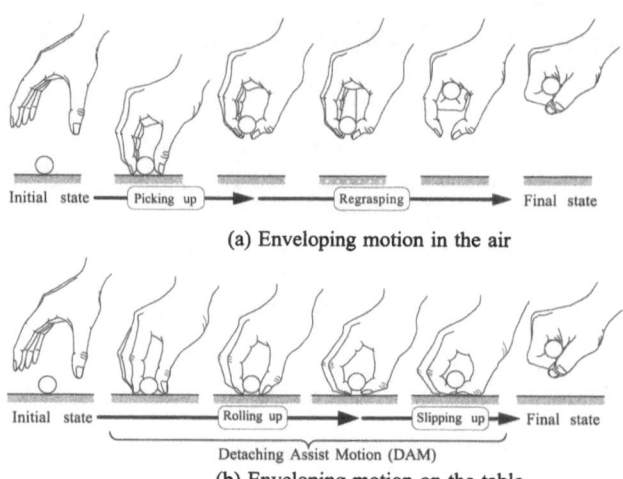

(a) Enveloping motion in the air

(b) Enveloping motion on the table

Figure 1. Two grasp strategies for enveloping a cylindrical object placed on a table.

grasps. The second advantage is that the *DAM* is achieved on the table in most phases and, therefore, it is not necessary to worry about dropping the object. Due to its simple motion planning, we can easily apply it to a multi-fingered robot hand[6]. In this paper, we further intend to implement the *DAM* not only to small objects but also to other size of objects, so that we can obtain a generalized grasp strategy which is applicable for various size of object, apart from the *Scale–Dependent Grasp.*

2. Related Works

Salisbury[7] has proposed the *Whole–Arm Manipulation (WAM)* capable of treating a big and heavy object by using one arm allowing multiple contacts with an object. Mirza and Orin[8] have applied a linear programming approach to solve the force distribution problem in power grasps. Bicchi[9] has showed that internal forces in power grasps can be decomposed into active and passive. Omata and Nagata[10] have considered the indeterminate contact force which does not influence on neither joint torque nor external wrench. Trinkle, Abel and Paul[11] have analyzed planning techniques for enveloping without friction. Trinkle et al.[12] have discussed the quasistatic, "whole-arm," dexterous manipulation of enveloped slippery workpieces. Kleinmann et al.[13] have showed a couple of approaches for finally achieving the power grasp from the fingertip grasp. There have been various papers discussing manipulation of object under enveloping style[14]–[16] or within the hand [17], [18].

3. Detaching Assist Motion (DAM)
3.1. What is DAM?

An enveloping grasp can be achieved by the following three fundamental phases: detaching an object from a table, lifting it up toward the palm, and firmly grasping. For detaching the object whose size is larger than fingertip, a human

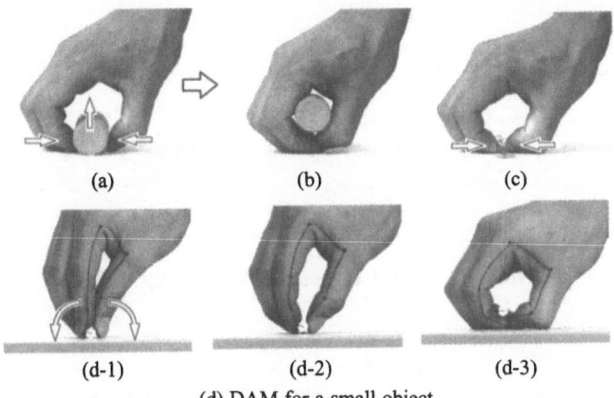

(a)   (b)   (c)

(d-1)   (d-2)   (d-3)

(d) DAM for a small object

Figure 2. Grasping motion by human.

often utilizes the *wedge-effect* where a simple pushing motion of the bottom part of object makes the object detach from the table as shown in Fig.2(a) and (b)[4]. Due to its simple motion planning, we can easily implement it into the grasping procedure of a multi–fingered robot hand. Either under significant friction or for an object with small diameter, we can not detach the object by using the *wedge–effect*, since the finger forces balance within the object and do not produce a lifting force any more as shown in Fig.2(c). Under such a situation, human can detach a small object without the *wedge-effect*. By changing finger posture from upright to crooked ones as shown in Fig.2(d), the object is automatically lifted up from the table. We call this grasping motion *DAM*. We would note that either a rolling or a sliding motion or perhaps both occur at the point of contact between the object and the fingertips.

3.2. Analysis of the Change of Finger Posture

Why does the *DAM* work effectively for detaching the object from a table? What kind of principle exists behind it? In this subsection, to clarify the basic working mechanism of *DAM*, we examine both finger posture and object position while human purposely applies the *DAM* as shown in Fig.2(d). The seven markers are attached at the side of object and each joint of index finger and thumb as shown in Fig.3(a). We measure the coordinates of markers from the video image sequences recorded by video camera system, where the sampling time is $1/30[sec]$. The absolute angle of tip of index finger θ_{ia} and thumb θ_{ta}, and the center of gravity of object $p_B = [p_{Bx}, p_{By}, 0]^t$ can be obtained from the image sequences.

Figs.3(b) through (d) show experimental results for a cylindrical object with the diameter of $8[mm]$, while human utilizes the *DAM* from the initial posture (Fig.2(d-1)) to the final posture (Fig.2(d-4)), where Figs.3(b),(c) and (d) show the changes of $\Delta\theta_{ia}$ and $\Delta\theta_{ta}$, the trajectory of p_B, and the change of Δp_{By} and angular displacement of object $\Delta\theta_B$, respectively, where $\Delta\theta_{ia} = \theta_{ia} - \theta_{ia0}$, $\Delta\theta_{ta} = \theta_{ta} - \theta_{ta0}$. $\Delta p_{By} = p_{By} - p_{By0}$ and $\Delta\theta_B = \theta_B - \theta_{B0}$, respectively, and subscript 0 denotes the value at initial posture ($0[sec]$). From Fig.3(b), it can be seen that both fingertips rotate uniformly with respect to

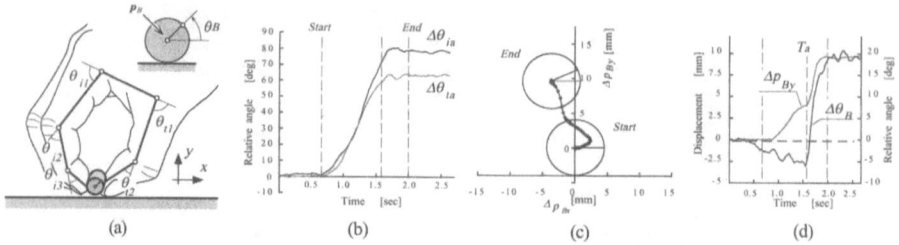

Figure 3. Visual observation during DAM.

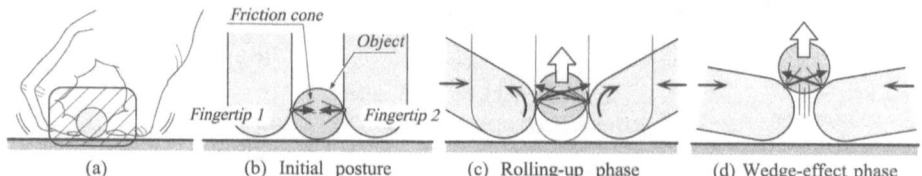

| (a) | (b) Initial posture | (c) Rolling-up phase | (d) Wedge-effect phase |

Figure 4. The basic mechanism of DAM.

time and finally keep constant in posture. An interesting behavior appears for the object motion when $t = 1.57[sec](= T_a)$. At the moment of T_a, the object suddenly start to move up with rotating motion as shown in Fig.3(d), while it slowly moves before T_a.

3.3. Basic Working Mechanism of DAM

Let us consider what happen during the *DAM* by using the fingertip model as shown in Fig.4. We assume that the object is small enough to ensure that a simple pushing motion in the horizontal direction can not lift up the object as shown in Fig.4(a). Further, we simplify the fingertip model as shown in Fig.4(b). If each fingertip does not slip on the surface of object, the object will be lifted up according to the geometrical constraint between the fingers and the object as shown in Fig.4(b) and (c), while both fingertips rotate from the initial to the final postures. We call this phase *Rolling–up phase*. As the object is lifted up, the normal direction of friction cone gradually changes upwards while the contact point moves towards the bottom of object. Finally, the moment the contact force is away from the friction cone, the object slips on the surface of fingertips. Once the contact force is away from the boundary, the *wedge–effect* effectively helps to move up the object as shown in Fig.4(d). We call the final phase *Wedge–effect phase*. These are the outline of the working mechanism of *DAM*. The phase from *Rolling–up* to *Wedge–effect* is automatically switched depending upon the contact friction as well as the finger rotating motion.

4. SPCM can Simulate DAM

While several strategies for robot hands which are equivalent to the human *DAM* can be considered, we utilize compliant motion of link system having one compliant joint (*s-th*) and one position–controlled joint (*p-th*) as shown in Fig.5(a). Now, suppose that we impart an arbitrary angular displacement $\Delta\theta_p$ at the position–controlled joint for such a link system contacting with an

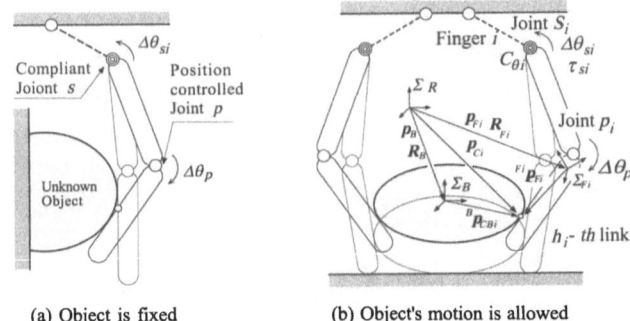

(a) Object is fixed   (b) Object's motion is allowed

Figure 5. Self-Posture Changing Motion.

environment. Under the condition, the link system will automatically change its posture while keeping contact between the environment and the link system, if $\Delta\theta_p$ is given appropriately. This series of motion is termed as *Self-Posture Changing Motion (SPCM)*[19]. *SPCM* has been conveniently utilized for detecting an approximate contact point between a link system and an unknown object under the assumption that the object does not move during sensing. For example, let us consider two different link postures during *SPCM*. Between two link postures, we can always find an intersection, which provides us with an approximate contact point. This approach allows us to detect an approximate contact point without implementing any tactile sensor, which is a great advantage. In this work, however, we allow the object to move according to the contact force imparted by the link as shown in Fig.5(b).

For an n-fingered robot hand with m-joints per finger, suppose that h_i-th ($h_i \geq 2$) link of each finger makes contact with an object, and the angular displacement, $|\Delta\theta_{pi}| \neq 0$ is applied at the position-controlled joint p_i ($h_i \geq p_i$) as shown in Fig.5(b). If we can find the vector $p_{ci} \in R^{3\times1}$ satisfying the following equations during a change of link posture, it is said that there exists *Self-Posture Changeability (SPC)*[19].

$$S_B(^Bp_{CBi}) = 0, \quad S_{Fi}(^{Fi}p_{CFi}) = 0 \tag{1}$$

$$p_B + R_B\,^Bp_{CBi} = p_{Fi} + R_{Fi}\,^{Fi}p_{CFi} = p_{Ci} \tag{2}$$

$$n_{CBi} = -n_{CFi} \tag{3}$$

where $S_B(^Bp)$ $(S_{Fi}(^{Fi}p))$ and n_{CBi} (n_{CFi}) are function representing the surface shape of object (i-th finger link) and unit normal vector directing outside at i-th contact point on the surface of object (i-th finger), respectively.

The series of motions bringing about a *SPC* is defined as the *Self-Posture Changing Motion (SPCM)* and we express it as $SPCM\{K_\theta, \Delta\theta_p\}$ where K_θ and $\Delta\theta_p$ are the stiffness matrix of compliant controlled joints and the angular displacement matrix of position controlled joints, respectively. Since the basic behavior of *DAM* is similar to that of *SPCM*, we can discuss the condition where the robot hand can lift up the object by utilizing the $SPCM\{K_\theta, \Delta\theta_p\}$. Now, let $f_c = [f_{c1}^t, \dots, f_{cn}^t]^t \in R^{3n\times1}$ and $W_{ext} \in R^{6\times1}$ be the contact force vector

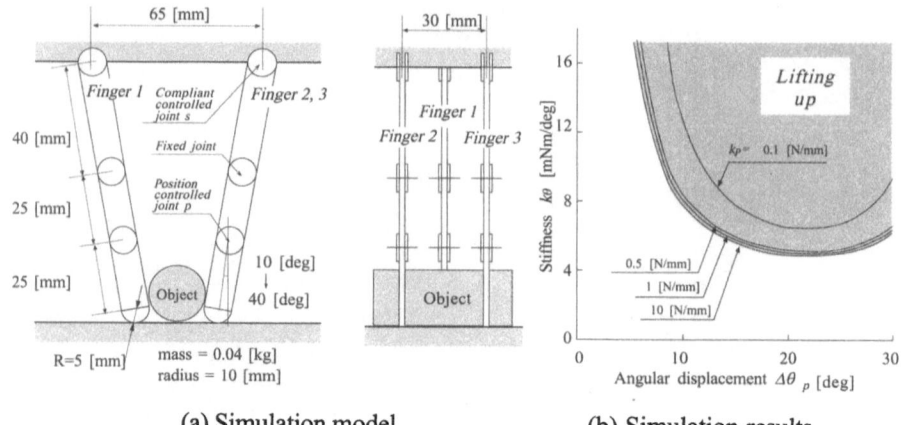

(a) Simulation model   (b) Simulation results

Figure 6. Analytical model and simulation result.

at each contact point and the load wrench, respectively. The equation of the force and the moment balancing on the object can be expressed as

$$W_{ext} = -G^t f_c,$$ (4)

where $G^t \in R^{6 \times 3n}$ is the grasp matrix and given by

$$G^t = \begin{bmatrix} I_3 & \cdots & I_3 \\ (R_B{}^B P_{CB1} \times) & \cdots & (R_B{}^B P_{CBn} \times) \end{bmatrix}.$$ (5)

Suppose that the load wrench is $W_{ext} = [0, 0, -(m_B g + f_{ez}), 0, 0, 0]^t$ where m_B, g, and f_{ez} are the mass of object, the gravitational acceleration, and the virtual force in the gravitational direction, respectively. The combination of K_θ and $\Delta\theta_p$ leading to DAM can be obtained by solving the following problem.

Search $SPCM\{K_\theta, \Delta\theta_p\}$ where f_c balances with W_{ext} and $f_{ez} \geq 0$.

The mathematical formulation for computing K_θ and $\Delta\theta_p$ has been discussed in [6]. Fig.6(a) shows the simulation model where parameters are chosen so that they are same as the robot hand we used for our experiment. Fig.6(b) shows the simulation result where we assume that the contact stiffness $k_p = 0.1[N/mm]$ through $10[N/mm]$, and the object is guaranteed for lifting up when we choose parameters with the hatched region. We would note that the lower boundary of the hatched region does not change largely as far as $0.5 \leq k_p \leq 10[N/mm]$, which means that the determination for both k_θ and $\Delta\theta_p$ can be easily achieved from Fig.6(b), if contact point is stiff enough.

5. Toward a Generalized Grasp Strategy
5.1. Outline
Since the motion planning of DAM (or $SPCM$) is simple enough, we now challenge to construct a generalized grasp strategy (GGS) which is applicable not only for small sized objects but also for other sized objects. Fig.7 shows an

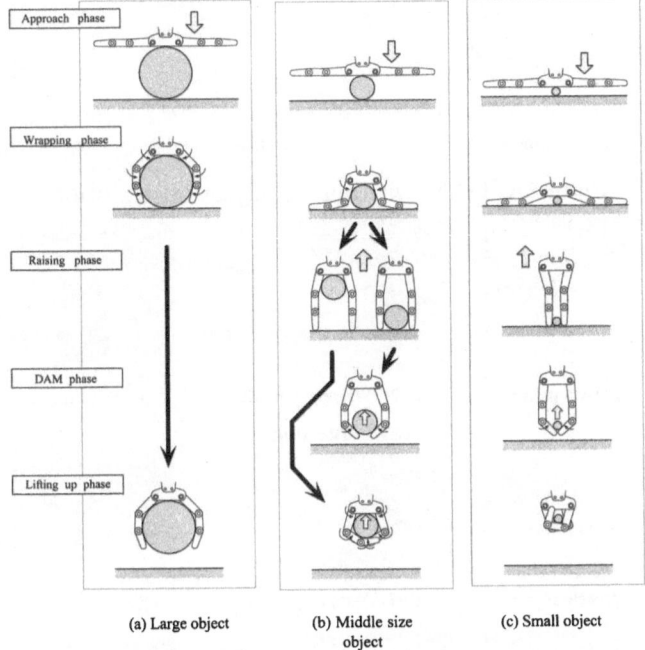

(a) Large object   (b) Middle size   (c) Small object
                              object

Figure 7. Generalized Grasp Strategy.

example of a *GGS* where it includes the *DAM* (or *SPCM*) phase in the central part. The *GGS* consists of the following five motions.

Approach phase : The robot hand approaches the target object until it makes contact with the object. By this contact, the robot can detect the height of the object H_{object}.

Wrapping phase : Each finger is closed in sequence from the 1st to the n-th joint under constant torque control, until the finger posture does not change any more. If an enveloping grasp is completed at the end of this phase as shown in Fig.7(a), the robot hand skips both *raising* and *DAM phases*, and switches to the *lifting up phase* directly.

Raising phase : The palm position is lifted under an appropriate compliance control where joint position is commanded so that the finger posture may be ready for starting the *DAM*. During this phase, there are two possible cases, the one is that the object is grasped by both palm and a couple of links, and the other one is that the object is stationary on the table. In case that the object is lifted, the target grasp can be achieved by simply closing each joint. In this case, we can skip the *DAM phase*.

DAM phase : The robot hand executes the *DAM* (or *SPCM*), so that the object can be captured within the hand. In case that the object needs the *initial adjustment motion*, the robot applies it before the *DAM*, as shown in

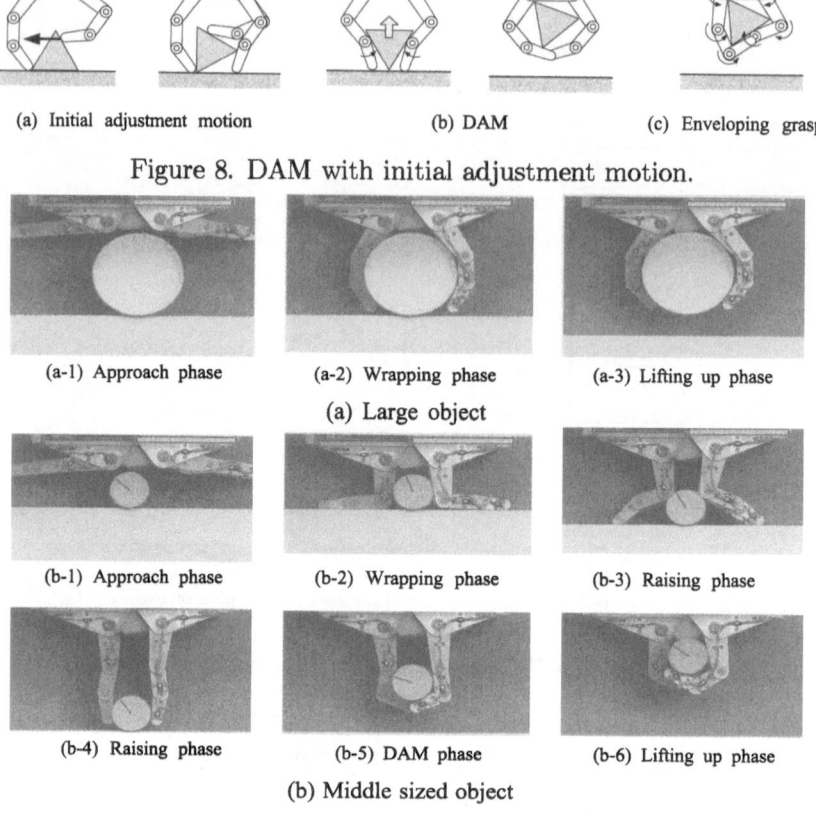

(a) Initial adjustment motion   (b) DAM   (c) Enveloping grasp

Figure 8. DAM with initial adjustment motion.

(a-1) Approach phase   (a-2) Wrapping phase   (a-3) Lifting up phase

(a) Large object

(b-1) Approach phase   (b-2) Wrapping phase   (b-3) Raising phase

(b-4) Raising phase   (b-5) DAM phase   (b-6) Lifting up phase

(b) Middle sized object

Figure 9. GGS by three–fingered robot hand.

Fig.8.

Lifting up phase : Each joint control is switched into constant torque control. The torque commands are chosen so that the grasped object may receive the upward resultant force under the given frictional coefficient[14].

In case that a vision system can observe the object continuously, we can immediately start either *wrapping phase* for a large object or *DAM phase* for other objects. Depending upon the contact friction, the object may stop during the *lifting up phase* due to so called jamming. When such a failure is detected, one feasible approach to recover is to apply small vibration signal (dither signal) to each joint, so that we can reduce the equivalent contact friction. We would note that according to the size, shape, and contact friction, there is a chance for detaching the object from a table in one of the three phases (*wrapping, raising and DAM phases*) during the whole grasp process.

5.2. Experiments

Fig.9 shows series of finger postures during the *GGS* by the robot hand. The robot hand consists of three same finger units and each finger has three links.

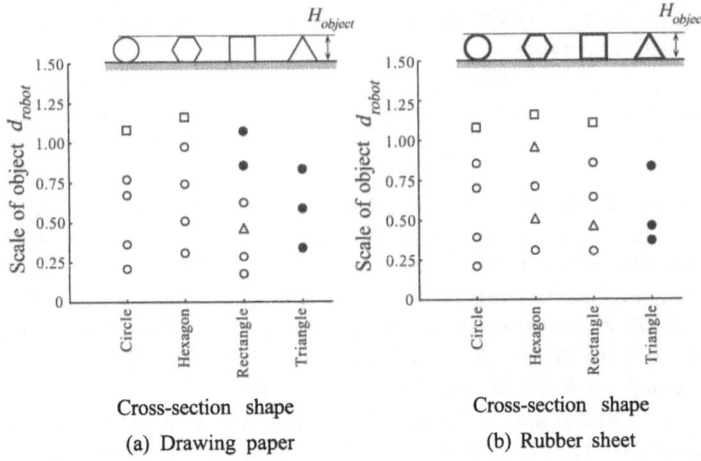

Figure 10. Experimental results for general column objects.

The lengths of each link are $l_1 = 40[mm]$, $l_2 = 25[mm]$, and $l_3 = 25[mm]$, respectively. Each finger link is driven by wire and a torque sensor is included in each joint. Rotary encoder is used as an angular sensor. The palm is equipped with ON/OFF type tactile sensor.

Figs. 10(a) and (b) show the experimental results where objects used in Figs. 10(a) and (b) are covered by drawing paper and by rubber, respectively. The horizontal and the vertical axes denote the shape of object and the normalized value $d_{robot} = L_o/L_r (L_r = 224[mm])$, respectively, where L_o and L_r denote the circumference of the object and the distance between fingertips, respectively. As the height of object increases, d_{robot} also increases. We prepare four types of object where cross–section of object are circle, hexagon, rectangle and triangle, respectively. "○", "□", "△" and "●" denote that the object can be detached from the table in *DAM phase*, *wrapping phase*, *raising phase* and by utilizing *initial adjustment motion*, respectively. The experimental results show that the robot hand can grasp various kinds of objects by utilizing the *GGS* proposed here.

6. Conclusion

We discussed the basic working mechanism of the *DAM* and examined the condition leading to the *DAM* by using *SPCM* which is easily implemented for robot hand. We have proposed a generalized grasping strategy (*GGS*) which is applicable for various sized (shaped) objects. We have experimentally shown that the robot hand can achieve the enveloping grasp for most kinds of objects which have various sizes, cross–section shapes, and contact friction by utilizing the *GGS*.

This work has been supported by CREST of JST (Japan Science and Technology).

References

[1] M. Cutkosky: "On Grasp Choice, Grasp Models, and the Design of Hands for

Manufacturing Tasks," *IEEE Trans. on Robotics and Automation*, Vol. 5, No. 3, JUNE, pp. 269–279, 1989.

[2] G.A. Bekey, H. Liu, R. Tomovic and W. Karplus: "Knowledge-Based Control of Grasping in Robot Hands Using Heuristics from Human Motor Skills," *IEEE Trans. on Robotics and Automation*, Vol. 9, No. 6, DECEMBER, pp. 709–722, 1993.

[3] S.B. Kang and K. Ikeuchi: "Toward Automatic Robot Instruction from Perception — Recognizing a Grasp from Observation," *IEEE Trans. on Robotics and Automation*, Vol. 9, No. 4, AUGUST, pp. 432–443, 1993.

[4] M. Kaneko, Y. Tanaka and T. Tsuji: "Scale–Dependent Grasp," *Proc. of the IEEE Int. Conf. on Robotics and Automation*, pp. 2131–2136, 1996.

[5] M. Kaneko and T. Tsuji: "Realization of Enveloping Grasp," *Video Proc. of the IEEE Int. Conf. on Robotics and Automation*, 1997.

[6] M. Kaneko, T. Shirai and T. Tsuji: "Detaching and Grasping Strategy Inspired by Human Behavior," *13th CISM-IFToMM Symp. on the Theory and Practice of Robots and Manipulators*, 2000 (Preprint).

[7] J.K. Salisbury: "Whole–Arm Manipulation," *Proc. of the 4th Int. Symp. of Robotics Research*, Santa Cruz, CA, 1987. Published by the MIT Press, Cambridge MA.

[8] K. Mirza and D.E. Orin: "Control of Force Distribution for Power Grasp in the DIGITS System," *Proc. of the IEEE 29th CDC Conf.* , pp. 1960–1965, 1990.

[9] A. Bicchi: "Force Distribution in Multiple Whole–Limb Manipulation," *Proc. of the IEEE Int. Conf. on Robotics and Automation*, pp. 196–201, 1993.

[10] T. Omata and K. Nagata: "Rigid Body Analysis of the Indeterminate Grasp Force in Power Grasp," *IEEE Trans. on Robotics and Automation*, Vol.16, No.1, pp. 46–54, 2000.

[11] J.C. Trinkle, J.M. Abel, and R.P. Paul: "Enveloping, Frictionless Planar Grasping," *Proc. of the IEEE Int. Conf. on Robotics and Automation*, pp. 246–251, 1987.

[12] J.C. Trinkle, R.C. Ram, A.O. Farahat, and P.F. Stiller: "Dexterous Manipulation Planning and Execution of an Enveloped Slippery Workpiece", *Proc. of the IEEE Int. Conf. on Robotics and Automation*, pp. 442–448, 1993.

[13] K.P. Kleinmann, J. Henning, C. Ruhm, and H. Tolle: "Object Manipulation by a Multifingered Gripper: On the Transition from Precision to Power Grasp," *Proc. of the IEEE Int. Conf. on Robotics and Automation*, pp. 2761–2766, 1996.

[14] M. Kaneko, M. Higashimori and T. Tsuji: "Transition Stability of Enveloping Grasps," *Proc. of the IEEE Int. Conf. on Robotics and Automation*, pp. 3040–3046, 1998.

[15] M. Kaneko, K. Harada and T. Tsuji: "A Sufficient Condition for Manipulation of Envelope Family," *Proc. of the IEEE Int. Conf. on Robotics and Automation*, pp. 1060–1067, 2000.

[16] S. Song, M. Yashima and V. Kumar: "Dynamic Simulation for Grasping and Whole Arm Manipulation," *Proc. of the IEEE Int. Conf. on Robotics and Automation*, pp. 1082–1088, 2000.

[17] A. Sudsang and J. Ponce: "In–Hand Manipulation: Geometry and Algorithms," *Proc. of the IEEE Int. Symp. on Intelligent Robots and Systems*, pp. 98–105, 1997.

[18] D. Rus: "In–Hand Dexterous Manipulation of Piecewise–Smooth 3D Objects," *Int. J. of Robotics Research*, Vol.18, No.4, pp.355–381, 1999.

[19] M. Kaneko and K. Tanie : "Contact Point Detection for Grasping an Unknown Object Using Self-Posture Changeability," *IEEE Trans. on Robotics and Automation*, Vol.10, No.3, pp.355–367, 1994.

Force-Based Interaction for Distributed Precision Assembly

Richard T. DeLuca, Alfred A. Rizzi, and Ralph L. Hollis

The Robotics Institute, Carnegie Mellon University
Pittsburgh, USA
{delucr,arizzi,rhollis}@ri.cmu.edu

Abstract: This paper documents our efforts to instantiate force-guided co-operative behaviors between robotic agents in the *minifactory* environment. Minifactory incorporates high-precision 2-DOF robotic agents to perform micron-level precision 4-DOF assembly tasks. Here we utilize two minifactory agents to perform compliant insertion. We present a custom force sensing device which has been developed as well as the control and communication systems used to coordinate the action of the agents. Finally, we conclude by presenting a set of experimental results which document the performance of the new force sensor as integrated in the minifactory system. These results document the first experimental confirmation of high-bandwidth ($> 100\,\mathrm{Hz}$) coordination between agents within the minifactory system.

1. Introduction

In the Microdynamic Systems Laboratory[1] we are developing a class of precision automated assembly systems based on collections of low-degree-of-freedom robots. These independent robotic devices (termed *agents*) each possess a set of actuators and sensors along with a control system and communications facilities to enable cooperative action [1, 2]. This collection of robotic agents forms *minifactory* – a high-precision, self-calibrating, agent based, distributed assembly system. It is a physical instantiation of a much broader philosophy for precision assembly systems called the Agile Assembly Architecture (AAA). The motivation behind AAA and the minifactory is to create a new standard for rapidly deployable automatic assembly systems capable of precision assembly of complex electro-mechanical products [3]. Developing and deploying suitably precise and flexible sensing capabilities on the individual robotic agents is critical to reach the level of inter-agent coordination necessary to achieve this goal. Toward this end our research group recently reported on the application of frame-rate visual sensing for inter-agent coordination within the minifactory [4]. This paper complements that earlier work and focuses on the development of force-based interaction strategies for coordinating minifactory agents.

The use of force sensors to enable the performance of contact tasks is a well-studied topic. The work presented here builds directly on many earlier efforts – most notably the work by Mason [5], Craig and Raibert [6], Hogan [7], Anderson and Spong [8], Khatib [9], and Siciliano [10]. Our challenge is to map these ideas effectively onto a flexible distributed automated assembly

---

[1]See http://www.cs.cmu.edu/~msl.

D. Rus and S. Singh (Eds.): Experimental Robotics VII, LNCIS 271, pp. 141–150, 2001.

(a)   (b)

Figure 1. (a) View of courier (lower robot), configured for an insertion task, collaborating with an overhead manipulator (upper robot). (b) Close up view of the mainpulator's end effector preparing to insert a peg into a hole.

system. The resulting control methodologies (described in Sec. 3.1) most closely resemble the class of impedance controllers originally proposed by Hogan.

The remainder of this paper describes the development of a force-based coordination strategy for the two major classes of agents in the minifactory system (couriers and manipulators). To achieve this level of inter-agent coordination requires deployment of both a suitable sensing system and a distributed control system. We document the design, implementation, deployment and experimental results obtained from force-sensing hardware designed expressly for distributed force-based assembly tasks and go on to detail the communication and control strategies required to coordinate the actions of two agents cooperatively performing such a task. Section 2 briefly describes the available minifactory infrastructure and the specific force sensor developed for this task. Section 3 documents the details of our distributed control system and presents experimental verification of its performance.

2. Force-Based Inter-Agent Coordination

To explore the applicability of force guided coordination in the minifactory environment, we have chosen to undertake a collection of insertion ("peg-in-hole") tasks. Briefly, the minifactory system (used to perform the experiments presented in this paper) uses two classes of robotic agents to perform automated assembly[1, 3]. The first class (courier agents) provides motion in the x, y plane – both translation and a limited range of rotation. It incorporates a magnetic position sensor device, providing $0.2\,\mu\mathrm{m}$ resolution ($1\,\sigma$) position measurements [11] and closed loop position control [12]. The second class (manipulator agents) provides vertical (z) and rotational (θ) movement. Motion resolution in z is approximately $2\,\mu\mathrm{m}$, while resolution in θ is roughly $0.0005°$. Our sample task requires the cooperative application of both types of agents: a courier to carry a plate bearing a chamfered hole, and a manipulator that will measure forces on a peg as it is inserted into the hole. By exchanging state, sensor, and command information between the two agents they will act as a single 4-DOF device to reliably perform the insertion task.

2.1. Communications Infrastructure

To facilitate cooperation between physically distinct agents in the minifactory, each agent is equipped with two 100 Mb Ethernet network interface devices.

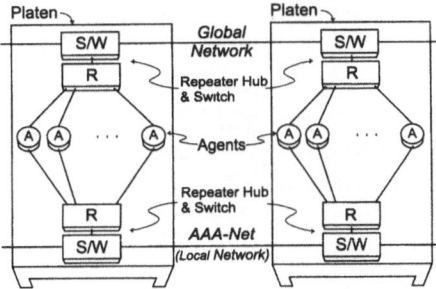

Figure 2. Physical network structure.

The first network carries non-latency-critical information, such as user commands or monitoring information – this network utilizes standard IP protocols. The second network (referred to as AAA-Net) carries real-time information critical to the timely coordination of activity between agents. This includes commands, state, and sensor information shared between the agents while performing the coordinated activities described in Section 3.

Fig. 2 depicts the physical communications infrastructure of the minifactory system. As can be seen, both the AAA-Net and the global IP network are configured as a chain-of-star topology local-networks, with a Fast Ethernet hub at the center of each star and Fast Ethernet switches forming the connections between the local-network segments. The switches serve to localize communications within the factory system by not transmitting data packets destined for local agents to the remainder of the factory and by selectively transmitting those packets destined for other network segments toward their destination, yielding a scalable communications infrastructure[2]. A more detailed description of the protocols used by this network and the resulting system performance can be found in [4] and [13].

2.2. Integrated 3-DOF Force Sensor

While there exists a wide selection of commercially available force sensors, the minifactory application presents several significant design constraints that led us to choose to fabricate a new force sensor. Precision assembly involving small parts requires small force resolution. The force sensor was designed such that manipulation devices could be interchanged easily and packaged such that vision, manipulation, and force sensing are integrated. The resulting sensor is capable of measuring two lateral moments and the applied force. It is packaged within the end effector of the manipulator agent which in turn directly interfaces with the manipulator through a single modular connector.

2.2.1. Sensor Design

The majority of applications under consideration for minifactory utilize vertical insertion assembly and single point vacuum gripping, making a three-axis force sensor sufficient. The sensor measures force along the z axis and torques about the x and y axes. Fig. 3(a) shows an assembly view of the CAD model for the

---

[2]Note that in AAA/minifactory the bulk of high-bandwidth communication will be between agents that are attached to the same network segment, while the remainder will involve agents that are typically attached to neighboring segments [3].

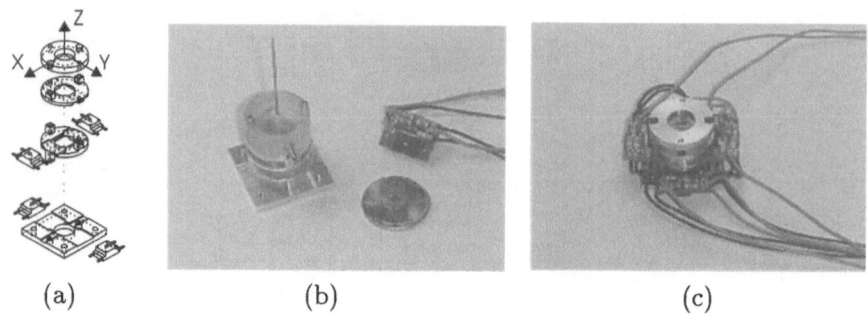

(a)   (b)   (c)

Figure 3. 3-Axis force sensor: (a) exploded assembly drawing of the flexure and single-axis load cells, (b) photograph of the flexure with clear Lexan™ vacuum chamber and one load cell, (c) flexure with four load cells in place.

sensor. The sensor was designed to provide sensitivity of at least 0.1 N, minimal hysteresis, and a force range of several Newtons. A wide range of sensing modalities were considered, including semiconductor and foil strain gauges, as well as capacitive, inductive, piezoelectric, and optical elements. The final design (shown in Fig. 3(b) and (c)) utilizes a set of four commercial single-axis piezoresistive load cells (Cooper Instruments LPM-562) in a mechanical flexure. The result is a low-cost, thermally stable, easily manufactured device.

The applied force and moments are transmitted to the four load cells through a flexure. Linear analysis of the flexure yields a model relating applied forces and moments to load cell readings of the form

$$
\begin{bmatrix} V_1 \\ V_2 \\ V_3 \\ V_4 \end{bmatrix} = \underbrace{\begin{bmatrix} \frac{\alpha_1}{\beta_1} & \frac{\epsilon_1}{\beta_1} & 0 \\ \frac{\alpha_2}{\beta_2} & \frac{\epsilon_2}{\beta_2} & 0 \\ \frac{\eta_1}{\delta_1} & 0 & \frac{\zeta_1}{\delta_1} \\ \frac{\eta_2}{\delta_2} & 0 & \frac{\zeta_2}{\delta_2} \end{bmatrix}}_{C} \begin{bmatrix} F \\ M_x \\ M_y \end{bmatrix},
\tag{1}
$$

where F, M_x, and M_y represent the applied force and moments respectively, while V_i represents the measured load cell voltages. This linear representation models the flexure as cantilever beams and the load cells as stiff springs. Ideally, if the load cells are symmetrically arranged within the flexure it will be the case that $\alpha_1 = \alpha_2$, $\beta_1 = \beta_2$, $\epsilon_1 = -\epsilon_2$, $\eta_1 = \eta_2$, $\delta_1 = \delta_2$, and $\zeta_1 = -\zeta_2$, leaving only four unknown parameters required to characterize the sensor.

2.2.2. Calibration

Our prototype force sensor was calibrated to both validate this model and to determine the appropriate parameters for our device as built. To accomplish this, the force sensor was mounted in the end effector and fitted with a flat plate. Forces ranging up to 2 N were applied at 49 points evenly spaced over the entire plate. Linear regression was performed on the resulting data to produce an estimate of the matrix C,

$$
\hat{C} = \begin{bmatrix} 0.594 & -121 & -9.66 \\ 0.791 & 129 & -5.31 \\ 0.0804 & -3.44 & 120 \\ 0.184 & -9.56 & -82.1 \end{bmatrix}.
$$

The absence of the expected zero elements and the lack of symmetry can be attributed to fabrication and assembly inaccuracies including load cell placement and machining tolerances. In particular, lateral displacement of the load cells was not considered in the idealized model. Based on this calibration of the sensor and observed noise properties of the sensing elements, we conclude that this force sensor is capable of resolving forces down to roughly 24 mN (1σ) and torques of approximately 0.13 mN $\cdot$ m (1σ).

3. Experimental Evaluation

We have undertaken a number of experiments to evaluate the performance of the distributed agent pair performing an insertion task. In order to accomplish this, we developed a control architecture which minimizes inter-agent communication by making use of algorithmically simple control schemes. Impedance control provides one such simple class of control policies and accomplishes stable interaction with an environment by converting the system to a form which naturally performs the task.

3.1. Control Architecture

To deploy an impedance control scheme we begin by specifying the desired behavior of the system. Given a desired mass (M_d), stiffness (K_d), damping (B_d), and force (F_d) the desired system behavior is given by

$$M_d\ddot{q} + K_d(q - q_0) + B_d\dot{q} + G_i \int_0^t (F_e - F_d)d\tau = F_e, \qquad (2)$$

where q represents the generalized configuration of the system and F_e the applied environmental force acting on the system. This system behaves like a mass attached to a spring and damper about a nominal target q_0, with the added integral term driving the system to the desired contact forces (F_d).

To realize the behavior defined by (2) we developed distributed control policies for the two independent robotic agents. The courier can be modeled as a mass with essentially ideal actuators in the x and y directions. This is a direct result of the simple actuator mechanism and the frictionless nature of the air bearing that supports the courier [12, 14]. The overhead manipulator's θ axis is directly driven, resulting in negligible friction and an equally simple model. The z axis is driven by a ball screw so a friction term was included to offset the effects of friction in that axis. Thus, the dynamic model of the system takes the form

$$M_a\ddot{q} + B_a\dot{q} + f(\dot{q}) = \tau_a - F_e, \qquad (3)$$

where M_a and B_a are 4×4 diagonal matrices that describe the overhead manipulator's and courier's mass and damping parameters; $f(\dot{q})$ contains the friction terms; τ_a represents the applied actuator forces; and F_e is the applied environmental forces. It is this last term that will be measured by the force sensor mounted on the manipulator's end effector described in Section 2.2.

Given the system model and desired impedance, applying inverse dynamics yields a control law of the form

$$\tau_a = M_a M_d^{-1}\left[F_e + K_d(q_0 - q) - B_d\dot{q} + G_i \int_0^t (F_e - F_d)d\tau\right] + B_a\dot{q} + f(\dot{q}) + F_e. \quad (4)$$

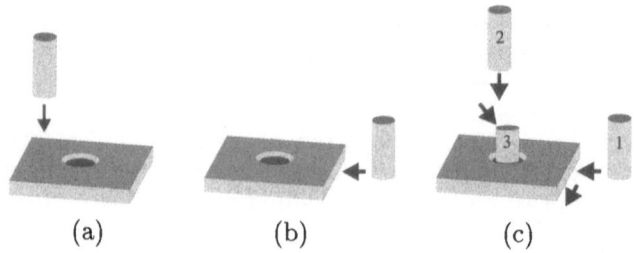

Figure 4. Graphic depiction of the three classes of experiments performed: (a) vertical contact, (b) lateral contact, and (c) peg-in-hole.

This control law is implemented separately on the two agents, with each agent responsible for its actuated degrees of freedom. By choosing to only allow diagonal matrices for M_d, K_d, B_d, and G_i this control policy is completely diagonal, which implies that the only information that must be shared between the agents are the sensed environmental forces.

Critical to implementing the above control law is the communication infrastructure described in Section 2.1. Messages sent to the courier from the overhead manipulator at real-time rates (roughly 500 Hz) contain force sensor information, velocity commands, and controller mode. The velocity commands and controller mode information allow the manipulator agent to command a variety of different behaviors from the courier agent.

3.2. Experimental Results

In the experiments described, the overhead manipulator performs contact tasks with a 0.813 mm (0.032 in.) diameter hypodermic tube attached to the force sensor (see Fig. 1(b) and 3(b)). Mounted on the courier is a plate containing several sets of holes 6.35 mm (0.25 in.) in depth and ranging in diameter from 2.54 mm (0.1 in.) to 0.838 mm (0.033 in.) (see Fig. 1). Each hole has a 45° chamfer.

Three types of tasks were performed to characterize overall system performance; vertical contact, lateral contact, and peg-in-hole insertions (see Fig. 4). Additionally, repeatability and reliability experiments were performed to verify overall system performance.

3.2.1. Vertical Contact

The vertical contact experiment involved the courier maintaining a fixed position while the overhead manipulator makes contact and maintains a constant force with the plate. Specifically, the courier performs a move (under position control) to place it under the manipulator and holds its position. Meanwhile, the manipulator finds the top of the plate on the courier by executing a constant velocity, force-guarded move. With the exact position of the plate top registered, the manipulator servos to a z position above the plate by a height equal to the depth of a hole. Once the tool tip has arrived at this position, an impedance controller is activated with the desired position located appropriately below the surface of the plate to obtain the desired contact force at equilibrium. For this experiment the desired z force was -1 N. Results from a typical experiment are shown in Fig. 5. Note the slightly underdamped response of the z force and the steady state value of -1 N. The observed high frequency noise in the force information is attributed to unmodeled dynamics in the gripper tube. The settling time from impact is approximately 1 s. Response rates faster than this were difficult to achieve without higher impact forces. Friction in z appears to be the limiting factor.

3.2.2. Lateral Contact

The goal of the lateral contact experiment is to position the tool tip below the plane of the top of the plate while the courier makes contact and maintains a constant lateral force with the side of the plate. The top of the plate is located with a constant velocity, force-guarded move as in the vertical contact experiment. Controllers are then deployed to reposition the courier and bring it into lateral contact with the tool tip. When contact is made, the manipulator executes an impedance controller which in turn issues desired position commands to the courier controller. The desired contact force was set to $-0.4\,\text{N}$ in y. Results of a typical experiment can be found in Fig. 6. Notice that the force applied in the y direction reaches a steady state value of $-0.4\,\text{N}$. A non-zero steady state value for the x force can be attributed to small misalignments of the gripper tube, misalignments of the courier plate, or compliance in the manipulator's θ axis. The observed settling time is roughly 0.3 s. It is expected that faster settling times can be achieved for lateral contact in which the stiffness of the θ axis of the manipulator is increased.

3.2.3. Peg-In-Hole

The peg-in-hole experiment consists of the courier bumping into and sliding along the manipulator gripper to register the plate corner and thus the entire plate geometry relative to the manipulator's tool tip. The initial bump and slide maneuver involves a hybrid position/force control sequence implemented through the use of the described impedance controllers. Once lateral contact is made, the manipulator commands a velocity and a mode change to the courier such that it maintains a constant force and slides along the plate edge. When the courier loses contact with the gripper, the courier's position is noted immediately and the plate geometry is registered to determine the location of the hole in tool tip coordinates. The plate top location is then calibrated as described earlier and the courier is positioned so that the gripper is directly over a chamfer of a hole. The manipulator servos to a position at the height of the plate top at which point an impedance controller activates with a desired position located at the center and appropriately below the bottom of the hole.

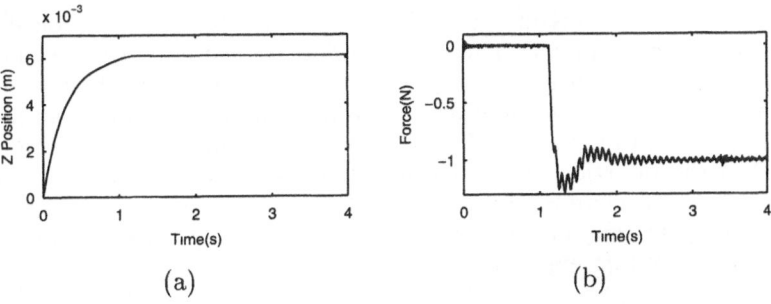

(a)   (b)

Figure 5. Position (a) and force (b) measurements during vertical contact.

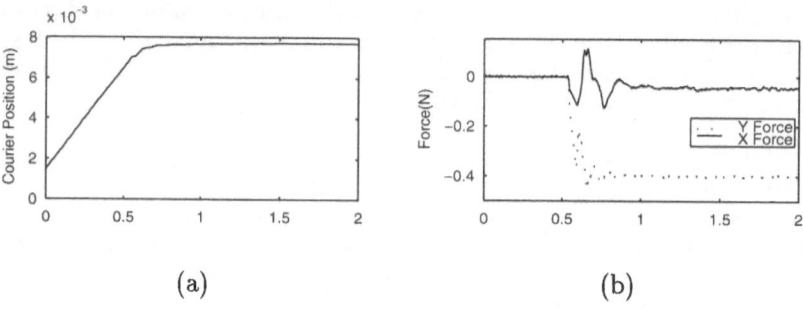

(a)   (b)

Figure 6. Position (a) and force (b) measurements during lateral contact.

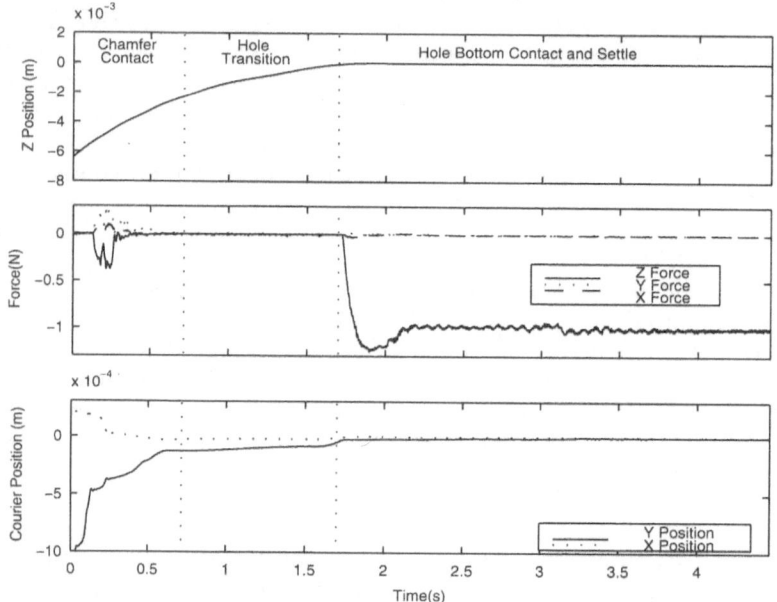

Figure 7. Overhead manipulator position, force, and courier position measurements during an insertion experiment.

The desired force is $-1\,\mathrm{N}$ in z and $0\,\mathrm{N}$ in x and y. Fig. 7 shows typical results for this type of experiment. The insertion data presented is from an insertion performed on a hole with an exaggerated chamfer of diameter 6.35 mm (0.250 in.). The oversized chamfer was used to produce data with longer chamfer contact regions to aid in analysis. The exaggerated chamfer explains the lengthy duration of the insertion event. During chamfer contact one can see that the tool tip follows the chamfer down into the hole as the courier moves in response to the x and y forces. The settling time for the z force is comparable to that of the vertical contact experiment.

3.2.4. Repeatability

Results are presented from two repeatability tests in which insertion tasks

Hole Clearance	Ave. Time	Attempts	Success	RMS z Force Error
0.2032 mm	2.43 s	1000	100%	0.0663 N
0.0254 mm	2.36 s	1000	93%	0.1091 N

Table 1. Repeatability experiment results.

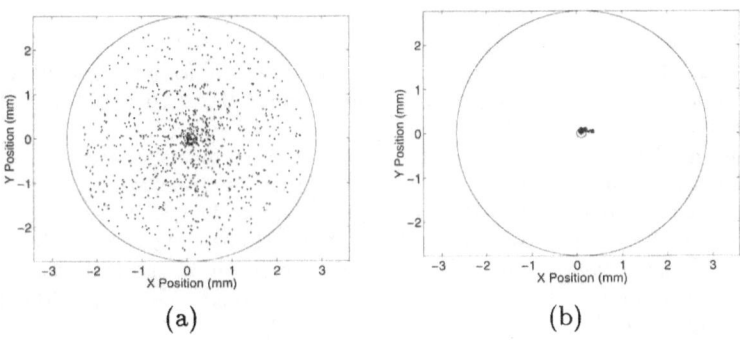

(a)   (b)

Figure 8. Tool tip start (a) and finish (b) positions plotted in configuration space for repeatability test involving a hole with 0.02032 mm (0.008 in.) clearance.

are initiated from uniformly distributed random points located above the chamfer. Repeatability was tested on a 1.016 mm (0.04 in.) hole with an exaggerated chamfer of diameter 6.35 mm (0.25 in.) and a 0.838 mm (0.033 in.) hole with a typical chamfer of diameter 1.52 mm (0.06 in.). Fig. 8 shows the start and stop positions of the tool tip in configuration space from a typical experiment. Finish positions shown outside of the hole in Fig. 8(b) are attributed to slight system miscalibration and our failure to account for small rotations of the courier. Table 1 summarizes these experiments. The difference in average insertion time is due to differences in the size of the chamfers. It is important to note that although only 93% of the 0.838 mm (0.033 in.) hole insertion attempts made it successfully into the hole, 63 of the failed attempts are attributed to system level errors. Discounting system level failures that prevented the experiment from beginning, the overall success rate rises to 99.3% for the 0.838 mm (0.033 in.) holes. Smaller clearance holes present a higher chance of wedging, which explains the larger RMS error in z force for the smaller clearance hole.

4. Conclusion

This paper serves to document the system-level communication, hardware, and control infrastructure used to perform force-based distributed precision assembly in the minifactory environment. A novel 3-axis force sensor design is presented for use in the prototype minifactory end effector. The force sensor incorporates compact design and suitable sensitivity for vertical insertion tasks. The performance measured from the results presented in this paper convincingly demonstrate the viability and potential of high-bandwidth cooperative manipulation tasks.

Acknowledgments

This work was supported in part by NSF grants DMI-9523156, and CDA-9503992. The authors would like to thank Arthur Quaid, Zack Butler, Jay Gowdy, Shinji Kume, Mike Chen, Ben Brown, and Patrick Muir for their invaluable work on the minifactory/AAA project and support for this paper.

References

[1] Ralph L. Hollis and Arthur E. Quaid. An architecture for agile assembly. In *Proc. Am. Soc. of Precision Engineering, 10th Annual Mtg.*, Austin, TX, October 15-19 1995.

[2] Alfred A. Rizzi and Ralph L. Hollis. Opportunities for increased intelligence and autonomy in robotic systems for manufacturing. In *International Symposium of Robotics Research*, pages 141-151, Hayama,Japan, October 1997.

[3] Alfred A. Rizzi, Jay Gowdy, and Ralph L. Hollis. Agile assembly architecture: An agent-based approach to modular precision assembly systems. In *Proc. IEEE Int'l. Conf. on Robotics and Automation*, pages Vol. 2, p. 1511-1516, Albuquerque, April 1997.

[4] Michael L. Chen, Shinji Kume, Alfred A. Rizzi, and Ralph L. Hollis. Visually guided coordination for distributed precision assembly. In *Proc. IEEE Int'l Conf. on Robotics and Automation*, pages 1651-1656, April 2000.

[5] Matt T. Mason. Compliance and force control for computer controlled manipulators. In *IEEE Transactions on Systems, Man, and Cybernetics*, pages 418-432, June 1981.

[6] Marc H. Raibert and John J. Craig. Hybrid position/force control of manipulators. In *Journal of Dynamic Systems, Measurement, and Control*, pages 126-133, 1981.

[7] Neville Hogan. Impedance control: An approach to manipulation: Part i - theory, part ii - implementation, part iii - applications. In *Journal of Dynamic Systems, Measurement, and Control*, pages 1-24, 1985.

[8] Robert J. Anderson and Mark. W. Spong. Hybrid impedance control of robotic manipulators. In *IEEE Journal of Robotics and Automation*, pages 549-556, Oct 1988.

[9] Oussama Khatib. A unified approach for motion and force control of robot manipulators: The operational space formulation. In *IEEE Journal of Robotics and Automation*, pages 43-53, Feb 1987.

[10] Stefano Chiaverini, Bruno Siciliano, and Luigi Villani. A survey of robot interaction control schemes with experimental comparison. In *IEEE Transactions on Mechatronics*, pages 273-285, Sept 1999.

[11] Zack J. Butler, Alfred A. Rizzi, and Ralph L. Hollis. Precision integrated 3-DOF position sensor for planar linear motors. In *Proc. IEEE Int'l. Conf. on Robotics and Automation*, pages 3109-14, 1998.

[12] Arthur E. Quaid and Ralph L. Hollis. 3-DOF closed-loop control for planar linear motors. In *Proc. IEEE Int'l Conf. on Robotics and Automation*, pages 2488-2493, May 1998.

[13] Shinji Kume and Alfred A. Rizzi. A high-performance network infrastructure and protocols for distributed automation. In *Proc. IEEE Int'l Conf. on Robotics and Automation*, 2001. (to appear).

[14] Bruce A. Sawyer. Linear magnetic drive system. U. S. Patent 3,735,231, May 22 1973.

Design and Implementation of a New Discretely-Actuated Manipulator

Jackrit Suthakorn
Department of Mechanical Engineering,
Johns Hopkins University,
Baltimore, Maryland, USA
song@jhu.edu

Gregory S. Chirikjian
Department of Mechanical Engineering,
Johns Hopkins University,
Baltimore, Maryland, USA
gregc@jhu.edu

Abstract: This paper presents a new 3-D design of a discretely-actuated robot manipulator powered by binary actuators. Binary actuators have two stable states, which are, for example, closed (0) and open (1). Major benefits of this kind of discretely-actuated manipulator are repeatability, relatively low cost, and no need for feedback control. In addition, hyper-redundancy allows tasks to be performed even when some actuators fail. A prototype of the new 3-D design is presented in this paper.

1. Introduction

Continuously-actuated robotic manipulators are the most common type of manipulators even though they require sophisticated and expensive control and sensor systems to increase their accuracy and repeatability. Discretely-actuated robot manipulators are potential candidates to be used in applications where high repeatability and reasonable accuracy are required. Such applications include pick-and-place, spot welding and assistants to people with disabilities.

A new design concept for this kind of manipulator in the spatial case is presented and hardware is demonstrated. The concept of a 3-bit planar binary Variable Geometry Truss (VGT) module is used in this implementation. Figure 1 illustrates all possible configurations of a 3-bit planar binary VGT module. The new design is developed from a

D. Rus and S. Singh (Eds.): Experimental Robotics VII, LNCIS 271, pp. 151–157, 2001.
© Springer-Verlag Berlin Heidelberg 2001

2-dimensional binary-actuated VGT manipulator and a 3-dimensional binary-actuated Stewart-platform-like manipulator which were previously designed and built by the authors and co-workers (shown in Figure 2 and 3 respectively). The new design uses 3-bit binary VGT modules stacked on top of each other with a discretely-actuated rotating joint between each module. As a result, the manipulator has the ability to reach many points and covers a full 3-dimensional sphere around the manipulator itself.

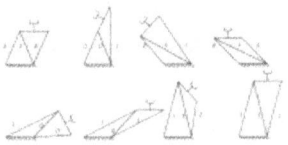

Figure 1: Eight-possible-configurations of a 3-bit binary VGT.

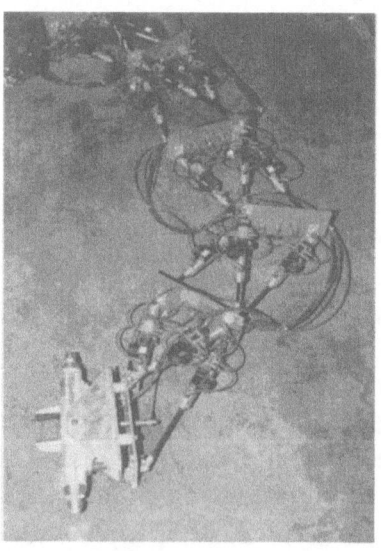

Figure 2: A 2-D binary VGT manipulator

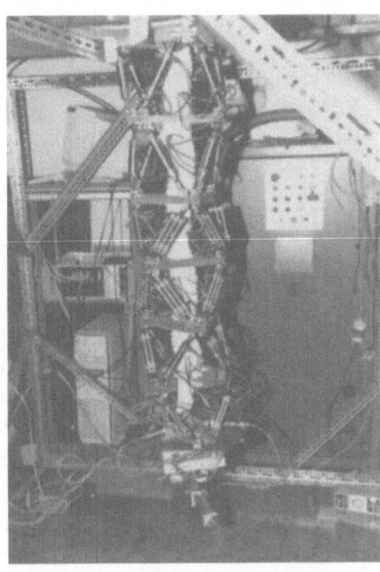

Figure 3: A spatial binary Stewart-platform-like manipulator.

2. Previous Work

Because of the high cost of traditional continuous manipulators with sophisticated sensing and control systems, a number of reduced complexity systems have become popular. Canny and Goldberg [1] discussed a reduced complexity paradigm for robotic manipulation. Furthermore, researchers such as Mason, Erdman and Goldberg have presented reliable sensorless manipulation schemes (See e.g. [2], [3] and references therein). A binary hyper-redundant manipulator is one kind of discretely-actuated manipulator. This manipulator does not require sensors or feedback control systems, which reduces its cost. Our concept of a binary hyper-redundant manipulator was influenced by research on snake-like robots [4], [5] and the design of variable geometry truss (VGT) manipulators [6], [7], [8]. One of the authors presented the concept of the binary manipulator in [9]. There have been several improvements in this manipulator concept. A planar binary VGT manipulator and a spatial binary Stewart-platform-like manipulator were designed and built by one of the authors and co-workers. The planar binary VGT manipulator consists of five modules of 3-bit binary VGTs stacked on top of each other. This manipulator has the ability to fold back to reach its own base. The spatial binary Stewart-platform-like manipulator consists of six modules of 6-bit binary Stewart-platforms stacked on the top of each other. Each module has $2^6 = 64$ configurations (a total number of 68.7 billion configurations for the whole manipulator). This 3-dimensional binary manipulator has a very limited hemispherical workspace. This problem occurred because of the structure of each Stewart-platform-like module. A number of papers present forward and inverse kinematics techniques and control for this kind of manipulator ([10], [11], [12], [13], [14], [15], [16]).

3. Design

The new manipulator design produces a hemispherical workspace. We have built a full size four-module prototype of this new 3-D manipulator using two-state pneumatic actuators. Furthermore, we have designed a new kind of discrete rotating joint controlled by three binary actuators.

3.1. Design of a discrete rotating joint

We have designed a discrete rotating joint using three binary actuators to control the angle and direction. We use a set of three gears. Two of these gears are mounted along with two rotary actuators on the base and the third is mounted on the turned module. One of two rotary actuators is used as a fluid-filled damper in order to decrease speed and reduce vibration. A series of holes aligned in a circular pattern on the turned module (rotor) acts as the position controller. While the rotary actuators and gear set are in operation, we use a compact linear binary actuator located between the base and rotor as the stopper. A cone shape is attached to the tip of the actuator, which is inserted into each hole to stop the rotor. The base and rotor are shown in Figure 4.

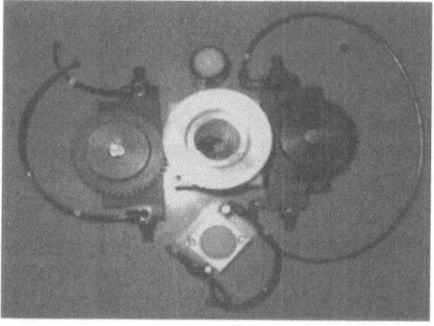

Figure 4 (a): A rotary joint comprising a rotary actuator, a rotary dashpot, linear actuator and a gear set.

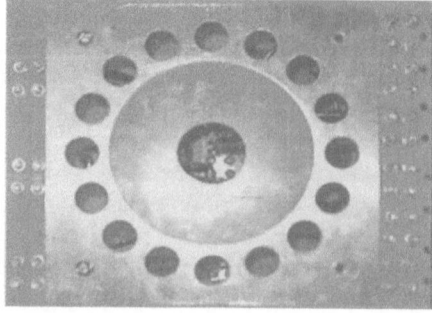

Figure 4 (b): A circular set of holes on the turned module (rotor).

3.2 Design of a new spatial binary-actuated manipulator

This design is influenced by the advantages and disadvantages of the previously built binary manipulators. The new design uses 3-bit binary VGT modules stacked on top of each other with a discretely actuated rotating joint between each module. Figure 5 illustrates some configurations of five concatenated modules. We have built a full size prototype using two-state pneumatic actuators. The prototype consists of three 3-bit binary VGT modules with discrete rotating joints between the modules. PC interfacing with a relay switchboard triggers the solenoid valves, and controls each actuator. In order to reduce the vibration occurring due to the high speed of the pneumatic actuators, dashpots have to be introduced in parallel to the actuators in each module.

Figure 5: Configurations of a five-binary-module.

4. Experiments and Results

The kinematics were successfully simulated before the prototype was built. A 3-bit planar VGT module has eight 2^3 possible reachable points. Therefore, the new discretely actuated manipulator can reach $2^{(3 \times N1) \times (N2 \times N3)}$ points. N1 is the number of 3-bit binary VGT modules, N2 is the number of rotating joints, and N3 is the number of states in each joint. We have built and tested the prototype manipulator. The load-bearing capacity of the manipulator depends on the actuators available on the market. Overall the experiments with the manipulator were very successful in demonstrate the range of motion.

5. Conclusion and future work

The results were very satisfactory. We have solved the vibration problem by using dashpots in parallel with the actuators in each module and each rotating joint. Figure 6 illustrates some configurations of the new spatial binary actuated manipulator. Attempts are currently being made to improve the performance of the structure and the rotating joints. We are interested in using a commercial magneto-rheolegical fluid, which is a kind of oil with magnetizable particles, to be filled in the dashpots to be placed in parallel with the actuators in each module. In this case, we will be able to control the stop positions by using electromagnetic systems, which will provide more states than a regular binary actuator. Our further experiments shall involve implementation of a new inverse kinematics algorithm [17].

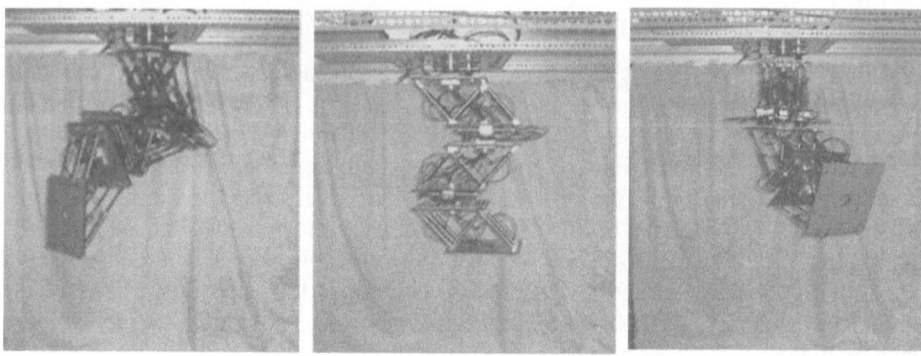

Figure 6: Configurations of the new spatial binary actuated manipulator.

Figure 7: The light weight of the new manipulator is illustrated in comparison
with traditional manipulators.

Acknowledgment

The authors would like to thank Mr. David Stein for design suggestions. We are also gratefully thank to Mr. Zheng Xu, Mr. Yong T. Kwon and Mr. Dory T. Lummer for their valuable technical assistance.

References

[1] Canny J, Goldberg K 1993 A risc paradigm for industrial robotics. *Tech Rep.ESRC 93-4/RAMP 93-2*. Engineering Systems Research Center, University of California at Berkeley
[2] Mason M T 1993 Kicking the sensing habit. *AI Magazine*.
[3] Goldberg K 1992 Orienting polygonal parts without sensors. *Algorithmica*. Special

robotics issue.

[4] Hirose S 1993 *Biologically Inspired Robots*. Oxford University Press.

[5] Chirikjian G, Burdick J 1995 Kinematically optimal hyper-redundant manipulator. *IEEE Transactions on Robotics and Automation*. 11(6): 794 –806

[6] Miura K, Furaya H 1985 Variable geometry truss and its applications to deployable truss and space crane arm. *Acta Astronautica*. 12(7): 599-607

[7] Hughes P, Sincarsin P, Carroll K 1991 Trussarm-a variable-geometry-Trusses. *J Intelligent Material Systems and Structures*. 1.2: 148-160

[8] Robertshaw H, Reinholtz C 1988 Varaible geometry Tresses. *Smart Materials, Structures, and Mathematical Issues*. 105-120

[9] Chirikjian G S 1994 A binary Paradigm for Robotic Manipulators. *Proceedings of the 1994 IEEE International Conference on Robotics and Automation*. 3063-3069

[10] Lees D, Chirikjian G S 1996 A combinatorial approach to trajectory planning for binary manipulators. *Proceedings of the 1996 IEEE International Conference on Robotics and Automation*. 2749-2754

[11] Lees D, Chirikjian G S 1996 An efficient method for computing the forward kinematics of binary manipulators. *Proceedings of the 1996 IEEE International Conference on Robotics and Automation*. 1012-1017

[12] Lees D, Chirikjian G S 1996 Inverse kinematics of binary manipulators with applications to service robotics. *Proceedings of the 1996 IEEE International Conference on Robotics and Automation*. 2749-2754

[13] Ebert-Uphoff I, Chirikjian G S 1996 Inverse kinematics of discretely actuated hyper redundant manipulators using workspace densities. *Proceedings of the 1996 IEEE International Conference on Robotics and Automation*. 1: 139 –145

[14] Ebert-Uphoff I, Chirikjian G S 1995 Efficient workspace generation for binary manipulators with many actuators. *J.Robotic Systems*. 12(6):383-400

[15] Chirikjian G S 1995 Kinematic synthesis of mechanisms and robotic manipulators with binary actuators. *J. Mechanical Design*. 117 (4): 573-580

[16] Chirikjian G S, Ebert-Uphoff I 1998 Discretely actuated manipulator workspace generation using numerical convolution on the Euclidean group. *Proceedings of the 1998 IEEE International Conference on Robotics and Automation*. 1: 742 -749

[17] Suthakorn J, Chirikjian G S A new inverse kinematics algorithm for binary manipulators with many Actuators. *Advanced Robotics*. (accepted for publication)

Design and Experiments on a Novel Biomechatronic Hand

P. Dario, M.C. Carrozza, S. Micera, B. Massa, M. Zecca
Mitech Lab, Scuola Superiore Sant'Anna
56127 Pisa, Italy
Centro INAIL RTR
55049 Viareggio (LU), Italy
E-mail: dario@arts.sssup.it

Abstract: An "ideal" upper limb prosthesis should be perceived as part of the natural body by the amputee and should replicate sensory-motor capabilities of the amputated limb. However, such an ideal "cybernetic" prosthesis is still far from reality: current prosthetic hands are simple grippers with one or two degrees of freedom, which barely restore the capability of the thumb-index pinch.
This paper describes the design and fabrication of a novel prosthetic hand based on a "biomechatronic" and cybernetic approach. Our approach is aimed at providing "natural" sensory-motor co-ordination to the amputee, by integrating biomimetic mechanisms, sensors, actuators and control, and by interfacing the hand with the peripheral nervous system.

1. Introduction

The development of an upper limb prosthesis that can be felt as a part of the body by the amputee (Extended Physiological Proprioception – EPP [1]), and that can substitute the amputated limb by closely replicating its sensory-motor capabilities ("cybernetic" prosthesis [2]), is far to become reality. In fact, current commercial prosthetic hands are unable to provide enough grasping functionality and to provide sensory-motor information to the user. One of the main problems of the current available devices is the lack of degrees of freedom (DOFs).

Commercially available prosthetic devices, such as Otto Bock SensorHand™, as well as multifunctional hand designs [3,4,5,6,7,8,9] are far from providing the manipulation capabilities of the human hand [10]. This is due to many different reasons. For example, in prosthetic hands active bending is restricted to two or three joints, which are actuated by a single motor drive acting simultaneously on the metacarpo-phalangeal (MP) joints of the thumb, of the index and of the middle finger, while other joints can bend only passively.

The way to overcome all these problems is to develop a "cybernetic" prosthesis following a *biomechatronic* approach, i.e. by designing a mechatronic system inspired by the biological world. A cybernetic prosthesis must solve the following problems of the commercial prostheses:

1. the reduced grasping capabilities;
2. the noncosmetic appearance;
3. the lack of sensory information given to the amputee;

D. Rus and S. Singh (Eds.): Experimental Robotics VII, LNCIS 271, pp. 159–168, 2001.

4. the lack of a "natural" command interface.

The first and the second problems can be solved by increasing the number of active and passive DOFs; this can be achieved by embedding a higher number of actuators in the hand structure and designing coupled joints.

The third and forth problems can be addressed by developing a "natural" interface between the Peripheral Nervous System (PNS) and the artificial device (i.e., a "natural" Neural Interface (NI)) to record and stimulate the PNS in a selective way. This can be useful in order to make possible the ENG-based control of the prosthesis (to solve the forth problem) and to give back some sensory feedback to the amputee by stimulating in an appropriate way his/her afferent nerves (after characterizing the afferent signals of the PNS in response to mechanical and proprioceptive stimuli) solving the third problem of the commercially-available hand prostheses. This approach is illustrated in Figure 1.

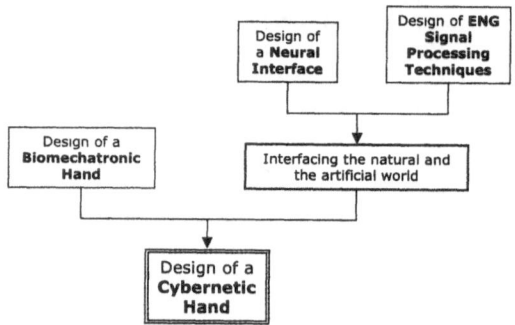

Figure 1. Approach for the development of a cybernetic hand

Current research activities at Scuola Superiore Sant'Anna aimed at the development of a biomechatronic prosthetic hand controlled through a "natural" NI are presented in this paper. In particular, preliminary results obtained in processing ENG signals from afferent nerves are illustrated and analyzed.

2. Design of the biomechatronic hand

The main requirements to be considered since the very beginning of a prosthetic hand design are the following: cosmetics, controllability, noiselessness, lightness and low energy consumption. These requirements can be fulfilled by implementing an integrated design approach aimed at embedding different functions (mechanisms, actuation, sensing and control) within a housing closely replicating the shape, size and appearance of the human hand. This approach can be synthesized by the term: "*biomechatronic*" design [11].

2.1. Architecture of the biomechatronic hand

The biomechatronic hand will be equipped with three actuator systems to provide a tripod grasping: two identical finger actuator systems and one thumb actuator system (see Figure 2).

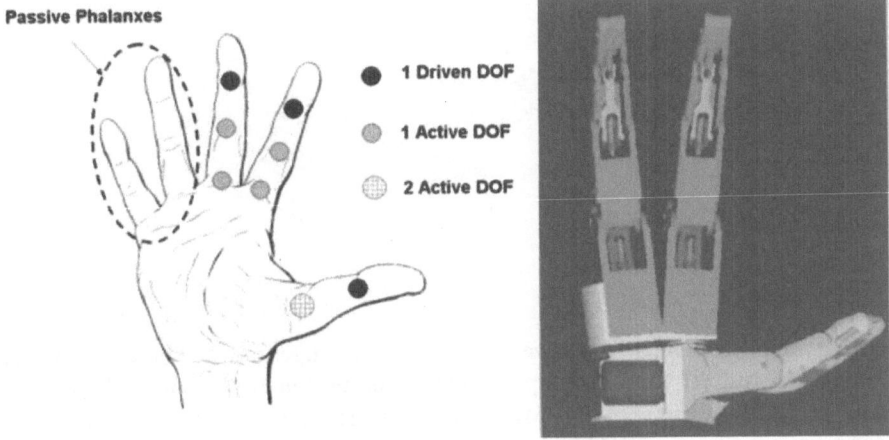

Figure 2. Architecture of the biomechatronic hand.

The finger actuator system is based on two micro-actuators, which drives the MP and the PIP joints respectively; for cosmetic reasons, both actuators are fully integrated in the hand structure: the first in the palm and the second within the proximal phalange. The DIP joint is passively driven by a four bars link connected to the PIP joint. The thumb is equipped with two active DOFs in the MP joint and one driven passive DOF in the IP joint.

The grasping task is divided in two subsequent phases in which the two different actuator systems are active:

1)   reaching and shape-adapting phases;
2)   grasping phase with thumb opposition.

In fact, in phase 1) the first actuator system allows the finger to adapt to the morphological characteristics of the grasped object by means of a low output torque motor. In phase two, the thumb actuator system provides a power opposition force, useful to manage critical grasps, especially in case of heavy or slippery objects.

It is important to point out that the most critical problem of the proposed configuration is related to the strength required to micro-actuators to withstand the high load applied during the grasping phase.

In order to demonstrate the feasibility of the described biomechatronic approach, we started by developing one finger (index or middle).

2.2. Design and development of the finger prototype

As outlined above, the two DOF finger prototype is designed by reproducing, as closely as possible, the size and kinematics of a human finger. It consists of three phalanges and of palm housing, which is the part of the palm needed to house the proximal actuator (see Figure 3).

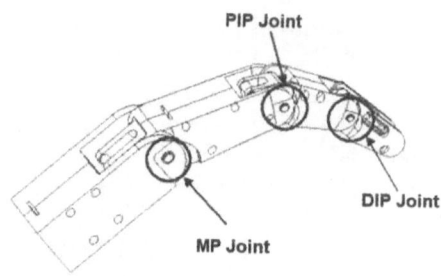

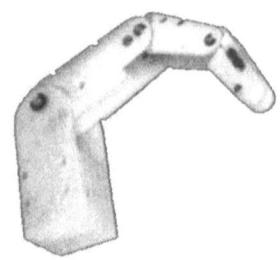

Figure 3. Finger design.   Figure 4. Finger prototype.

In order to match the size of a human finger, two micro-motors are incorporated, respectively, in the palm and in the proximal phalange. The actuator system is based on Smoovy™ (RMB, Eckweg, CH) micro-drivers (5 mm diameter) linear actuators based on DC brushless motors.

The output force resulting from motor activation is sufficient to move the phalanges for achieving adaptive grasp. In addition, the shell housing provides mechanical resistance of the shaft to both axial and radial loads. This turns out to be essential during grasping tasks, where loads, derived from the thumb opposition, act both on the actuator system and on the whole finger structure.

A first prototype of the finger was fabricated using the Fused Deposition Modeling [FDM] process (see Figure 4). This process allows the fabrication in a single process of three-dimensional objects, made out of acrylonitrile/butadiene/styrene [ABS] resin, directly from CAD-generated solid models.

2.3. Fingertip force characterization

A first set of experimental tests has been performed in order to evaluate the force that the finger is able to exert on an external object [12]. To this aim we have measured the force resulting when the finger is pressing directly on a force sensor (3-axial piezoelectric load cell 9251 A, Piezo-Instrumentation KISTLER, Kiwag, CH), corresponding to different configurations of the joints.

Two "pressing" tasks were identified in order to evaluate separately and independently force obtained by the two actuators incorporated in the finger:

TASK 1: the pushing action was exerted only by the distal actuator.

TASK 2: the pushing action was exerted only by the proximal actuator.

Corresponding to each task, two subtasks were identified according to the position of the non-active joint (extended, flexed). The different values of joint rotation angles corresponding to each subtask are illustrated in Figure 5.

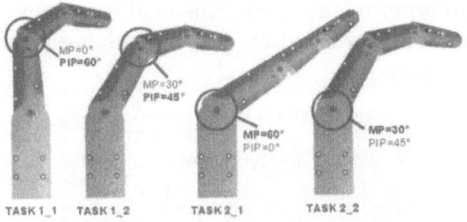

Figure 5. Positions of finger joints for each task.The active joint for each task and position is indicated by a small circle.

During the force characterization the fingertip pushed on the force sensor. The Z force component was recorded, the X and Y outputs of the load cell were monitored. This was obtained by adjusting the finger position for obtaining a force parallel to the Z-axis of the load cell. A first set of experimental tests was done on the finger prototype, with the aim of evaluating how much force the finger is able to apply on an object.

2.4. Results and discussion

Ten tests were performed for each subtask. The obtained results are illustrated in Figure 6. These force values are comparable with force exerted by "natural" human finger during fine manipulation, thus demonstrating the feasibility of the biomechatronic approach, at least for this class of manipulation tasks. The output force resulting from motor activation is sufficient to move the phalanges for achieving adaptive grasp [13].

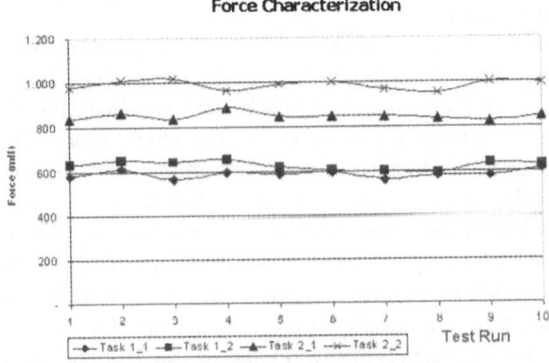

Figure 6. Experimental results.

3. Development of an intelligent Neural Interface (NI)

3.1. Microelectrodes Array Fabrication

Different microelectrode arrays on silicon substrates (named "dice") were designed and fabricated using various microfabrication technologies. Three different dice designs were fabricated with different dimensions, electrode size, and through-holes dimensions. The designs mainly utilized for *in vivo* experiments were the so called "Active Die 1" and "Active Die 2". A photograph of Active Die 2 is shown in Figure 7. Some dice did not incorporate electrodes ("passive dice") and were used for control purposes. In order to achieve mechanical robustness, each silicon die was mounted in a titanuim ring, fabricated by laser machining purposely to host the die and the electrical connections.

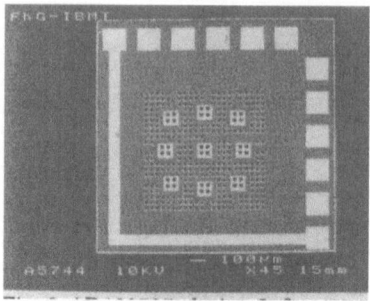

Figure 7. Active Die 2. Electrode size of 9x7,696 μm^2, through holes dimensions of 24x24 μm^2; center spacing 36 μm; total number of through-holes: 20x20; number of Ti-Pt electrodes is 10.

3.2. Electrophysiological Results

3.2.1. Electrophysiological Set-up and Methods

In vivo tests were performed on adult female New Zealand rabbits. For control purposes a first set of experiments was performed using first empty guidance channels and then NI based on "passive" dice (i.e. dice with through-holes but no electrodes) with no interconnects. A second set of experiments was performed using complete NI incorporating active dice. The same surgical procedure was adopted for the two sets of experiments.

Different set-up configurations were used depending on the experiment performed. For recording from the intact nerve, from nerve regenerated in empty guidance channels and from nerve through passive dice, the sciatic nerve was stimulated proximally through a pair of electrodes (FHC Hook Electrodes # 06-11-2) with constant current pulses (duration = 0.08-0.1 ms) of different intensities. Responses for both anodic and cathodic stimulations were obtained. The electroneurogram (ENG) analysis was performed by measuring the compound action potential (CAP) from the recording electrodes (placed at a distance of 50 mm from the stimulation site). In animals implanted with the NI the connector was plugged to a standard electrophysiological set-up for nerve recording and stimulation. A mechanical switch allowed each of the 10 neural interface electrodes to be connected to the electronic apparatus.

Single pulses (amplitude range = 40 μA-1 mA, duration = 0.2 ms) were delivered at each of the 10 electrodes of the active interface. When muscle contraction was observed, thus indicating the presence of functional axons, the same channel of the neural interface used for stimulation was connected to the recording apparatus and the spontaneous nerve activity was monitored. Neural activity was also recorded in response to passive leg movements (i.e., stretching).

3.2.2. Computer Analysis of Electrophysiological Signals

Signals from the amplifier were stored on a tape recorder and then played back and digitized on a computer hard disk (using a MIO-16 A/D converter). Suitable programs, based on LabView programming techniques, were used in order to acquire experimental data and synchronize the electrical stimulation. The amplitude and delay of compound action potentials were measured in order to assess the functional recovery of the regenerated nerve.

3.2.3. Stimulation and Recording from Nerve Axons using NI#6

The most interesting results were those obtained on rabbit #inter18 using NI #6. In this specific animal, stimulation and recording were performed after 48 days from implant. At this time nerve regenerated through the NI. No signs of nerve damage and/or device failure were visible. Tissue reaction seemed similar to that found in the control experiments. The flat connecting cable was still intact and it could be easily twisted in order to attach the signal conditioning plug.

The action potential duration is about 1-1.5 ms and its amplitude is about 110 µV. An intense electrical activity with respect to the resting level in response to an imposed leg movement (extension of the leg and foot) is shown in Figure 8.

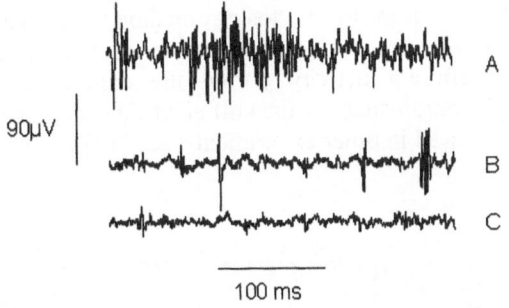

Figure 8. Recording of electrical activity during a leg/foot movement (A) and after the movement (B). The control signal at rest is shown in (C).

3.2.4. Nerve Stimulation Using NI #6

Nerve stimulation was obtained through the NI #6, as demonstrated by a leg/foot contraction for each pulse delivered to the nerve through the microelectrode array. The current threshold which induced a contraction was of the order of tenth of µA, lower than the current needed to excite the nerve using the control experiment extracellular electrodes. Clear EMG signals were recorded by muscle groups as illustrated in Figure 9.

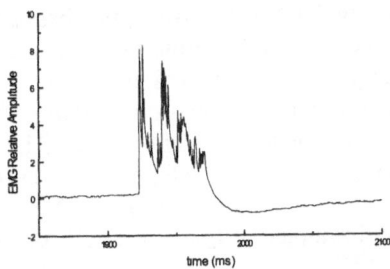

Figure 9. EMG signal recorded during nerve stimulation.

4.   ENG Signal Processing Techniques

In order to verify the feasbility of extracting sensory-motor information from the ENG signals recorded from afferent nerves we carried out some preliminary experiments in collaboration with the Center for Sensory-Motor Interaction (Aalborg University, Aalborg – DK). In this Section the results of these experiments are briefly described.

4.1. Experimental Set-up

Acute experiments were performed using four femal adult New Zealand rabbits (identified by progressive numbers). The Danish Committee for Ethical use of Animals in Research approved all procedures used in the experiments. Two tripolar, whole nerve cuff electrodes were implanted around the tibial and peroneal nerves (which are major branches of the sciatic nerve) in the rabbit's left leg (cuff lengths were approx. 20 mm; the inner diameters were 2 mm and 1.8 mm, respectively). The cuff electrodes were produced according to the procedure described in [14] except that a straight cut was used as a closing method. The sural nerve was cut immediately distal to the peroneal cuff electrode to minimize the recording of unwanted cutaneous afferent activity during the experiments. In Figure 10, a schematic of the implantation sites for the cuff electrodes is presented (similar rabbit preparations have been used in other experiments, see [14]).

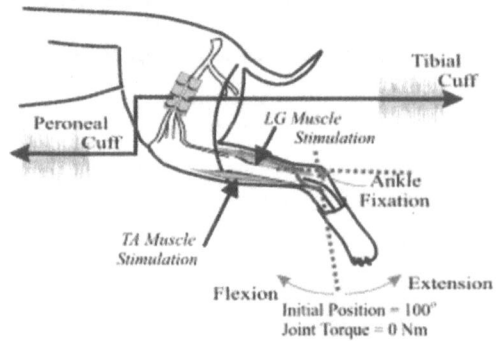

Figure 10. Schematic illustration of the implantation sites.

The equipment used during the experiments consisted of a computer controlled servomotor used to passively rotate the rabbit's ankle in the sagittal plane.

A support and fixation device equipped with four strain gauges was used as torque transducer (sensitivity 10 Nm/V). An optics-based rotation transducer was used to record the position of the ankle during the movements (sensitivity = 10°/V). The position and torque signals were sampled at 500 Hz. The rabbit was placed on its right side, and the left foot was mounted on a cradle. The knee and ankle joint were fixated during the experiment. An elaborate description of the experimental equipment can be found in [14]. The whole nerve cuff recordings were pre-amplified 200,000 times, bandpass filtered using a 2nd order Butterworth analog filter (500 Hz - 5 kHz), and sampled at f_S =10 kHz (12 bit National Intruments A/D board).

The ankle angle of a normal human subject was recorded during quite standing and this signal was used as a template to move the ankle of a rabbit preparation. The whole nerve activity of the tibial and peroneal nerves were recorded as described in [14]. All ENG recordings were rectified and bin integrated during a 9 ms window. The position and torque data were low-pass filtered at $f_{SP} = 100$ Hz (12th order digital Butterworth filter).

4.2. Fuzzy Models

Three fuzzy models were implemented with characteristics as follows:

1. The Modified FCRM Fuzzy Model is a Takagi-Sugeno (TS) fuzzy system. To obtain the rules directly from the data, a fuzzy clustering algorithm named fuzzy C-regression model (FCRM) is implemented [15].

2. The Adaptive Network-based Fuzzy Inference System (ANFIS) model is a TS fuzzy system implemented as a feed-forward neural network. The principal characteristic of this network is the hybrid learning procedure described in [16].

3. The Dynamic Non-Singleton Fuzzy Logic System (DNSFLS) is a Mamdani fuzzy system, implemented in the framework of recurrent neural networks [17].

4.3. Results of the prediction

To compare the performances of the different fuzzy models, the root mean square (RMS) of the prediction error has been introduced as a figure of merit. In Tables 1 the RMS of the prediction error is presented.

Table 1: RMS of the Prediction Error for the Different Fuzzy Models

Fuzzy model	Training Traject RMS	Test Traject RMS	Rule Number
FCRM	0.0216	0.0480	49
ANFIS	0.0047	0.0079	49
DNSFLS	0.0013	0.0057	45

5. Conclusions

This paper show research activities carried out at Scuola Superiore Sant'Anna for the development of a cybernetic prosthesis. We are currently designing of a new prototype of the biomechatronic hand and developing ENG signal processing techniques to characterize the afferent response of the PNS. Moreover, the Consortium of the GRIP EU Project (coordinated by Scuola Superiore Sant'Anna) is currently developing an implantable system to record and stimulate the PNS with a telemetry connection with an external control system. This device can be used in experiments to design the cybernetic prosthesis.

6. Acknowledgements

This work has been supported by a research project entitled "Design and development of innovative components for sensorized prosthetic systems" currently ongoing at the "Applied Research Center on Rehabilitation Engineering" funded by INAIL (National Institute for Insurance of Injured Workers), and originated by a joint initiative promoted by INAIL and by Scuola Superiore Sant'Anna. This work has been partly funded by the GRIP EU Project ("An integrated system for the neuroelectric control of grasp in disabled persons", ESPRIT LTR #26322).

The authors are also grateful to Mr. Carlo Filippeschi and Mr. Gabriele Favati for their valuable technical assistance. The authors would also thank Mr. Rinaldo

Sacchetti for helpful discussions and criticism on the biomechatronic prosthetic hand concept.

References

[1] D. C. Simpson, The Choice of Control System for multimovement prostheses: Extended Physiological Proprioception (EPP), in *The Control of Upper-Extremity Prostheses and Orthoses*, P. Herberts *et al.*, Eds., 1974.

[2] J. A. Doeringer, N. Hogan, Performance of above elbow body-powered prostheses in visually guided unconstrained motion task, *IEEE Trans. Rehab. Eng.* 42 (1995), 621-631.

[3] P.J. Agnew, Functional effectiveness of a myoelectric prosthesis compared with a functional split hook prosthesis: a single subject experiment, Prost. & Orth. Int. 5 (1981), 92–96.

[4] S.-E. Baek, S.-H. Lee, J.-H. Chang, Design and control of a robotic finger for prosthetic hands, *Proc. Int. Conf .Intelligent Robots and Systems* (1999), 113-117.

[5] M. E. Cupo, S. J. Sheredos, Clinical Evaluation of a new, above elbow, body powered prosthetic arm: a final report, *J. Rehab. Res. Dev.* 35 (1998), 431-446.

[6] R. Doshi, C. Yeh, M. LeBlanc, The design and development of a gloveless endoskeletal prosthetic hand, *J. Rehab. Res. Dev.* 35 (1998), 388–395.

[7] P. J. Kyberd, O. E. Holland, P. H. Chappel, S. Smith, R. Tregidgo, P. J. Bagwell, and M. Snaith, MARCUS: a two degree of freedom hand prosthesis with hierarchical grip control, *IEEE Trans. Rehab. Eng.* 3 (1995), 70–6.

[8] D. H. Silcox, M. D. Rooks, R. R. Vogel, L. L. Fleming, Myoelectric Prostheses, *J. Bone & Joint Surg.*, 75 (1993), 1781–1789.

[9] R. Vinet, Y. Lozac'h, N. Beaundry, G. Drouin, Design methodology for a multifunctional hand prosthesis, *J. Rehab. Res. Dev.* 32 (1995), 316–324.

[10] M. R. Cutkosky, Robotic Grasping and Fine Manipulation, Boston: Kluwer Academic Publishers, 1985.

[11] Carrozza M.C., Dario P., Lazzarini R., *et al*, An actuator system for a novel biomechatronic prosthetic hand. In Proceedings of Actuator 2000 Bremen, Germany (2000), 276-280.

[12] K. Nagai, Y. Eto, D. Asai, M. Yazaki, Development of a three-fingered robotic hand-wrist for compliant motion, *Proc. Int. Conf. Intelligent Robots and Systems* (1998), 476-481.

[13] D.T.V. Pawlock, R.D.Howe, Dynamic contact of the human fingerpad against a flat surface, *ASME J Biomech Eng* 121 (1999), 605-611.

[14] R.R. Riso and Farhad K. Mosallaie and Winnie Jensen and Thomas Sinkjaer, Nerve Cuff recordings of muscle afferent activity from tibial and peroneal nerves in rabbit during passive ankle motion, IEEE Trans Rehab Eng, 2000.

[15] E. Kim, M.Park, S.Ji, and M.Park, "A new approach to fuzzy modeling", IEEE Trans Fuzzy Sys, vol. 5, pp. 328-337, August 1997.

[16] J.-S. R. Jang, "ANFIS: Adaptive-network-based fuzzy inference system", IEEE Trans Sys Man Cyber, vol. 23, pp. 665-685, May/June 1993.

[17] G.C. Mouzouris, J.M. Mendel, "Dynamic non-singleton fuzzy logic systems for nonlinear modeling", IEEE Trans Fuzzy Sys, vol. 5, pp.199-208, May 1997.

Autonomous Injection of Biological Cells Using Visual Servoing

Sun Yu   Bradley J. Nelson
Advanced Microsystems Lab
Department of Mechanical Engineering
University of Minnesota, Twin Cities
Minneapolis, MN 55455
yus, nelson@me.umn.edu

Abstract: The ability to analyze individual cells rather than averaged properties over a population is a major step towards understanding the fundamental elements of biological systems. Recent advances in microbiology such as cloning demonstrate that increasingly complex micromanipulation strategies for manipulating individual biological cells are required. In this paper, a microrobotic system capable of conducting automatic embryo pronuclei DNA injection is presented. Both embryo pronuclei DNA injection and intracytoplasmic injection (cell injection) are methods of introducing foreign genetic material into cells. Conventionally, cell injection has been conducted manually, however, long training, disappointingly low success rates from poor reproducibility in manual operations, and contamination all call for the elimination of direct human involvement. The system presented is capable of performing automatic embryo pronuclei DNA injection autonomously and semi-autonomously through a hybrid visual servoing control scheme. MEMS-based cell holders were designed and fabricated to aid in injection. Upon the completion of injection, the DNA injected embryos were transferred into a pseudopregnant foster female mouse to reproduce transgenic mice for cancer studies. Experiment result shows that the injection success rate is 100%.

1. Introduction

The ability to analyze individual cells rather than averaged properties over a population is a major step towards understanding the fundamental elements of biological systems. Studies on single cells are a key component in the development of highly selective cell-based sensors, the identification of genes, and bacterial synthesis of specific DNA. Treatments for severe male infertility and the production of transgenic organisms require that individual cells are isolated and individually injected. These recent advances in microbiology as well as other significant research efforts such as cloning, demonstrate that increasingly complex micromanipulation strategies for manipulating individual biological cells are required.

Microrobotics and microsystems technology can play important roles in manipulating cells, a field referred to as biomanipulation. In this paper we present a visually servoed microrobotic system capable of performing automatic pronuclei DNA injection, which is a method for introducing DNA into embryos in order to create transgenic organisms. In Figure 1, a holding pipette holds a mouse embryo and an injection pipette performs the injection task. The objective of pronuclei injection is,

D. Rus and S. Singh (Eds.): Experimental Robotics VII, LNCIS 271, pp. 169–178, 2001.

in this case, to produce transgenic mice for use in cancer studies.

Conventionally, cell injection is conducted manually. Operators often require at least a year of full time training to become proficient at the task, and success rates are disappointingly low. One reason for this is that successful injections are not precisely reproducible. A successful injection is determined greatly by injection speed and trajectory [8]. Automated cell injection can be highly reproducible with precise control of pipette motion. The second reason for the low success rate of conventional cell injection is due to contamination. This also calls for the elimination of direct human involvement. Therefore, the main advantages of automated cell injection are that it reduces the need for extended training, reduces the risk of contamination, and is highly reproducible which greatly increases the success rate.

The autonomous microrobotic system described in this paper is being developed to conduct autonomous pronuclei DNA injection of mouse embryos. In our pronuclei DNA injection experiments, we developed both autonomous and semi-autonomous injection strategies through a hybrid control scheme. A third injection strategy aided by microfabricated cell holders was also investigated. Upon the completion of injection, the DNA injected embryos were transferred into the ampulla of a pseudo pregnant foster female mouse to reproduce transgenic mice. Results show that the injection success rate is 100%.

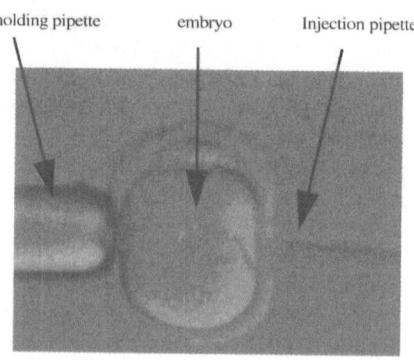

Figure 1. Cell injection of a mouse embryo. The embryo is approximately 50μm in diameter.

2. Biomanipulation

Biomanipulation entails such operations as positioning, grasping, and injecting material into various locations in cells. Existing biomanipulation techniques can be classified as non-contact manipulation including laser trapping [3][4][5][15] and electrorotation [1][12][14], and contact manipulation referred to as mechanical micromanipulation [8]. When laser trapping [3][4][5][15] is used for non-contact biomanipulation, a laser beam is focused through a large numerical aperture objective lens, converging to form an optical trap in which the lateral trapping force moves a cell in suspension toward the center of the beam. The longitudinal trapping force moves the cell in the direction of the focal point. The optical trap levitates the cell and holds it in position. Laser traps can work in a well controlled manner. However, two reasons make laser trapping techniques undesirable for automated cell injection. The high dissipation of visible light in aqueous solutions requires the use of high energy light close to the UV spectrum, raising the possibility of damage to the cell. Even though some researchers claim that such concerns could be overcome using wavelengths in the near infrared (IR) spectrum [5], the question as to whether the incident laser beam might induce abnormalities in the cells' genetic material still exists. One alternative to using laser beams is the electro-rotation technique. Electric-field-induced rotation of cells was demonstrated by Mischel [9], Arnold [2] and Washizu [14]. This non-contact cell manipulation technique is based on controlling the phase shift and magnitude of electric fields. These fields, appropriately applied, produce a torque on the cell. Different system configurations have been established for cell manipulation based on this principle [1][12], which can achieve high accuracy in cell positioning. However, it lacks a means to hold the cell in place for further manipulation, such as injection, since the magnitude of the electric fields has to be kept low to ensure the viability of cells. The limits of non-contact biomanipulation in the laser trapping and electro-rotation techniques make mechanical micro-manipulation desirable. The damage caused by laser beams in the laser trapping technique and the lack of a holding mechanism in the electro-rotation technique can be overcome by mechanical micro-manipulation.

3. Embryo Preparation

The embryos used in our experiments are collected in the Cancer Center at the University of Minnesota in accordance with standard embryo preparation procedure [6]. Three week old FVB/N female mice are injected with pregnant mare serum (PSM) to promote oval maturation. After approximately 45 hours the mice are injected with human chorionic gonadotropin (hCG) to promote synchronized ovulation. Then the superovulated female mice are mated to fertile male mice. Finally embryos are collected from the ampulla of female mice. A typical embryo is shown in Figure 2. The average diameter of the embryos is 50 μm.

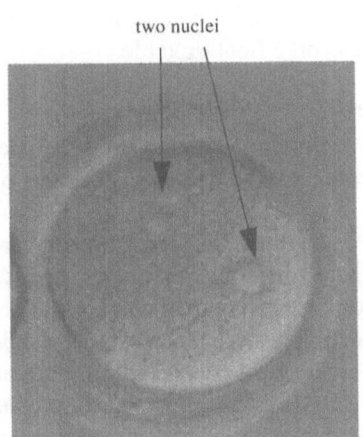

Figure 2. Embryo with two nuclei.

For embryo pronuclei DNA injection, only embryos with two visible pronuclei can be selected for injection. DNA is deposited in one of the two nuclei. From the perspective of control, intracytoplasmic injection into embryos is less demanding than pronuclei DNA injection because intracytoplasmic injection only requires that a foreign genetic material be deposited within the embryo membrane, not necessarily in the nucleus.

4. System Setup

The autonomous embryo injection system is composed of an injection unit, an imaging unit, a vacuum unit, a microfabricated device, and a software unit. Figure 3 shows the system setup.

For embryo injection vibration must be well controlled. Vibration not only causes difficulty in visually tracking features but also produces permanent and fatal harm in

coarse manipulator   3-DOF microrobot

inverted microscope

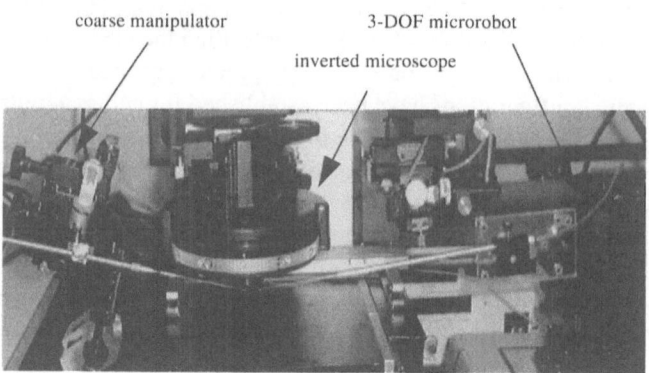

Figure 3. Autonomous embryo injection system.

the injected location and the surrounding area. To avoid vibration, all units except the host computer and the vacuum units of our embryo pronuclei DNA injection system are placed on a floating table.

4.1. Injection Unit

The injection unit of the system includes a holding pipette, an injection pipette, two standard pipette holders, a high precision 3 DOF microrobot, and a coarse manipulator.

The injection and holding pipettes are both processed using a micropipette puller. The dimensions of the pipette tips are $1\mu m$ in inner diameter for the injection pipettes and $20\mu m$ in outside diameter for the holding pipettes. Both the holding pipettes and injection pipettes are held by pipette holders.

Extremely high precision motion control is required for successful embryo injection. A 3 DOF microrobot is used in which the XYZ axes each has a travel of 2.54 cm with a step resolution of 40nm. An injection pipette with a pipette holder is installed on the microrobot as shown in Figure 4.

3-DOF microrobot

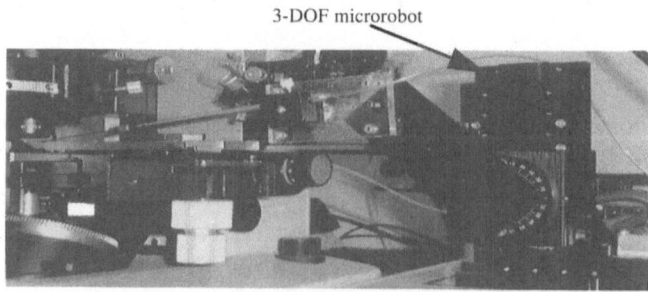

Figure 4. 3-DOF high precision microrobot.

The holding pipette is installed on a micromanipulator that is a manually operated three dimensional coarse manipulator. The holding pipette holder and the injection pipette holder are both connected with Teflon tubing such that negative and positive pressure is provided to the tips of the pipettes for holding embryos and depositing DNA.

4.2. Imaging Unit

The imaging unit of the embryo injection system includes an inverted microscope, a CCD camera, a PCI framegrabber, and a host computer. An inverted microscope is used with a 400x objective. The CCD camera is mounted on port of the microscope. The framegrabber captures thirty frames per second. The tracking of image features, which is required for semi-autonomous teleoperation and autonomous injection, is performed on the host computer (a 450MHz Celeron) at 30Hz.

5. Embryo Injection

5.1. Automatic Embryo Injection

For embryo pronuclei DNA injection, focusing needs to be done precisely on the central plane of one of the two nuclei, the tip of the holding pipette, and the tip of the injection pipette. Failure to do so will cause the injection pipette to "slide" over the top of the embryo failing to puncture the nucleus membrane and possibly causing serious injure to the cell membrane.

5.1.1. Hybrid Control Scheme for Embryo Injection

Our hybrid control scheme consists of image-based visual servo control and precise position control. In image-based visual servo control, the error signal is defined directly in terms of image feature parameters. The motion of the microrobot causes changes to the image observed by the vision system. Although the error signal is defined on the image parameter space, the microrobot control input is typically defined either in joint coordinates or in task space coordinates. In formulating our visual servo system, task space coordinates are mapped into sensor space coordinates through a Jacobian mapping. Let x_T represent coordinates of the end-effector of the microrobot on the task space, and $\dot{x}_T$ represent the corresponding end effector velocity. Let x_I represent a vector of image feature parameters and $\dot{x}_I$ the corresponding vector of image feature parameter rates of change. The image Jacobian, $J_v(x_T)$, is a linear transformation from the tangent space of task space T at x_T to the tangent space of image space I at x_I.

$$\dot{x}_I = J_v(x_T)\dot{x}_T \tag{1}$$

where $J_v(x_T) \in \Re^{k \times m}$, and

$$J_v(x_T) = \left[\frac{\partial x_I}{\partial x_T}\right] = \begin{bmatrix} \dfrac{\partial x_{I1}(x_T)}{\partial x_{T1}} & \cdots & \dfrac{\partial x_{I1}(x_T)}{\partial x_{Tm}} \\ \cdots & \cdots & \cdots \\ \dfrac{\partial x_{Ik}(x_T)}{\partial x_{T1}} & \cdots & \dfrac{\partial x_{Ik}(x_T)}{\partial x_{Tm}} \end{bmatrix} \tag{2}$$

m is the dimension of the task space T. The derivation of the model can be found in [13].

The state equation for the visual servo control system is as follows.

$$x(k+1) = x(k) + TJ_v(k)u(k) \qquad (3)$$

where $x(k) \in \Re^{2M}$ (M is the number of features being tracked); T is the sampling period of the vision system; and $u(k) = \begin{bmatrix} \dot{X}_T & \dot{Y}_T \end{bmatrix}$ is the microrobot's end-effector velocity.

The control objective of the system is to control the motion of the end-effector, i.e. the injection pipette, in order to place the image plane coordinates of the feature on the target in the switch area shown in Figure 6(a). The control strategy used to achieve the control objective is based on the minimization of an objective function that places a cost on errors in feature positions, $[x(k+1) - x_{switch}]$, and a cost on providing a visual control input $u(k)$.

$$E(k+1) = [x(k+1) - x_{switch}]^T Q[x(k+1) - x_{switch}] + u^T(k)Lu(k) \qquad (4)$$

This expression is minimized with respect to the current control input $u(k)$. The result is the expression for the visual control input.

$$u(k) = -[TJ^T_v(k)QTJ_v(k) + L]^{-1} TJ^T_v(k)Q[x(k) - x_{switch}] \qquad (5)$$

The weighting matrices Q and L allow the user to place more or less emphasis on the feature error and the control input. Methods for selecting these matrices can be found in [10].

When the visual servo controller guides the end-effector of the microrobot into the switching area, the control scheme switches to precise position control. The visual servo control and the precision position control jointly form the hybrid control scheme for automatic embryo pronuclei DNA injection. The complete hybrid control scheme is

$$U(k) = F_\sigma \begin{bmatrix} x(k) - x_{switch} \\ x_T(k) - x_{TD} \end{bmatrix} \qquad (6)$$

where

$$F_\sigma = \begin{bmatrix} -\sigma_1[TJ^T_v(k)QTJ^T_v(k) + L]TJ^T_v(k)Q & \sigma_2 I \end{bmatrix} \qquad (7)$$

I is a 2×2 unit matrix; and x_{TD} is the desired position on the task space T.

The switching condition is $\sigma_1\sigma_2 = 0$, and

$$\sigma_1 = 1 \quad \text{when } x(k) \notin (c, r);$$
$$\sigma_2 = 1 \quad \text{when } x(k) \in (c, r).$$

where (c, r) is the switching area shown in Figure 6(a).

Figure 5 shows the block diagram of the hybrid control system for embryo pronuclei DNA injection.

The hybrid controller $c_{switch}(t)$ selects the controller in the hybrid control system based on visual feedback and the switching conditions.

The injection pipette is originally positioned away from the embryo shown in Figure 6(a). The hybrid controller guides the microrobot with the injection pipette into the nucleus of the embryo where DNA is deposited. Figure 6(b) shows the injection process.

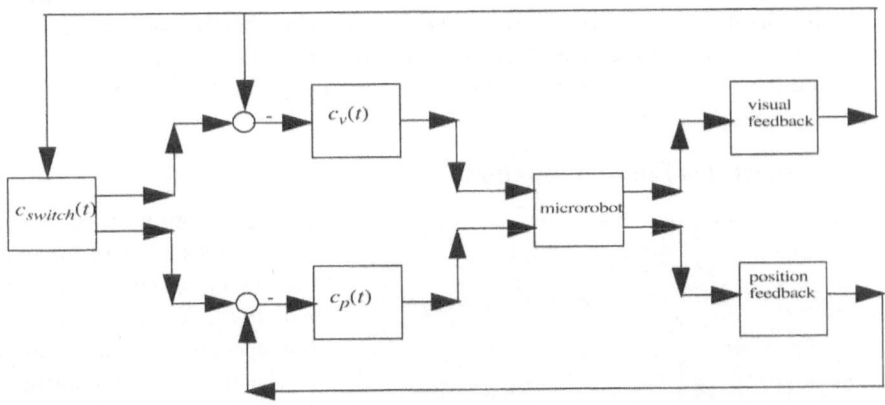

Figure 5. Hybrid control scheme for embryo pronuclei DNA injection.

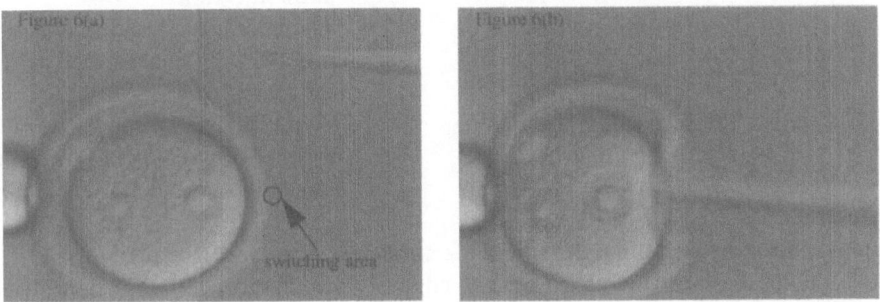

Figure 6. Embryo pronuclei DNA injection using hybrid control.

5.1.2. SSD Tracking Algorithm

For visual servo control, we adopt the sum-of-squared-differences optical flow (SSD) tracking algorithm [11]. SSD is an effective method for tracking in a structured environment where image patterns do not change considerably between successive frames of images. In visual servoing of microrobot under a microscope, the predictable environment and controlled illumination make SSD a robust tracking method. It is desirable to select features with high gradients, such as edges and corners that are distinct from their neighboring regions. In embryo injections, we select the tip of the injection pipette as a feature. The basic assumption of SSD tracking is that intensity patterns $I(x, y, t)$ in a sequence of images do not change rapidly between successive images $I(x, y, t + 1)$. In implementing the algorithm, we acquire a template of 20x20 pixels $T_{20 \times 20}$ around the feature, i.e. the tip of the injection pipette. An SSD correlation measure is calculated for each possible displacement (dx, dy) within a 20x20 pixel search window in the new image $I(x, y, t + 1)$

$$SSD(dx, dy) = \sum_{i, j \in N} [I(x_1 + dx + i, y_1 + dy + j) - T(x_1 + dx + i, y_1 + dy + j)]^2 \qquad (8)$$

The distance (dx, dy) having the minimum SSD measure shown in (8) is assumed to be the displacement of the feature. The amount of processing depends greatly on the template size and the size of the search window. A large template will increase robustness, while a larger search window will handle larger displacements, provided frames of images can be processed in real time.

5.2. Teleoperated Embryo Injection

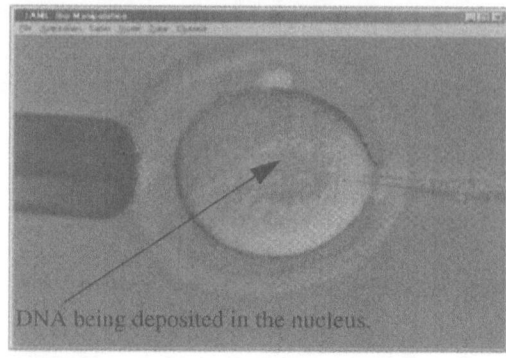

Figure 7. Teleoperated embryo injection.

In teleoperation mode, a supervisor guides a cursor on a monitor (using a computer mouse) which the visual servoing system accepts as a control input to a visual servoing control law. Figure 7 shows the program interface that performs teleoperation injection.

The main teleoperation injection process flow is described as follows:

Step1: Focus on the nucleus of the embryo, the tips of the holding pipette and the injection pipette.

Step2: Control the vacuum unit to hold the embryo.

Step3: Guide the injection pipette to the edge of the embryo, well aligned with the larger nucleus of the embryo.

Step4: Control the injection pipette to move into the nucleus of the embryo.

Step5: Deposit DNA inside the nucleus of the embryo.

Step6: Move the injection pipette out of the embryo. This completes the teleoperation process.

6. Injection with Cell Holders

The cell holder shown in Figure 8 is a device fabricated in the Microtechnology Laboratory at University of Minnesota for cell manipulation. The wells on the device are of different dimensions, which are 100μm, 80μm, and so on, down to 30μm.

The fabrication process of the cell holder is described as follows.

Step1: PECVD. Deposit SiO_2.

Step2: Deep trench etching. Etch through the silicon wafer.

Step3: Anodic bonding. Bond the

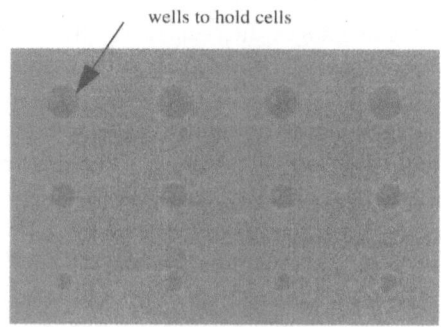

Figure 8. Microfabricated cell holder.

wafer with a pyrex glass substrate.

Step4: Polishing. Remove 400μm from the silicon wafer. The remaining thickness of the silicon wafer shown as the top layer in Figure 9 is 100μm, which is appropriate for holding both media and an embryo.

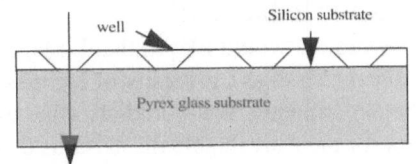

Figure 9. Schematic of the cell holder.

The vertical arrow in Figure 9 represents the transparent property of the active part of the device.

There are two main purposes in designing and fabricating the cell holder:

1. When well calibrated, the system with the cell holder makes it possible to inject large numbers of cells using only position control. For cell injection, the operation can be conducted in a move-inject-move manner.

2. This cell holder is not only used for embryo injection. It can also be used in other biological applications, such as loading embryos for producing embryonic stem (ES) cell derived fetuses [7].

7. Experiment Results

Eight mouse embryos were collected for embryo pronuclei DNA injection. Three of the eight embryos were discarded due to abnormalities in the nuclei. The other five embryos were injected by the autonomous system and were then placed in an incubator for 45 minutes. The injected embryos were visually inspected for viability. All five injected embryos proved to be viable and were transferred into a foster female mouse to reproduce transgenic mice. In nineteen days, transgenic mice were reproduced. Experimental results demonstrate that the success rate for automatic injection is 100%.

8. Conclusions

An autonomous embryo pronuclei DNA injection system was developed. It is not only capable of conducting pronuclei DNA injection, but it is also suitable for performing intracytoplasmic injection. The task of embryo pronuclei DNA injection is more demanding than that of intracytoplasmic injection into embryos. In embryo pronuclei DNA injection, DNA must be deposited in one of the two nuclei, while for intracytoplasmic injection foreign genetic material such as sperm only needs to be deposited within the embryo. Experimental results show that our success rate for embryo pronuclei DNA injection is 100%. With the aid of the microfabricated cell holder, the ability to inject large numbers of cells becomes possible. The complete system including the microfabricated cell holder demonstrate that microrobotics and microfabrication technology can play important roles in automating and facilitating biomanipulation.

Acknowledgements

The authors would like to thank Ms. Sandra Horn and Dr. David Largaespada of the Cancer Center at University of Minnesota for assistance in embryo preparation and in the experiments. We would also like to thank Prof. John Bischof and Prof. Ken Roberts for their invaluable discussions concerning this research.

References

[1]   F. Arai, K. Morishima, T. Kasugai, T. Fukuda, "Bio-Micromanipulation (new direction for operation improvement)," Proceedings of the 1997 IEEE/RSJ International Conference on Intelligent Robot and Systems, IROS 1997, New York, 1300-1305.

[2]   W.M. Arnold, U. Zimmermann, "Electro-Rotation: Development of a Technique for Dielectric Measurements on Individual Cells and Particles," Journal of Electrostatics, Vol. 21, No. 2-3, pp. 151-191, 1988.

[3]   A. Ashkin, "Acceleration and Trapping of Particles by Radiation Pressure," Physical Review Letters, Vol. 24, No. 4, pp. 156-159, 1970.

[4]   T.N. Bruican, M.J. Smyth, H.A. Crissman, G.C. Salzman, C.C. Stewart, J.C. Martin, "Automated Single-Cell Manipulation and Sorting by Light Trapping," Applied Optics, Vol. 26, No. 24, pp. 5311-5316, 1987.

[5]   J. Conia, B.S. Edwards, S. Voelkel, "The Micro-robotic Laboratory: Optical Trapping and Scissing for the Biologist," Journal of Clinical Laboratory Analysis, Vol. 11, No. 1, pp. 28-38, 1997.

[6]   B. Hogan, R. Beddington, F. Costantini, E. Lacey, *Manipulating the Mouse Embryo: A Laboratory Manual*,   Second Edition, Cold Spring Harbor Laboratory Press, 1994.

[7]   A.L. Joyner, *Gene Targeting - A Practical Approach*, Oxford University Press, Oxford, UK. 1993.

[8]   Y. Kimura, R. Yanagimachi, "Intracytoplasmic Sperm Injection in the Mouse," Biology of Reproduction, Vol. 52, No. 4, pp. 709-720, 1995.

[9]   M. Mischel, A. Voss, H.A. Pohl, "Cellular Spin Resonance in Rotating Electric Fields," Journal of Biological Physics, Vol. 10, No. 4, pp. 223-226, 1982.

[10] N.P. Papanikolopoulos, B.J. Nelson, P.K. Khosla, "Full 3-D tracking using the controlled active vision paradigm," IEEE Int. Symp. Intell. Contr., ISIC-92, 1992, pp. 267-274.

[11] N.P. Papanikolopoulos, "Selection of Features and Evaluation of Visual Measurements During Robotic Visual Servoing Tasks," Journal of Intelligent & Robotic Systems: Theory and Applications, Vol. 13, pp 279-304, 1995.

[12] M. Nishioka, S. Katsura, K. Hirano, A. Mizuno, "Evaluation of Cell Characteristics by Step-Wise Orientational Rotation Using Optoelectrostatic Micromanipulation," IEEE Transactions on Industry Applications, Vol. 33, No. 5, pp. 1381-1388, 1997.

[13] B. Vikramaditya, B.J. Nelson, "Visually Guided Microassembly using optical microscopes and active vision techniques," IEEE Int. Conference on Robot and Automation, Albuquerque, NM, Apr. 21-27, 1997, pp. 3172-3177.

[14] M. Washizu, Y. Kurahashi, H. Iochi, O. Kurosawa, S. Aizawa, S. Kudo, Y. Magariyama, H. Hotani, "Dielectrophoretic Measurement of Bacterial Motor Characteristics," IEEE Transactions on Industry Applications, Vol. 29, No. 2, pp.286-294, 1993.

[15] W.H. Wright, G.J. Sonek, Y. Tadir, M.W. Berns, "Laser trapping in cell biology," IEEE Journal of Quantum Electronics, Vol. 26, No. 12, pp. 2148-2157, 1990.

An active tubular polyarticulated micro-system for flexible endoscope

J. Szewczyk*
V. de Sars*
Ph. Bidaud*
G. Dumont**
Laboratoire de Robotique de Paris*
10-12, av. de l'Europe, 78140 Vélizy-Villacoublay
FRANCE
Email: philippe@robot.uvsq.fr
ENS de Cachan - Antenne de Bretagne**
Campus de Ker Lann, 35170 Bruz
FRANCE

Abstract: This paper describes an original active steering device for endoscopes and boroscopes. Its mechanical structure is based on a tubular hyper-redundant mechanism. Distributed SMA actuators with their own local controller are integrated in this structure for producing bending forces in reaction to the interaction detected between the instrument and its environment.

The SMA actuators are two thin NiTi springs in an antagonist configuration. Joint actuation relies on martensite/austenite phase transformation in NiTi alloys. The global behavior of the endoscope is controlled through a multi-agent approach.

1. Introduction

Current instruments for endoscopy suffer from limitations mainly caused by the lack of mobility and ability to perform maneuvers into very small and geometrically complex 3D spaces.

An endoscope is a long thin tubular device for non-invasive inspection in interior cavities, canals, vessels, etc... inserted through a natural or surgically produced orifice. A typical outer diameter of endoscopes is 10 mm and their length varies from 70 to 180 mm. The endoscope body contains several light guides (typically 2), tool channels (biopsy grippers, snare, cytology brush) and optics or electronics for the image transmission. Endoscopes can be rigid or flexible. The latter use optical fiber bundles for image transmission to the headset. The user can view the image transmitted by the instrument directly

D. Rus and S. Singh (Eds.): Experimental Robotics VII, LNCIS 271, pp. 179–188, 2001.

through an eyepiece or, when a camera is connected to the headset, the image displayed on a monitor. This arrangement is called indirect video endoscopy. When a CCD chip is integrated in the distal part, the image is electronically transmitted, it is direct (or distal) video endoscopy.

A steerable tip can be mounted on most of these instruments. The change of the tip orientation facilitates the endoscope progression in cavities and also modifies the viewing direction. This passively bendable part is generally deflected by one or two pairs of cables (depending upon the number of planes of bending) connected to a remote control mechanism located close to the headset.

Similar non-medical devices (boroscopes) are used for internal visual inspection of highly integrated mechanical systems such as jet engines or satellites. The boroscope must be insertable into narrow cooked passageways, the associated progression difficulties paralleling those of endoscopes.

These devices, while highly flexible, have limited steering ability. They cannot traverse tight bends nor negotiate complexe interior structures. As a consequence, for instance about 60% of the gastro-intestinal track is unreachable with current endoscope technology. Moreover, one of the major risk with these instruments is the perforation of the patient's tissues due to their substantial stiffness. There are also problems which results from excessive stresses applied on the operating cables, they frequently break or acquire a permanent strain.

In addition, the trend is to move towards smaller and smaller diameter endoscopes required by applications such as neuro-surgery, cardio-vascular-surgery or obstetrical procedures and this can not be tackled by a scale reduction of the current technologies.

This paper describes the design of a mechanical system and a distributed actuation system with a self-guiding control strategy for a scalable steering device able to dexterously maneuver through small and geometrically complex 3D structures.

Very few devices of this kind can be found in the scientific literature [1, 2, 3] but some are described in patents [4, 5]. They relate generally on shape memory alloy (SMA) distributed films integrated control drivers deposited on a flexible substrate and integrated using VLSI techniques.

2. Design principles and system description

The device has been designed to give the user more dexterity in endoscopic procedures than with current instruments. It means to provide an endoscope system which can be easily inserted deep into the body cavity to be inspected while minimizing the risk for perforation of organ walls or damaging the instrument (cable stretching or breaking).

Here, technology is a key issue. It is clear that cables is no longer a solution for getting tight bends in 3D space. The outer diameter has to be as small as possible considering that the room needed for optical fibers bundles and chanels for surgical tools and fluids, which defines the inner diameter, can not be reduced. Another important feature is that the bending force required for a given stiffness of the inner components increases when reducing the endoscope diameter. Moreover, the technology selection must withstand the sterilization process (140°C during 20 minutes). It has to make the system as simple as possible and to facilitate its manufacturing at small scale.

The controllably bendable portion of the instrument must be able to adapt

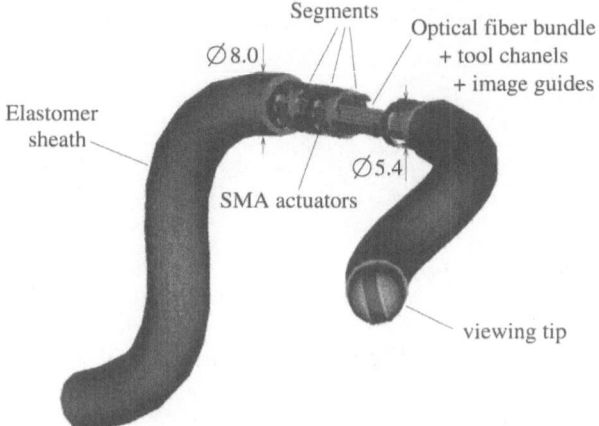

Figure 1. Schematic view of the steering device

its local curvature to the interior geometry by a spontaneous reaction to the interactions with the environment while the viewing tip follows a track in a vessel or in a cavity.

Figure 1 schematically illustrated the endoscopic system we designed. The mechanical structure of the device can be viewed as an hyper-redundant manipulator which embrace the endoscope components (optic bundle, light guides, tool chanels). It is a serial arrangement of tubular segments articulated to each other by pin joints. This design is modular, the number of segments can be adjusted to the application and is in theory infinite. On the actual design, the segment length is 4 mm, the inner diameter 5.4 mm and the outer diameter (including the outer elastomer cover) is 8 mm. The distal viewing tip integrates a variable field optical system comprising an image guide, objective and rotatable prism actuated by a polymer gel actuator. This enable to obverse continuously over a wide range inside narrow body cavities or tubes without removing the endoscope for replacing the distal end type.

This system is usually protected by a metallic sheath in industrial endoscopes and/or by a flexible polymer sheath in medical endoscopes. Notice that this sheath significantly increases the strength requirements for the bending actuators.

3. A tubular hyper-redundant manipulator

The hyper-redundant manipulator is composed of identical modules (iron rings) linked together by pin joints whose axes are alternatively oriented at 90° in the same plane. This mechanical design allows to bend the endoscope body along complex curves in the 3-D space.

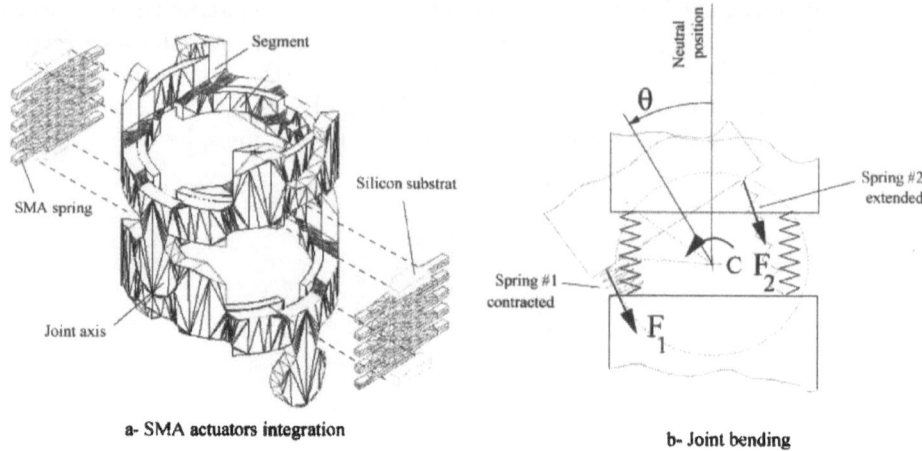

a- SMA actuators integration   b- Joint bending

Figure 2. Joint description

The modules are obtained by an electro-erosive processing technique. By using this substractive manufacturing method, joints and links are made in one piece. The pin and the hinge are respectively the positive and the negative cutting in the cylindrical shell of the segment. The relative translation of two consecutive modules along the joint axis is suppressed by inserting a very thin internal ring has shown on figure 2-a. The steering mechanism is assembled by simply plugging these segments whose length can be reduce to 4 mm such that a 15 mm curvature radius can be achieved.

Two spring-like actuators are integrated with their own control circuit in each module to change the relative orientation of two consecutive segments. These actuator are Shape Memory Alloy (SMA) springs mounted in an antagonist configuration.

The actuator is controlled by the electrical power supplied to the SMA. A control circuit is associated to each actuator. It is integrated on a substrate of alumina by using hybrid electronic technologies.

4. SMA actuator design

The SMA actuator elements are springs cut out by photochemical etching process in a NiTi (Nickel-Titanium) ribbon (Figure 3-a).

Basically, these SMA actuators undergo a micro-structural transformation from their austenite phase to their martensite phase [6]. This phase transformation can be activated by heating and cooling the material or by applying an external stress.

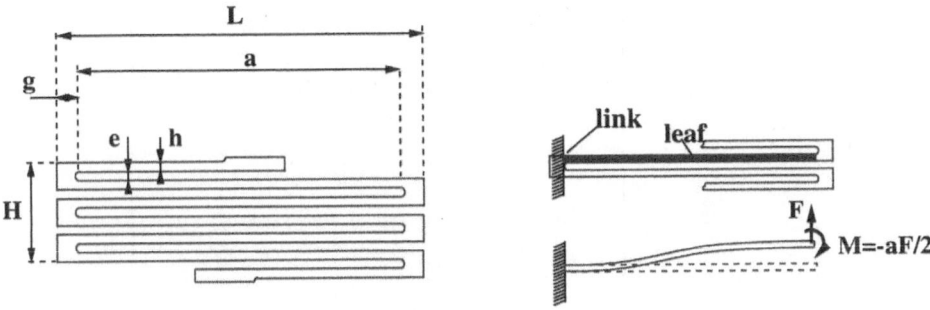

Figure 3. SMA spring description

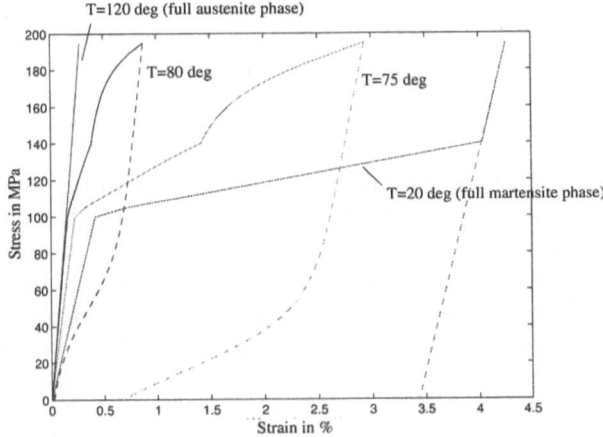

Figure 4. Strain-stress diagram of the NiTi (50%-50%) alloy

For a 50%-50% NiTi alloy, as the one used here, the transition tempera-
tures are $A_s = 40°$ and $M_s = 70°$ for a null applied constraint to the material.

This phase transformation also induces a large modification in the material Young's modulus (Figure 4). This property is exploited here to create
unballanced pulling forces applied by the two antagonist SMA springs. This
results in an output joint torque (Figure 3-b). The mapping between the spring
pulling forces and the output torque is joint configuration dependent and high-
tly non-linear.

SMA actuators design has to take into account dimensional constraints
and deflexion resistance of the endoscope.

Resistance to deflexion, due to the endoscope body and the outer elastomer sheath, was experimentally evaluated. On Figure 5, we represented the
necessary output joint torque to bend a rotoïd axis in an existing endoscope.
For deflexions smaller than 15°, the resistive torque is about 0.01Nm.

The SMA spring on Figure 3-a can be approximated by an assembly of
flexible parts (leaves of length a) linked by rigid parts. For symetrical reasons,

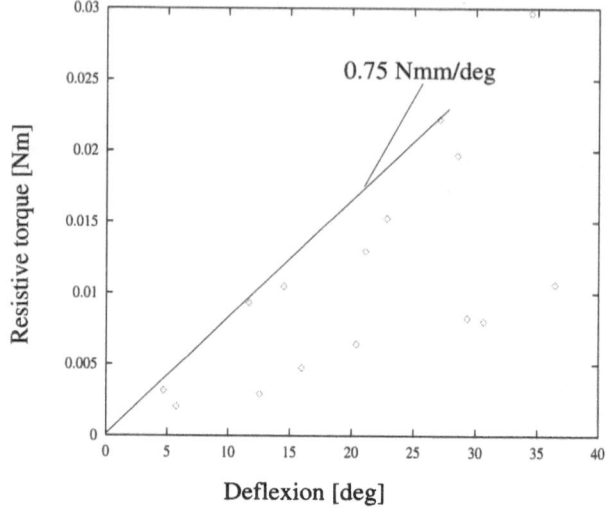

Figure 5. Resistive torque [Nm] vs deflexion [deg]

the orientation of each link remains constant while the spring is stretched.

Thus, a leaf can be modelized as a flexible beam rigidly fixed at one extremity and submitted to a combination of a force F and a moment $M(F)$ with $M(F) = -a\,F/2$ such that a null inclination remains at the other extremity.

The material Young's modulus E is the slope of the corresponding curve on Figure 4 considering that the stretched spring is at temperature $20°C$ (E varying from 1 to $20GPa$) while the contracted one is at temperature $120°C$ ($E = 70GPa$).

The desired maximal flexion is $15°$. In this configuration, for $g = 0.135mm$, $e = 0.25mm$ and a number of leaves set to 6, we represent on Figure 6 the normalized output joint torque (the reference value is $0.01Nm$) and the normalized maximal stress in the material (the reference value is $135MPa$).

The best trade-off corresponds to $l = 2.35mm$ and $H = 2.0mm$ (i.e. $h = 0.125mm$). The output joint torque in this configuration is greater than $0.008Nm$ and maximal stress is less than $145MPa$.

5. Joint control

The distal end of the system is remotly driven. The endoscope configuration is self-guided in such a way that the local interactions between the instrument and its environment are minimized.

Changes in configuration are controlled at joint level by switching between position and a temperature control loops. The resulting controller for the antagonist actuators is represented on Figure 7. Only one spring is actuated at once for producing a displacement in the desired direction. When the error is large, position feedback is used (see below case 1 or 2). A switching on the temperature loop occurs when the static error is less than ϵ (see below case 3 or 4). The temperature input is the one memorized just before switching. If $\tilde{\theta}$

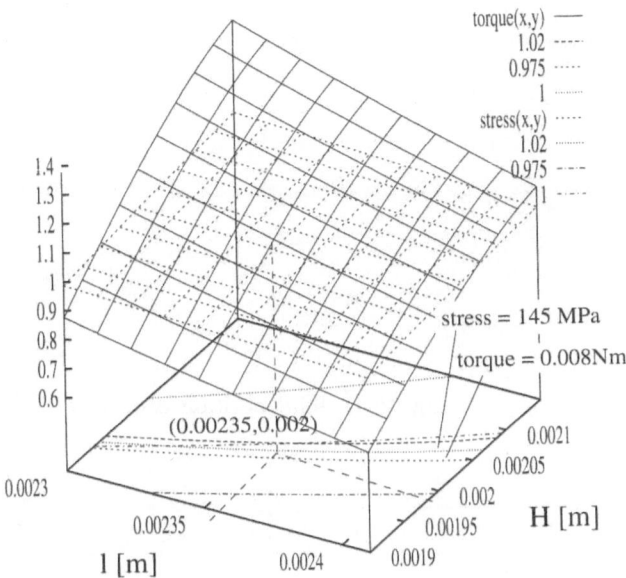

Figure 6. Normalized joint output torque and maximal internal stress for different SMA spring geometries

is the desired joint position and $\Delta\theta = \tilde{\theta} - \theta$ the joint position error, then the swiching rules are the following :

$$\text{Case 1}: \quad \Delta\theta \geq 0 \ \& \ |\Delta\theta| \geq \epsilon$$
$$\text{Case 2}: \quad \Delta\theta \leq 0 \ \& \ |\Delta\theta| \geq \epsilon$$
$$\text{Case 3}: \quad \tilde{\theta} \geq 0 \ \& \ |\Delta\theta| \leq \epsilon$$
$$\text{Case 4}: \quad \tilde{\theta} \leq 0 \ \& \ |\Delta\theta| \leq \epsilon$$

Figure 8 shows experimental results obtained in a position step response of a SMA actuator using this kind of switching controller. For the implementation of the local controller, we have developed a specialized micro-system based on hybrid electronic technologies. The temperature and position sensors which are electrical resistances, as well as the electrical connections are obtained by ink serigraphy on an alumina substrate. The power for SMA actuators and its electronics are transmitted by a two wires bus. The bending information is transmitted by modulation on the same bus.

6. Configuration behavior

Controlling the endoscope configuration aims at positionning and orienting correctly the tip of the structure while limiting forces coming from interactions with the environment.

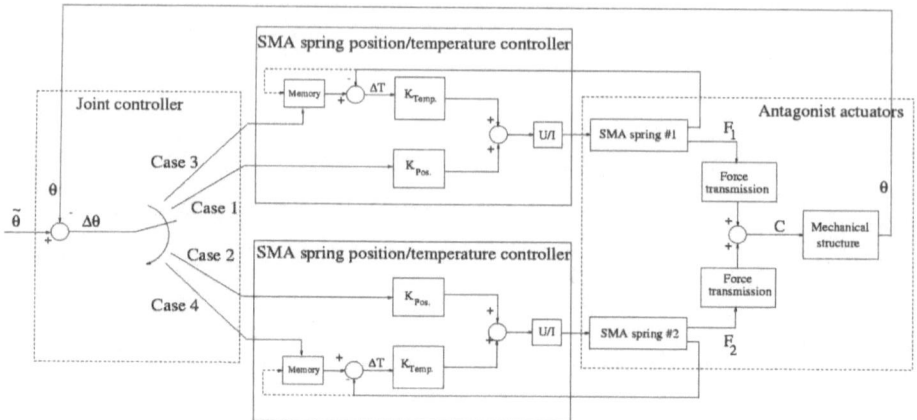

Figure 7. Joint position control

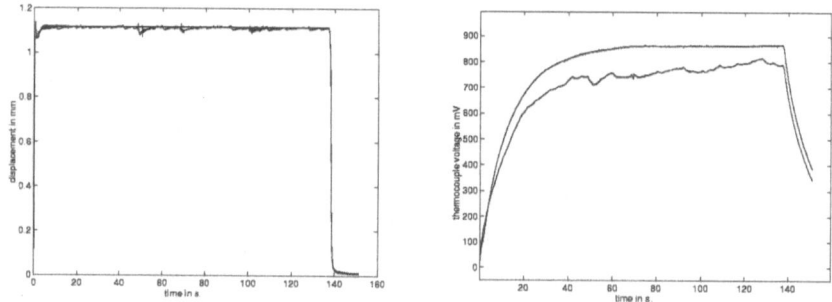

Figure 8. Step response with the position feedback controller (filled line) and the controllers combination (dotted line) (on the left-hand side) and the thermocouple output signal (on the right-hand side)

An algebraic resolution for this problem is extremely complex and the explored environment is a-priori unknown so a reactive resolution method is preferable.

The solution we propose relies on the virtual split of the steering mechanism into independent sub-systems and by considering them as agents [7]. Each agent is able to detect a contact with the environment and to accordantly modify the endoscope local configuration.

This is a very simple and modular solution indepedent from the length of the structure. Moreover, it is a strictly distributed approach minimizing the quantity of informations exchanged between the agents.

Three different kinds of local behaviors are described on Figure 9. They correspond to sub-systems composed of 1, 2 or 3 segments. The first behavior is very simple but it significantly disturbs the global configuration of the endoscope. The last one requiers information exchanges over three consecutive segments but preserve the global configuration.

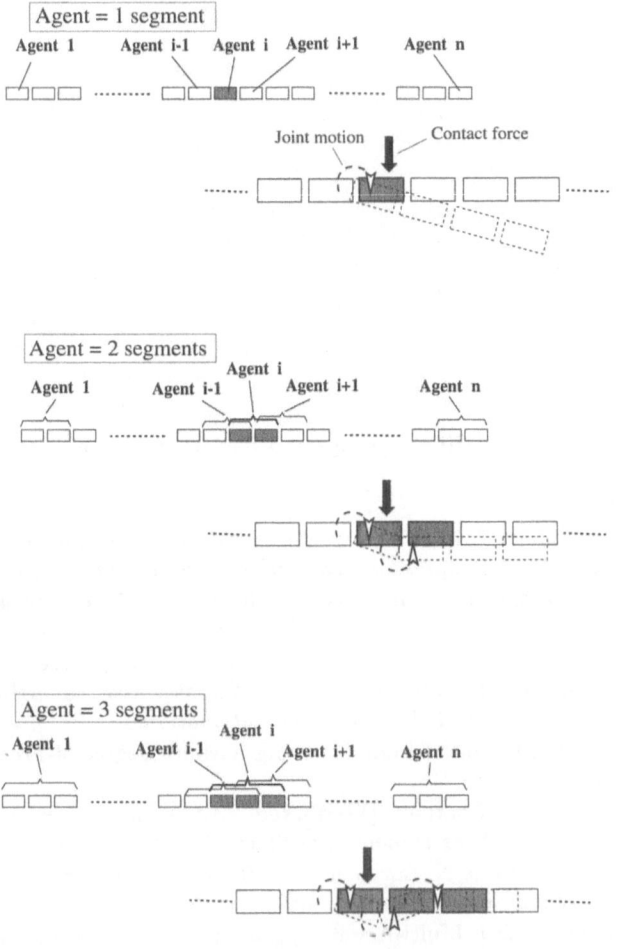

Figure 9. Three different kind of local behavior

The global behavior of a planar structure (20 segments) progressing into a pipe and guided by the local reaction of the agents only, has been tested in simulation. As shown on figure 10-b, ultra-local approch (agent $\equiv$ 1 segment), leads to instability. The second and third solutions (agent $\equiv$ 2 or 3 segments) keep the endoscope stable while minimizing interactions (Figures 10-c and 10-d).

7. Conclusion

This paper proposed a new design concept for actively guided steerable endoscope. At this point, the mechanical structure and the associated actuators have been manufactured.

The local controller has to be experimentaly improved and the whole integration has to be done for testing the proposed behavior control strategies.

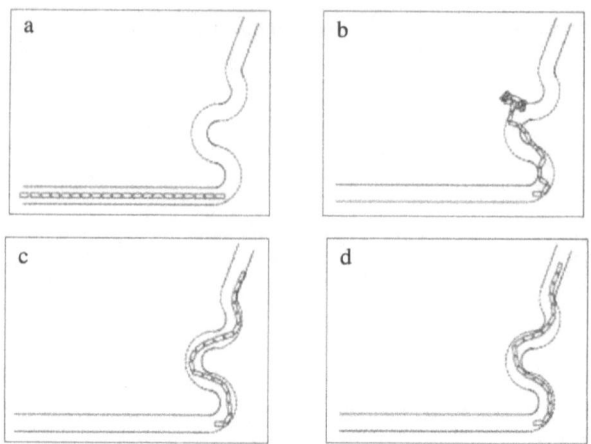

Figure 10. Multi-agent behavior simulation

References

[1] Müglitz J., Schönherr J. 1999 Miniaturized Mechanisms - Joint design, Modeling, Example *Tenth World Congress on Theory of Machines and Mechanisms.*

[2] Yusa A. 1998 Tubular manipulator with multi-degrees of freedom *Micro Machine Center Journal* 19.

[3] Dario P, Paggetti C, Troisfontaine N et al. 1997 A miniature steerable end-effector for application in an integrated system for computer-assisted arthroscopy *IEEE Internaltional Conference on Robotics and Automation.*

[4] Takayama S 1997 Method of Manufacturing a Multi-Degree-of-Freedom Manipulator, US Patent No 5,679,216.

[5] Maynard R et al. 1995 Spatially Distributed SMA Actuator Film providing unrestricted Movement in Three Dimensional Space, US Patent No 5,405,337.

[6] Bidaud Ph, Troisfontaine N, Larnicol M 1999 Optimal Design of Micro-actuators based on SMA Wires *Smart Materials and Structures* 8.

[7] Duhaut D 1993 Using a Multi-Agent Approach to Solve the Inverse Kinematics *Intelligent Robot and System Conference IROS.*

Towards semi-autonomy in laparoscopic surgery through vision and force feedback control

Alexandre Krupa*, Christophe Doignon*, Jacques Gangloff*
Michel de Mathelin*, Luc Soler[†] and Guillaume Morel[‡]

*University of Strasbourg I, [†]IRCAD, [‡]EDF R&D, France

Abstract: This paper shows ongoing research results on the development of automatic control modes for robotized laparoscopic surgery. We show how both force feedback and visual feedback can be used in an hybrid control scheme to autonomously perform basic surgical subtasks. Preliminary experimental results on an example clamping tasks are given.

Introduction

In laparoscopic surgery, small incision points are made in the human abdomen contrary to open surgery where a large incision is made. The surgeon puts a trocar at these incision points. Then, surgical instruments and an endoscopic camera are inserted through trocars. Looking at the video signal, the surgeon can move the tools in order to perform the desired surgical task. The advantages of laparoscopic surgery are obvious: reduced pain and hospital stay, as well as a quicker recovery. The main inconvenience of this surgical technique is due to the stand of the surgeon which is very tiring and limits the duration of the surgical procedure as well as the surgeon's performance.

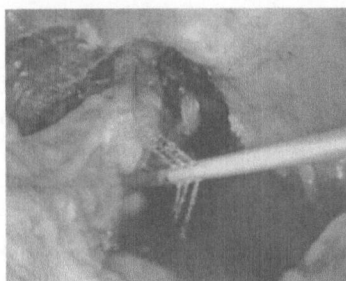 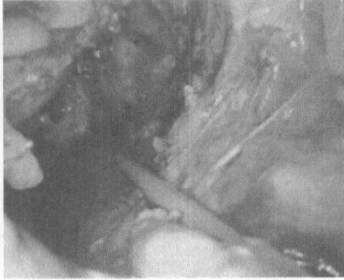

Figure 1. Example task : cleaning-suction process in laparoscopic surgery

Laparoscopic surgical robots have appeared recently. Several commercial systems for laparoscopic surgery exist today, e.g., ZEUS (Computer Motion, Inc.) or DaVinci (Intuitive Surgical, Inc.). With these systems, robot arms are used to manipulate the instruments and the camera. The surgeon teleoperates the robots through master arms using the visual feedback from the laparoscope. This reduces the surgeon tiredness, and potentially increases accuracy by the use of a high master/slave motion ratio.

Our research in this field is aimed at expanding the potentialities of such systems by providing "automatic modes" in which the system autonomously performs simple subtasks. In this case, the robot controller uses the visual feedback from the laparoscope to automatically drive instruments, through a visual servo

D. Rus and S. Singh (Eds.): Experimental Robotics VII, LNCIS 271, pp. 189–198, 2001.
© Springer-Verlag Berlin Heidelberg 2001

loop, towards their desired location. Pioneer research in this field can be found in [1] or [2] for which the laparoscope is planned to automatically track a surgical instrument in its field of view, or in [3] where a 3 dof surgical robot is automatically placed with vision feedback.

In cooperation with IRCAD (Institut de Recherche sur le Cancer de l'Appareil Digestif, Strasbourg, France), we particularly focus on liver surgery. This surgery involves a number of repetitive gesture, such as the cleaning-suction process (Fig. 1): first, the surgeon has to sweep a liquid projecting instrument over the surface to be cleaned up ; then, the same instrument (with the pump reversed) is used to suck up the remaining liquid. It shall be noticed that although this cleaning process is rather simple as compared to critical surgical gestures, it involves repetitive movements for the surgeon who drives master devices. Providing semi-autonomy to the robot by the use of vision based control will then relieve the surgeon of tiredness induced by this kind of simple and repetitive tasks and allow a maximized concentration for the delicate phases of the surgical operation. In this case, the surface to be cleaned up on the screen is bounded by the surgeon and the robot autonomously performs the cleaning-suction process.

This paper first shows how force feedback control can be used in order to limit the forces applied on the patient through the trocar. Then, an application combining vision and force control to automatic clamping is given, with preliminary experimental results.

1. The use of force control in laparoscopic manipulation

A particularity of laparoscopic manipulation lies in the presence of a trocar, which limits the surgical instruments motion to 4 dof. This problem has been addressed over the past by the mean of mechanical design [4], [5], [6]. The proposed solutions use either a remote rotation center device, which requires a precise positioning of the robot in the trocar prior to the surgical operation, or a 6 DOF robot with two passive joints in the wrist. This last solution suffers from a lack of precision due to backlash that may occur between the instrument and the trocar. Our proposition is to use a fully actuated 6 DOF robot, that provides both functional flexibility and accuracy. In order to avoid large forces to be apply on the patient body through the trocar, the robot tip is equipped with a force sensor.

1.1. Kinematics

The kinematics of the laparoscopic manipulation is depicted in figure 2, where F_c is a frame attached to the tip of the tool handler, such that the z_c axis is colinear to the tool penetration axis ; F_s is the F/T sensor frame, with z_s colinear to z_c ; P is the point of the tool handler that instantaneously coincides with the trocar ; l is the distance between the origins of F_s and F_c and d is the distance from P to the origin of F_c.

As only 4 dof are available, we choose to use the following parametrization for the operational space velocity vector:

$$\dot{W} = \begin{pmatrix} \dot{d} & \omega_x & \omega_y & \omega_z \end{pmatrix}^T \tag{1}$$

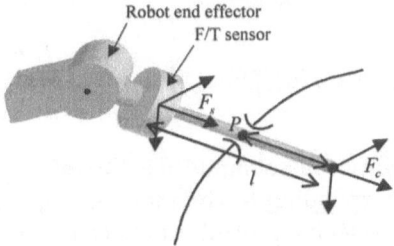

Figure 2. Manipulation through a trocar

where $\dot{d}$ is the translational velocity of the tool handler along z_c, and ω_x, ω_y and ω_z are projections of the absolute rotational velocity over x_c, y_c and z_c respectively. Let $\dot{r}_p$ be a vector, expressed in F_c, of the twist components describing the absolute velocity of the tool handler at point P. Since the trocar does not allow any tangential motion of the tool handler at point P, we have:

$$\dot{r}_p = \left(\begin{array}{c} v(P)_{c/0} \\ \omega_{c/0} \end{array} \right) = \left(\begin{array}{c} 0_{2\times4} \\ I_{4\times4} \end{array} \right) \dot{W} = A\dot{W} \qquad (2)$$

Then, we can express the same twist at the origin O_c of F_c:

$$\dot{r}_c = \left(\begin{array}{c} v(O_c)_{c/0} \\ \omega_{c/0} \end{array} \right) = M(d)\dot{r}_p \quad \text{with:} \quad M(d) = \left(\begin{array}{cc} I_3 & \left(\begin{array}{ccc} 0 & d & 0 \\ -d & 0 & 0 \\ 0 & 0 & 0 \end{array} \right) \\ 0_{3\times3} & I_3 \end{array} \right)$$
$$(3)$$

where both $\dot{r}_p$ and $\dot{r}_c$ are expressed in F_c. Furthermore, as F_c is rigidly attached to the robot end-effector, standard kinematics can be used to provide the robot *natural* jacobian matrix J_N, such that $\dot{r}_c = J_N \dot{q}$, where $\dot{q}$ is the robot joint velocity. Finally, the inverse kinematic model is :

$$\dot{q} = J_N^{-1}(q)M(d)A\dot{W} \qquad (4)$$

where the robot kinematics is supposed to be nonsingular.

1.2. Force feedback

If the penetration depth d was perfectly known, equation (4) could be used to drive the robot joint velocities while respecting the trocar constraint. However, in practice, due to the experimental conditions of a surgical operation, this assumption is not realistic, $M(d)$ is not perfectly known. As a consequence, using equation (4) to control the robot motion will induce a lateral motion of P and forces could be applied on the patient through the trocar. To cope with this problem, a force feedback controller is added to the system in order to limit the lateral forces applied at the incision point.

Let f_x and f_y be the measured forces along x_s and y_s, respectively. Hybrid position/force control can be used to servo these measures to zero. Assuming that robot joints are velocity controlled, we get a conventional motion rate control :

$$\dot{q}^* = J_N^{-1}(q)M(\hat{d})\dot{r}_p^* = J_N^{-1}(q)M(\hat{d}) \begin{pmatrix} k\ f_x \\ k\ f_y \\ \dot{W}^* \end{pmatrix} \qquad (5)$$

where $\dot{q}^*$ is the joint velocity control input, $\hat{d}$ is the estimation of the penetration depth d, k is a gain corresponding to the planned apparent damping, and $\dot{W}^*$ is the operational space velocity control input, that can be provided either by the surgeon through master devices (teleoperation mode) or by a vision based controller (automatic mode). This control strategy is known to be very robust providing that the force loop bandwidth is low enough as compared to the joint velocity loop bandwidth.

Actually, the first experiments with small depth estimation errors exhibited good results in lateral force limitations. However, for larger estimation errors, the force loop was not fast enough to efficiently compensate for lateral motions at the trocar, and large forces occurred (see experimental results).

In order to increase the overall controller performance, it was then necessary to provide online estimation of the penetration depth d. Two drastically different strategies were experimented:

- In the first one, we use the measured forces f_x, f_y and torques T_x, T_y in the xy plane to estimate the distance. Algorithm is based on a robust least square identification involving a forgetting factor, a sliding window and a threshold. If we define m as the distance between the force/torque sensor's center and the incision point, i.e., $m = l - d$, then:

$$m = \frac{\sqrt{T_x^2 + T_y^2}}{\sqrt{f_x^2 + f_y^2}} = \frac{T_r}{f_r} \qquad (6)$$

In order to provide a robust identification the following cost function was used with sliding window and forgetting factor:

$$J(t, t_0) = \int_{max(t-T, t_0)}^{t} e^{-\lambda(t-\tau)} (T_r(\tau) - f_r(\tau)\widehat{m}(t))^2 d\tau \qquad (7)$$

where $\lambda > 0$ is a forgetting factor and $T > 0$ is the size of the sliding window. The least-squares estimate $\widehat{m}(t)$ that minimizes $J(t, t_0)$ is equal to:

$$\widehat{m}(t) = R(t, t_0)^{-1} Q(t, t_0) \qquad (8)$$

$$\text{with} : \begin{cases} R(t, t_0) = \int_{max(t-T, t_0)}^{t} e^{-\lambda(t-\tau)} f_r^2(\tau) d\tau \\ Q(t, t_0) = \int_{max(t-T, t_0)}^{t} e^{-\lambda(t-\tau)} f_r(\tau) T_r(\tau) d\tau \end{cases} \qquad (9)$$

However, equation (8) cannot be used directly to estimate m, particularily when the force signals are close to zero, bringing a high noise to signal ratio. Therefore, dead-zone is added (cf. [7]). If $f_r(t)$ or $T_r(t)$ decrease below some threshold value f_{th} or T_{th}, the computation of the least-squares estimate is frozen, i.e., only the reference tool velocity $\dot{d}^*$ along z_c is taken into account. Consequently, the robust estimation algorithm is defined as follows:

$$\hat{d}(t) = \begin{cases} l - R(t,t_k)^{-1}Q(t,t_k) & \text{if } f_r(t) > f_{th} \\ & \text{and } T_r(t) > T_{th} \\ & \text{and } t \geq t_k + T_0 \\ \hat{d}(T_k) + \int_{T_k}^{t} \dot{d}^*(\tau)d\tau & \text{otherwise} \end{cases} \tag{10}$$

where t_k is the last time instant when f_r and T_r left the dead-zone area, and T_k is the last time instant when f_r or T_r entered the dead-zone area.

- In the second one (see appendix), an adaptive approach is proposed to estimate $\hat{d}$. It uses as inputs both rotational velocities (ω_x^*, ω_y^*), and the measured tangential forces (f_x, f_y).

1.3. Experimental results

In order to evaluate the efficiency of the force feedback controller and depth estimation strategies, an experimental testbed was built, consisting of an 6 dof robot equipped with a force sensor, manipulating a rigid 40 cm bar, through a trocar. During these experiments, we apply square reference signals ω_x^* and ω_y^* with a magnitude of ± 1.5 deg/s. In the first set of experiments, the penetration depth $d = 0.1m$ and its initial estimate is $d_0 = 0.3m$. The corresponding results are shown in the first plot column of Fig. 3. In the second set of experiments, the penetration depth $d = 0.2m$ and its initial estimate is $d_0 = 0.02m$. The corresponding results are shown in the second plot column of Fig. 3. Three strategies are compared: The first one considers a constant estimation of d, that is $\hat{d} = d_0$. The results (first line of Fig. 3) show significant forces (about ± 5 N) in both xy directions. For faster motions, these forces would clearly become unacceptable for the patient. The second strategy uses the direct estimation of d with forces and torques measurement. One can see (second line of Fig. 3) that forces are limited to $\pm 2N$ and that estimation of d quickly converges towards the actual value. A small fluctuation of $\hat{d}$ remains, which does not significantly affect the force controller performance. Finally, the third strategy (last line, Fig. 3) uses the adaptive algorithm. This method exhibits a rather slower convergence of the depth estimation as compared to direct estimation of d. However, the estimated signal $\hat{d}$ is smoother. This is due to the dynamics of the gradient law that acts on $\hat{d}$, whereas the least-squares algorithm gives instantaneously the estimation of $\hat{d}$. Once the estimation convergence is obtained, the force control performance is similar for the two on-line estimation approaches.

2. Vision based control of the instrument movements

In "conventional" laparoscopic telemanipulation, the four degrees of freedom of the instrument are directly controlled by the surgeon through master devices,

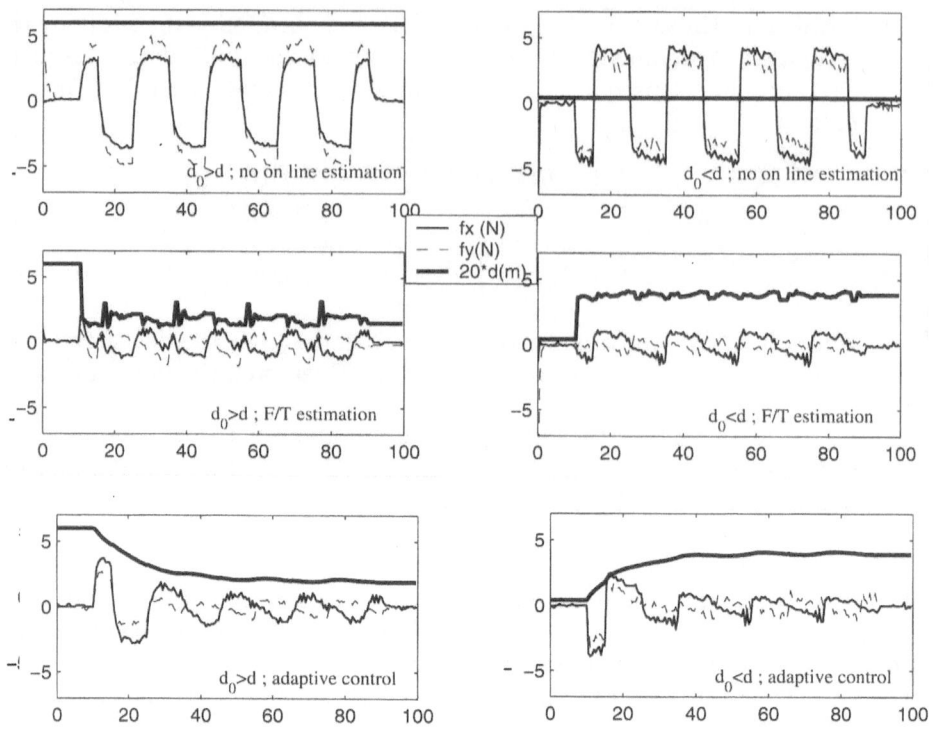

Figure 3. Force feedback experimental results *(x axis : time in seconds)*

using the video feedback of the laparoscope. Rather, we want to provide autonomy to the system, that is to use the video feedback in a control algorithm in order to achieve surgical tasks.

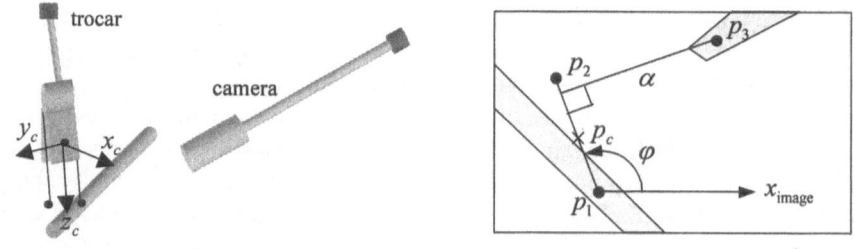

Figure 4. Vessel clamping task

2.1. The use of optical markers

Of course, a major difficulty lies in the poor structuration of the observed scene. To cope with this problem, structured lightening can be added by the mean of laser pointers attached to the tool and/or the camera. A first attempt was shown in [3] where a laser spot was used for depth estimation in a surgical robot. In this work, only one degree of freedom of the surgical instrument was driven with this optical marker. Here, we propose here to generalize this approach.

We consider a clamping task, depicted in figure 4. In this experiment, we use two laser pointers mounted on a surgical tool. The laser beams are parallel, colinear to z_c, in the (z_c, y_c) plane. We also add an optical marker on the tip of the tool. This marker lies in the x_c axis. Note that the camera is fixed during this experiment, although in the future, we intend to mount it on a second robot, in the ZEUS system. An interesting property of this setup is that robust extraction of the optical marker images can be obtained by comparing an image with and without the markers on.

From the image coordinates of three spots (p_1, p_2, p_3), we built the following image feature vector (see figure 4):

$$s = (u_c, \ v_c, \ \alpha, \ \varphi)^T \tag{11}$$

where (u_c, v_c) are the coordinates of $p_c = \frac{1}{2}(p_1 + p_2)$, φ is angle from the x image axis to the vector joining p_1 to p_2, and α is the distance from p_3 to the line $p_1 p_2$. The desired value s^* of s is supposed to be known. In the final scenario, the surgeon will indicate on a tactile screen the location of both the vessel to be clamped and the point where the clamping should be done. This will automatically set u_c^*, v_c^* and φ^*. Also, the clamping position is characterized by a known value of α^*.

2.2. Visual servoing

Due to the complexity of the scene, there is no way to precisely model the exact image jacobian matrix, that maps the instrument velocity into $\dot{s}$. Rather, we built a simplified jacobian matrix J_i for a nominal configuration, for which the lightened scene is planar and z_c is perpendicular to this plane. The following properties can be demonstrated: $i)$ the velocity $\dot{d}$ only affects $\dot{\alpha}$; $ii)$ the velocity ω_z does not affect $\dot{u}_c$ and $\dot{v}_c$; $iii)$ the velocities ω_x and ω_y do not affect $\dot{\varphi}$; Thus the image jacobian matrix J_i is given by:

$$\dot{s} = J_i \dot{W} \text{ with, } J_i = \begin{pmatrix} 0 & J_{i12} & J_{i13} & 0 \\ 0 & J_{i22} & J_{i23} & 0 \\ J_{i31} & J_{i32} & J_{i33} & J_{i34} \\ 0 & 0 & 0 & J_{i44} \end{pmatrix} \tag{12}$$

It is intersting to notice that we have a quasi-triangular system, apart from the bloc mapping (ω_x, ω_y) into $(\dot{u}_c, \dot{v}_c)$.

As a number of unknown geometrical and optical parameters is involved in the computation of J_i, a first identification stage can be run at the beginning of any experiment. Constant velocities ω_x, ω_y, ω_z and $\dot{d}$ are applied independently during a short time T. The variations of s are computed and the jacobian matrix components are estimated by:

$$\begin{cases} \widehat{J}_{i12} = \frac{\Delta u_c}{\omega_x T} & \widehat{J}_{i13} = \frac{\Delta u_c}{\omega_y T} & \widehat{J}_{i22} = \frac{\Delta v_c}{\omega_x T} & \widehat{J}_{i23} = \frac{\Delta v_c}{\omega_y T} \\ \widehat{J}_{i31} = \frac{\Delta \alpha}{\dot{d} T} & \widehat{J}_{i32} = \frac{\Delta \alpha}{\omega_x T} & \widehat{J}_{i33} = \frac{\Delta \alpha}{\omega_y T} & \widehat{J}_{i34} = \frac{\Delta \alpha}{\omega_z T} & \widehat{J}_{i44} = \frac{\Delta \varphi}{\omega_z T} \end{cases} \tag{13}$$

The visual servoing then consists of the quasi decoupling control law:

$$\dot{W}^* = \lambda \widehat{J}_i^{-1} (s^* - s) \tag{14}$$

The remaining degrees of freedom of the robot are still controlled using force feedback, and the overall controller consists of an hybrid vision/force controller that combines equation (5) and (14), together with an on-line estimation of $\hat{d}$. The well known stability condition is that $J_i \widehat{J}_i^{-1}$ remains positive definite. Due to the complexity of the scene, the stability properties cannot be formally derived. Notice that a number of techniques have been proposed in the past using constant image jacobian matrices, exhibiting good experimental stability robustness properties (see e.g. [9]).

2.3. Planning

Vessel clamping, is decomposed in three stages (Fig. 5):

1. the angle φ is servoed to its desired value φ^*. Other components are not servoed. Then, the resulting motion is a pure rotation of the instrument around z_c axis.

2. the image coordinates u_c, v_c are servoed towards u_c^*, v_c^*, while φ is maintained at φ^*. This step involves mainly ω_x and ω_y motions.

3. the distance α is servoed to α^*, while other components of s are kept to their desired value. This phase mainly consist of a final z_c translation.

With such a decomposition, the image jacobian matrix components are not all identified at the beginning of the experiment. Rather, they are estimated progressively during the clamping, when they are required. $\widehat{J}_{i_{44}}$ is identified before the phase 1, $\widehat{J}_{i_{12}}$, $\widehat{J}_{i_{13}}$, $\widehat{J}_{i_{22}}$ and $\widehat{J}_{i_{23}}$, before the phase 2 and $\widehat{J}_{i_{3k}}$, $k = 1..4$, before phase 3. Note that in practice, only $J_{i_{31}}$ is actually identified, since, in the final approach configuration, $J_{i_{32}} \approx 0$, $J_{i_{33}} \approx 0$ and $J_{i_{34}} \approx 0$.

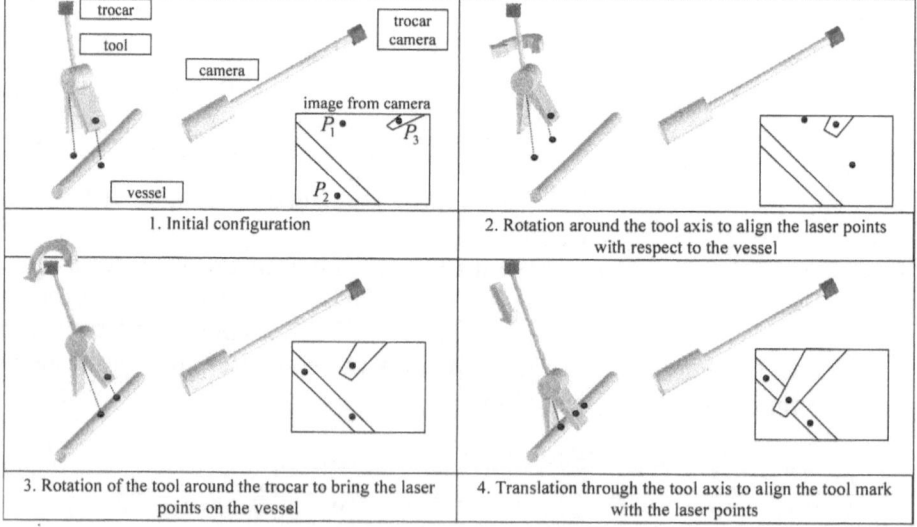

Figure 5. Planning a clamping task

2.4. Experimental results

This strategy was experimented on a lab testbed. A surgical instrument was equipped with to laser beams and an LED. In these preliminary experiments, the desired values of s are set manually to a value determined during a learning phase: $u^* = 350$ pixels, $v^* = 300$ pixels, $\varphi^* = -20$ deg. and $\alpha^* = 30$ pixels. A number of experiments with different initial configurations has been conducted. Experimental results are depicted in Figure (7), where the different phases are described. During the first 3 seconds, an open loop motion aimed at identifying $\frac{\dot{\varphi}}{\omega_z}$ is performed. Then the first phase is running, providing convergence of φ. Note that in presence of large initial errors, the controller output ω_z^* is saturated, which explains the linear convergence (during this first phase, the three other components of s are not controlled). At $t \approx 15$s, after the convergence of φ, a second identification phase is running to estimate the mapping between (ω_x, ω_y) and $(\dot{u}_c, \dot{v}_c)$. This produces a variation of the different components of s. The exponential convergence of u_c, v_c is then observed (from $t \approx 20$s to $t \approx 40$s, while φ is maintained to its desired final value. Then the final phase consists of the exponential decrease of α, that is the final clamping z_c translational motion.

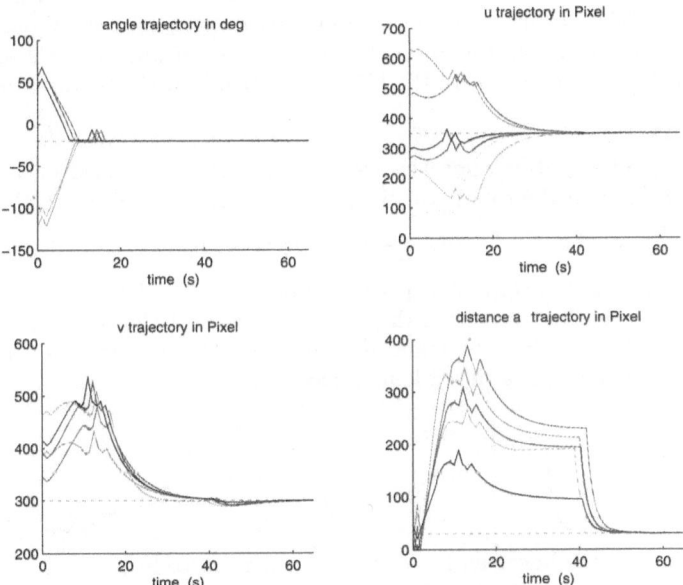

Figure 6. Visual servoing results.

References

[1] G.-Q. Wei, K. Arbter and G. Hirzinger. Real-Time Visual Servoing for Laparoscopic Surgery. *IEEE Engineering in Medicine and Biology*, 16(1), pp. 40-45, 1997.

[2] A. Casals, J. Amat, D. Prats and E. Laporte. Vision Guided Robotic System for Laparoscopic Surgery. *Proc. of the IFAC Int. Congress on Advanced Robotics*. Barcelona, Spain, 1995.

[3] M. Mitsuishi, S. Tomasaki, T. Yoshidome, H. Hashizume and K. Fujiyara Tele-micro-surgery system with intelligent user interface. *Proc. of the 2000 IEEE Int. Conf. on Robotics and Automation*. pp 1607-1614, San Francisco, CA, 2000.

[4] A. Faraz and S. Payandeh. A Robotic Case Study: Optimal Design for Laparo-scopic Positioning Stands. *Proc. of the 1997 IEEE Int. Conf. on Robotics and Automation* ,pp. 1553-1560. Albuquerque, New Mexico, April 1997.

[5] A. Madhani, G. Niemeyer and J.K. Salisbury. The Black Falcon: A Teleoperated Surgical Instrument for Minimally Invasive Surgery. *Proc. of the IEEE/RSJ Int. Conf. on Intelligent Robots and Systems*, Victoria B.C., Canada, October, 1998.

[6] V.F. Munoz, C. Vara-Thorbeck, J.C. DeGabriel, J.F. Lozano, E. Sanchez-Badajoz, A. Garcia-Cerezo, R. Toscano and A. Jimenez-Garrido. A Medical Robotic Assistant for Minimally Invasive Surgery. *Proc. of the 25th IEEE Int. Conf. on Robotics and Automation*. San Francisco, CA, pp. 2901-2906, April 2000.

[7] M. de Mathelin and R. Lozano. Robust adaptive identification of slowly time-varying parameters with bounded disturbances. *Automatica*, vol. 35, pp. 1291-1305, July 1999.

[8] F. Chaumette, P. Rives, B. Espiau. Classification and realization of the differ-ent vision-based tasks. *Visual servoing*, Koichi Hashimoto, pp. 199-228. World Scientific Press, 1993.

[9] B. Espiau, F. Chaumette, P. Rives, *A New Approach to Visual Servoing in Robotics*. IEEE Trans. on Robotics and Automation, vol 8 no 3, june 1992

[10] A. Krupa, M. de Mathelin, G. Morel. The use of force control in laparoscopic surgery *Technical Report*, LSIIT GRAViR, http://gravir.u-strasbg.fr/, 2000

Appendix

The alternative adaptive approach used to estimate $\hat{d}$ is based on a model of the interaction between the robot and the patient at the incision point, that is:

$$f_x = -g.x_p \quad f_y = -g.y_p \tag{15}$$

where x_p and y_p are the lateral displacement of point P with respect to its equilibrium position and g is the stiffness of the abdominal wall of the patient. Neglecting the robot joint dynamics (which, again, is supposed to be very fast as compared to the force loop dynamics), the closed loop behavior is :

$$\begin{cases} -kf_x &= (\hat{d} - d)\omega_y^* + \frac{\dot{f_x}}{g} \\ -kf_y &= -(\hat{d} - d)\omega_x^* + \frac{\dot{f_y}}{g} \\ \dot{d} &= \dot{d}^* \end{cases} \tag{16}$$

These equations are linear with respect to parametrization error $(d - \hat{d})$.
Given a good estimate of g, the following normalized gradient algorithm can be used to estimate d:

$$\dot{\hat{d}} = \dot{d}^* + k_1(\dot{f}_x + g\ kf_x)\frac{\omega_y^*}{\epsilon + \omega_x^{*2} + \omega_y^{*2}} - k_1(\dot{f}_y + g\ kf_y)\frac{\omega_x^*}{\epsilon + \omega_x^{*2} + \omega_y^{*2}} \tag{17}$$

where $k_1 > 0$ is the gain of this gradient algorithm and $\epsilon > 0$ is a normalization coefficient. The stability and convergence properties of this estimation algorithm are given in [10]: the previous algorithm is stable and the convergence of the parameter error to zero is obtained if there is enough excitation (i.e., enough rotational velocities around axis x_c and y_c).

Optimized Port Placement for the Totally Endoscopic Coronary Artery Bypass Grafting using the da Vinci Robotic System*

Ève Coste-Manière[1], Louaï Adhami[1], Renaud Severac-Bastide[2],
Adrian Lobontiu[3], J. Kenneth Salisbury Jr.[3], Jean-Daniel Boissonnat[1],
Nick Swarup[3], Gary Guthart[3], Élie Mousseaux[2], Alain Carpentier[2]

INRIA Sophia-Antipolis (www.inria.fr/chir)[1]
Hôpital Européen Georges Pompidou[2]
Intuitive Surgical Inc. (www.intusurg.com)[3]

Abstract. This work presents the first experimental results of an ongoing cooperation between medical, robotics and computer science teams aimed at optimizing the use of robotic systems in minimally invasive surgical interventions. The targeted intervention is the totally endoscopic coronary artery bypass graft (TECAB), performed using the daVinci^{TM} system (by Intuitive Surgical, Inc.). An integrated and formalized planning and simulation tool for medical robotics is proposed, and experimental validation results on an artificial skeleton and heart are presented.

1 Introduction

The introduction of robots in cardio-vascular surgery came from the limitations imposed on the surgeon by manually controlled minimally invasive surgery (MIS) instruments. Namely, the surgeon finds his movements, vision and tactile sensing reduced and severely altered [8, 3]. The use of a robotic manipulator can remedy the loss of dexterity by incorporating additional degrees of freedom at the end of the tools, as is the case with the EndoWrist^{TM} movement of the daVinci^{TM} system (see [4] for details about the system). In addition, a robotic system offers an increased precision and stability of the movement.

However, this innovation has its own limitations and problems. Beginning with the limitations, and despite the increased dexterity, the region reached from a set of incision sites will remain restrained. Therefore these sites have to be carefully chosen for each patient, depending on his anatomy and the requirements of the intervention. Moreover, the forces that can be delivered by a robotic manipulator may vary significantly with the position of the latter, which stresses even more on the choice of the *ports*. Now moving to the problems introduced by the use of a robot, and setting aside classical control and liability concerns, the main handicap of such systems is the issue of potential collisions with the

---

* This work is partially supported by the Télémédecine project of the French Ministry of Research.

D. Rus and S. Singh (Eds.): Experimental Robotics VII, LNCIS 271, pp. 199–208, 2001.

manipulator arms. Again this stresses on the proper positioning of both the incision ports and the robot. Finally, and no matter how intuitive the controlling device is made, the surgeon will need time and proper training before using his new "hands" in the most efficient way. Therefore simulation would be used to rehearse the intervention, validating the planned ports and helping the surgeon get accustomed to both his tools and his patient. The simulation can also be very useful as an educational tool.

Experiment driven guidelines for optimal port placement had already been proposed by leading surgeons in the field (e.g. [8]); however, it should be clear that the above enumerated criteria cannot be mentally pictured and taken into consideration for each patient. Such a quantity of high precision information cannot be handled by the surgeon alone. A computerized approach offers an efficient fusion of the anatomical data of the patient, the robot characteristics and the requirements set out by the surgeon, by translating the aforementioned into mathematical criteria suitable for optimization. This can be done through an integrated planning and simulation interactive system presented in this paper.

2   General Approach

We propose an integrated approach in which all the processing and user interaction are lumped in a single interface 1, schematically summarized in figure 2.

With the patient's pre-operative data, we formulate the needs of the surgeon and the characteristics of the robot as mathematical criteria, in order to optimize the settings of the intervention. Then we automatically reproduce expected surgeons' movements and guarantee their feasibility. We also simulate the intervention in real-time, paying particular attention to potential collisions between the robotic arms. This paper focuses on the planning and experimental testing steps. Details about the rest of the steps can be found in [1].

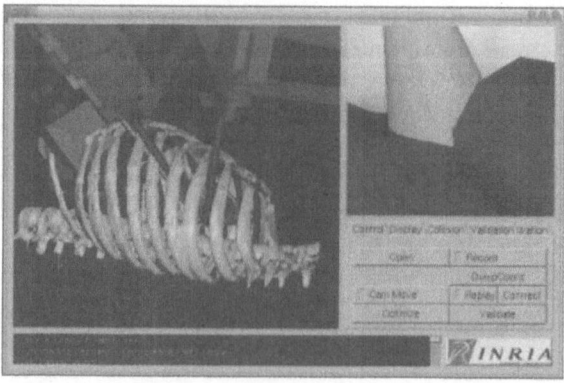

Fig. 1. The planning and simulation interface.

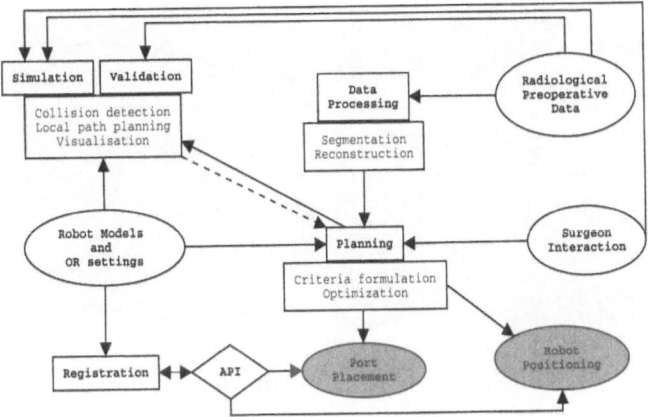

Fig. 2. A modular view of the overall approach.

3   Planning

The planning step can further be broken down into:

1. Finding an optimal triplet for the incision sites.
2. Finding an adequate positioning of the robot.

A two step approach simplifies considerably the problem, and enables a better distinction between the limitations imposed by the minimal invasive access, and those imposed by the robot. Clearly, the two steps are not totally independent, as will be seen in the rest of this section.

3.1   Triplet Optimization

After having identified a set of points that can be used for the access of the robotic tools (the intercostal spaces as explained in [1], see 3 (b)), referred to as admissible points, an exhaustive search for a triplet that insures the best accessibility of the tools is carried out. Moreover, other requirements have to be present such as a "comfortable" position for the surgeon, in addition to some anticipation for the robot positioning step.

Targets: Target points represent the area on which the surgeon would work, and the direction along which he would be able to achieve his task, which is dictated by the physiological of the patient. A typical example is shown in figure 3 (a).

Criteria: Each admissible point goes through a series of tests to characterize its adequacy for use as an entry point for the robotic tool or the endoscope. There are qualitative and quantitative tests as described next:

Qualitative tests concern the reachability from an admissible point to the target areas, where the point is eliminated if any of the following conditions holds:

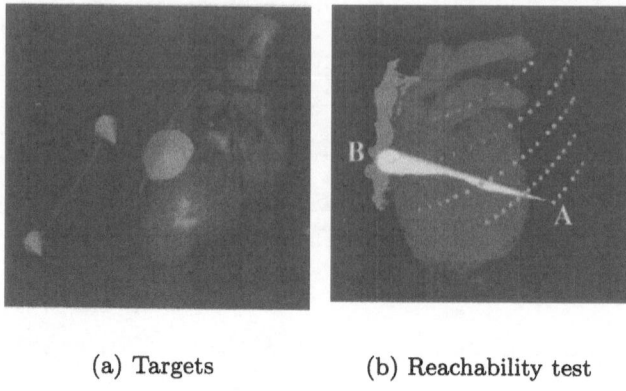

(a) Targets   (b) Reachability test

Fig. 3. Targets and collision test (see text).

- The length of the tool between target and admissible is outside a given range, which simply means that the area cannot be operated using the concerned instrument. (d in figure 4 (a), not used for endoscope)
- The angle between the admissible direction and the line relating the target to the admissible is too big, in which case the tool may damage the adjacent ribs (β in figure 4 (a)).
- The path between the admissible point and the target area is not clear; i.e., it is hindered by an anatomical structure as is shown for instance in figure 3 (b). The graphics hardware is used to perform this test in a way similar to the work described in [7].

Quantitative tests concern the dexterity of the robot, where each admissible point is graded based on the angle between the target direction and the line relating the target to the admissible (α in figure 4 (a)). This measure translates the ease with which the surgeon will be able to operate the concerned target areas from a given port in the case of a robotic tool, or the quality of viewing these areas for an endoscope.

Optimization: Finding the best triplet of ports is done in two steps: First the best endoscope position is chosen based on the above listed criteria, then all possible pairs are ranked according to their combined quantitative grade and their position with respect to the endoscope. More precisely, the triplet is ranked in a way that insures a symmetry of the left and right arms with respect of the endoscope, and favors positions further away from the endoscope to give a clear field of view (formally this corresponds to maximizing ϕ and θ defined in figure 4). Moreover, and in order to anticipate on the next step which is the robot positioning, port that are too close are not considered in the optimization, as they would most certainly result in a colliding state.

This optimization is exhaustive; however, it does not cause a performance problem since the search is hierarchical and thus only a small number of admissible points are left for ranking after all the tests are performed.

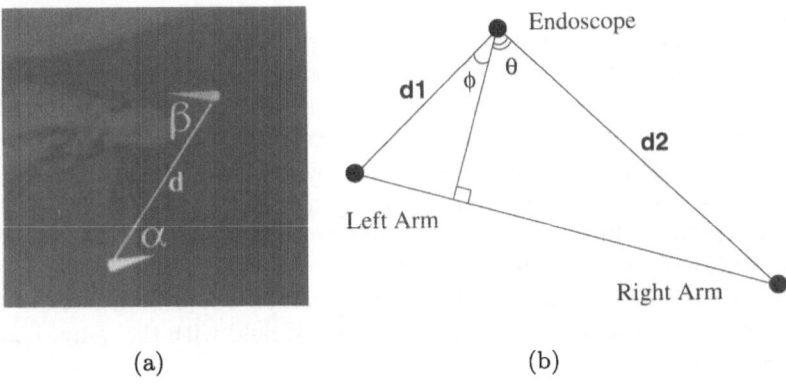

(a) (b)

Fig. 4. Parameter definition for triplet optimization (see text).

3.2   Robot Positioning

Once a suitable port placement has been found, the robot has to be positioned in a way that avoids collisions between its arms, in addition to other constraints. The following section describes the problem in more detail and discusses the proposed solution.

Problem Description The positioning problem consists of finding an initial pose of the robot with respect to the patient, such that **all targets** are reachable without violating any of the constraint enumerated in the next section. In addition, the robot can have degrees of freedom (dofs) that are not teleoperated (passive), and that serve for the set-up of the active joints (see 4.1 for the case of the daVinci^{TM} robot which increases the total number available dofs to 34).

Constraints: The following list captures the constraints imposed upon the robot.

1. Ports: 9 dofs corresponding to 3 positions in space
2. Collisions between the robot arms
3. Collisions between an arm and the patient (e.g. shoulder)
4. Other collisions (e.g. with anesthesia equipment or OR table)
5. Miscellaneous constraints (e.g. endoscope orientation for assistant surgeon)

Certain constraints are more difficult to express than others, especially when it comes to subjective measures such as the preferences of the surgeon or the settings of the OR. In addition, not all constraints can always be expressed; for instance, if the CT scan does not incorporate the shoulder, then the corresponding collision constraint cannot be formulated.

Solution Although impressive theoretical and practical results can be found in the literature on the solution of path planning problems solved both in the robot articular [5, 9] and cartesian space [6, 2], it should be understood that this is not a path planning problem in two respects:

First the start and goal points are not defined, since there are infinitely many solutions to place the robot over the desired ports. In terms of the robot articular space, this means that there are a start subspace and a goal subspace that can have dimensions as high as half the dimension of the articular space.

Then we should keep in mind that even if, for instance, a path is identified to go from one position to another given a certain configuration, then this path may be too complex for the surgeon who will be teleoperating the robot to reproduce. In other words, there is no point in planning in the articular space of the active joints, which leaves us with passive joints that in turn are supposed to stay stationary during the intervention.

A potential field method on the passive joints would be a natural solution for such a problem; however, formulating a good field with the above constraints does not seem to be feasible.

At the present time we use a combined probabilistic and gradient descent approach, in which configurations of the passive joints (including a translation of the base) are randomly drawn in robot articular space. To each configuration, a cost function is associated that depends on the above enumerated constraints. These configurations are then used to get new ones, for which a low cost function gives its corresponding configuration a high selection probability. This process is repeated until convergence to a cost function that is less than a given tolerance.

Clearly, the formulation of the constraints plays a key role in the solution. For a given configuration, the cost function is infinite if a collision is detected, else it is inversely proportional to the distance from the desired port. Moreover, and once the cost function is low enough (i.e., the passive joints are close enough to the desired port), the active joints are moved over all the targets (using inverse kinematics), and collisions are checked.

The main advantage of this method is its flexibility in formulating constraints. For instance, we can easily favor a low opening of the left passive joints in order to avoid conflicts with the anesthesia equipments, by giving these joints a positive effect in the cost function. However, much tuning is required in setting the cost function and the different parameters, which may make the use of such an approach tedious. A more systematic method based on a local planning method is being prepared and will be presented in future works.

4   Experimentation and Results

The validation of the approach is carried out on a plastic skeleton and heart that were CT scanned, and on which the usual pre-processing, planning, validation and simulation steps were performed. The results obtained, as well as the registration and the "clinical" assessment of the positioning, is presented in this section.

4.1   Experimental Setup

The TECAB intervention: The TECAB intervention consists of grafting on a damaged coronary artery, such as the left anterior descending (LAD), another artery (typically the left internal mammary artery or LIMA) that would be used as a bypass to irrigate the affected portion of the heart. The entire intervention is performed through incisions in the chest, where carbon dioxide is insufflated to collapse the left lung, thus enabling the movement of the instruments.

OR Setup: The setup is shown in figure 5, where a plastic skeleton and a heart fixed inside it are used. Moreover, two radio opaque metal strips were used inside the skeleton to represent the LIMA and LAD.

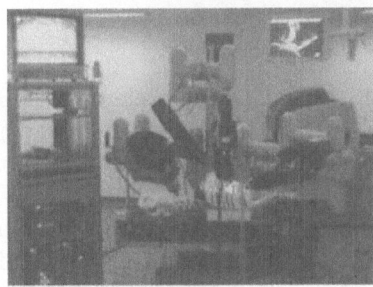

Fig. 5. Experimental setup, the simulation view can be seen on the top right corner of the left image.

The daVinci^{TM} *System and API:* The daVinci^{TM} system is a teleoperated robot composed of two arms to hold the surgical tools and a third to hold the endo-scope, all mounted in the same base. Each arm comprises an active (4 dofs) and a passive part (6 dofs for a tool arms, and 4 for the endoscope arm). In addition, the tool arms incorporate an active end effector that has 3 dofs and a gripper. The passive joints, referred to as set-up joints, are used to position the remote center (fixed point) of the active joints on the ports before the intervention begins. More details about the system can be found in [4].

The API connects a client software to the robot, enabling a real-time reading of the active joints, and a cartesian position of the remote centers with respect to the base of the robot.

4.2 Planning results

The optimized port placement and robot positioning are shown in figure 6, which are the result of the targets shown in 3 (a). Referring to 3.2, only the first two constraints where incorporated in the planning. These may be described as being on the 3rd and 5th intercostal space below the anterior auxiliary line, and on the 7th at the cartilage limit for the left, endoscope and right arm respectively.

4.3 Registration

The positioning of the robot according to the planned results is achieved in two steps: first register the robot to the skeleton, then place the port. This is a preliminary method used as a first attempt to quantify the difficulty of the registration, and should not be considered as a standard approach.

Fig. 6. The planned positioning and port placement.

Patient-Robot The pose of the robot with respect to the skeleton is subject to 6 dofs (3 translations and 3 orientations). The rotational dofs are not registered because we assume that the robot base will be parallel to the OR table, and that the relative tilt between the skeleton and the OR table is the same as the one in the CT scan. The translational pose is registered by pointing an easily identified point (the tip of the sternum) *in the simulator* using the endoscope arm, and reading the corresponding articular values. Then robot base and its first translational joint (up/down) are moved so that the articular values read through the API match the computed ones. This was done manually with a positioning error of 1.5°.

Ports Once the robot is registered to the skeleton, positioning the ports can simply be achieved by moving the robot arms according to the precomputed articular values that correspond to having the remote center on the port. On the other hand, the results of the planning can also be expressed as a quantitative description of the positions of the port; e.g., endoscope arm at 3rd intercostal space at the limit of the cartilage. This is a relatively accurate description since the port are known to be in the intercostal spacing; therefore, only 1 dof (inside the spacing) remains to be set, which is easily achieved since the skeleton has many metal fiducials used to bind it together.

This placement method was successfully carried out on the endoscope arm with a relative error in the positioning of 0.6°, and an absolute error in the port location of 22 mm. However, positioning the 6 dofs setup joints was not possible by manually moving every articulation. Therefore, the quantitative description was used to place the ports. In other words, a 6 dofs arm is positioned using a 3 dofs constraint. The errors on the joint values was very high as expected (more than 85°); however, the absolute port position error was 12 mm.

4.4   Clinical Assessments

Although the registration step is not yet satisfactory, the clinical assessment of the positioning turned out to be very encouraging are described next.

Reachability and Dexterity The steps of the TECAB intervention were performed on the skeleton, and the surgeon described the configuration as very satisfactory in terms of reach and the available dexterity at the tool tips.

Collisions   No collisions occurred between the arms during the entire "intervention". However, it should be noted that the robot arm would have been in conflict with the shoulder and the diaphragm of the skeleton, which means that these constraints would most probably have had a big impact on the port placement. These will be included in the next set of tests by adding a shoulder and diaphragm to the skeleton.

4.5   Comparison with Clinical Solution

The positioning problem of the robot is largely due to the non-invasive nature of the intervention. More precisely, the surgeon has two major problems to face: he does not have an accurate idea about the location of the areas he wants to operate on with respect to the patient's chest, and he cannot try out different port configurations to avoid collisions with the robot. In the case of the plastic phantom; however, all the anatomical entities are visible from the outside, and there is no harm in re-positioning the ports according to the needs of the intervention. Therefore, and given enough time, a conventional port placement on a plastic skeleton can be used as a gold standard for the proposed automatic port placement.

In terms of anatomical landmarks, the two results differ in the endoscope position which is one rib higher, and in the right arm position which is one rib lower and closer to the sternum. The similarity between the results is encouraging, especially if we keep in mind that the clinical solution was driven by experience on real patients; therefore, for instance, the surgeon would not use the 7th intercostal space for the right arm (as proposed by the automatic planning), since it may be too close to the diaphragm.

Alternatively, we believe the proposed planning will ultimately change the way the clinical procedure is performed, were the final verdict will be given by the goings of the intervention. Finally, comparing the robot positioning does not yield any useful information, since both positions were collision free.

4.6   Recording

An additional useful feature of the system is that the entire intervention can be recorded and played back at a later time, or at a remote location. This was done successfully; however, the replay in simulation was difficult to assess since registration errors were too large, thus the reproduced gestures did not correspond to the clinical ones.

4.7   Assessments

The results obtained indicate that the proposed approach can considerably simplify and improve the positioning of the robot on the patient. The registration remains the weak chain in the process, although its effects can be minimized when complemented with anatomical landmarks for the port locations.

Potential problems due to an incomplete modeling of the plastic phantom and the OR settings were identified, these should be incorporated into future experimentations.

5 Conclusion

An integrated planning and simulation system for minimal invasive robotically assisted surgery was presented, along with a preliminary registration method. The approach was validated on a plastic phantom, and preliminary results were very encouraging. We believe that this approach is poised to bring significant advanced in the way robot will be used in the operating room. In addition to being a useful visualization and simulation tool, it opens new possibilities for testing prototype robotic tools and unconventional interventions in simulation.

Future efforts will be directed towards more validation of the approach, and improving the robot positioning algorithm and the registration process.

References

1. L. Adhami, E. Coste-Manière, and J.-D. Boissonnat. Planning and simulation of robotically assisted minimal invasive surgery. In *Proc. Medical Image Computing and Compter Assisted Intervention (MICCAI'00)*, volume 1935 of *Lect. Notes in Comp. Sc. 1954*. Springer, Oct. 2000.
2. O. Brock and O. Khatib. Executing motion plans for robots with many degrees of freedom in dynamic environments. In *Proceedings of the IEEE International Conference on Robotics and Automation (ICRA-98)*, Piscataway, 1998. IEEE.
3. G. B. Cadière and J. Leroy. *Principes généraux de la chirurgie laparoscopique. Encycl Méd Chir (Techniques chirurgicales - Appareil digestif)*, volume 40, page 9. Elsevier-Paris, 1999.
4. G. Guthart and J. K. Salisbury Jr. The intuitive telesurgery system: Overview and application. In *Proceedings of the 2000 IEEE International Conference on Robotics and Automation*, pages 618–622, San Francisco, CA, 2000.
5. L. Kavraki, P. Svestka, J.-C. Latombe, and M. Overmars. Probabilistic roadmaps for path planning in high-dimentional configuration spaces. *IEEE transactions on Robotics and Automation*, 1999.
6. O. Khatib. The potential field approach and operational space formulation in robot control. In K. S. Narendra, editor, *Adaptive and Learning Systems - Theory and Applications*, pages 367–378, New York and London, 1986. Plenum Press.
7. J.-C. Lombardo, M.-P. Cani, and F. Neyret. Real-time collision detection for virtual surgery. In *Computer Animation, Geneva*, May 1999.
8. D. Loulmet, A. Carpentier, N. d'Attellis, A. Berrebi, C. Cardon, O. Ponzio, B. Aupècle, and J. Y. M. Relland. Endoscopic coronary artery bypass grafting with the aid of robotic assisted instruments. *The journal of thoracic and cardiovascular surgery*, 118(1), July 1999.
9. R. Z. Tombropoulos, J. R. Adler, and J. C. Latombe. Carabeamer: A treatment planner for a robotic radiosurgical system with general kinematics. *Medical Image Analysis*, 3(3):237–264, 1999.

ETS-VII Flight Experiments
For Space Robot Dynamics and Control

Theories on laboratory test beds ten years ago, Now in orbit

Kazuya Yoshida
Dept. of Aeronautics and Space Engineering
Tohoku Unoversity
Sendai, 980-8579, Japan
yoshida@astro.mech.tohoku.ac.jp

Abstract: The Engineering Test Satellite VII (ETS-VII), developed and launched by National Space Development Agency of Japan (NASDA) has been successfully flown and carried out significant experiments on orbital robotics with a 2 meter-long, 6 DOF manipulator arm mounted on this un-manned spacecraft. The ETS-VII should be noted as one of remarkable outcomes of the research effort on space robots, particularly characterized as an orbital free-flying robot. This paper highlights a story how the theories have been developed mathematically, studied with laboratory test bed, then demonstrated in orbit.

1. Introduction

The ideas for the rescue and service to a malfunctioning satellite by a free-flying space robot has been discussed since early 80s (for example [1]), but very few attempts have ever done in orbit. The maintenance missions of the Hubble Space Telescope and the retrieval of the Space Flyer Unit are such important examples carried out with the Space Shuttle Remote Manipulator System. However, in these missions the manipulator was manually operated by a well-trained flight crew. Autonomous target capture by an un-manned space robot is a big challenge for space robotics community for many years, and very recently, essential parts of this technology have been successfully verified and demonstrated in orbit by a Japanese free-flying space robot, ETS-VII.

The Engineering Test Satellite VII (ETS-VII, Figure 1), developed and launched by National Space Development Agency of Japan (NASDA) in November 1997, has been successfully flown and carried out a lot of interesting orbital robotics experiments with a 2 meter-long, 6 DOF manipulator arm mounted on this un-manned spacecraft. The mission objective of ETS-VII is to test robotics technology and demonstrate its utility for un-manned orbital operation and servicing tasks. The mission consists of two subtasks, autonomous rendezvous/docking (RVD) and robot experiments (RBT). The robot experiments include a variety of topics such as: (1) teleoperation from the ground with large time-delay, (2) robotic servicing task demonstrations such as ORU exchange and deployment of a space structure, (3) dynamically

D. Rus and S. Singh (Eds.): Experimental Robotics VII, LNCIS 271, pp. 209–218, 2001.

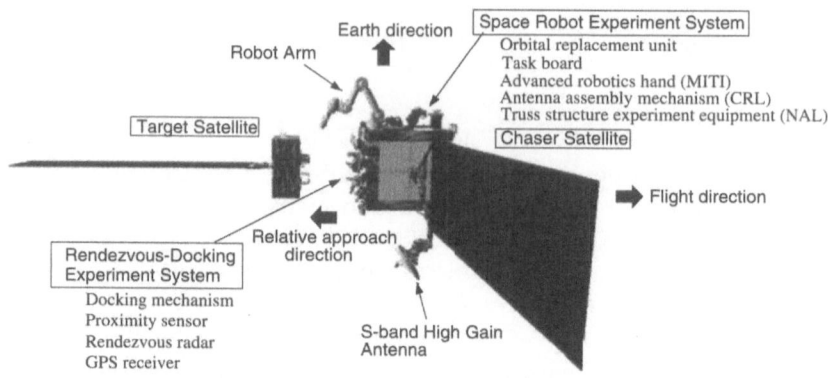

Figure 1. The Engineering Test Satellite VII

coordinated control between the manipulator reaction and the satellite attitude, and (4) capture and berthing of a target satellite. Early reports on some
of these experiments were made in the paper [2] and in the proceedings [3], for
example.

The initially planned flight experiments were successfully completed by
the end of May 1999. But since the spacecraft was still operational in a good
condition, an extensive mission period was set up till the end of December 1999.
In this period the opportunity was opened for academic proposals and research
groups of Japanese universities were given the time to do their flight experiments. Using this precious opportunity, the present author has proposed and
carried out the experiments to highlight the motion dynamics of a free-flying
space robot, verifying the theories on the coupling and coordination between
the manipulator reaction and the base spacecraft attitude. Some details are
already reported in [4][5][6][7].

In this paper, the focus is made on the evolution of experimental robotics
effoert from simplified laboratory test beds to a real flight system in orbit. The
paper is organized as follows. In section 2, the motion dynamics to characterize
a free-flying space robot and concepts for the manipulation in space, originally
discussed in late 80s and early 90s, are reviewed. In section 3, the laboratory
test beds having been developed by the present author since 1987 are recalled.
In section 4, the results of the corresponding ETS-VII flight experiments carried
out in 1999 are presented. In section 5, a complementary discussion is made
for future practical satellite servicing missions.

2. Dynamics of a Free-Flying Space Robot

A unique characteristics of a free-flying space robot is found in its motion dynamics. According to the motion of the manipulator arm, the base spacecraft
moves due to the action-to-reaction principle or the momentum conservation.

The reaction of the arm disturbs its footing base, then the coupling and co-ordination between the arm and the base becomes an important issue. This is a main difference from a terrestrially based robot manipulator and a draw-back to make the control of a space manipulator difficult. Earlier studies for the modeling and control of such a free-flying robot are collected in the book [8]. Here the basic modeling is reviewed to derive the key concepts named the generalized Jacobian matrix and the reaction null-space.

2.1. Basic equations

The equation of motion of a free-flying space robot as a multibody system is, in general, expressed in the following form:

$$\begin{bmatrix} H_b & H_{bm} \\ H_{bm}^T & H_m \end{bmatrix} \begin{bmatrix} \ddot{x}_b \\ \ddot{\phi} \end{bmatrix} + \begin{bmatrix} c_b \\ c_m \end{bmatrix} = \begin{bmatrix} \mathcal{F}_b \\ \tau \end{bmatrix} + \begin{bmatrix} J_b^T \\ J_m^T \end{bmatrix} \mathcal{F}_h \qquad (1)$$

where we choose the linear and angular velocity of the base satellite (reference body) $\dot{x}_b = (v_b^T, \omega_b^T)^T$ and the motion rate of the manipulator joints $\dot{\phi}$ as generalized coordinates. The symbols used here are defined as follows:

$H_b \in R^{6 \times 6}$: inertia matrix of the base.

$H_m \in R^{n \times n}$: inertia matrix for the manipulator arms (the links except the base.)

$H_{bm} \in R^{6 \times n}$: coupling inertia matrix.

$c_b \in R^6$: velocity dependent non-linear term for the base.

$c_m \in R^6$: that for the manipulator arms.

$\mathcal{F}_b \in R^6$: force and moment exert on the centroid of the base.

$\mathcal{F}_h \in R^6$: those exert on the manipulator hand.

$\tau \in R^n$: torque on the manipulator joints.

Especially in the free-*floating* situation, the external force/moment on the base which can be generated by gas-jet thrusters, and those on the manipulator hand are assumed zero; i.e. $\mathcal{F}_b = 0$, $\mathcal{F}_h = 0$. The motion of the robot is governed by only internal torque on the manipulator joints τ, and hence the linear and angular momenta of the system $(\mathcal{P}^T, \mathcal{L}^T)^T$ remain constant.

$$\begin{bmatrix} \mathcal{P} \\ \mathcal{L} \end{bmatrix} = H_b \dot{x}_b + H_{bm} \dot{\phi} \qquad (2)$$

2.2. Angular momentum

The integral of the upper set of the equation (1) gives the momentum conser-vation, as shown in Equation (2), which is composed of the linear and angular momenta. The linear momentum has further integral to yield the principle that the mass centroid stays stationary or linearly moves with a constant velocity.

The angular momentum equation, however, does not have the second-order integral hence provides the first-order non-holonomic constraint. The equation is expressed in the form with the angular velocity of the base $\boldsymbol{\omega}_b$ and the motion rate of the manipulator arm $\dot{\boldsymbol{\phi}}$ as:

$$\tilde{\boldsymbol{H}}_b \boldsymbol{\omega}_b + \tilde{\boldsymbol{H}}_{bm} \dot{\boldsymbol{\phi}} = \mathcal{L} \tag{3}$$

where $\mathcal{L}$ is the initial constant of the angular momentum, and the inertia matrices with tilde are those modified from Equation (2). $\tilde{\boldsymbol{H}}_{bm}\dot{\boldsymbol{\phi}}$ represents the angular momentum generated by the manipulator motion. These equations of (2) and (3) provide a basis for further discussion.

2.3. Generalized Jacobian Matrix (GJM)

The velocity of the manipulator hand in the inertial frame is expressed as:

$$^{(i)}\dot{\boldsymbol{x}}_h = \boldsymbol{J}_m \dot{\boldsymbol{\phi}} + \boldsymbol{J}_b \dot{\boldsymbol{x}}_b \tag{4}$$

Then an idea came to combine it with (2), to yield the equation directly connect the manipulator joints and hand with canceling out the base variables:

$$^{(i)}\dot{\boldsymbol{x}}_h = \boldsymbol{J}_g \dot{\boldsymbol{\phi}} \tag{5}$$

$$\boldsymbol{J}_g = \boldsymbol{J}_m - \boldsymbol{J}_b \boldsymbol{H}_b^{-1} \boldsymbol{H}_{bm} \tag{6}$$

where $(\mathcal{P}^T, \mathcal{L}^T)^T = \boldsymbol{0}$ is assumed for simplification. The matrix $\boldsymbol{J}_g$ is termed *Generalized Jacobian Matrix (GJM)*[9], and with using it the manipulator hand can be operated by resolved motion-rate control or resolved acceleration control properly in the inertial space, while allowing the base reaction but not disturbed by it.

2.4. Reaction Null-Space (RNS)

From a practical point of view, the attitude change is not desirable, then the manipulator motion planning methods to have minimum attitude disturbance on the base are also studied. An ultimate goal of those approaches is completely zero disturbance, and such operation is possible from the insight of the angular momentum equation.

The angular momentum equation with zero initial constant $\mathcal{L} = \boldsymbol{0}$ and zero attitude disturbance $\boldsymbol{\omega}_b = \boldsymbol{0}$:

$$\tilde{\boldsymbol{H}}_{bm} \dot{\boldsymbol{\phi}} = \boldsymbol{0} \tag{7}$$

yields the following null-space solution:

$$\dot{\boldsymbol{\phi}} = (\boldsymbol{I} - \tilde{\boldsymbol{H}}_{bm}^+ \tilde{\boldsymbol{H}}_{bm}) \dot{\boldsymbol{\zeta}} \tag{8}$$

The joint motion given by this equation is guaranteed to make zero disturbance on the base attitude. Here the vector $\dot{\boldsymbol{\zeta}}$ is arbitrary and the null-space of the inertia matrix $\tilde{\boldsymbol{H}}_{bm}$ is termed *Reaction Null-Space (RNS)* [10].

The degrees of freedom for $\dot{\boldsymbol{\zeta}}$ is $n-3$, and in ETS-VII the manipulator arm has 6 DOF, i.e. $n = 6$, then there remains 3 DOF for the reaction null-space

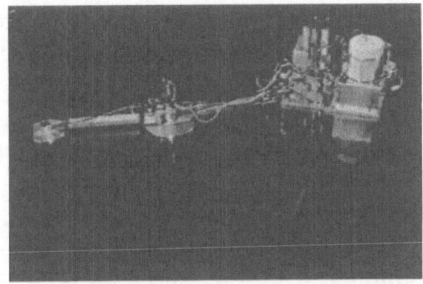

(a) The experimental free-floating robot simulator, EFFORTS (1987)

(b) The experimental flaxible-base manipulator TREP (1996)

Figure 2. Laboratory test beds for space manipulator systems

to be specified by additional criteria. In the flight experiment presented later, a criterion is chosen that the orientation of the manipulator hand (3 DOF) is constraint, while the translation of the hand (3 DOF) is allowed in realizing the zero reaction on the base. This manipulator motion is termed the *reactionless manipulation*.

3. Laboratory Test Beds

3.1. EFFORTS

EFFORTS is the **E**xperimental **F**ree-**Fl**O**ating **R**obo**T** **S**atellite simulator developed since 1987 (Figure 2(a)) [11], which is one of the earliest attempts of air-cushion type microgravity test beds. The robot model is floated by pressurized air on a horizontal planar table. The motion is constraint on a plane though, the test bed is very useful to study the reaction dynamics of an articulated link systems and effective to demonstrate the performance of the GJM based inertial manipulation in the microgravity environment. The test bed well contributed to appeal the importance of the reaction effect in space manipulation.

3.2. TREP

TREP is a test bed comprising rigid articulated manipulator arm(s) mounted at the end of a flexible beam (Figure 2(b))[10]. The flexible beam has a parallel double-beam structure to allow the bending deformation on a horizontal plane, while the deformations in other directions are restricted. Using this test bed the manipulation that yields zero reaction to a certain direction of the base, as well as the effective vibration suppression control by the manipulator motion have been studied. A number of primitive experiments for the RNS based reactionless manipulation have been carried out with this test bed [12].

4. ETS-VII Flight Experiments

The robotic flight experiment proposed by the present author was carried out on September 30, 1999, using three successive flight paths. The flight path is a communication window between Tsukuba Space Center, NASDA, and ETS-VII via TDRS, a US data relay satellite located in the geosynchronous orbit above the pacific ocean. In each path almost net 20 minutes operation (command uplink) and dense telemetry (including video downlink) are established.

The purpose and also the advantage of the flight experiments using a real space system is to obtain the proof of the theories that are suggested only by confined laboratory test beds, and to demonstrate the practical availability of the methods in real world. However the difficulty is that practical systems always have many constraints and restrictions from design specification and safety point of view. For example, the manipulator arm mounted on ETS-VII is a 6 DOF non-redundant arm. If it would be a redundant one, we could demonstrate a wide range of interesting performances. But the present system is still good enough to show the basic performance of the proposed concepts. In operation, it is very difficult to implement a newly proposed control algorithm into the existing on-board computers and test the closed-loop performance. But the open-loop type control experiments were accepted in ETS-VII.

For the proposed experiment, the manipulator motion trajectories were carefully prepared in a motion file and the safety was preliminary checked on an offline simulator. During the experiment, the motion file is uploaded to ETS-VII at 4 Hz frequency as an isochronous command and the manipulator arm is controlled to follow the prepared motion profile. In preparing the motion profile, best knowledge on the inertia parameters of the system was used, which were elaborately identified with the flight telemetry data obtained until May 1999.

4.1. GJM Based Inertial Manipulation

For the GJM based inertial manipulation, the experiment was carried out under the free-floating environment without any base attitude control actions or zero initial momentum in the system.

Figure 3 depicts the experimental flight data for the GJM based manipulation. The motion demonstrated here is straight-path tracking in the inertial frame, under the resolved motion-rate control with the generalized Jacobian using the inversion of Equation (5). The top graph shows a profile for the pitch angle of the manipulator hand in the satellite base frame. The middle shows the pitch attitude of the base satellite, disturbed by the manipulator reaction. The bottom is the summation of the top and the middle graphs, then represents the attitude of the manipulator hand in the inertial frame. By means of the control with the generalized Jacobian, the attitude of the hand in the inertial frame is kept almost zero against a non-negligible satellite attitude disturbance.

Figure 4 campares the performance to reach a given point in the inertial frame such as a free-floating target. The graph depicts the error distance between the hand and the target. The broken line shows the flight data to capture a floating target by the manipulator control without GJM. Without GJM the hand moves to incorrect direction, but the error is corrected by means of visual servo-tracking, then finally the capture is attained. However with GJM, as shown in the solid line, the target capture is attained earlier because

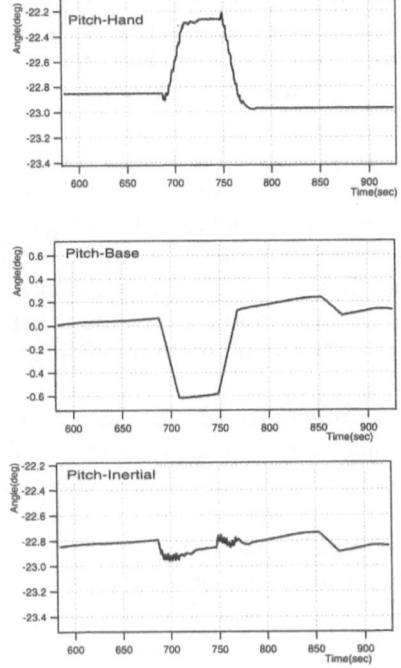

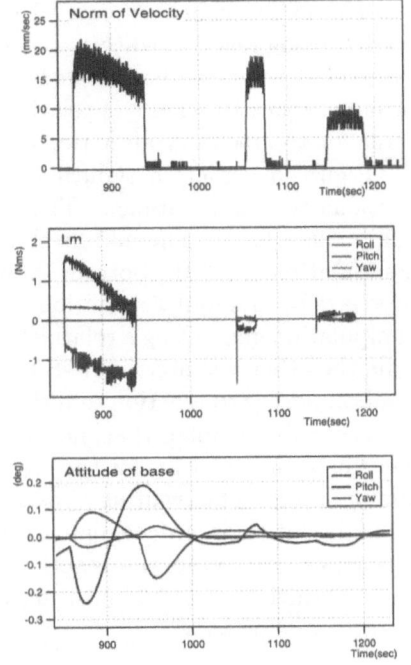

Figure 3. ETS-VII flight data for the GJM based inertial manipulation

Figure 5. ETS-VII flight data for the RNS based reactionless manipulation

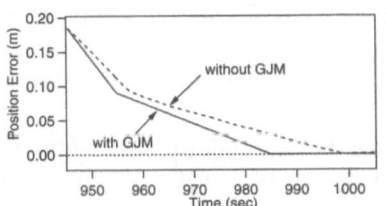

Figure 4. Comparison of the approaching performance to an inertially-floating target, between with and without GJM

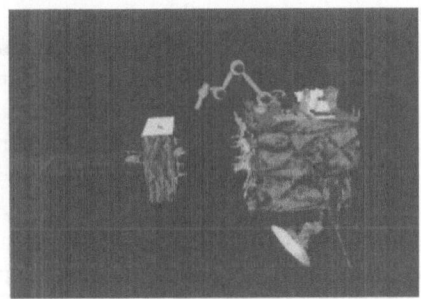

Figure 6. A computer graphic image for the target capture simulation

the hand moves toward the target exactly straightly.

4.2. RNS based reactionless manipulation

In the RNS experiment, several sets of reactionless trajectories were prepared based on the reaction null-space formulation. We prepare the trajectories to go to or from useful control points such as an onboard ORU or a target satellite, and compared with the motion by conventional PTP trajectories.

The experiment was carried out under the attitude control of the base satellite using reaction wheels. Even under the control, the attitude disturbance is observed when the base receives the manipulator reaction, since the control torque of reaction wheels is relatively small. The attitude control here mainly works for the recovery after the attitude disturbed.

Figure 5 depicts a typical flight data to compare the conventional and reactionless manipulations. The top graph shows the velocity norm of the manipulator hand. The middle shows the reaction momentum induced by the manipulation. And the bottom shows the attitude motion. The graphs include three sets of manipulator motion, where the first one is the conventional PTP manipulation generating a relatively large momentum and attitude disturbance, while the other two are the RNS based reactionless manipulation yielding very small, almost zero reaction and disturbance.

It should be noted that, not only the maximum attitude change is remarkably small, but the time for the recovery is also very short with the reactionless manipulation. This waiting time for the attitude recovery in the conventional manipulation is not negligible and degrade the efficiency of the operation in practice. However, the reactionless manipulation provides almost zero attitude disturbance and almost zero recovery time, thus assuring a very high operational efficiency.

5. Toward Practical Satellite Servicing

In the ETS-VII flight experiments, a whole sequence of the capture of a free-floating target was not demonstrated continuously, however, all necessary key elements were verified each by each. The technology necessary for the orbital maneuver in approaching to a target, rendezvous and precise proximity flight control has been successfully demonstrated with the ETS-VII's main satellite (chaser) and a target satellite separated from the chaser. Three different flight paths called FP-1, 2 and 6, including contingency maneuvers, were tested and all resulted in autonomous soft-docking safely [13].

For the target capture, not for docking, a manipulator arm should be operated to track and grasp a fixture mounted on the target satellite, while maintaining the proximity flight with the target. One of the key technology is the visual servo-tracking of the grasping point by the manipulator arm. This was successfully tested with a set of CCD camera mounted on the manipulator hand and an optical marker located at the target fixture, using on-board real-time video signal processing.

Further control technology to improve the manipulator control, and thereby increase the fidelity and safe margin of the successful capture, were also tested and important flight data were obtained. They are GJM based inertial manipulation and RNS based reactionless manipulation, as presented above.

The reactionless manipulation is an idea to operate the manipulator arm while not disturbing the base satellite attitude. Such manipulator trajectories are very limited in case of a 6 DOF arm although, they are proven very effective to minimize the base attitude disturbance and save time to wait for the attitude recovery. The RNS based reactionless manipulation should be useful for coarse approach to the target under the proximity flight, where the attitude disturbance due to the manipulator reaction is highly undesirable.

During the final approach with visual servoing, the manipulator hand must be controlled for the target floating in the inertial frame. In this phase the GJM based inertial manipulation, which is also proven effective in practical situation, should be particularly useful.

For practical satellite capture operation expected in near furure, it will be therefore a best combination of technology to make the coarse approach with the proximity flight control of the spacecraft and the RNS based reactionless manipulation of the arm, then switch to the final approach with the visual-servo tracking and the GJM based inertial manipulation of the hand while the spacecraft is left under free-drift until finally captured.

Post flight analysis and computer simulatuions (Figure 6) are now extensively continuing with possible assumptions for a practical satellite servicing mission. The analysis suggests that the introduction of a redundant arm, with 7 DOF or more, will enhance the advantage and performance of the RNS based reactionless manipulation [14].

6. Conclusions

This paper presents a story about the development of the space robot dynamics and control technology. The Generalized Jacobian Matrix (GJM) based inertial manipulation and the Reaction Null-Space (RNS) based reactionless manipulation have been theoretically studied since middle 80s and their preliminary verifications have been done with simplified and confined laboratory test beds. Finally the theories are evaluated in the practical performance by the flight demonstration of a real space robot in orbit.

Both of GJM and RNS based control concepts are successfully verified on ETS-VII, as well as other technology relevant to the target capture such as the rendezvous control and the visual servo-tracking of the manipulator hand. A whole sequence of the autonomous target capture was not demonstrated in orbit although, it is analytically inferred by post flight analysis. From the flght data, the GJM based inertial manipulation is verified with higher performance to reach a given point in the inertial frame such as a free-floating target. Also, the RNS based reactionless manipulation verified with almost zero attitude disturbance and almost zero recovery time, thus assuring a very high operational efficiency.

The ETS-VII opened a very solid way to the autonomous target capture. We hope that these technologies are further improved and applied to practical missions for satellite servicing or robotic rescue.

Acknowledgements

The present author acknowledges his special thanks to Dr. Mitsushige Oda, Mr. Noriyasu Inaba and other ETS-VII operation team of NASDA, Japan, for kind help an assistance to prepare and carry out the collaborative flight experiments. The flight data of ETS-VII are obtained under the collaboration between Tohoku University and NASDA.

References

[1] D. L. Akin, M. L. Minsky, E. D. Thiel and C. R. Curtzman, "Space Applications of Automation, Robotics and Machine Intelligence Systems (ARAMIS) phase II," *NASA-CR-3734 – 3736*, 1983.

[2] M. Oda et al, "ETS-VII, Space Robot In-Orbit Experiment Satellite," *Proc. 1996 IEEE Int. Conf. on Robotics and Automation*, pp.739–744, 1996.

[3] *Proc. 5th Int. Symp. on AI, Robotics and Automation in Space, iSAIRAS'99*, June 1999, ESTEC, Netherlands.

[4] K. Yoshida, "Space Robot Dynamics and Control: To Orbit, From Orbit, and Future," *Robotics Research, The Ninth International Symposium*, ed. by J. Hollerbach and D. Koditschek, Springer, pp.449-456, 1999.

[5] K. Yoshida, D. N. Nenchev, N. Inaba and M. Oda, "Extended ETS-VII Experiments for Space Robot Dynamics and Attitude Disturbance Control," *22nd Int. Symp. on Space Technology and Science*, Morioka, Japan, ISTS2000-d-29, May 25-29, 2000.

[6] K. Yoshida, K. Hashizume, D. N. Nenchev, N. Inaba and M. Oda, "Control of a Space Manipulator for Autonomous Target Capture - ETS-VII Flight Experiments and Analysis -," *AIAA Guidance, Navigation, and Control Conference*, Denver, CO., AIAA2000-4376, Aug. 14-17, 2000.

[7] K. Yoshida, "Space Robot Dynamics and Control: a Historical Perspective," *J. of Robotics and Mechatronics*, vol.12, no.4, pp.402-410, 2000.

[8] *Space Robotics: Dynamics and Control*, edited by Xu and Kanade, Kluwer Academic Publishers, 1993.

[9] Y.Umetani and K.Yoshida, "Continuous Path Control of Space Manipulators Mounted on OMV," *Acta Astronautica*, vol.15, No.12, pp.981-986, 1987. (Presented at the 37th IAF Conf, Oct. 1986)

[10] K. Yoshida, D. N. Nenchev and M. Uchiyama, "Moving base robotics and reaction management control," *Robotics Research: The Seventh International Symposium*, Ed. by G. Giralt and G. Hirzinger, Springer-Verlag, 1996, pp. 101–109.

[11] K. Yoshida, "Space Robotics Research Activity with Experimental Free-Floating Robot Satellite (EFFORTS) Simulators," *Experimental Robotics III, The 3rd International Symposium*, ed. by T. Yoshikawa and F. Miyazaki, Springer-Verkag, pp.561-578, 1993.

[12] A. Gouo, D.N.Nenchev, K.Yoshida, M.Uchiyama, "Dual-Arm Lomg-Reach Manipulators: Noncontact MOtion Control Strategies," *Proc. 1998 IEEE/RSJ Int. Conf. on Intelligent Robots and Systems*, Victoria, Canada, October, 1998, pp.449-454.

[13] I. Kawano et al, "First Result of Autonomous Rendezvous Docking Experiments on NASDA's ETS-VII Satellite," Proc. of 49th Int. Astronautical Congress, *IAF-98-A3.09*, 1998.

[14] K. Yoshida and K. Hashizume, "Zero Reaction Maneuver: Flight Velification with ETS-VII Space Robot and Extention to Kinematically Redundant Arm," *Proc. 2001 IEEE Int. Conf. on Robotics and Automation*, Seoul, Korea, 2001.

Experimental Demonstrations of a New Design Paradigm in Space Robotics

Matthew D. Lichter, Vivek A. Sujan, and Steven Dubowsky
Department of Mechanical Engineering
Massachusetts Institute of Technology
Cambridge, MA 02139 USA
{lichter | vasujan | dubowsky} @mit.edu

Abstract: This paper presents a study to experimentally evaluate a new design paradigm for robotic components, with emphasis on space robotics applications. In this design paradigm, robotic components are made from embedded binary actuators and compliant mechanisms in order to reduce weight and complexity. This paper presents a series of five experiments that demonstrate the concept. These studies include a reconfigurable rocker suspension for rocker-bogie rovers, a sample acquisition gripper, a pantograph mechanism for robotic legs, an articulated binary limb, and a hyper-degree-of-freedom mechanism building block.

1. Introduction

Future missions to planets such as Mars will require explorer/worker robots to perform tasks of increased complexity such as exploring, mining, conducting science experiments, constructing facilities, and preparing for human explorers. To meet the objectives of missions in the year 2010 to 2040 timeframe, planetary robots will need to work faster, travel larger distances, and perform highly complex tasks with a high degree of autonomy [1]. They will also need to cooperate in teams and reconfigure themselves to meet their mission objectives. Current electromechanical technologies of motors, optical encoders, gears, bearings, etc. will not be lightweight and robust enough for these robots.

In this research, a new class of building blocks for planetary robots is being studied. These devices, called Articulated Binary Elements (ABEs), consist of compliant mechanisms with large numbers of embedded actuators and sensors. Figure 1 shows two concepts of the ABE elements. These actuators would be binary (on/off) in nature and could be made from conducting polymer, electrostrictive polymer artificial muscle, shape memory alloy, etc. [2,3]. By using hundreds or thousands of very simple but reliable actuators, one can approximate a continuous robotic system in dexterity and utility. This is analogous to the leap from analog to digital computing. The advantage of binary actuation is the fact that systems can be controlled in the absence of feedback sensing and using only very simple digital electronics, due the robust nature of a two-state actuator. In this work, the feasibility of the ABE concept for planetary robots is being studied analytically and experimentally.

This paper presents an overview of a series of five experimental developments performed to establish credibility and feasibility of the concepts of embedded binary

D. Rus and S. Singh (Eds.): Experimental Robotics VII, LNCIS 271, pp. 219–228, 2001.
© Springer-Verlag Berlin Heidelberg 2001

(a) mating two rovers   (b) coring rock samples

Figure 1. Two ABE robotic potential applications.

actuation and elastic robotic elements. They are a reconfigurable rocker suspension for rocker-bogie rovers, a science sample gripper, a pantograph mechanism for a robot leg, an articulated binary limb, and a hyper-degree-of-freedom mechanism building block. Each of these case studies suggests that light weight and simple robotic elements can be constructed with this design paradigm. The work also highlights several limitations that will need to be overcome for the concept to reach its full potential.

2. Background and Literature

The concept of binary robotics is not altogether new. In the 1960's and 70's sporadic work was done in areas of binary actuation and sensorless robotics [4,5]. Deeper study of this area did not occur until recently [6-10], when computation power made the analysis, control, and planning for binary robots feasible.

Solving the forward and inverse kinematics, in order to follow trajectories for a binary robot, is a very different problem than that for a continuous robot. Because the actuators can achieve only finite displacements, a binary robot can reach only finite locations and orientations in space. The workspace of a binary robot is thus a point cloud and the inverse kinematics problem becomes simply a search through this cloud. However, for large numbers of actuators this search space explodes rapidly (the number of search points is 2^N, where N is the number of actuators). Methods for solving these computational challenges have been developed [8-10]. The most common methods reduce the search space through a combinatorial approach or use stochastic searches such as genetic algorithms.

Some experimental work has been done on binary redundant manipulators [8]. An example is a variable-geometry truss (VGT) manipulator that was constructed using pneumatic actuators which were overpowered and held at their stops. These studies showed the fundamental ability of binary robots to perform some tasks. Such devices constructed using conventional components cannot demonstrate the full advantages of binary robotic devices, such as achieving large redundancy with simple implementation. To date, little work has been done to experimentally demonstrate simple, lightweight, robust binary designs that are able to perform effectively.

The focus of this work is to study the issues within the context of the application of binary elements in space robots. Binary-actuated elements have great appeal for this area because they can be designed at a fraction of the weight, complexity, and volume of conventional continuously-actuated elements (i.e. motors, hydraulics, etc.) [2,3]. Incorporating the concept of compliant mechanisms enables further design simplifications through the use of elastic flexures in place of mechanical hinges, bearings, and lubrication. The results of this paper suggest that given the development of high-performance smart materials and artificial muscle technology, a whole new realm of robotics design will emerge.

3. Technology Demonstrations

3.1. Approach

The goal of this research is to investigate and demonstrate the concept and technologies of achieving high degree of freedom binary systems with physically simple and robust implementations. In this concept the actuators are assumed to be polymer-based materials. These materials include conducting polymers [2] and electrostrictive polymers [3]. However, while these materials are anticipated to meet the needs of the concept in the future, they have not yet reached a sufficient state of development to perform practical experimental demonstrations in devices today. For example, conducting polymers require immersion in an ion solution and provide only small displacements. So for this research shape memory alloys (SMAs) were chosen as surrogate actuator material.

In all the devices studied, the challenge was to amplify the very small displacements provided by the muscle-type actuators. The ABE systems described in this paper run the range from one-DOF devices to many-DOF systems which use embedded actuators and continuously compliant structures.

3.2. Case Studies

3.2.1. Reconfigurable Rocker Suspension

When traversing slopes, a rover can greatly increase its tip over stability by changing the spread angle of its rockers θ (see Figure 2) [11]. To examine the concept of achieving this with binary actuation, an SMA-actuated rocker was designed and developed (see Figure 3) [12]. The working prototype has demonstrated the feasibility and simplicity of such a design. While this first experiment uses conventional bearings and structural materials, it shows that binary muscle-type actuation in its simplest form can be effective.

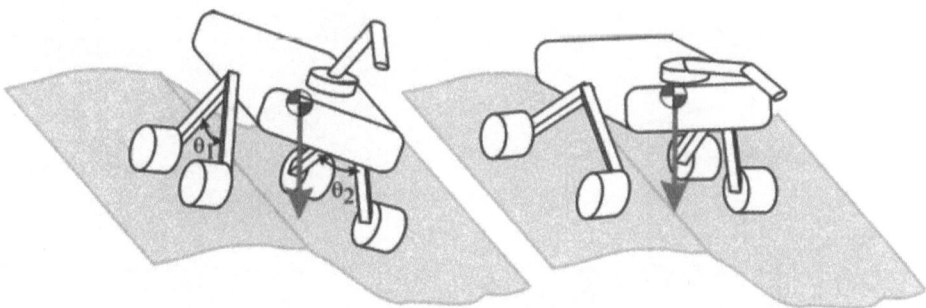

Figure 2. Rover increases tip over stability by reconfiguring its rocker angles θ_1, θ_2.

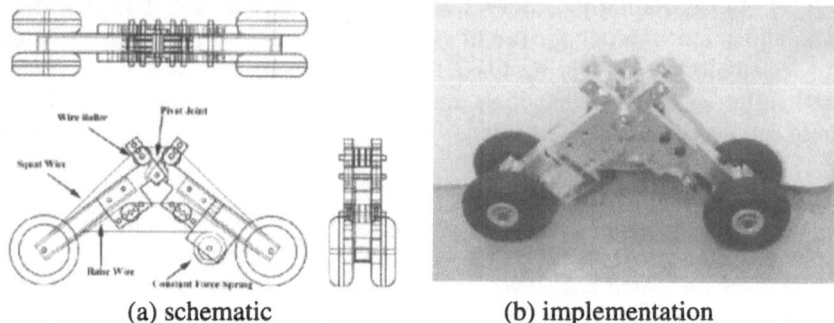

| (a) schematic | (b) implementation |

Figure 3. Reconfigurable rocker.

The working prototype has the performance capabilities required for a real rover design. The rocker angle has approximately a 60 degree range of motion (expanding the rocker angle from 90 degrees to about 150 degrees). This results in a change in height of the pivot point of 2.75 cm. The lightweight SMAs allow the rocker to lift a 10 kg payload, half the weight of the rover it was designed for. Actuation takes 0.5 seconds and consumes less than 1.9 W peak. PID control of the SMAs was also studied to examine continuous (non-binary) operation, with good results.

3.2.2. Binary Gripper

In this second device the conventional bearings were replaced by elastic flexure hinges. A one degree-of-freedom binary gripper was designed to show the light weight, simplicity, and utility of the design concept [12]. This gripper could be used for rock sample collection on a space explorer. The gripper has been used for this purpose on a laboratory rover test bed [13].

When the single SMA wire is actuated, the fingers of the gripper close around the object to be grasped (see Figure 4). The key of the gripper design was to amplify the small (5%) deformations of the SMA actuator to be large enough to grasp rock

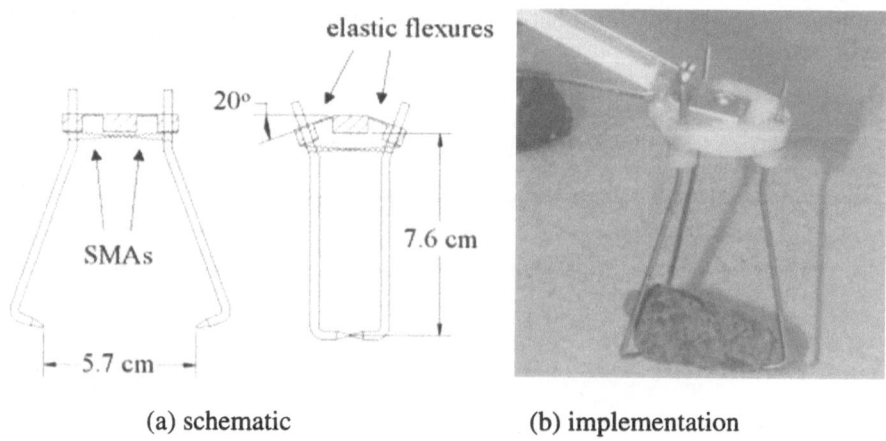

| (a) schematic | (b) implementation |

Figure 4. Binary gripper.

samples, while keeping forces at a usable level. Elastic flexures are incorporated at all three finger pivots. The compliance of these flexures acts as a return spring to open the fingers upon release of the object. This device uses no sensors, bearings, or gears.

The gripper with its needle-like fingers and simple binary action is able to reliably grasp rocks of varying size. The mounting base dimension is 2.5 cm in diameter. Each finger is 7.6 cm long, with a motion range of 20 degrees, and can pick up objects up to 5.7 cm in diameter. It can provide a normal force of up to 0.110 N and 0.0422 N at each finger tip, with a 250 micron and 150 micron SMA wire respectively. The expected normal force is 0.149 N and 0.0536 N for the 250 micron and 150 micron wires respectively. The difference in observed verses expected forces can be attributed to wire slip at mounting points, slop in the wire, etc. These observed forces result in reliably lifting objects weighing up to 330 g, six times its own weight of 55 g.

3.2.3. Pantograph Mechanism

A single DOF 4-bar pantograph mechanism was designed to demonstrate large amplification of binary muscle actuator motion (see Figure 5). This amplification is required by some ABE applications, such as legged walking machines.

The pantograph mechanism exemplifies some of the advantages of using compliant members and is an advance over the previous devices considered. The mechanism is simple, since the flexures and bars can be molded or cut from a single piece of material. In this case, a laser cutter was used to cut the part from sheet plastic (PETG — polyethylene terephthalate, glycol modified), although other methods coulld have been used. Like the previous two studies, this mechanism demonstrates how very small actuator motions can be amplified to usable scales in designs. Although SMA actuator technology can achieve usable elongations of only a few percent, this design can amplify this displacement by a factor of 8 in theory. In the working prototype, the endpoint deflected 29 mm given an actuator contraction of only about 4.5 mm, yielding an amplification of 6.5. The difference between theory and implementation is due mostly to un-modeled compliance in the bars and in irregularities in the SMA actuators.

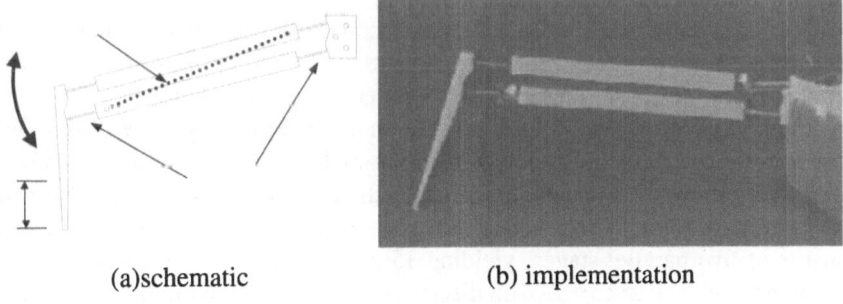

(a)schematic   (b) implementation

Figure 5. Binary pantograph mechanism.

3.2.4. Articulated Binary Limb (ABL)

A more general Articulated Binary Limb (ABL) was designed to study some of the issues involved with multi-DOF robotic systems [14]. The ABL could be used in a wide variety of space robotic applications, as a dexterous manipulator, as a cam-

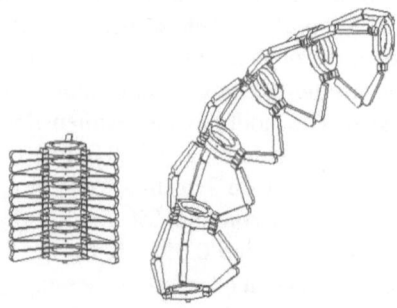

(a) full system stowed and deployed

Flexure to allow lateral motion of the links ⤏

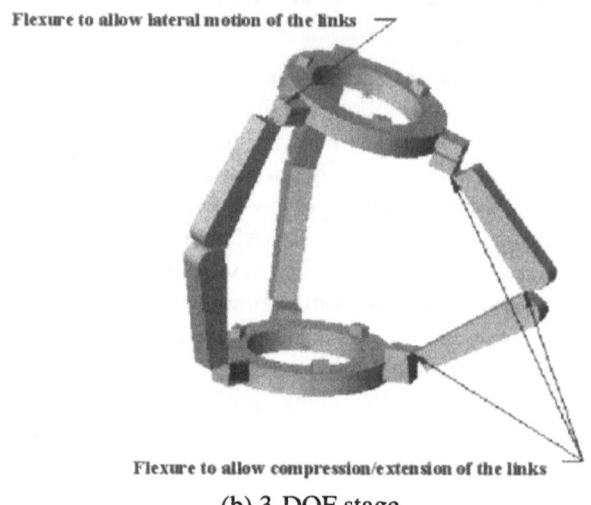

Flexure to allow compression/extension of the links ⤐

(b) 3-DOF stage

Figure 6. Articulated Binary Limb (ABL).

era or instrument mount, as an articulated connection between cooperating robotic explorer/workers, etc. (see Figure 1). This device is comprised of a serial chain of parallel stages (see Figure 6). In the experimental implementation, each 3-DOF stage is composed of three 1-DOF links, each with a shape memory alloy (SMA) actuator. The links are fabricated from polyethylene using high precision water jet cutting. The link shape provides for an elastic hinge at the end of each element. The actuation scheme allows for only binary operation of each link. The experimental structure built consists of five parallel stages, yielding 15 binary degrees of freedom. The ABL is thus able to achieve $2^{15} = 32,768$ discrete configurations. With this many configurations, the device can approximate a continuous system in dexterity and utility. By its polymer construction and binary actuation the design is very lightweight and simple.

In order to actuate and control such a system, based on the desired kinematics, power must be applied to each actuator by a central controller. The large number of actuators can rapidly make the physical realization of such a system difficult, if each actuator requires an individual set of power supply lines from the central controller

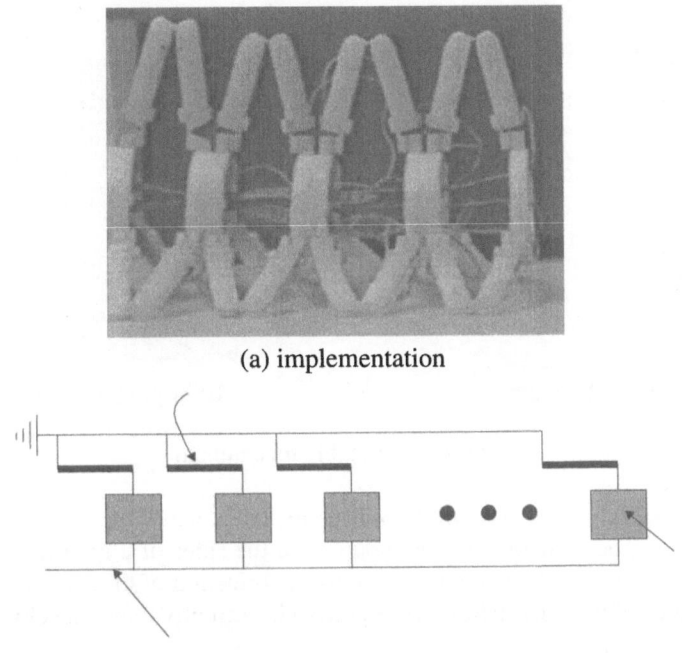

(a) implementation

(b) power bus architecture

Figure 7. Articulated Binary Limb implementation and control.

(see Figure 7a). This multitude of wires adds unacceptable weight and complexity. In this study, a more effective control architecture has been designed (see Figure 7b). A common power line and ground runs between all of the actuators. The power signal is encoded with a high-frequency component that instructs which actuators to turn on or off. Each actuator has a small decoder chip that can be triggered into either binary state by the carrier signal piggybacked on the power line. The carrier signal is a sequence of pulses that identifies a unique address in the form of a binary word. Once triggered, the decoder chip stays latched to that state until triggered otherwise. This power bus architecture reduces the wiring of the entire system to only two wires (power and ground) and can easily be implemented in the form of conducting paint/tape (on the non-conducting polyethylene substrate) to minimize the structural disturbance forces due to wiring.

A key element of the ABE concept is the bi-stable structure. The devices above do not have true bi-stable character in that they depend on actuator forces to hold them in position. To keep a binary actuator in a fixed state often requires power. In order to eliminate this need for continuous power supply and save energy, bi-stable or locking mechanisms can be employed in binary mechanisms to lock the mechanism into each state. In this study methods for bi-stability have been developed. One of the more successful designs is shown below (see Figure 8). This design is comprised of bi-stable mechanisms sandwiched by flexure beams. The bi-stable mechanisms use

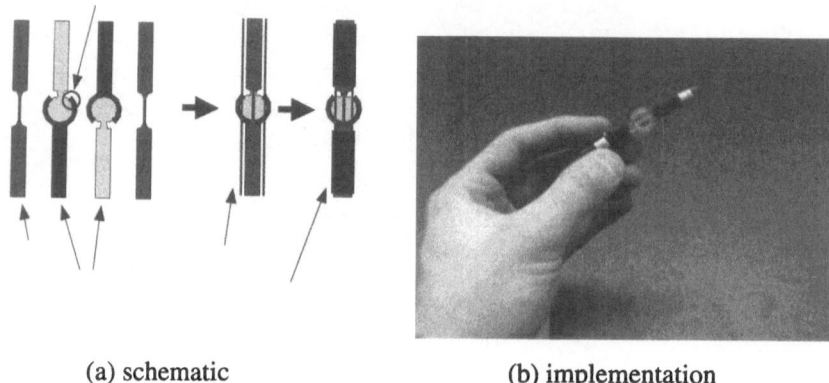

(a) schematic   (b) implementation

Figure 8. Bi-stable mechanism.

detents to passively lock the joint into discrete states, while the flexures add out-of-plane rigidity. The actuators can be mounted to the sides of a joint like this, similar to the musculature of a human elbow or knee. Fabrication of these bi-stable elements into the hinges of the ABL (shown in Figure 6) is currently under development.

3.2.5. Hyper-DOF Mechanisms

In the progress of developing the ABE concept, continuously compliant structures with embedded actuators are being studied with the aim of understanding a basic building block for more complex designs [15]. Embedding a large number of actuators into a continuously compliant structure can achieve unique motion amplifications and shape deformations (see Figure 9). A mechanism with a cellular structure of hundreds of small deformable cells may provide high performance and large ranges of motion.

Structural finite element analyses have been done of various shapes for individual structures, such as beam elements with voided interior volumes, diamond shaped elements, dog legged elements, elbow elements, and hexagonal elements (see Figure 10). This analysis shows the strong relationship between the structure shape and deformation of the elements.

The manufacture of these elastic elements with embedded actuation is not trivial. In our studies, compliant structures have been cast using different elastomeric materials (see Figure 11). Future work will consider such methods as stereolithography or selective laser sintering as possible alternative fabrication methods.

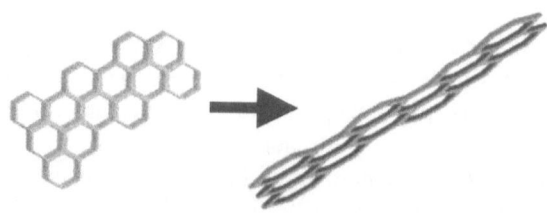

Figure 9. Hyper-DOF network of cells achieving large displacements.

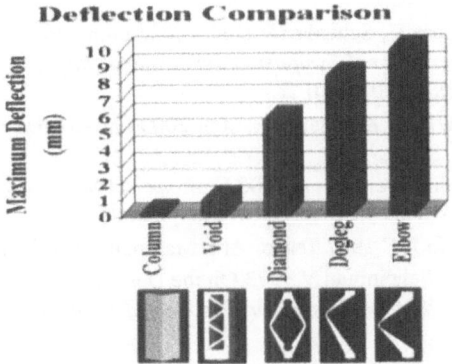

Figure 10. Expected results for various structures.

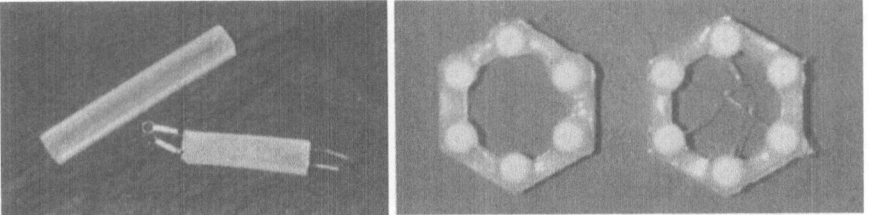

Figure 11. Hyper-DOF systems: flexible structures with embedded SMAs.

4. Summary

This paper presents an experimental study of a new paradigm for the design and fabrication of robotic components, with attention to potential space robotics applications. These robotic components would use embedded binary actuators and elastic members with bi-stable elements. These components have the potential to be implemented with a fraction of the weight, volume, cost, and complexity of conventional systems composed of motors, bearings, geartrains, and encoders. The results of this study suggest that these devices would be effective for robotic systems that are capable of accomplishing space missions of substantial complexity, with flexibility and robustness. The concept is currently limited by polymer actuator development, where present technologies possess practical limitations such as high power requirements, low displacements, or the need for immersion in solutions.

Acknowledgements

The authors would like to acknowledge the NASA Institute for Advanced Concepts (NIAC) for supporting this research. Also the important contributions of Kate Andrews, Robert Burn, and Guillermo Oropeza (students at MIT) and Sharon Lin (research engineer) are acknowledged.

References

[1] Weisbin C, Rodriguez G, Schenker P, et al 1999 Autonomous Rover Technology for Mars Sample Return. *International Symposium on Artificial Intelligence, Robotics and Automation in Space (i-SAIRAS'99)*, 1-10

[2] Madden J D, Cush R A, Kanigan T S, et al 2000 Fast-contracting Polypyrrole Actuators. *Synthetic Metals*, 113: 185-193

[3] Pelrine R, Kornbluh R, Pei Q, et al 2000 High-speed Electrically Actuated Elastomers with Over 100% Strain. *Science*, Vol. 287, No. 5454, 836-839

[4] Anderson V C, Horn R C 1967 Tensor Arm Manipulator Design. *ASME paper*, 67-DE-57

[5] Roth B, Rastegar J, Scheinman V 1973 On the Design of Computer Controlled Manipulators. *First CISM-IFTMM Symposium on Theory and Practice of Robots and Manipulators*, 93-113

[6] Goldberg K 1992 Orienting Polygonal Parts Without Sensors. *Algorithmica*, special robotics issue

[7] Canny J, Goldberg K 1993 A RISC Paradigm for Industrial Robotics. *Tech. Rep. ESCR 93-4/RAMP 93-2*, Engineering Systems Research Center, University of California at Berkeley

[8] Chirikjian G S 1994 A Binary Paradigm for Robotic Manipulators. *Proc. IEEE International Conf. on Robotics and Automation*, 3063-3069

[9] Ebert-Uphoff I, Chirikjian G S 1996 Inverse Kinematics of Discretely Actuated Hyper-Redundant Manipulators Using Workspace Densities. *Proc. IEEE International Conf. on Robotics and Automation*, 139-145

[10] Lees D S, Chirikjian G S 1996 A Combinatorial Approach to Trajectory Planning for Binary Manipulators. *Proc. IEEE International Conf. on Robotics and Automation*, 2749-2754

[11] Iagnemma K, Rzepniewski A, Dubowsky S, et al 2000 Mobile Robot Kinematic Reconfigurability for Rough-Terrain. *Proceedings of the SPIE Symposium on Sensor Fusion and Decentralized Control in Robotic Systems III*

[12] Burn R D 1998 Design of a Laboratory Test Bed for Planetary Rover Systems. Master Thesis, Department of Mechanical Engineering, MIT

[13] Iagnemma K, Burn R, Wilhelm E, et al 1999 Experimental Validation of Physics-Based Planning and Control Algorithms for Planetary Robotic Rovers. *Proceedings of the Sixth International Symposium on Experimental Robotics, ISER '99*

[14] Oropeza G 1999 The Design of Lightweight Deployable Structures for Space Applications. Bachelor Thesis, Department of Mechanical Engineering, MIT

[15] Andrews E K 2000 Elastic Elements with Embedded Actuation and Sensing for Use in Self-Transforming Robotic Planetary Explorers. Master Thesis, Department of Mechanical Engineering, MIT.

A first-stage experiment of long term activity of autonomous mobile robot — result of repetitive base-docking over a week

Yasushi Hada   Shin'ichi Yuta

Intelligent Robot Laboratory
University of Tsukuba
Tsukuba 305–8573, Japan
E–mail:{had, yuta}@roboken.esys.tsukuba.ac.jp

Abstract

The aim of this research is the realization of an autonomous mobile robot which can perform many kind of tasks in a real environment for a long duration without supports by human operators. We call such autonomy long term activity.

A robot, which works for long term autonomously should be able to get re-charging the battery without failure, at least. As the first-step of long term activity, we made a basic experiment in which our mobile robot are presented in this paper. In the experiment, our robot "Yamabico-Liv" survived over a week while it had repetitively performed going in and out of the battery charge station every 10 minutes. In this paper, the implementation and result of this experiment, and next one which is now undergoing are presented.

1. Long term activity of autonomous robots

1.1. Problem - long term activity

We are investigating on the long term activity of autonomous mobile robots. Long term activity means that the robot performs given tasks continuously without any human's supports.

Most of the currently existing intelligent robots can work only for a limited time, and require to be reset and maintained their hardware and software at each task. For example, when we try to perform a navigation experiment, we reset the hardware and software of the robot at first, and bring it to the starting point by pushing or by manual control before an experiment. And when the experiment is finished at the goal destination, we reset the system again. The robot is autonomous only during the exact duration of the experiment. Our aims are to automate these troublesome procedures and develop a robot which can work for long term.

1.2. Elementary technologies for long term activity

For long term activity, to realize high-level robustness, durability of hardware, rich basic functions, energy supply function and system level software which manage them are needed. Among them, it is one of the most important abilities to get energy from the environment to survive. It has to deal with many kinds

D. Rus and S. Singh (Eds.): Experimental Robotics VII, LNCIS 271, pp. 229–238, 2001.

of possible obstacles in its tasks and go back to its home position to supply its battery.

To get energy time to time autonomously, a special energy supply base is prepared and placed in the environment where robot works. We call it a robot base.

1.3. Target long term activity

We realize long term activity of a robot by taking the following steps. First, realize basic long term activity to charge batteries autonomously while going in and out of the base for a week. Next, realize a long term activity in which robot performs one task during basic long term activity. And then, realize long term activity in which robot performs multiple tasks during basic long term activity.

In this work, we put a concrete goal that our experimental mobile robot "Yamabico-Liv" can perform several kinds of tasks autonomously in the corridor and our laboratory for a week (10080 minutes). [1]

2. A first-stage experiment

2.1. Motion in experiment

As a first basic trial of long term activity, we have implemented hardware and software of an autonomous robot to make repetitive motion of going out from the base and going into dock with base. The experimental environment is placed in our laboratory(3L302 room in Univ. of Tsukuba), where is usually busy with students and not prepared for the robot specially.

The robot moves for 150[cm] from the base to a fixed point and goes back from the point to the base to re-charge the batteries. The robot repeats with this navigating motion every 10 minutes continuously for a week.

2.2. Experimental platform "Yamabico-Liv"

We are trying to realize long term activity on "Yamabico-Liv" which is based on our experimental platforms Yamabico for robotics research (Figure 1). We have been developing a series of small size wheeled mobile robots as the research platforms which are named Yamabico, and we have more than 30 Yamabico robots in our laboratory and are using them for various experiments on autonomous mobile robots. In addition to the standard functions of Yamabico robots, "Yamabico-Liv" has some extra hardware and software integrated on it for long term activity.

2.2.1. Basic function of Yamabico

Yamabico is a small size (40cm cubed) mobile robot platform with functionally distributed controller architecture.

It is using DC motors, and PWS (Power Wheeled Steering) driving mechanism. It has a dead-reckoning function using integration of pulse from encoders attached to the motors. Standard Yamabico has four ultrasonic sensors which can measure the distance to the objects in the environment. The seven segments of touch sensors on the bumper detect the collision with obstacles. It also has a function to emit synthesized or recorded voice sounds. These functions are distributedly implemented as the function modules , and on master module there is a task level program which makes a total motion of the robot. Operator who uses the Yamabico has to develop this level program.[2]

"Yamabico" uses two 12[V]4[Ah] lead acid batteries, and its life time is about 1 hour. To make Yamabico work longer, it is necessary to supply energy from environment.

 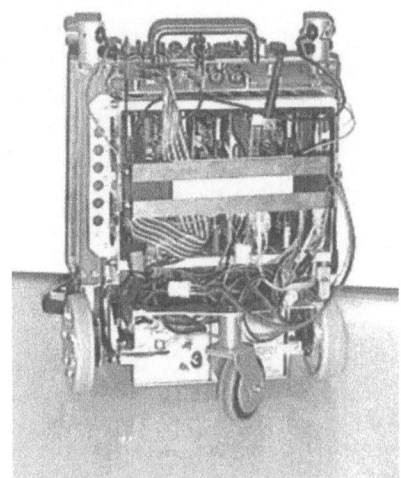

Figure 1. Yamabico-Liv(left:front view, right:back view)

2.2.2. Additional functions for Yamabico-Liv

Yamabico-Liv has some additional functions for long term activity.

Battery management system   Besides the electric terminal to get energy, Yamabico-Liv has a sensor which measures the input/output electric current of the batteries, and a software which computes the remaining power energy of the batteries. It can also recognize the connection to the energy supply dock.

Optical position sensor for docking   We installed two photo-micro-sensors facing to the floor, along the axis of the wheels. This sensors detect a reflecting tape stuck in front of the base, and know the accurate position for connecting to the base.

2.3. An energy supply base and terminals

The energy supply base is placed in the environment in order to charge Yamabico-Liv batteries. Yamabico-Liv and the supply base have special electric terminals. By butting them into contact, Yamabico-Liv can get electricity to charge the batteries.

The supply base is shown in Figure 2. It is boxed shape, and contains the terminal which is a 22[cm]*6[cm] sheet copper connected to a DC power supply. A delicate sloped floorboard is provided in the entrance of the base in order that Yamabico-Liv can connect to the base terminal without motors driving. The terminal is supported by a spring to ease an impact during the connection and provides a margin for positioning error of Yamabico-Liv. The terminal of Yamabico-Liv is also the same size sheet copper, and connected to the batteries.

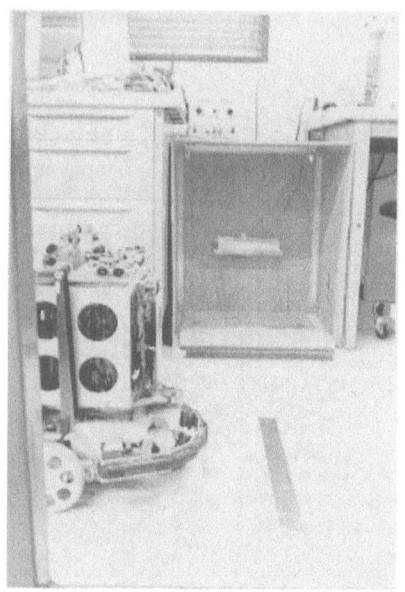 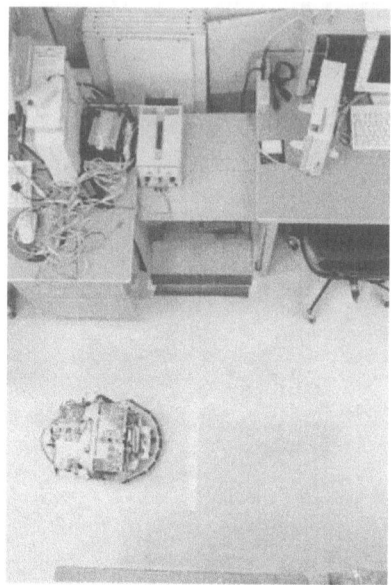

Figure 2. Energy supply base(left:front view, right:top view)

2.4. Base-docking motion

In this experiment, Yamabico-Liv goes out from the base, moves to a fixed point and navigates back to connect to the base.

This is very simple motion, however, we considered various possibilities of the fail and made a navigation program which can cope with them. It is written using our state transition language "ROBOL/0".[3] The number of states are 28 in this program.

When the robot comes back into the base, it starts from position A in Figure 3 and moves straight and searches the reflecting tape. Using the difference of time two sensors detects the tape, the angle between the robot and the tape can be calculated. Using this information, the right angle to the station can be obtained at the position B. Then it goes into the base and connects the terminal at C. The connection can be detected by the current to the batteries. If the navigation fails for some reasons, the robot goes back to the point A, and tries the same procedure again.

2.5. Result

The experiment was carried out from May 8 to May 15, 2000, over a week. In this duration, Yamabico-Liv performed the in-and-out navigation 1,080 times and it run 3,391[m] in total distance. Although the robot met many types of obstacles during experiment, the program coped with all of them and worked

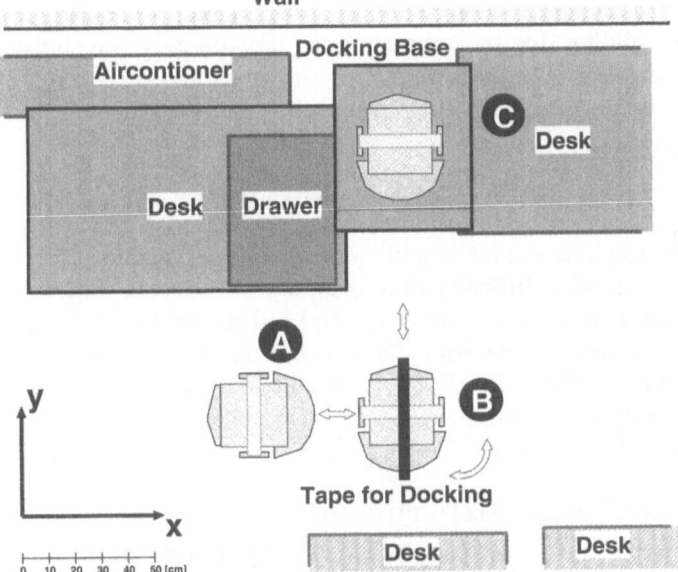

Figure 3. Algorithm to connect to the base

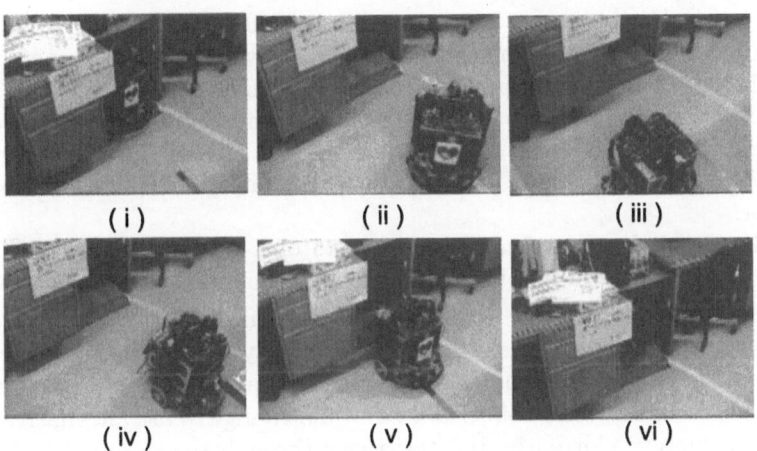

Figure 4. Docking motion

continuously.

Among the 1,080 in-and-out navigations, 1,018 times are done smoothly, however, it could not detect and run over the tape on the floor for 64 times. Moreover, for 20 times it got wrong angle to the station because of the sensor noise. The reason of these accidents are the dirt of the tape caused by passengers, or scattered paper on the ground. The collision with obstacles occurred 18 times in the experiment, where the obstacles were the leg of chairs or people. Other accidents were little slippage caused by a scattered paper.

However, the robot could recover these errors autonomously by detecting a failure and re-trying the navigation. A really unexpected accident was a power cut of our building due to a thunderbolt. Fortunately the shutdown lasted a few minutes and it was no problem for the continuation of experiment except for our PC for logging data was turned off.

2.6. Consideration

This realized experimental system had already been improved for about some possible obstacles based on past experiments. Through in this experiment, we checked the robustness and reliability of this base-docking system.

We also found a difficulty of debugging. Because, it continued to run day and night, and the robot might stack and fail the motion behind us.

Since our system employs odometry base positioning, the slippage is a serious problem potentially. Even though little slippage can be corrected using external sensors, large slippage potentially makes the robot lose its position and not be able to continue the motion.

3. A second-stage experiment

We are now trying to next step in which a robot survives in the corridor with navigating longer distance for a week.

3.1. A task in a corridor for next experiment

In the previous experiment, our robot didn't perform any meaningful tasks. It continued only basic motion repetitively. Next, we give a task to the robot.

The task is a patrol navigation in the corridor. It starts from the fixed point near the base in front of the 3L402 room and it goes around the corridor for 50[m]. The task is done every 1 hour and 192 tasks will be done in a week. The width of the corridor is about 3[m] and many passengers go through the corridor.

The target long term activity consists of a docking motion part and a patrol navigation part. We improved the experimental system as follows:

Improvement of energy supply system   In last experiment, we found one of the most serious error is the bad quality of the electric terminal to connect to the base. It becomes dirty caused by a little electric spark and it is dangerous for human and robot. So, we would change to use alternative current to supply energy. Currently, the electricity through the terminal is AC100[V] and Yamabico change to DC28[V] using internal power regulator (Figure 5).

3.2. New energy supply base

We designed a new energy supply base. In addition to the power supplying function, it has an TV cameras and VCR in the upper space (Figure 6) to record scenery of experiment. This videotape would be used for debugging in case of failure. Old base is also available to supply energy and Yamabico-Liv can navigates between them.

3.3. Image recording system for debugging

Since the robot navigates in the corridor day and night for several days, we cannot watch the whole of experiment. For this reason, it becomes harder to

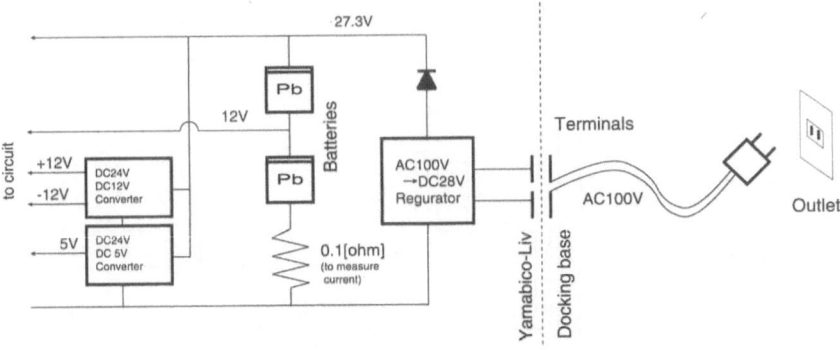

Figure 5. Electric circuit for battery re-charging

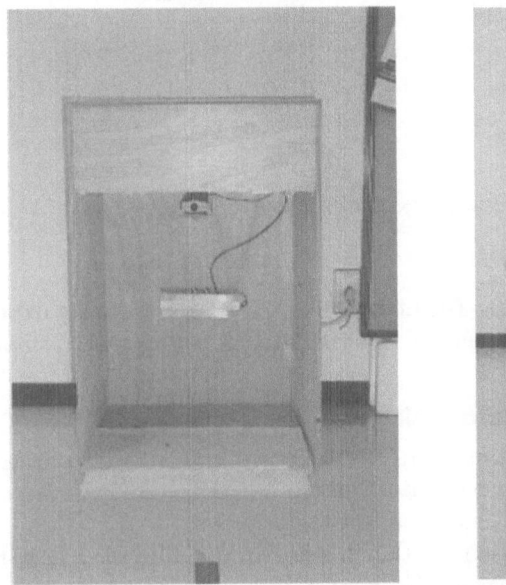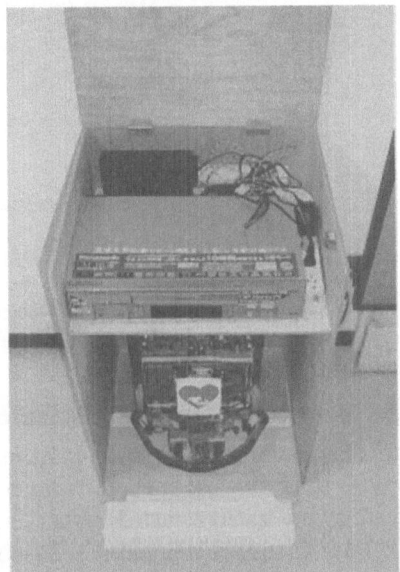

Figure 6. Energy supply base(front view, top view)

develop and debug the experimental system than usual.

To solve the problem, we establish a system to monitor the motion of the robot and its environment over while the robot lives. This system consists of two CCD cameras and VCR. These work independently from the robot and its base.

cameras   We installed 2 cameras to take motion of the robot during experiment. Since the most important motion is connection to the base, one camera is installed inside of the base. The other camera is installed on the ceiling of the corridor to overlook the experiment.

picture mixing machine   To record images from cameras, we use picture mixing machine(PQS-2000C) which mingle 4 NTSC signals(max) into one

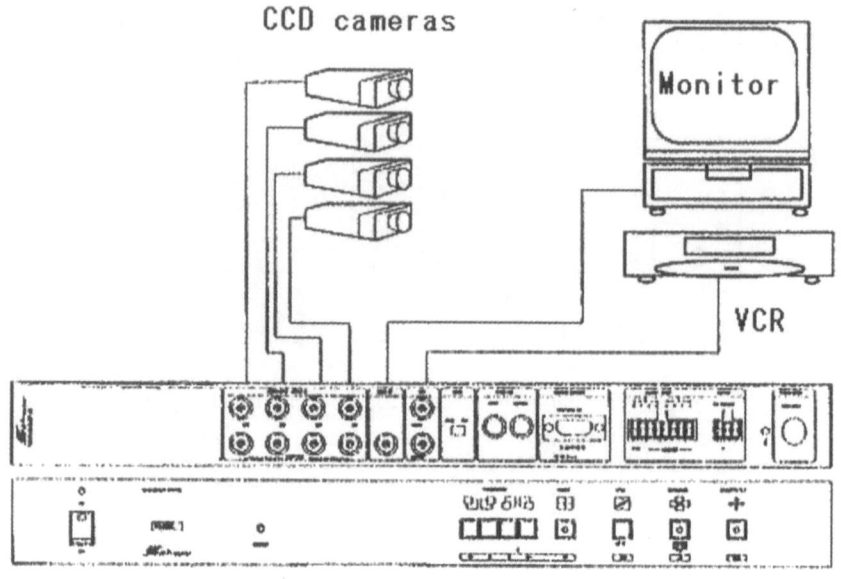

Figure 7. Image recording system

NTSC signal.

VHS VCR with long recording function   A picture taken during experiment is recorded to VHS VCR(Panasonic NV-HSB20). It can record 17.5hours with a tape.

3.4. Navigation software with a task

The navigation program is written in ROBOL/0 and the number of states are 38. It is divided into docking navigation part and patrol navigation part. Docking navigation part is almost same as last experiment.

Patrol navigation part is based on the System Route Runner. It is the map-based robust navigation program for Yamabico, which has been already implemented by our group. It plans the route between any two point in the environment map thanks to the MaP(Map management and Planning) function module and it makes the robot trace the planed route[5]. System Route Runner also has position estimation function and obstacle detection function using ultrasonic sensor[4].

3.5. Current progress

Most of the system is accomplished. Up-to-now, Yamabico-Liv could success the navigation only 20-30 times continuously. We are trying to fix some problem which potentially makes the failure and improve the robustness.

4. Summary

In this paper we discussed the concept of the long term activity and its necessary elementary technologies. Then we described our experimental system and

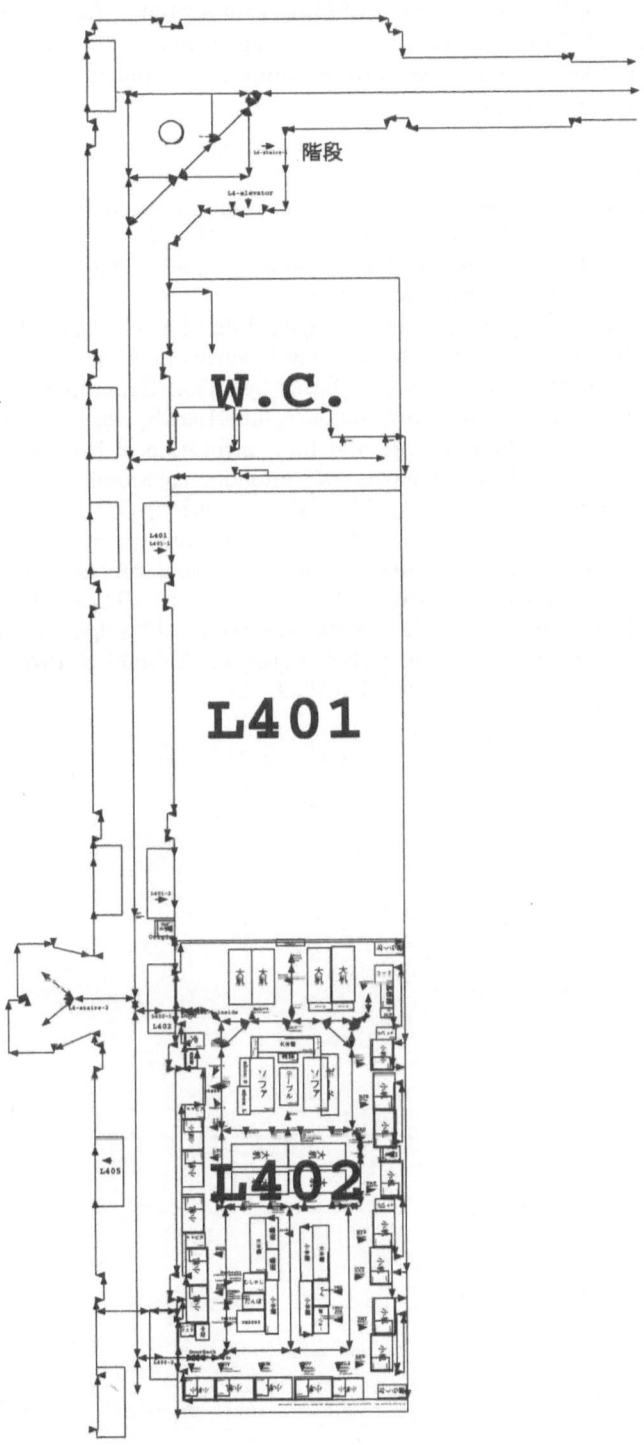

Figure 8. Environment Map in the corridor and 3L402

the results of basic experiment of long term activity. As a result, Yamabico-Liv successed base-docking motion 1'080 times repetitively. We are now in the second step of the research. We are planning to further it to be more autonomous in the real environment.

References

[1] Shin'ichi Yuta and Yasushi Hada: Long term activity of the autonomous robot -Proposal of a bench-mark problem for the autonomy,Proceedings of the 1998 IEEE/RSJ International Conference on Intelligent Robots and Systems (IROS'98),Oct.1998,pp.1871-1878

[2] Shin'ichi Yuta, Sho'ji Suzuki, Shigeki Iida Shin'ichi Yuta: Implementation of a small size experimental self-contained autonomous robot – sensors,vehicle control, and description of sensor based behavior, Experimental Robotics(The 2nd International Symposium),Toulouse,June.1991,Springer-Verlag,pp.344-359

[3] S. Suzuki, S. Yuta: Design and Implementation of Programming Environment for a Sensor Based Behavior of Autonomous Mobile Robots, Intelligent Autonomous Systems(IAS-3),Feb.1993,Pittsburg,pp.407-416

[4] Yasushi Hada and Shin'ichi Yuta: Robust navigation and battery re-charging system for Long Term Activity of autonomous mobile robot,The 9th International Conference on Advanced Robotics('99 ICAR),Oct.1999,pp.297-302

[5] Tomoki Ogura and Shin'ichi Yuta: Autonomous Navigation Function as System-level Software in Mobile Robot Platform 'Yamabico' Proceedings of the 4th Robotics Symposia, pp.359-366,1999

Comparing the Locomotion Dynamics of the Cockroach and a Shape Deposition Manufactured Biomimetic Hexapod

Sean A. Bailey, Jorge G. Cham, Mark R. Cutkosky
Center for Design Research
Stanford University
Stanford, CA 94305 USA
baileys@cdr.stanford.edu

Robert J. Full
Department of Integrative Biology
University of California at Berkeley
Berkeley, CA 94720 USA
rjfull@socrates.berkeley.edu

Abstract: We describe the locomotion dynamics of a biomimetic robot and compare them with those of its exemplar: the cockroach. The robot is a small (0.275kg) hexapod created using a layered manufacturing technique that allows us to tailor the compliance and damping of the limbs to achieve passive stabilization similar to that observed in insects. The robot runs at over 3 body-lengths per second (55 cm/s) and easily traverses hip-height obstacles. However, high-speed video and force data reveal differences between the robot's locomotion dynamics and the inverted spring-pendulum model that characterizes most running animals, including cockroaches. Closer examination of the individual leg forces shows that these differences stem from the behavior of the middle and rear legs and points to suggestions for future designs and further experimentation.

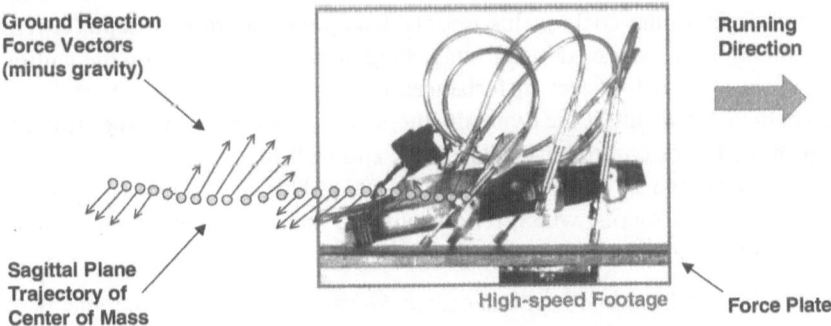

Figure 1. *Sprawlita*, a dynamically-stable running hexapod, and the force plate used for measuring ground reaction forces. Reflective markers were used for high speed video motion capture.

D. Rus and S. Singh (Eds.): Experimental Robotics VII, LNCIS 271, pp. 239–248, 2001.

1. Introduction

The basic behaviors of running and walking in animals have been reproduced in legged robots [1][2]. Some have attempted to copy the morphology of biological systems in their design [3], while others implement controllers based on observed animal behavior [4]. More recent work has focused on emulating the mechanical properties of biological structures. As described in [5], a novel rapid-prototyping technique called Shape Deposition Manufacturing allows robots to be built with soft, visco-elastic materials integral to the structure that provide functional compliance and damping [6]. Using this technique, we have built a biomimetic robot named *Sprawlita* intended for fast robust locomotion through uncertain terrain.

In this paper, we describe the cockroach-inspired robot design and the resulting robot performance. We then present results using two experimental measures traditionally used in biomechanics to characterize running: pendulum-like energy recovery and ground reaction forces. Finally, we draw conclusions about the difference in locomotion styles and discuss the changes in design and future experiments that these findings suggest.

2. Biomimetic Design: The Animal and the Robot

For its size, *Periplaneta americana* is among the fastest known animals with maximum speeds of over 50 body lengths per second [7]. Although it is significantly slower at 10 body lengths per second, the *Blaberus discoidalis* cockroach [8] is still far faster for its size than legged running machines built to date.

The cockroach's physical robustness is widely recognized, but its performance over extremely rough terrain is less well-known. The *Blaberus discoidalis* cockroach can easily traverse a fractal surface containing obstacles of up to three times the height of its center of mass [9], a feat only recently achieved in legged robots [10]. Of particular interest is that this fast and robust performance is thought to be achieved by a relatively simple motor control pattern. Preliminary results suggest that there are only minor changes in the cockroach's muscle activation pattern as it rapidly transitions from smooth to uneven terrain [9], suggesting heavy dependence on the ability of the mechanical system to reject disturbances [11]. In essence, fast robust locomotion appears to be the result of the dynamic interaction between sprawled posture, a timed feedforward motor controller [10][12][13], and well-tuned passive visco-elastic elements, also known as "preflexes" [14][15][16]. This is in contrast to the control schemes of many robots, which rely heavily on active feedback control rather than passive components.

We have built a biomimetic robot named *Sprawlita* which incorporates these suggested components for fast robust locomotion: passive visco-elastic mechanical properties tuned to a timed feedforward motor controller. This robot was fabricated using a rapid-prototyping technique called Shape Deposition Manufacturing (SDM) [5][17] which allows for integrated structures with soft, viscoelastic materials that provide compliance and damping. The ability of the SDM process to embed active components such as actuators inside the structure of the robot also allows us to approach the

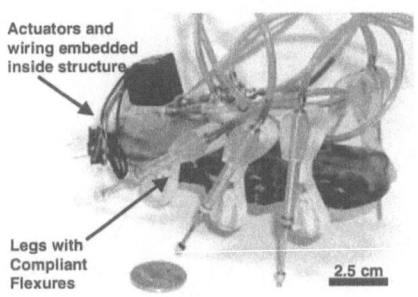

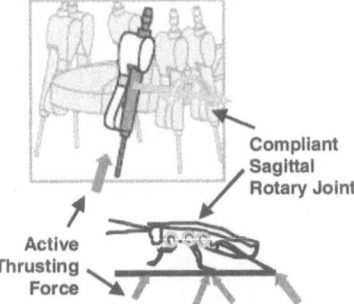

Figure 2. *Sprawlita* was designed based on functional principles from biomechanical studies of the cockroach. The prototype was fabricated using Shape Deposition Manufacturing and is capable of speeds of over 3 body-lengths per second. Studies of ground reaction forces in cockroach locomotion show that forces are directed towards the hip joints, essentially acting as thrusters.

physical robustness of the mechanisms found in nature.

Rather than directly copying the morphology of the cockroach leg, the robot was designed through "functional biomimesis," drawing from studies of leg function, arrangement and passive properties [17]. As shown in Figure 2, the robot's legs are arranged in a sprawled posture in the sagittal plane and consist of a simple mechanism that incorporates a pneumatic piston attached to the body through a viscoelastic hip joint. The compliant hip joint is designed to mimic the function of the trochanter-femur joint which is believed to be a mostly passive, viscoelastic element [6] rotating about an axis perpendicular to the sagittal plane. The thrusting piston is designed to mimic the function of the coxa-femur-tibia linkage.

Our robot is controlled by alternately activating each of the leg tripods in an open loop fashion at fixed time intervals. Each tripod is pressurized by separate 3-way solenoid valves. This simple, open-loop control scheme is the extreme of the minimal feedback control hypothesized for the cockroach. Therefore, the robotic system relies heavily upon the passive, self-stabilizing properties of the robotic mechanical system[1].

For the results presented here, the tripods are alternately activated over a 130 ms stride period at a 35% duty cycle, with 50% corresponding to a half-stride, or 65 ms. Despite the binary pneumatic actuation scheme, the force output at the pistons is surprisingly muscle-like in form as shown in Figure 3. The tubing lengths, valve porting, and small piston orifices conspire to transform the square wave valve input into a smooth force output.

Despite this simple mechanical arrangement and motor controller, *Sprawlita* achieves speeds of over 3 body lengths per second, or 0.55 m/s, and can overcome hip-height obstacles with little difficulty. This performance, though humble in light of the cockroach's, begins to compare to that seen in nature.

[1]The same degrees of freedom were designed into another biomimetic robot called *Rhex* [10], although the active and passive degrees of freedom are reversed resulting in a different functional mapping.

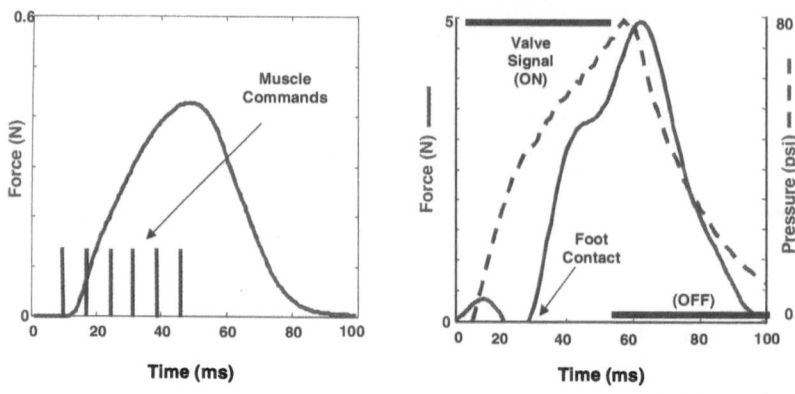

Figure 3. A comparison of isometric muscle force output [18] in response to motor commands and pneumatic piston force output in response to a solenoid valve input.

3. Basis for Comparison: Walking and Running Models of Animals

In animals there are very distinct patterns of force and motion when walking or running [19][20]. During walking, the kinetic and potential energy of the center of mass fluctuate sinusoidally and out of phase. Theoretically, the potential and kinetic energies can be exchanged via a pendulum-like energy recovery mechanism.

In contrast, running in animals is characterized by the kinetic energy and potential energy being in-phase, eliminating the possibility of pendulum-like energy exchanges. This type of motion is characterized by what is called the spring loaded inverted pendulum (SLIP) model. In addition, this model produces a characteristic set of ground reaction patterns, with the vertical force leading the horizontal force by 90 degrees [19].

As we will see, ground reaction force patterns and pendulum-like energy recovery measures help qualitatively determine how much each basic mechanism of locomotion is utilized.

4. Comparison Testing: Equipment and Methods

4.1 Cockroach Measurements

Position, velocity and ground reaction force measurements for the *Blaberus discoidalis* cockroach (mean mass 0.0026kg) were originally obtained in [8]. In summary, the cockroaches were run along a track with a force platform while a high-speed video system captured the locomotion at 60 frames/second. Velocity, position and kinetic and potential energy data were calculated by integrating the force signals. Stride beginnings and endings were determined by vertical ground reaction force patterns and verified using video information.

4.2 Robot measurements

Sprawlita (mass 0.275kg) was run along a plywood surface, with reflective markers attached to nose, back, each leg, and each foot. A high-speed video system captured the locomotion at 250 frames per second.

The force platform was a modified 6-axis force sensitive robotic wrist. An aluminum plate covered with a thin rubber layer to prevent slippage, as shown in Figure 1, was attached to the force wrist and placed flush with the plywood surface. The natural frequency of the force plate was 143Hz. Forces were filtered by an analog 4th order Butterworth filter at 100Hz, and then sampled at 1000Hz and converted to a digital signal. Forces were then digitally filtered at 50Hz by a Butterworth filter with zero phase shift. The minimum resolution of the force plate is approximately 0.1N in the vertical and fore-aft directions.

Center of mass position data were calculated by tracking the reflective markers attached to the body. The accuracy of this method is approximately 0.0001m. Velocity was calculated by taking the derivative of the position data. As with the cockroach, stride beginnings and endings were determined by vertical ground reaction force patterns, and verified using video information.

5. Biomimetic Comparison: Pendulum-like Energy Recovery

As discussed previously, a significant amount of energy may be available for recovery during walking via a pendulum-like energy recovery mechanism. In animals, this mechanism is used extensively, as energy recovery values approach 70% in walking humans [19] and 50% in crabs[20]. This measure can be calculated by:

$$\frac{(\Sigma HKE + \Sigma GPE) - \Sigma TE}{\Sigma HKE + \Sigma GPE} x 100\%$$

Here, ΣHKE is the sum of the positive changes in horizontal kinetic energy during one stride, ΣGPE is the sum of the positive changes in gravitational potential energy during one stride, and ΣTE is the sum of the positive changes in the total mechanical energy of the center of mass during one stride [19]. If there is only one peak in the given energy measure per stride, then the sum of the positive changes is simply the amplitude. In addition, vertical kinetic energy is typically excluded from these calculations as it is generally negligible in comparison to the other energies. Typical pendulum-like energy recovery is about 2% in running animals [19]. Thus, this metric is a quantitative indication of whether the observed locomotion is well represented by an inverted pendulum model, indicating walking dynamics.

5.1 Energy Recovery - *Blaberus discoidalis*

The pendulum-like energy recovery values for a cockroach during running are quite low, with a mean of 15.7%. This is a result of the kinetic energy leading the potential energy by only 7.6 degrees as shown in Figure 4 (**P**) [8]. While this is not surprising for the animal during fast locomotion, it is interesting that even at one-quarter the

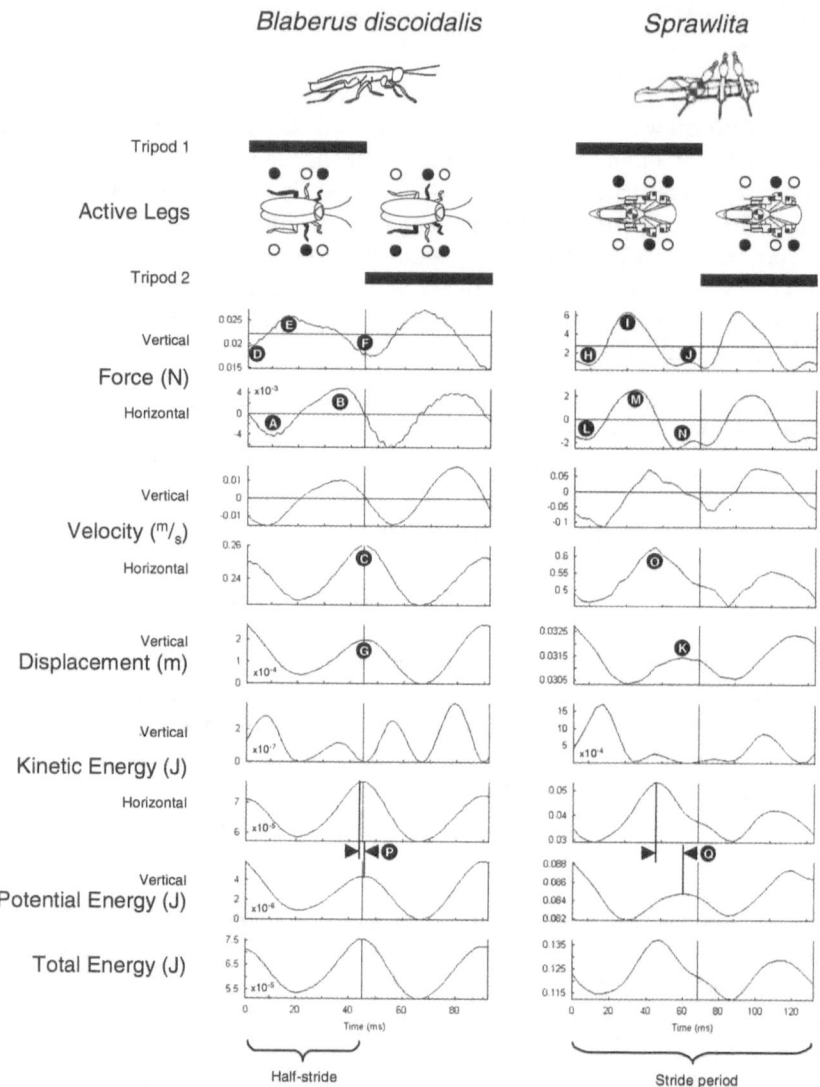

Figure 4. The results of force plate and high-speed video experiments described in Section 3 show differences in the locomotion of *Blaberus discoidalis* [8] and *Sprawlita*. The respective amounts of pendulum-like energy recovery, calculated from the center-of-mass energetics, indicate that neither hexapod is "walking." The respective ground reaction force plots show that the standard model of animal running, the spring-loaded inverted pendulum (SLIP) model, fits the cockroach well but the robot poorly. Labels (A) - (Q) correspond to features discussed in Sections 5 and 6.

maximum stride frequency (3Hz), the amount of pendulum-like energy recovery is low. At very low speeds, locomotion becomes intermittent, taking only a few quick strides at a time. Thus it seems that this animal actually *prefers* a running gait.

5.2 Energy Recovery - *Sprawlita*

The phasing as shown in Figure 4 (**Q**) between the kinetic and potential energies in our robot seem to place it closer to the inverted pendulum model observed in walking animals than to the running observed in the cockroach, as the kinetic energy leads the potential energy by 60 degrees. However, when the actual pendulum-like energy recovery is calculated, the value for *Sprawlita* is surprisingly low at 10.2%. This low value is due to the non-sinusoidal shapes of the energetics and the almost one order of magnitude difference between the magnitudes. Thus, while the robot, like the cockroach, does not exhibit the pendulum-like energy recovery associated with walking, it is dynamically dissimilar to the cockroach. The dynamic dissimilarity is underscored by examination of the ground reaction forces.

6. Biomimetic Comparison: Ground Reaction Forces

6.1 Ground Reaction Forces - *Blaberus discoidalis*

The ground reaction forces produced by *Blaberus discoidalis* are what one would expect for a running animal with bouncing dynamics. During the first part of a half-stride, the fore-aft horizontal force applies a braking force, slowing the body down as shown in (**A**) of Figure 4. As the half-stride progresses, the fore-aft force changes direction and an accelerating force is produced (**B**), causing the body to increase speed, with maximum horizontal velocity attained at the end of half-stride (**C**). In short, there is a clear *brake-propel* pattern over the course of each half-stride.

As shown in Figure 4, the vertical force pattern is just as distinctive. The vertical force is a minimum at the beginning of a half-stride (**D**) and increases to a maximum that occurs during in the middle of the half-stride (**E**). The vertical force then returns to the minimum by the end of the half-stride (**F**), resulting in a maximum vertical displacement as the cockroach switches from one tripod of legs to another (**G**). In short, the vertical force oscillates about the weight of the body in a *minimum-maximum-minimum* pattern over the course of the half-stride.

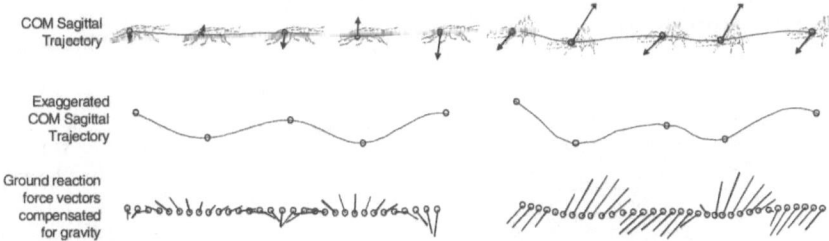

Figure 5. Ground reaction force vectors superimposed onto position data for an entire stride period. The vertical axis of the middle plot is exaggerated for detail. The ground reaction force vectors shown have been compensated for gravity by subtracting the weight of the robot from the vertical force measurement.

The aggregate of these fore-aft and vertical force patterns as shown in Figure 5 verify that the overall body motion is well characterized by the spring-loaded inverted pendulum model and is dynamically similar to other animals during running [8][21].

6.2 Ground Reaction Forces - *Sprawlita*

As shown in Figure 4, the vertical force patterns generated by the robot are quite similar to the cockroach. At the beginning of the half-stride, the vertical force is a minimum (**H**), very close to zero. Midway through the half-stride, the vertical force peaks (**I**) and then decreases back towards the minimum by the end of the half-stride (**J**), resulting in a maximum displacement near the tripod switch (**K**). As with the cockroach, there is a clear minimum-maximum-minimum pattern over the half-stride.

The fore-aft horizontal forces, on the other hand, are not as similar. As in the cockroach, the fore-aft forces begin the half-stride at a minimum (**L**), decelerating the body, and increase to a maximum (**M**), accelerating the body. Considering only this portion of the half-stride, there is a *brake-propel* cycle in both the animal and the robot. However, the latter part of the half-stride shows a pattern of light vertical forces (**J**) and decelerating fore-aft forces (**N**), resulting in an early horizontal velocity peak (**O**). This difference in the horizontal forces explains the large phase difference between the kinetic and potential energies as discussed earlier and is the key dynamic dissimilarity between the robot and the cockroach.

Examination of the video data reveals that the robot assumes a "pseudo-flight" phase during this part of the half-stride. Unlike a true flight phase, the middle and rear feet never leave the ground. Instead, they drag along in light contact, which accounts for the differing force patterns. The phenomenon is a result of the thrusting pistons reaching the end of their stroke before the stride is complete. At the same time, the torsional elements in the hips apply torques to the legs which keep the feet in contact

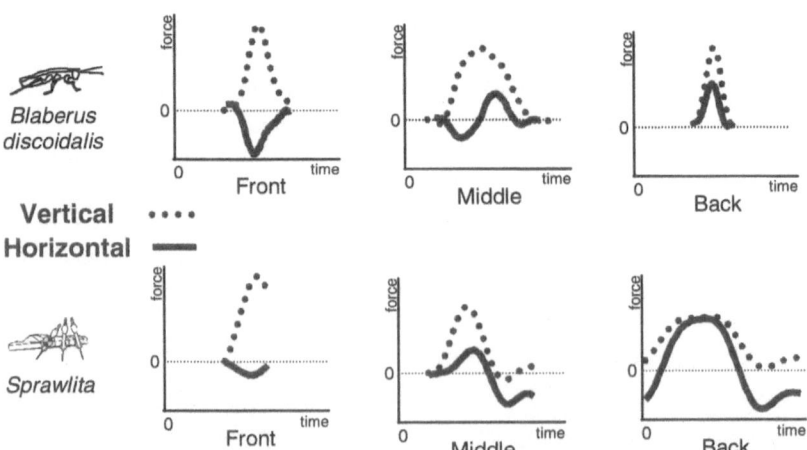

Figure 6. Plots of the individual leg ground reaction forces for *Blaberus discoidalis* [22] and *Sprawlita*. As indicated, dragging occurs in the middle and rear legs of *Sprawlita* during locomotion. This and the relative lack of deceleration provided by the front legs account for differences in locomotion dynamics.

with the ground.

While roughly similar in form, the comparison in Figure 5 shows that *Sprawlita* is not well characterized by the SLIP model.

7. Biomimetic Comparison: Individual Leg Ground Reaction Forces

7.1 Individual Leg Ground Reaction Forces - *Blaberus discoidalis*

There are many ways in which the SLIP model ground reaction force patterns can be produced by a system with multiple legs. In contrast to the Raibert approach of running with symmetry [1], the cockroach legs carry out very different functions in producing the SLIP-like behavior. While the vertical force patterns for individual limbs are similar, forces in the fore-aft direction are quite different. In general, the front legs decelerate, the rear legs accelerate, and the middle legs do both, as shown in Figure 6.

7.2 Individual Leg Ground Reaction Forces - *Sprawlita*

When we examine the plots in Figure 6, we see that there are differences between the individual leg functions in the cockroach and the robot. While the rear legs accelerate during the first part of the half-stride, there is a significant amount negative fore-aft force during the latter part of the half-stride due to dragging. When we consider the middle leg force profile, we see that it is actually the opposite of the cockroach's. The middle legs initially provide acceleration, and then deceleration. Finally, unlike the cockroach, the front legs provide little deceleration.

8. Discussion and Conclusions

Sprawlita has demonstrated the feasibility of small, biomimetic robots that exploit passive properties in combination with an open-loop controller to achieve fast, stable locomotion over obstacles.

However, while *Sprawlita's* scurrying is insect-like, a comparison of the ground reaction forces reveals significant differences, particularly in the horizontal direction. A closer inspection of the individual leg forces shows that the front, middle and rear legs behave differently than they do for a cockroach (or other running hexapedal animals). Instead of being decelerated primarily by the front and middle legs at the end of each stride, *Sprawlita* is decelerated substantially by foot dragging in the rear legs. As a consequence, the robot does not display the typical phasing of horizontal and vertical forces associated with the SLIP model found in running animals. In essence, the rear legs are "running out of stroke length," resulting in a pseudo-flight phase with dragging feet.

These observations suggest modifications for incorporation into the next generation of biomimetic hexapods. In particular, we can increase the stroke length of the middle, and especially the rear legs by embedding custom pistons with a longer stroke length or by fabricating a compliant SDM linkage that multiplies the pistons' motion. We anticipate that if we can prevent the pseudo-flight phase and foot dragging, the front and middle legs will be able to take on the role of compliantly decelerating the robot at the end of each stride, and a more SLIP-like motion will be observed. Whether this motion will truly be faster or more robust remains to be verified, but given that it is ubiquitous in running animals it is certainly worth investigating.

Acknowledgments

Special thanks to the members of the U. C. Berkeley PolyPEDAL Lab for their enthusiasm, open communication and technical contributions to this work. The authors also thank the members of the Stanford CDR and RPL teams for their contributions to the work documented in this paper, especially Jonathan Clark for his role in building Sprawlita and Arthur McClung for individual leg force measurements. Sean Bailey is supported by an NDSEG fellowship. This work is supported by the National Science Foundation under grant MIP9617994 and by the Office of Naval Research under N00014-98-1-0669.

References

[1] Raibert, M. H., "Legged robots that balance," MIT Press, Cambridge, MA, 1986.
[2] McGeer, T., "Passive dynamic walking," IJRR, Volume 9, 62-82, April 1990.
[3] Nelson, G. M., Quinn, R. D., Bachmann, R. J., Flannigan, W. C., Ritzmann, R. E., and Watson, J. T., "Design and Simulation of a Cockroach-like Hexapod Robot," IEEE ICRA Proceedings, Albuquerque, NM, 1106-1111, 1997.
[4] Cruse, H., Dean, J., Muller, U., and Schmitz, J., "The stick insect as a walking robot," Proceeding of the 5th International Conference on Advanced Robotics, Pisa, Italy, 936-940, 1991.
[5] Bailey, S. A., Cham, J. G., Cutkosky, M. R., and Full, R. J., "Biomimetic Mechanisms via Shape Deposition Manufacturing," in Robotics Research: the 9th International Symposium, J. Hollerbach and D. Koditschek (Eds), Springer-Verlag, London, 2000.
[6] Xu, X., Cheng, W., Dudek, D., Hatanaka, M., Cutkosky, M. R., and Full, R. J., "Material Modeling for Shape Deposition Manufacturing of Biomimetic Components," ASME Proceedings of DETC/DFM 2000, Baltimore, Maryland, September 10-14, 2000.
[7] Full, R. J., and Tu, M. S., "Mechanics of a Rapid Running Insect: Two-, Four-, and Six-legged Locomotion," Journal of Experimental Biology, 156, 215-231, 1991.
[8] Full, R. J., and Tu, M. S., "Mechanics of Six-legged Runners," J. Exp. Bio., 148, 129-146, 1990.
[9] Full, R. J., Autumn, K., Chung, J. I., Ahn, A., "Rapid negotiation of rough terrain by the death-head cockroach," American Zoologist, 38:81A, 1998.
[10] Saranli, U., Buehler, M., and Koditschek, D. E., "Design, Modeling, and Control of a Compliant Hexapod Robot," IEEE ICRA Proceedings, 2589-2596, San Francisco, USA, April 2000.
[11] Full, R.J. and Koditschek, D. E., "Templates and Anchors - Neuromechanical hypotheses of legged locomotion on land," J. Exp. Bio. 202, 3325-3332, 1999.
[12] Cham, J. G., Bailey, S. A., and Cutkosky, M. R., "Robust Dynamic Locomotion Through Feedforward-Preflex Interaction," ASME IMECE Proceedings, Orlando, Florida, November 5-10, 2000
[13] Ringrose, R., "Self-stabilizing running," IEEE ICRA Proceedings, Albuquerque, NM, 1997.
[14] Brown, I. E., and Loeb, G. E., "A reductionist approach to creating and using neuromusculoskeletal models," in Biomechanics and Neural Control of Movement, Winters, J. M. and Crago. P. E. (Eds.), Springer-Verlag, New York, 2000.
[15] Kubow, T. M., and Full, R. J., "The role of the mechanical system in control: A hypothesis of self-stabilization in hexapedal runners," Phil. Trans. Roy. Soc. London B. 354, 849-862, 1999.
[16] Meijer, K., Full, R. J., "Stabilizing properties of invertebrate skeletal muscle," American Zoologist, Volume 39, Number 5, 117A, 1999.
[17] Clark, J. E., Cham, J. G., Bailey, S. A., Froehlich, E. M., Full, R. J., and Cutkosky, M. R., "Biomimetic Design and Fabrication of a Hexapedal Running Robot," Submitted to IEEE ICRA, Seoul, Korea, 2001.
[18] Full, R. J. and Meijer, K., "Artificial muscles versus natural actuators from frogs to flies," In Smart Structures and Materials 2000: Electroactive Polymer Actuators and Devices (EAPAD) (ed. Yoseph Bar-Cohen). Proc. SPIE Vol. 3987, p. 2-9, 2000.
[19] Cavagna, G. A., Heglund, N. C., and Taylor, C. R., "Walking, Running, and Galloping: Mechanical Similarities between Different Animals," Scale Effects in Animal Locomotion, Proceedings of an International Symposium, T. J. Pedley (Ed.), 111-125, Academic Press, New York, USA, 1975.
[20] Blickhan, R., and Full, R.J., "Locomotion energetics of the ghost crab. II. Mechanics of the centre of masss during running and walking," J. exp. Biol., 130, 155-174, 1987.
[21] Full, R.J. and Farley, C. T., "Musculoskeletal dynamics in rhythmic systems - a comparative approach to legged locomotion," in Biomechanics and Neural Control of Posture and Movement, J. M. Winters and P. E. Crago (Eds.), Springer-Verlag, New York, 2000.
[22] Full, R.J. and Blickhan, R. and Ting, L.H., "Leg design in hexapedal runners," J. Exp. Bio. 158, 369-390, 1991.

Super Mechano-System: New Perspective for Versatile Robotic System

Shigeo Hirose

Tokyo Institute of Technology
Department of Mechanical and Aerospace Engineering
2-12-1 Ookayama Meguro-ku, Tokyo 152-8552 Japan
hirose@mes.titech.ac.jp   http://mozu.mes.titech.ac.jp

Abstract: This paper discusses the general concept of the Super Mechano-System, or SMS and its anticipated applications. The SMS is a new type of robot system, which can transform its shape and also its internal control system in accordance with the given tasks. Snake robots, running and walking robot with manipulation function, group robots, and unified and reconfigurable robots are the three main examples of the SMS. Demining task, rescue operations, and planetary exploration are the expected applications of the projects. Design and new features of constructed robot systems, such as 3D snake-like robots, demining quadruped walking robots, and group type robots named as Super Mechano-Colony will be explained.

1. Introduction

The group of researchers in control and mechanical engineering of Tokyo Institute of Technology (TITech) has started COE projects sponsored by the Japanese Ministry of Education, Science and Culture. In the COE project, we are developing Super Mechano-System (SMS), a mechanical system with the function of task-adaptive-self- organization of its configuration so as to realize both of high mechanical performance and versatility.

Let me explain the basic concept of SMS. We think that human body is regarded as one of the typical examples of SMS [1]. Human body has hundreds of motion degrees of freedom. These motion degrees of freedom automatically reconfigure themselves in accordance with the task to be performed. For example, when a person throws a ball as shown in Fig.1 (a), the body configures itself as a serial link, composing the legs, body, arm and fingers connected in series. This configuration of serial linkage enables to accumulate the swing velocity of body segments up to the fingertip and allow the ball to be thrown with fast speed. To the contrary, when the person lifts a heavy weight, the body reconfigures itself as a parallel linkage, as shown in Fig.1 (b). The parallel linkage formed by the body enables to accumulate force of muscles to lift the heavy object. As we observed like this, human body skillfully changes its posture in accordance with the task and show high versatility just as SMS.

D. Rus and S. Singh (Eds.): Experimental Robotics VII, LNCIS 271, pp. 249–258, 2001.
© Springer-Verlag Berlin Heidelberg 2001

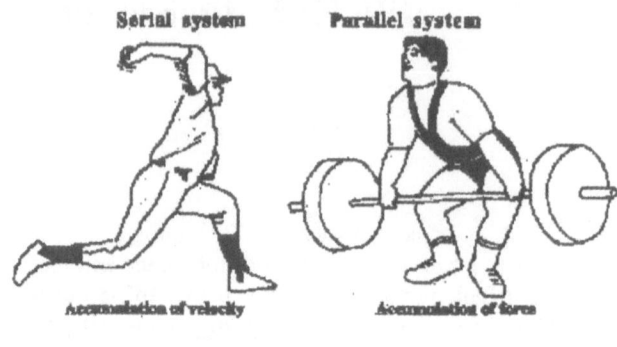

(a)   (b)
Fig.1 Human body as a typical example of SMS

The objective of TITech/COE research is to realize a mechanical system with versatility just like a human body. Of course, many attempts have been done to realize versatile mechanical system like this until now, but most of the former trials have been, I suppose, not always successful.

In our SMS project, we will emphasize close cooperation between the researchers of mechanical and control engineering within Tokyo Institute of Technology, and thus expecting to realize practical versatile mechanical system, which can really be called as Super Mechano-System. Several examples of SMS including snake-like robots, walking robots and group robots will be explained in the following chapters.

2. Snake-like SMS

Snake-like robot is composed of several articulated body segments linked linearly like a snake. By using unified train-like configuration and the active coordinated motion of the segments, the snake-like robot is expected to exhibit following characteristics [2][3][4]:

1) It can pass over uneven terrain and narrow paths by actively adapting its long trunk to the ground topography.

2) It can cross a ditch by stiffening the joint servomechanisms to bridge the ditch. At the same time, it can steadily wade a marsh by softening the joint servomechanisms and distributing its weight to all the segments.

3) The cross section of the path required for the locomotion is small if it compared with the total volume of the body. Furthermore, the driving mechanism of each unit can be comparatively simple as they have unified mechanisms.

4) Thanks to the effect of unified structure, the reliability and maintainability can be high. The malfunctioning segment will easily be detached and replaced.

5) The body can be an excellent manipulator as well as a locomotor by transforming the hyper redundant degrees of freedom just as a trunk of an elephant.

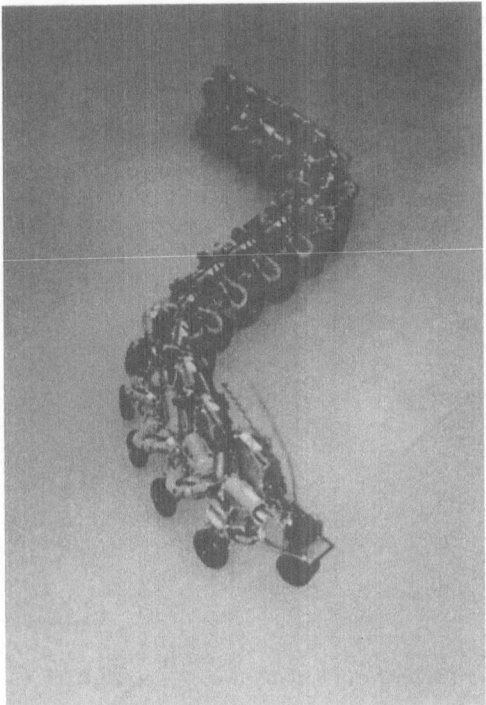

Fig.2 Snake like robot ACM R1

The author has been developing many types of snake-like robots. Fig.2 is the Active Cord Mechanism revised model 1(ACM-R1) [5]. The ACM-R1 consists of 16 segments. By using the coordinated control of the joints, it moved just like real snakes with the velocity of about 50 cm/sec.

By attaching skate blade for each joint, ACM-R1 has also demonstrated to glide on the ice skating rink. Fig. 3 is the future view of new types of snake-like robots with 3D-motion freedom [6]. There are so many interesting research topics on the control of this highly redundant robotic mechanism.

Fig. 3 Future view of the 3D snake robot ACM R2

Fig.4 is "Soryu-I", three-tracked snake like robot specially designed for the rescue operation. Soryu-I is designed to be as slender as possible and capable to make active bending on both of the joints. The bending is done symmetrical. It installed small TV camera on the head, and going to mount various sensors. It is driven by only three motors to minimize the weight. It at present weighs 10 kg. It can move inside narrow and winding path. It can also recover the posture even if it is tumbled upside down by using the twist motion of the segments. By using these features we are going to establish rescue system to inspect under the debris of building just after big earthquake.

Fig.4 Rescue and inspection robot "Soryu-I" which can move inside narrow and winding path under the debris of building after the earthquake

3. Walking Type SMS

Generally speaking, legged mobile robots are very difficult to use for practical applications. The mechanism usually weighs too much since large numbers of actuators are required for the drive of multi-DOF legs, and, at present, actuators are bulky and heavy. But legged locomotion has intrinsic characteristics derived from its specific configuration as follow [7]:

1) The legs can be utilized not only for locomotion, but also for posture stabilization in standstill operation. For example, in a stationary posture, the four legs serve as outriggers to hold the upper body stable even on uneven ground. At the same time, the upper body can be actively driven while the feet are fixed to the ground. Thus the legs form active supporting base to assist the motion of a manipulator mounted on the body.

2) The foot contacts the ground at discrete points, and the contact points can be arbitrarily selected according to the terrain condition. At the same time, it can move all directions in non-holonomic omni-directional motion. Therefore, the legs are inherently suitable for highly maneuverable locomotion in uneven terrain.

3) The ground contact area of the sole of the foot can be made as large as needed, so the pressure over the area of contact can be reduced more than that of a wheel makes contact over only a small area. Moreover by using the multiple DOFs of the leg joint, a legged robot can be steered without slippage on the ground even if the sole contacts the ground over a large area. Therefore, the legs are suitable for locomotion on soft surfaces, e.g. on sandy soil, on damp-dry concrete, and on the grated floor of construction sites.

3.1 Steep slope walking quadruped walking robot

These advantages are especially important for the realization of SMS. Some of the examples will be shown here. Fig.5 is the steep-slope-climbing robot TITAN VII for construction tasks [8]. TITAN VII can walk on a steep slope even of 70 degrees while supported by a pair of force-controlled wires and it could make omni-directional motion on the slope. It also performs several operational tasks such as a drilling task by using powerful up and down motion of the legs. The versatility of the legs exhibited in the TITAN VII is typical example of SMS concept.

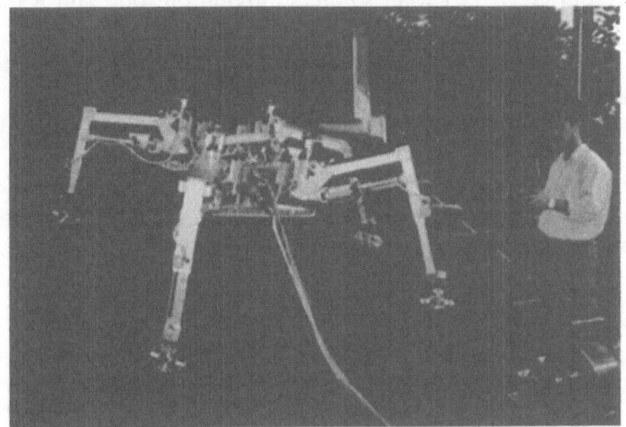

Fig.5 Steep slope climbing robot TITAN VII

Fig.6 Future view of the walking type demining robot

3.2 Walking type humanitarian demining robot

Fig.6 illustrates an anticipated application of a walking robot for humanitarian mine detection and removal task [9]. The walking robot is planned to have the function to change its end-effectors being attached on the foot. It will make grass cutting, mine detection, and mine retrieval task by attaching different end-effectors. Fig.7 shows an experiment using small walking vehicle model TITAN VIII. The reason why the walking configuration with insect type legs is selected for this specific application are as follows:

(1)   Mines are buried by targeting soldiers moving on foot, and walking robots are suitable to approach the site.
(2)   The leg can also be a powerful manipulator to dig out and handle the mine on the ground.
(3)   Legs form stable, and active base for operation even on uneven terrain, swamp and sandy ground.
(4)   It is possible to prevent total loss of the robot in the case of an unexpected explosion of the mine, for a posture to maintain the body low and stretching the legs away from the body outside like an insect.

Fig.7 Experiment of the demining task using the modified model of TITAN VIII

3.3 Leg and wheel hybrid vehicle Roller Walker

Fig.8 is the leg and wheel hybrid mobile robot named "Roller-Walker" [10]. Until now many studies have been done about leg-wheel hybrid mobile robot, but most of then attaches active wheels, wheels driven and steered by electric motor. But installation of active wheels greatly reduces the mobility of walking machine, because they are usual heavy and bulky. Proposed hybrid mobile robot "Roller-Walker" is a vehicle with a special foot mechanism, which can be changed from the sole of the walking vehicle to passive wheels for wheeled vehicle by the mechanism shown in Fig.9. Fig.8 shows the trajectory of the foot and the body in roller walking motion. There are many interesting control problems too for this walking vehicle, which makes roller-skating.

Fig.8 Wheel-leg hybrid robot "Roller Walker

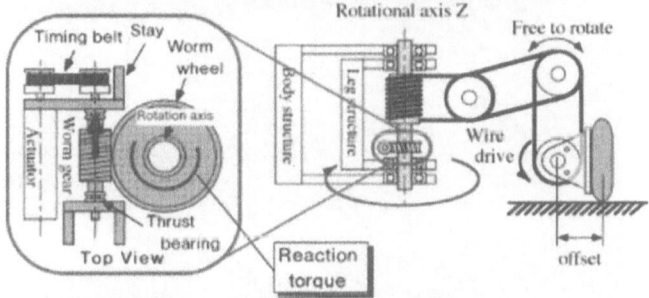

Fig.9 Leg mechanism of Roller Walker

3.4 Dinosaur-like running and task-performing walking robot TITRUS

To balance the posture of biped robot is by no means an easy task. The swing motion of the legs may generate undesirable shaking to the body, and canceling motion is required. Especially the rotational motion around yaw angle is usually large. Human uses the swing motion of hands to cancel this effect. If the walking vehicle has long neck and long tail just like dinosaur, this shape may increase the inertia of the body around yaw axis and makes it easy to depress the yaw axis rotation. The tail will be effectively used as the third leg to have static stability. The neck, if it is designed as flexible structure, may be used as the hand as shown in Fig. 10.

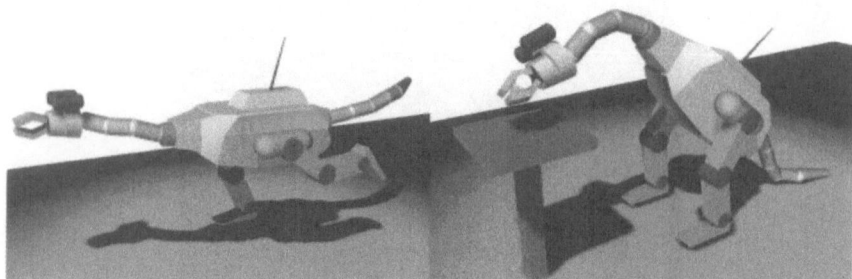

Fig.10 Dinosaur-like running and task-performing walking robot TITRUS

From these standpoints, the dinosaur-like walking robot is said to have unconventional utilities as a robotic system, and it can also be considered as typical example of the SMS. We will thus going to investigate this type of running robot.

4. Super Mechano-Colony

Although there have been quite a few researches on the robotic system consisting of multiple agents, in our SMS project special interest is paid for the heterogeneous system of the decentralized autonomous agents with leader. We are going to call such a system as "Super Mechano-Colony (SMC)". One of the typical examples of SMC is shown in Fig.11 a new type of planetary rover based on the SMC. It is named as "SMC rover" [11].

Fig.11 Expected view of the SMC rover

The SMC rover consists of a main body and multiple detachable wheel units. Each wheel unit of the SMC rover is composed of wheel and a single arm as shown in Fig.12. Design of versatile, reliable and robust mechanism for the wheel unit is crucial for the realization of the SMC rover. We thus paid special attention of the mechanical design of the wheel unit One of the most important points of the wheel unit is psudo-multi-DOF driving mechanism. It is to use only two main driving motors and five braking mechanisms to drive five joints in a psudo-5-DOF motion. The arm has active shoulder joint, elbow joint, wrist joint, and a gripper. The arm can also be rotated around the wheel axis. Therefore when the wheel unit is in "locomotion mode" as shown in Fig.12 (a), the wheel can move around on the ground with high mobility. In this propulsive motion, the arm is used to sustain reaction moment generated by the propulsive motion of the wheel rotation by pressing the arm against the ground. To reduce the friction of the arm against the ground, a caster is installed around the wrist. Changing the orientation (yaw angle) of the caster attached on the wrist will steer the wheel unit.

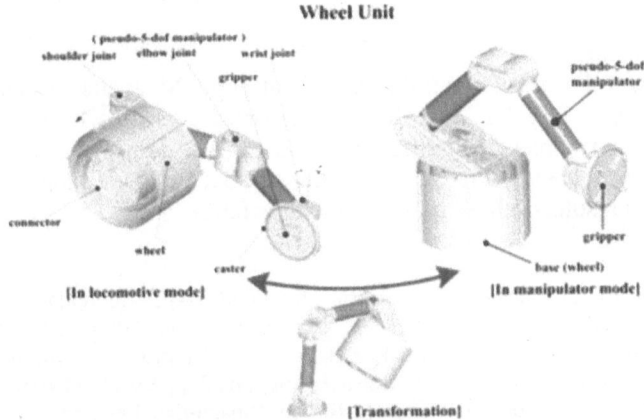

Fig.12 Single wheeled rover

Fig.13 Connected single wheeled rover

The wheel unit can also be transformed into a "manipulator mode" as shown in Fig.12(b). In this posture, the arm acts as a 4-DOF manipulator with a gripper. Pushing motion of the arm against the ground enable to exchange the mode form wheel to manipulation and vice versa.

The wheel units can be connected as shown in Fig.13. If more wheel units are connected in series it will form a snake-like Active Cord Mechanism [3]. This snake-like formation is expected to demonstrate highly efficient mobility over irregular terrain. SMC rover will have unprecedented functions. When it is required to make a scouting of the route on the unknown planet, SMC rover will detach some of the wheel units and let them survey on surrounding terrain. When the SMC rover is going to forage the rocks of the planet, detached wheel units will also execute the task effectively by the coordinated group operation. Fig.11 depicts the expected view of the foraging mission.

The swarm intelligence control of SMC for these applications is left unsolved for the future as one of the most interesting controlling problem of our project.

5. Conclusion

In this paper, I discussed about the concept of Super-Mechano System on which our Tokyo Institute of Technology COE researches group are working, and illustrated some of the examples. Other group members are doing intensive study on the controlling problems of these robots in parallel. We are hoping to report the experimental results of these robots in the near future.

References

[1] Shigeo Hirose: Considerations on the Design of Hyper-Redundant Versatile Robotic System, Proc. TITech COE/Super Mechano-Systems workshop '98, pp.12-17 (1998)
[2] Shigeo Hirose, and Akio Morishima: Design and Control of a Mobile Robot with an Articulated Body, Int. J. of Robotics Research, vol.9, no.2, pp.99-114 (1990)
[3] Shigeo Hirose: Biologically Inspired Robots (Snake-like Locomotor and Manipulator), Oxford University Press (1993)
[4] Edwardo F. Fukushima, Shigeo Hirose: How to Steer the Long Articulated Body Mobile Robot "KR-II"; Proc. Int. Conf. on Advanced Robotics, pp.729-735 (1995)
[5] Shigeo Hirose, Gen Endo: Development of Autonomous Snake-Like Robot ACM R-1, Proc. Annual Conf. Robotics & Mechatronics '97, pp.309-310 (1997)(in Japanese)
[6] Shigeo Hirose; Considerations on the Design of Hyper-Redundant Versatile Robotic System, Proc. TITech COE/Super Mechano-Systems workshop '98, , , pp.12-17 (1998)
[7] Shigeo Hirose, Kan Yoneda, Kazuhiro Arai, Tomoyoshi Ibe :Design of a Quadruped Walking Vehicle for Dynamic Walking and Stair Climbing; Advanced Robotics, 9[2] pp.107-124 (1995)
[8] Shigeo Hirose, Kan Yoneda, Hideyuki Tsukagoshi; TITAN VII: Quadruped Walking and Manipulating Robot on a Steep Slope, Proc. Int. Conf. on Robotics and Automation, Albuquerque, New Mexico, pp.494-500 (1997)
[9] Shigeo Hirose, Keisuke Kato: Quadruped Walking Robot to Perform Mine Detection and Removal Task, Proc. 1st Int. Symp. Clawar '98,   pp.261-266 (1998)
[10] Gen Endo, Shigeo Hirose: Study on Roller-Walker (Basic Experiments on Self-Contained Vehicle System), Proc. TITech COE/Super Mechano-Systems workshop '99, pp.153-160 (1999)
[11] Shigeo Hirose: Super-Mechano-Colony and SMC Rover with Detachable Wheel Units, Proc. TITech COE/Super Mechano-Systems workshop '99, pp.67-72 (1999)

Using Modular Self-reconfiguring Robots for Locomotion

Keith Kotay, Daniela Rus, Marsette Vona
Department of Computer Science
Dartmouth
Hanover NH 03755, USA
{rus,kotay,mav}@cs.dartmouth.edu

Abstract: We discuss the applications of modular self-reconfigurable robots to navigation. We show that greedy algorithms are complete for motion planning over a class of modular reconfigurable robots. We illustrate the application of this result on two self-reconfigurable robot systems we designed and built in our lab: the *robotic molecule* and the *atom*. We describe the modules and our locomotion experiments.

1. Introduction

Self-reconfiguring robots have the ability to adapt to the operating environment and the required functionality by changing shape. They consist of a set of identical robotic modules that can autonomously and dynamically change their aggregate geometric structure to suit different locomotion, manipulation, and sensing tasks. A primary design goal for a self-reconfiguring robot is to allow the robot to assume any geometric shape. For example, a self-reconfiguring robot system could self-organize as a snake shape to pass through a narrow tunnel and reorganize as a multi-legged walker upon exit to traverse rough terrain. Self-reconfiguring robots are suited for a range of applications that require the geometric modification of a part and are characterized by incomplete a priori task knowledge. Such a robot could match its geometric structure to the shape of the surrounding terrain for versatile locomotion. This can be achieved by requiring the robot to metamorphose from one shape to another to best match the shape of the terrain in a statically stable gait, as illustrated in Figure 1.

Figure 1. This figure demonstrates using shape metamorphosis for locomotion. A statically stable gait is used to translate the robot from left to right.

In our previous work [7, 5, 17, 18, 19] we describe two different robot systems capable of self-reconfiguration: the *Robotic Molecule* system and the *Robotic Crystal* system. In this paper we examine using self-reconfiguration for locomotion an we describe our experimental results in simulation and on the hardware units we built in our lab. These results have the flavor of [8], where

D. Rus and S. Singh (Eds.): Experimental Robotics VII, LNCIS 271, pp. 259–269, 2001.

we examine locomotion with Inchworm robots.

2. Two Self-reconfigurable Robot Systems

2.1. The Molecule

 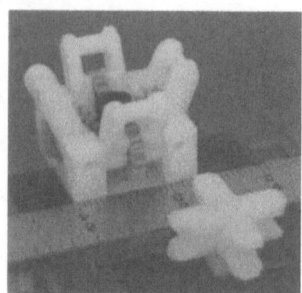

Figure 2. (**Left**) The robotic Molecule. The Molecule is composed of two atoms, connected by an right-angle rigid bond. The Molecule has 4 degrees of freedom: two rotational degrees of freedom about the bond and one rotational degree of freedom per atom about a single inter-Molecule connector. The connectors have been implemented with electromagnets. (**Right**) The prototype gripper connection mechanism. The gripper is a male-female design. The male component is in the upper left and the female component is in the lower right. Molecules will either have all male components or all female components as connectors. This does not cause a problem because the Molecule design naturally partitions 3D space into two regions. A single Molecule can only occupy one of the regions and can only connect to Molecules in the other region.

A Molecule robot [6] consists of multiple units called *Molecules*; each Molecule consists of two *atoms* linked by a rigid connection called a *bond* (see Figure 2). Each atom has five inter-Molecule connection points and two degrees of freedom. One degree of freedom allows the atom to rotate 180 degrees relative to its bond connection, and the other degree of freedom allows the atom (thus the entire Molecule) to rotate relative 180 degrees relative to one of the inter-Molecule connectors at a right angle to the bond connection. We have already prototyped the Molecule (see Figure 2.)

Our current design uses R/C servomotors for the rotational degrees of freedom. A new feature of our prototype is the use of a gripper-type connection mechanism (see Figure 2). In our previous design we used electromagnets as the connection mechanism, but electromagnets have several disadvantages including continuous power consumption to maintain connections and requiring a sheath to prevent unwanted rotation about the axis of connection. Since a sheath must extend beyond the bounding sphere of the atom to allow it to interlock with its mating sheath, a binding condition in introduced restricting mating motion to a face-to-face approach (a sliding approach, in which the two mating faces come into contact by sliding past each other is not possible because of sheath collisions). A gripper-type connection mechanism, in which the gripper arms can retract into the bounding sphere of the atom allows sliding face-to-face approaches and atom rotations in place. Also, since the gripper arms are driven by a non-backdrivable worm gear mechanism, they will maintain their grip when electrical power is no longer applied, decreasing the power consumption of Molecule self-reconfiguration.

The rotating connection points on each atom are the only connection points required for Molecule motion. The other connection points are used for attachment to other Molecules to create stable 3D structures. Each Molecule also contains a microprocessor and the circuitry needed to control the servomotors and connectors. The diameter of each atom is 4 inches (10.2 cm.), making the atom–atom distance in the Molecule approximately 5.7 inches (14.4 cm.). The weight of the Molecule is 3 pounds (1.4 kg.).

2.2. Molecule Motion

An individual Molecule has the following basic motion capabilities: (1) linear motion in a plane on top of a lattice of identical Molecules, irrespective of the absolute orientation of the plane; (2) convex 90-degree transitions between two planar surfaces composed of Molecules; and (3) concave 90-degree transitions between two planar surfaces composed of Molecules.

The details of controlling these motions using the 4 molecular DOFs are provided in [6]. Figure 3 illustrates the linear walk algorithm.

Figure 3. A linear walk sequence. The checkered surface represents a plane of Molecules. An atom with a black dot is attached to the Molecule below it. The left image represents the initial configuration. A clockwise rotation of 90 degrees about the connected atom produces the next image. The atoms then swap attachment as indicated by the movement of the black dot. Finally, a counterclockwise rotation of 90 degrees produces the right image. Another attachment swap would return the Molecule to its initial pose, translated by two squares in the vertical direction. A similar sequence could be used to translate the Molecule horizontally. Thus, these sequences of moves are sufficient for Molecular translation to any pair of white squares in the plane.

2.3. The Atom

The Crystalline robot consists of a set of modules called *Atoms*; each Atom is a mechanism that has some of the motive properties of muscles, that can be closely packed in 3D space, and that can attach itself to similar units. We chose a design based on cubes with connectors to other modules in the middle of each face. The idea is to build a cube that can contract by a factor of two and expand

Figure 4. Three snapshots from a simulation of locomotion using Crystalline robots. The left image shows the initial state. The middle image shows the robot after shrinking two modules in the direction of motion. The right image shows the robot after relaxing the shrunk modules in the direction of motion. Notice that the entire structure moved forward one unit, in an inchworm-like fashion.

to the original size (see Figure 4). We wish to effect compression along all three principal directions (e.g., x, y, z) individually or in parallel. We call the module an *Atom*, and each connector a *bond*. Figure 5 shows a design for the mechanics of a two-dimensional (square rather than cubic) implementation of the Atom and Figure 6 shows the physical prototype. We use complimentary rack and pinion mechanisms to implement the contraction and expansion actuation for the two-dimensional prototype. The connection mechanisms are based on a *channel and key* concept. When fully contracted, the Atom is a square with a 2 inch side. When fully expanded, the Atom is a square with a 4 inch side. The height of the Atom is 7 inches and its weight is 12 ounces.

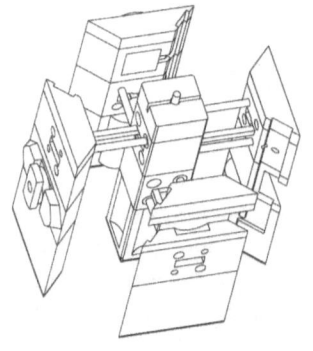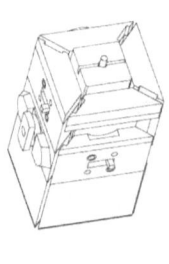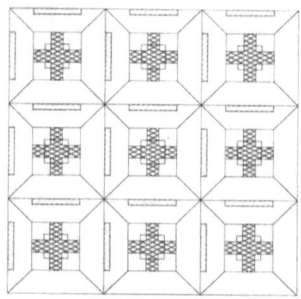

Figure 5. The mechanics of a 2D Atom actuated by complimentary rack-and-pinion mechanisms. The Atom is 4 inches tall (not including electronics, which are not shown). When expanded (left), the Atom occupies a 4 inch square; when contracted (middle) the Atom occupies a 2 inch square. The right figure shows a tiling of nine compressed 3DOF Atoms. Note that every inter-Atomic interface contains exactly one active connection mechanism.

The two-dimensional version of the Crystalline Atomic module (see Figure 6) was created based on the CAD designs shown in Figure 5 (left and middle). The module has an expansion/contraction ration of 2. All the faces of the Atom have to be fully extended or fully contracted. Each face of the

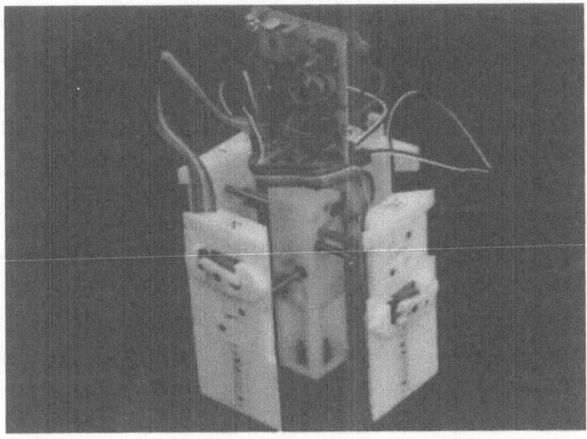

Figure 6. The physical prototype for the Crystalline Atom.

module has a connection slot. However, only two out of the four faces[1] have active connection slots (see Figure 5 (right)). These active slots provide a key-and-lock mechanism for forming rigid connections with adjacent modules. Thus, the entire unit can be realized with three degrees of freedom: one to expand/contract the faces of the Atom, and two for the active connectors. All three degrees of freedom can be implemented with binary actuators. Since Atoms can never rotate relative to one another, the use of two rather than four connectivity degrees of freedom leads to no mechanical limitations. Every inter-Atomic interface of a structure will have one active connection mechanism. The module has on-board electronics and four 3V 2/3A size Lithium batteries, so that it can function untethered.

A Crystalline Atom can connect with identical modules to create Crystalline robot systems. Only lattices whose faces are normal to the x, y, and z axes can be created using Crystalline robots. By manipulating the size of the Atom, it is possible to approximate any finite solid shape to an arbitrary precision using Crystalline modules[2].

Each Atom contains an on-board processor (Amtel AT89C2051 microcontroller), power supply (four 2/3 A Lithium batteries), and support circuitry, which allows both fully untethered and tethered operations. Atoms are connected by a wired serial link to a host computer to download programs. For untethered operations, an experiment specific operating program specified as a state sequence is first downloaded over a tether. When the tether is removed, an on-board IR receiver is used to detect synchronization beacons from the host.

---

[1] The active connection slots are situated on adjacent faces, which allows any lattice of Crystalline Atoms to be fully connected.

[2] The aliasing error for any shape on a raster display can be arbitrarily reduced by increasing the resolution of the display.

3. Locomotion with Modular Self-reconfigurable Robots

In this section we examine locomotion capabilities for modular self-reconfiguring robot systems.

Self-reconfigurable robots consist of modules that can move relative to each other. The modules can climb on top of each other[3], slide relative to each other, etc. This enables the modules of the robot to move outside the plane supporting the robot. Thus, a modular self-reconfigurable robot is capable of climbing on top of obstacles. In this section we examine the power of this capability to motion planning in the absence of maps. For example, we may consider a factory floor where dynamic obstacles make it impossible to supply the robot with an accurate map. The model for this problem is a self-reconfiguring robot that starts at a known location and is to find its way to a goal location, identifiable by a beacon.

In this section we do not assume any specific design for the unit module. To preserve generality, we assume a unit-modular system where an individual module can move linearly and make convex and concave transitions relative to a collection of modules. These capabilities enable an entire class of robots to move linearly and make convex and concave transitions in the environment[4].

A simple strategy, such as the right hand rule, or the on-line navigation algorithms proposed by [9], may be employed to find a path to the goal. However, because modular self-reconfiguring robots have the ability to move out of plane, we propose a simpler algorithm for this problem. The basic idea is to move the robot greedily in the direction of the goal. When an obstacle is reached, instead of going around the obstacle, which is the technique employed by robots confined to move in the plane (for example wheel-based robots), self-reconfiguring robots can simply climb over the obstacles, maintaining their original heading. The algorithm described in Figure 7 summarizes this intuition.

```
(define (greedy-navigation start goal gravitation-direction)
   (align-robot (make-path start goal))
   (loop (cond ((at-goal?)
                 'stop)
               ((obstacle? 'front-IR)
                (concave-transition))
               ((free-space? 'front-foot-IR)
                (convex-transition))
               (else (linear-step)))))
```

Figure 7. A greedy algorithm for on-line navigation. The robot moves in the direction of the goal using step. If an obstacle is encountered, the robot uses convex-transition and concave-transition to climb over the obstacle.

Theorem 1 *Suppose a self-reconfiguring robot has to travel from an initial*

---

[3]Each system implements this operation in a different way, using its own specific actuation capabilities.

[4]We have shown these capabilities for Molecule robots in [7] and for Atoms in [19]. The Experiments section details how the two systems accomplish such motions.

location S to a goal location G in an unknown environment with piecewise-planar segments. The greedy algorithm in Figure 7 is complete and takes $O(1)$ time to compute a path to the goal, provided each segment is wide enough to allow the robot to step on it.

Proof:

The path followed by the self-reconfiguring is obtained by intersecting the environment with the plane defined by S, G, and the direction of gravitation. This plane can be computed in $O(1)$ time. Starting in this plane, the robot will move greedily towards the goal. At each step, the robot will use its heading and sense the direction of gravitation[5] to ensure that its motion stays confined to the motion plane.

This greedy algorithm operates as hill climbing towards the goal and it is complete. The actual path is a simple polygon that connects the S to G; thus the Inchworm is guaranteed to reach G. The total length of the path is of length at most $2\sum H + D$, where D is the straight line distance from S to G and $\sum H$ sums the heights of the obstacles in the space.

Note that this on-line algorithm (see Figure 7) allows the robot to reach the goal provided the robot can place itself completely on each edge of the path. This condition translates into the assumption that all the polygonal edges of its path are of length at least $k + 1$ (where k is the size of one module) for robots with discrete orientations for their modules such as ours. $\square$

This on-line motion planning algorithm will not always move the robot on the shortest path to the goal. For example, if the environment has very high but skinny obstacles, the robot will do extra work to reach the goal. The advantage of on-line navigation with self-reconfiguring robots relies on the capability of such robots to move out of plane. The resulting algorithm is very simple and it only requires computing the direction of motion for each step. This algorithm is significantly simpler than strategies such as [9] where the robot has to completely surround an obstacle to compute the best way to move.

4. Experiments

4.1. Experiments with Molecules

We have constructed two prototype modules and used them to perform experiments to evaluate the feasibility of implementing locomotion using self-reconfiguration. Two other modules are currently under construction. We have also developed a simulator that allowed us to experiment with locomotion algorithms applied to systems consisting of many more modules.

A two-molecule system can walk on a floor of connectors. A four-Molecule Robot can use the individual Molecule actuation to translate and rotate in the plane without needing any additional support from external connectors [7]. The four-Molecule linear translation can be extended to a $2k$-Molecule linear chain of Molecule pairs An eight-Molecule system can climb stairs. Several

---

[5]This part requires a sensor such as a potentiometer to sense the direction of gravitation.

algorithms can implement this motion. One possible algorithm to implement this task is a *Slinky*-style motion: the modules get organized into a tower on the plane of one step. The tower is then reversed from the top on the next step. How do we insure that all the modules of the tower (including those at the base, that provide support for the tower) get moved to the next step? Figure 8 illustrates this issue using robots that consist of *Robotic Molecules*. To remove the bottom modules from the lower step the entire structure needs to be supported and balanced by the robotic molecules on the upper step. This poses constraints on the number of modules that need to be located on the upper step prior to breaking ground contact on the bottom step.

Self-reconfiguring robots can adapt to the geometry of the terrain, but planning for such versatility requires additional constraints about the dynamics of the robot structure—our algorithm for stair climbing guarantees dynamic stability during motion.

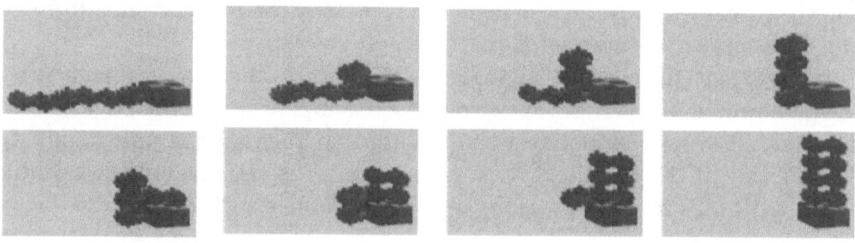

Figure 8. A stair climbing sequence using Robotic Molecules [7]. The light and dark gray cubes on the right of each image represents the first step in a staircase.

We are implementing the four molecule walk on a plane. We have already experimented with a two-molecule system that can successfully climb on top of each other, thus implementing the basic rolling primitive necessary for stepping forward. This two-molecule system requires the support of connectors in the environment.

4.2. Experiments with Atoms

We have constructed ten prototype modules and used them to perform experiments to evaluate the feasibility of using multiple Atoms to demonstrate reconfiguration.

To facilitate experimentation, a row of 8 fixed passive connectors was constructed to simulate the surface of a Crystal. This arrangement not only frees us from having to construct many units at the outset, but it also allows us to perform experiments that are focused narrowly on the specific activities under study. The fixed connectors are placed as they would be for a flat Crystal surface composed of 8 contracted Atoms. In the descriptions that follow, we will refer to two of the prototype Atoms as **a** and **b**, and we will number the fixed connectors **0–7**. The North and West faces of **a** and **b** (those that contain active connection mechanisms) will be referred to as **a.n/b.n** and **a.w/b.w**, respectively, and the South and East faces will be similarly named. **a** and **b** are always oriented so that **a.n** and **b.n** are facing the row of fixed connectors.

Our locomotion experiment was designed to evaluate whether Atoms could work together to effect a reconfiguration. Initially, both **a** and **b** were contracted. **a** was connected to **0** (at **a.n**) and **b** was connected to **1** (at **b.n**). **a** and **b** were connected together at **b.w**. The Atoms were programmed with state sequences designed to perform a variant of inchworm translation along the fixed connectors:

> 1. **free b.n** from **1**
> 2. **expand a**
> 3. **expand b**
> 4. **connect b.n** to **2**
> 5. **disconnect a.n** from **0**
> 6. **contract a** and **b**
> 7. **connect a.n** to **1**
> 8. **repeat**

This sequence is illustrated in Figure 9, and Figure 10 presents several photographs of the Atom prototype hardware performing the experiment.

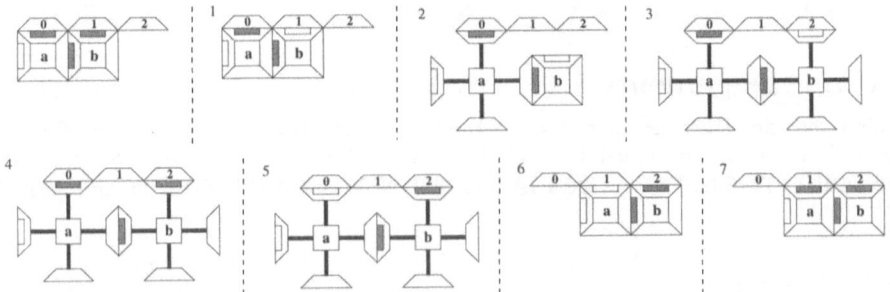

Figure 9. The second experiment tests an inchworm propagation algorithm.

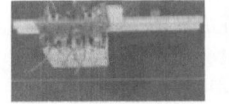   

Figure 10. Several snapshots of the Atom prototype hardware performing the inchworm experiment.

5. Related Work

We are inspired by pioneering work in self-reconfiguring robotics. In [2], Fukuda et al propose a cellular robotic system to coordinate a set of specialized modules. Several specialized modules and ways of composing them were proposed. In [21] Yim studies multiple modes of locomotion that are achieved physically by manually composing a few basic elements in different ways. This work also presents extensive examples of locomotion and self-reconfiguration in simulation. In [10, 23, 20, 11], Murata et al consider a system of modules that can achieve planar motion by walking over one another. The reconfiguration motion is actuated by varying the polarity of electromagnets that are embedded in each module. More recently [12] this group developed a twelve DOF module capable of three-dimensional motion. In [13] Chirikjian et al describe metamorphic robots that can aggregate as two-dimensional structures with varying geometry. The modules are deformable hexagons. This work also examines theoretical bounds for planning the self-reconfiguring motion of such modules.

6. Discussion

We have discussed how modular self-reconfiguring robots can be used to implement versatile locomotion. The ability of such robots to move out of the plane supporting the robot enables them to climb on top of obstacles. Thus, greedy algorithms that move the robot on a straight line to the goal (which might involve climbing over obstacles) are complete for a class of environments where the size of the obstacles is compatible with the size of the robots' discrete steps. We have illustrated our point on two very different robot systems we have designed and built in our lab. We have reported on our experiments with these robots.

Acknowledgements

This paper describes research done in the Dartmouth Robotics Laboratory. Support for this work was provided through the NSF CAREER award IRI-9624286, NSF award IRI-9714332, NSF award EIA-9901589, NSF award IIS-98-18299, and NSF IIS 9912193

References

[1] G. Chirikjian, A. Pamecha, and I. Ebert-Uphoff. Evaluating Efficiency of Self-Reconfiguration in a Class of Modular Robots. *Journal of Robotic Systems*, Vol. 13., No. 5., May 1996.

[2] T. Fukuda and Y. Kawauchi. Cellular robotic system (CEBOT) as one of the realization of self-organizing intelligent universal manipulator. In *Proceedings of the IEEE Conference on Robotics and Automation*, pp. 662-667, 1990.

[3] G. Hamlin and A. Sanderson. Tetrabot modular robotics: prototype and experiments. In *Proceedings of the IEEE/RSJ International Symposium of Robotics Research*, pp 390-395, Osaka, Japan, 1996.

[4] Kazuo Hosokawa, Isao Shimoyama, and Hirofumi Miura. Dynamics of self-assembling systems — analogy with chemical kinetics. *Artificial Life*, 1(4), 1995.

[5] K. Kotay, D. Rus, M. Vona, and C. McGray. The self-reconfigurable robotic molecule. In *Proceedings of the 1998 International Conference on Robotics and Automation*, 1998.

[6] K. Kotay and D. Rus. Motion Synthesis for the Self-reconfiguring Robotic Molecule. In *Proceedings of the 1998 International Conference on Intelligent Robots and Systems*, 1998.

[7] K. Kotay and D. Rus. Locomotion Versatility through Self-reconfiguration In *Robotics and Autonomous Systems*, 26 (1999), pp 217–232.

[8] K. Kotay and D. Rus. The Inchworm Robot: A Multi-functional System In *Autonomous Robots*, Vol. 8, no. 2, pp 53–69, 2000.

[9] V. Lumelsky. Algorithmic issues of sensor-based robot motion planning. in *Proceedings of the 26th Conference on Decision and Control*, pp 1796–1801, Los Angeles, CA 1987.

[10] S. Murata, H. Kurokawa, and Shigeru Kokaji. Self-assembling machine. In *Proceedings of the 1994 IEEE International Conference on Robotics and Automation*, San Diego, 1994.

[11] S. Murata, H. Kurokawa, K. Tomita, and Shigeru Kokaji. Self-assembling method for mechanical structure. In *Artif. Life Robotics*, 1:111–115, 1997.

[12] S. Murata, H. Kurokawa, E. Yoshida, K. Tomita, and S. Kokaji. A 3-D Self-Reconfigurable Structure. In *Proceedings of the 1998 IEEE International Conference on Robotics and Automation*, Leuven, 1998.

[13] A. Pamecha, C-J. Chiang, D. Stein, and G. Chirikjian. Design and implementation of metamorphic robots. In *Proceedings of the 1996 ASME Design Engineering Technical Conference and Computers in Engineering Conference*, Irvine, CA 1996.

[14] A. Pamecha, I. Ebert-Uphoff, and G. Chirikjian. Useful Metrics for Modular Robot Motion Planning. *IEEE Transactions on Robotics and Automation* pp531-545, Vol.13, No.4, August 1997.

[15] C. Paredis and P. Khosla. Design of Modular Fault Tolerant Manipulators. In *The First Workshop on the Algorithmic Foundations of Robotics*, eds. K. Goldberg, D. Halperin, J.-C. Latombe, and R. Wilson, pp 371-383, 19 95.

[16] D. Rus. Self-Reconfiguring Robots. *IEEE Intelligent Systems*, 13(4), 2-5, July/August 1998

[17] D. Rus and M. Vona. Self-Reconfiguration Planning with Unit Compressible Modules. *Proceedings of the 1999 IEEE International Conference on Robotics and Automation*, pp 2513–2520, Detroit, MI, 1999.

[18] D. Rus and M. Vona. A Physical Implementation of the Crystalline Robot. In *Proceedings of the 2000 IEEE International Conference on Robotics and Automation*, San Frncisco, CA, 2000.

[19] D. Rus and M. Vona. Crystalline Robots: Self-reconfiguration with Unit-compressible Modules. Autonomous Robots Vol. 10, no. 1, pp 107–124, 2001.

[20] K. Tomita, S. Murata, E. Yoshida, H. Kurokawa, and S. Kokaji. Reconfiguration method for a distributed mechanical system. In *Distributed Autonomous Robotic Systems 2*, pp 17–25, Springer Verlag 1996.

[21] M. Yim. A reconfigurable modular robot with multiple modes of locomotion. In *Proceedings of the 1993 JSME Conference on Advanced Mechatronics*, Tokyo, Japan 1993.

[22] M. Yim. Polypod II. http://www.parc.xerox.com/spl/members/yim/

[23] E. Yoshida, S. Murata, K. Tomita, H. Kurokawa, and S. Kokaji. Distributed Formation Control of a Modular Mechanical System. In *Proceedings of the 1997 International Conference on Intelligent Robots and Systems*, 1997.

Open-loop Verification of Motion Planning for an Underwater Eel-like Robot

Kenneth A. McIsaac and James P. Ostrowski
General Robotics Automation, Sensing and Perception Laboratory (GRASP)
University of Pennsylvania
Philadelphia PA 19104
{kamcisaa,jpo}@grip.cis.upenn.edu

Abstract: In this paper, we perform experimental verification of open-loop motion planning for a biomimetic robotic system using our underwater eel-like robot. Our results from past work provide theoretical justification for proposed gaits for forward/backward swimming, circular swimming, sideways swimming and turning in place. We have developed a five-link, underwater eel-like robot focusing on modularity, reliability and rapid prototyping, to verify our theoretical predictions. Results from experiments performed with this robot using visual position sensing in an aquatic environment show good agreement with theory.

1. Introduction and Background

Mobile robots continue to challenge researchers with new applications in a variety of environments. Of recent interest has been the application of robotic technology to underwater exploration, monitoring, and surveillance. In this paper, we explore the modeling, simulation, and design of controllers for snake-like robotic systems capable of both crawling overland and underwater swimming.

In recent research, *biomimetic* (biologically based) approaches to underwater locomotion have been pursued. The biomimetic approach to locomotion systems has several potential advantages, including increased underwater efficiency and agility. Recent research has explored various size ranges of robots, including the RoboTuna [15], and smaller fish-like projects [2, 5]. Less work has been done in the area of *anguilliform* (eel-like) locomotion, though recently Ekeberg [11] has simulated the motion of such systems when controlled by biologically based neural networks.

A central issue in studying mobile robots is how to enable a robot to move from one location to another, the *motion planning problem*. There is an extensive literature on motion planning [1, 7], though the vast majority has focused on kinematic systems in which the robot's motion can be described by a differential equation that is linear in the inputs, e.g., nonholonomic, car-like robots. More recently, however, researchers have begun to explore algorithms for developing the motion plans for robots with more complex governing equations, for example, in flexible part handling [4] or mobile manipulators with dynamics [12, 14, 16]

Our interest in this area has emerged from the study of underactuated dynamic mobile robots, ranging from the snakeboard [13] to a vision-guided blimp [17] to an *anguilliform* (eel-like) robot [8, 10], discussed in this paper.

D. Rus and S. Singh (Eds.): Experimental Robotics VII, LNCIS 271, pp. 271–280, 2001.
© Springer-Verlag Berlin Heidelberg 2001

In prior work, we have presented a proposed solution to the motion planning problem for the anguilliform robot. We use a geometric analysis of the dynamics of the robotic system to determine gaits for momentum generation [9], and a sampled feedback/feedforward approach to perform closed-loop steering control [10]. In this work, we verify our open-loop approach to momentum generation with our robotic eel using visual feedback to extract position information in an aquatic environment, and compare results from the robotic system to our simulated dynamic model. We present gaits for forward/backward and circular swimming, as well as novel gaits for sideways swimming and turning in place which were predicted in [9]. We also propose a simple extension to our open-loop system that will allow verification of our closed-loop steering control and accomplish full motion planning in the plane for the underwater, anguilliform robot.

2. Mathematical Model and Gait Selection

In [8], we studied anguilliform locomotion using a simplified physical model of a snake (we use the term "snake" interchangeably with "eel") to be used as a platform to test various locomotive gaits (see also [11]). We model the snake as a planar, serial chain of identical links with mass m, length $2d$, and inertia J. We assume full control of the internal shape of the snake (the joint angles ϕ_i) which allows us to solve the dynamic equations in terms of the unknown configuration variables (x, y, θ)– the position and orientation of the middle link. The mechanical robot used in this work is a five link model. It is advantageous to choose an odd number of links to gain symmetry about the central link. Figure 1(A) shows the three link case.

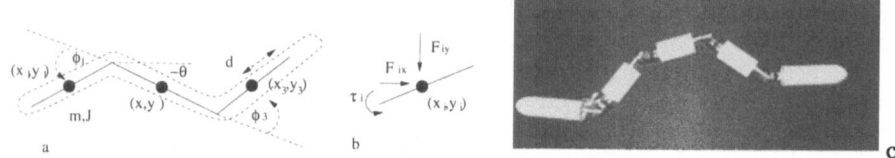

Figure 1. (a) Model of snake, (b) Forces and torques on link i, (c) The REEL II eel robot.

2.1. Friction Models

The crucial elements in this model are the drag force terms, which generate the locomotion. To simulate the forces in the water, we adopt a simple fluid mechanical model. We assume that the Reynolds number is high enough that inertial forces dominate over viscous effects—a reasonable approximation for smooth bodies in water. We also assume that the fluid is stationary, so the force of the fluid on a given link is due only to the motion of that link. The pressure differential created by an object moving in a fluid causes a drag force opposing the motion. Under the assumptions above, the drag force developed takes the form $F \propto \mu_w v^2$. Here, v is the forward speed of the link and μ_w is a drag coefficient for the water, determined by the formula $\mu_w = \rho AC/2$, where A is the effective area of the object, ρ is the density of water, and C is a shape coefficient.

In our simulations, we assume that pressure differentials in the directions parallel to the moving body are decoupled from pressure differentials perpendicular to the body, to yield:

$$F_i^\perp = -\mu_w \text{sgn}(v_i^\perp) \cdot (v_i^\perp)^2 \tag{1}$$

where $v_i^\perp$ is the projection of the vector $(\dot{x}_i, \dot{y}_i)$ along a direction perpendicular to the link. We exclude drag forces parallel to the link because they were determined in simulation to have negligible effects. The discontinuity in $\text{sgn}(v)$ means that this expression is not tractable for use in calculations. Therefore, in our analytical derivations we use an approximation to this function, which turns out to be linear in v:

$$F_{\text{approx}} = \mu v^\perp \tag{2}$$

where μ is defined by a least squares fit over a small range around $v = 0$. We also note that this force model can be interpreted as a viscous damping model, as might be encountered with a snake moving over soft sand.

2.2. Equations of Motion

The derivation of the system of equations governing the time evolution of the system momentum is involved, and presented in detail elsewhere [8]. We present a brief synopsis here.

The system state is governed by the time-evolution of the (body-referenced) velocity vector ξ, which describes forward, sideways and turning velocities, and by the time-evolution of the generalized momentum vector p of momentum in the three body-referenced dimensions. Starting from the Lagrangian formulation of the equations of motion and taking advantage of the invariance of the system with respect to changes in position and orientation, we are able to express the equations of motion by:

$$\xi = -\mathbb{A}(r)\dot{r} + I^{-1}(r)p, \tag{3}$$
$$\dot{p} = \dot{r}^T \sigma_{\dot{r}\dot{r}}(r)\dot{r} + p^T \sigma_{p\dot{r}}(r)\dot{r} + p^T \sigma_{pp}(r)p + \tau_g \tag{4}$$

where $\mathbb{A}$, I, $\sigma_{\dot{r}\dot{r}}$, $\sigma_{p\dot{r}}$ and σ_{pp} depend on the system geometry. Using the force approximation given in Equation 2, the external forces τ can be expressed:

$$\tau_i = \alpha(r)_i^j p_j + \eta(r)_{ij}\dot{r}^j \tag{5}$$

2.3. Perturbation Analysis

In order to gait some insight into the effect of gait selection on the time evolution of the momentum, we use a perturbation approach, making the assumtion that the joint angles vary sinusoidally around some initial value. We set $r^j(t) = r_0^j + \epsilon r_1^j(t)$ (introducing a scaling parameter ϵ) in Equation 5, and solve for the momentum in ascending orders of ϵ ($p_i = p_{i0} + \epsilon p_{i1} + \epsilon^2 p_{i2} + ...$) using Equation 4. With the assumption that $p(0) = 0$, we see immediately that $p_{i0} = 0$ for all time. We can also show that p_{i1} will have zero average using cyclic inputs. We conclude that, to second order in ϵ the momentum is dominated by:

$$\dot{p}_{i2} = \alpha(r_0)_i^j p_{j2} + r_1^j \kappa(r_0)_{ijk} \dot{r}_1^k \tag{6}$$

where the tensor $\kappa_{ijk} = \frac{\partial \eta_{ik}}{\partial r^j}$ has been introduced. (Other terms appear in this equation–for example, a term of the form $\frac{\partial \alpha}{\partial r} p_1 r_1$, but their affect was determined to be negligible using numerical solution of the equation, so they have been ignored.) This is a simple, first-order dynamic equation, parametrized by the gait parameter r_0. Using this equation, we are able to propose gait selection criteria for the eel.

2.4. Gait Selection Criteria

Equation 6 allows us to select appropriate gait waveforms to accomplish arbitrary maneuvers in the water. The α and κ tensors, affected by the choice of r^i, determines the coupling between momenta in the three planar dimensions. By careful choice of gait inputs, we can decouple forward motion, sideways motion and rotary motion, or couple them to accomplish, for example, circular trajectories.

We limit our attention to travelling wave gaits of the form:

$$r^i(t) = \epsilon \sin(\omega t + A(i)\phi_s) + B(i)\phi_{\text{offs}} \tag{7}$$

for some amplitude ϵ, frequency ω, phase-shift ϕ_s and turning offset ϕ_{offs} (the functions $A(i)$ and $B(i)$ are link-dependent parameters controlling phasing and steering offset). Even using this simple, restricted gait model, we are able to accomplish the following modes of locomotion:

- Forward motion, using $A(i) = -i$ and $\phi_{\text{offs}} = 0$ as above.

- Backward motion, using $A(i) = i$ and $\phi_{\text{offs}} = 0$.

- Circular motion, forwards or backwards, using $\phi_{\text{offs}} \neq 0$. The sign of ϕ_{offs} will determine the direction of the turn, and the turning radius will be inversely related to its magnitude.

- Sideways motion, using $A > 0$ for $i < N/2$, $A < 0$ for $i > N/2$, $B > 0$ for $i < N/2$ and $B > 0$ for $i > N/2$. (Pairs of signs may be reversed, to change direction).

- Turning in place, using $A > 0$ for $i < N/2$, $A < 0$ for $i > N/2$, $B > 0$ for $i < N/2$ and $B < 0$ for $i > N/2$. (Pairs of signs may be reversed, to change direction).

3. Robot Design and Experimental Setup

In [8] we presented the design of the REEL (Robotic EEL), a radio controlled robot used to perform preliminary, qualititative verification of our control algorithm. The REEL consisted of four identical links, and used a rubber tube as its waterproof "skin". In this work, we developed a second generation version of this robot (the REEL II–see Figure 1), which addressed some difficulties in the first design and has performed admirably in experimental testing. The rubber "skin" used in the original design was determined to be inappropriate, since replacement of a malfunctioning part (or discharged battery) required removal of the entire skin–a difficult procedure. The use of external skin as a waterproofing agent is also a poor choice. The skin must have a re-sealable opening (to allow access for maintenance), which becomes a point of failure since re-sealing is often done haphazardly, especially during field trials. Finally, it is difficult to find waterproof materials which are flexible enough to permit shape changes without resistance, but rigid enough to hold a hydrodyamic cross-section.

The solution adopted in REEL II (see Figure 1) was to design plastic links in the shape of elliptical cylinders (rounded nose cones were attached to the head and tail links to achieve a streamlined hydrodynamic profile). We manufactured these links using a fused deposition modeling (FDM) machine. The electronic parts (servo-motors, radio receiver and battery) were individually, permanently waterproofed using an epoxy resin sealant. A waterproof connector capable of "wet" insertion was used for the power supply connection, to enable easy battery replacement in the field. Because our links are made of interchangeable, identical parts, we have a robust, modular design for the robot. During testing, the robot was dropped, breaking one of the servo-motors. The motor was replaced without any modifications being needed to the rest of the robot.

The robot shape is radio-controlled. A PC ground station calculates the shape variables (joint angles), which are transmitted using an off-the-shelf radio controller to a receiver in the nose of the robot. The joint actuators are position controlled, medium-torque servo-motors with a specified maximum angular velocity of $315°/sec$, and an maximum angular velocity in water (observed) of $45°/sec$, which enables 0.5 Hz operation for the robot. The robot operates for approximately 20 minutes using a 600mAh battery.

4. Experimental Results and Analysis

4.1. Visual position sensing

We perform our open-loop experiments using a fixed, digital camera to record the behaviour of the robot in a pool of still water. Image processing is performed off-line, in real-time using the Matrox Imagining Library software. We use edge detection, followed by a closing operation to locate the robot as a single blob in the image. The robot's position and orientation in the image plane are determined from the centroid and orientation of this blob. Figure 2(a) shows a sample frame of captured video, and Figure 2(b) shows the post-processed image used in feature extraction.

Once data is available in the image plane, we are able to extract real-world coordinates by unwarping the data. We use the position of fixed reference points in the image to determine the appropriate transformation to yield the robot's true position and orientation in the pool.

 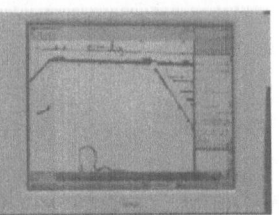

a   b

Figure 2. Experimental images (a) Raw image. (b) Post-processed image.

Our mathematical model of the eel in water leads to two non-dimensional parameters that influence the dynamics in water: an inertia parameter, $\hat{J} =$

J/md^2, and a drag parameter $\hat{\mu} = \mu/m$. Using data from one experiment, we tuned the drag parameter until our simulations matched observed data for all experiments, except for one case (sideways swimming) which we will describe below. We can conclude that our friction model has sufficient predictive power to be used for the development of open loop controls. We also demonstrate the necessity for feedback, since the match between observed and predicted open-loop paths is only approximate.

4.2. Forward motion

The first gait we explored is a simple travelling wave gait used for forward motion. This gait is a discrete approximation to the *serpenoid curve*, which Hirose [3] proposed as the true gait used by crawling snakes, and has also been justified by our perturbation analysis of the dynamics of the eel [9]. We use:

$$\phi_i(t) = A\sin(\omega t + i\phi_s) \tag{8}$$

for amplitude $A = \pi/6$, frequency $\omega = \pi$ and phase shift $\phi_s = \pi/3$. Our choices of amplitude and frequency were motivated by physical limits in the robot, while the choice of phase ϕ_s was motivated by our perturbation analysis [9] which predicted that 60° is the optimal gait phase for a five link snake. To compare observed data to predictions, we used our full dynamic model with the quadratic fluid drag approximation introduced in Section 2.1 and tuned the fluid drag parameter μ_w. The best match to observed data was found when $\mu_w = 0.15$, which differs from the theoretical value taken from Ekeberg [11] ($\mu_w = 0.45$) but is of the same order of magnitude. Figure 3(a) is a plot of observed vs. simulated data for 20 seconds using these parameter values. The path followed by the robot in the water (solid curve) does not display the oscillations of the simulated data because the vision algorithm samples a center of mass position, whereas the simulated data is the position of the center of one link. We also note that the observed path deviates from a straight line. This justifies our expectation that feedback will be needed to perform true motion planning for the eel robot—open loop operation is not sufficient to follow planned trajectories.

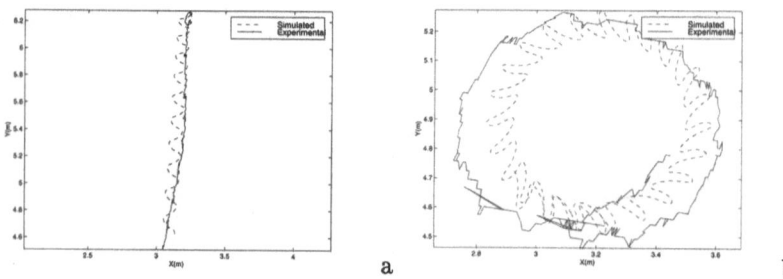

Figure 3. Simulated and experimental data for the (a)forward and (b) circular drive gaits. Simulated data is shown dashed, and experimental data solid.

4.3. Circular paths

We next explored our proposed gait for circular motion. The circular gait is another travelling wave gait, with a steering offset introduced to create a bias in the momentum generated during each cycle. We had qualitative justification of this gait in our experiments with the original REEL I robot, as well as theoretical justification in our perturbation analysis of the dynamics of the eel [9]. The joint angles ϕ_i take the form:

$$\phi_i(t) = A\sin(\omega t + i\phi_s) + \phi_{\text{offs}} \tag{9}$$

for amplitude $A = \pi/6$, frequency $\omega = \pi$ and phase shift $\phi_s = \pi/3$ and offset $\phi_{\text{offs}} = \pi/12$. Our choices of amplitude, frequency and offset were motivated by physical limits in the robot, while the choice of phase ϕ_s was motivated by our perturbation analysis as in Section 4.2. We compared observed data to predicted data using our full dynamic model with the quadratic fluid drag approximation and the μ_w parameter as tuned in our forward gait experiment. Figure 3(b) is a plot of observed versus simulated data for approximately 40 seconds using these parameter values. As in the case of forward swimming, the actual path deviates from a perfect circle, however the robot follows a closed elliptical path with only minor eccentricity, validating our choice of circular turning gait.

4.4. Turning in place

The third gait tested was a novel gait, predicted by our perturbation analysis (see Section 2.3) for turning in place. This gait essentially consists of two reinforcing travelling waves propagating outwards from the central link.

In this case, we use joint angles of the following form:

$$\phi_i = \begin{cases} A\sin(\omega t - (N/2 - i)\phi_s) - \phi_{\text{offs}} & 1 \le i \le N/2 \\ A\sin(\omega t - (i - 1 - N/2)\phi_s) + \phi_{\text{offs}} & N/2 < i \le N \end{cases}$$

for $N = 4$, the number of joints, for amplitude $A = \pi/6$, frequency $\omega = \pi$ and phase shift $\phi_s = \pi/2$ and offset $\phi_{\text{offs}} = \pi/12$. Our choices of amplitude, frequency and offset were motivated by physical limits in the robot, while the choice of phase ϕ_s was motivated by our perturbation analysis [9] which predicted that 90° is the optimal gait phase for a three-link snake. (The turn-in-place gait can be viewed as two three-link snakes connected at the tail, each attempting to turn in the same direction.) We compared observed data to predicted data using our full dynamic model with the quadratic fluid drag approximation and the μ_w parameter as tuned in our forward gait experiment (see Section 4.2). Figure 4(a) is a plot of observed versus simulated orientation for 20 seconds using these parameter values. Figure 4(b) plots observed robot position versus time for the same 20 seconds, to show that the robot is not moving.

4.5. Sideways swimming

The final gait tested in our experiments was another novel gait, predicted by our perturbation analysis [9] for sideways swimming. This gait essentially consists of two opposing travelling waves propagating outwards from the central link, and is reminiscent of the backstroke used by human swimmers.

In this case, we use joint angles of the following form:

$$\phi_i = \begin{cases} A\sin(\omega t - (N/2 - i)\phi_s) + \phi_{\text{offs}} & 1 \le i \le N/2 \\ A\sin(\omega t - (i - 1 - N/2)\phi_s) + \phi_{\text{offs}} & N/2 < i \le N \end{cases}$$

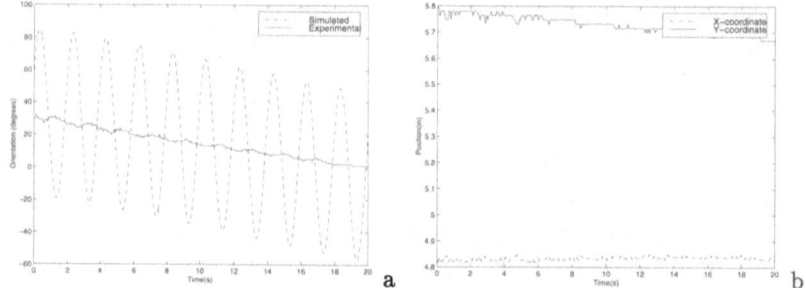

Figure 4. (a)Simulated and experimental data for the turn in place drive gait. Simulated data is shown dashed, and experimental data solid. (b)Position data for the turn in place gait. Note that the robot only drifts a total of 10cm.

for $N = 4$, the number of joints, for amplitude $A = \pi/6$, frequency $\omega = \pi$, phase shift $\phi_s = \pi/2$ and offset $\phi_{\text{offs}} = \pi/12$. Our choices of amplitude, frequency and offset were motivated by physical limits in the robot, while the choice of phase ϕ_s was motivated by our perturbation analysis which predicted that 90° is the optimal gait phase for a three-link snake. (The sideways gait can be viewed as two three-link snakes connected at the tail, attempting to turn in opposing directions.) We compared observed data to predicted data using our full dynamic model with the quadratic fluid drag approximation and the μ_w parameter as tuned in our forward gait experiment (see Section 4.2). As is evident in Figure 5, there is considerable discrepancy in the observed and predicted behaviour of the robot using this gait. In fact, even attempting to tune the fluid drag coefficient parameter μ_w, we were unable to match our observed data. We provide a possible explanation below in Section 5.

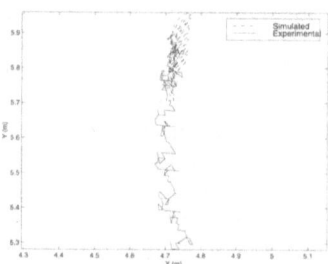

Figure 5. Simulated and experimental data for the sideways gait. Simulated data is shown dashed, and experimental data solid. Notice that there is a large discrepancy between predicted and actual performance.

5. Discussion of Experimental Results

In general, our experimental data show good agreement with predictions. Gaits for forward motion, motion in circular paths and turning in place have been verified experimentally, and we see that we are able to tune our simulations to match experimental data simply with modifications to one fluid drag parameter.

Our closed loop control algorithm presented in [10] used a feedforward approach to improve turning performance by precalculating necessary steering offsets. Since we are able to tune our simulation to match experimental results, it should be possible to determine these open-loop steering commands in simulation with a reasonable expectation of success in closed-loop control.

We have also verified our choice of vision-based position sensing for closed-loop control. Open-loop motion planning will not suffice in our real-world setting since there are deviations from predicted paths. Visual feedback provides a cheap position sensor that does not need to be waterproofed. We are able to capture data in real-time (15 frames/second), which meets our closed-loop requirements, since our control algorithm [10] is based on averaged sampling over one gait cycle (0.5Hz).

We observed minor errors in our experimental trajectory. The most likely sources of these discrepancies is that our simulations are based on an idealized model of the discrete eel that does not perfectly reflect some of the features of the robot. We have assumed that the robot is symmetric for modelling and simulation. In fact, it is not symmetric—the first and last links are approximately 50% longer than middle links, and the first link is both heavier than other links (due to the presence of the battery), and has increased drag, due to the presence of the battery cable which drags outside the link. It is also possible that small currents exist in the pool due to filtration. Open-loop control will not correct for such disturbances. Finally, we believe there are infrequent errors in the radio-communication protocol that lead to "glitches" or "spasms" in the gait waveform as motors receive discontinuous angular commands. These have the effect of causing one-time disturbances or "kinks" in the trajectory.

The only major discrepancy between predicted and observed data comes in the case of the sideways gait. The gait performed considerably *better* than expected—robot velocities were almost an order of magnitude higher than predictions! We are able to propose one possible explanation for this. Our model and simulations are based on a simple fluid dynamic approach that considers *only* drag forces on the links and completely neglects the phenomenon of momentum shedding and thrust in the wake. Under conditions of a slender body undergoing small oscillations, this is a reasonable approximation, however in the case of our proposed sideways gait, these conditions are not met, since the robot has a very large cross section in the direction of motion and is undergoing large shape changes. The excess momentum generated is likely caused by some sort of thrust or wake effect that would have to be modelled by some form of carangiform (fish-like) swimming [5, 6].

6. Conclusions and Future Work

We have developed an experimental system to evaluate the open-loop performance of a biomimetic, underwater eel-like robot. Using predictions from theory made in prior work on the analysis of the system dynamics [9], we tested five drive gaits for various modes of locomotion. Our robot performed well in these tests, despite some experimental errors due to asymmetry in the robot implementation. We have a robust, modular robotic design that will be a useful platform for future experiments.

In future work, we plan to extend our experimental analysis of the system. In prior work [10], we proposed an algorithm for closed-loop control of the eel, using visual feedback. We can implement this control scheme in our test platform with only software modifications. We also plan to add new degrees of freedom to the robot to enable motion in the vertical dimension, so we will not be restricted to the plane.

References

[1] R. W. Brockett and L. Dai. Nonholonomic kinematics and the role of elliptic functions in constructive controllability. In Z. Li and J. F. Canny, editors, *Nonholonomic Motion Planning*, pages 1–21. Kluwer, 1993.

[2] K. Harper, M. Berkemeier, and S. Grace. Decreasing energy costs of swimming robots through passive elastic elements. In *Proc. IEEE Int. Conf. Robotics and Automation*, pages 1839–1844, Albuquerque, NM, April 1997.

[3] S. Hirose. *Biologically Inspired Robots: Snake-like Locomotors and Manipulators*. Oxford University Press, Oxford, 1993. Translated by Peter Cave and Charles Goulden.

[4] L. Kavraki and F. Lamiraux. A general framework for planning paths for elastic objects. Submitted to the *International Journal of Robotics Research*, 1999.

[5] S. D. Kelly, R. J. Mason, C. T. Anhalt, R. M. Murray, and J. W. Burdick. Modelling and experimental investigation of carangiform locomotion for control. In *Proc. of the American Control Conference (ACC)*, 1998. (submitted).

[6] S. D. Kelly and R. M. Murray. Lagrangian mechanics and carangiform locomotion. In *Nonlinear Control Systems Design (NOLCOS)*, Enschede, The Netherlands, July 1998.

[7] J.-C. Latombe. *Robot Motion Planning*. Kluwer, Boston, 1991.

[8] K. A. McIsaac and J. P. Ostrowski. A geometric approach to anguilliform locomotion: Simulation and experiments with an underwater eel robot. In *Proc. IEEE Int. Conf. Robotics and Automation*, volume 1, pages 2843–2848, Detroit, MI, 1999.

[9] K. A. McIsaac and J. P. Ostrowski. A geometric approach to gait generation for the anguilliform robot. In *Proc. of Int. Conf. on Intelligent Robots and Systems (IROS 2000)*, volume 1, pages 2230–2235, Tokyo, October 2000.

[10] K. A. McIsaac and J. P. Ostrowski. Motion planning for dynamic eel-like robots. In *Proc. IEEE Int. Conf. Robotics and Automation*, volume 1, pages 1695–1700, San Francisco, CA, 2000.

[11] Örjan Ekeberg. A combined neuronal and mechanical model of fish swimming. *Biological Cybernetics*, 69:363–374, 1993.

[12] J. P. Ostrowski. Steering for a class of dynamic nonholonomic systems. *IEEE Transactions on Automatic Control*, 45(8):1492–1498, August 2000.

[13] J. P. Ostrowski and J. W. Burdick. The geometric mechanics of undulatory robotic locomotion. *International Journal of Robotics Research*, 17(7):683–702, July 1998.

[14] J. P. Ostrowski, J. P. Desai, and V. Kumar. Optimal gait selection for nonholonomic locomotion systems. *International Journal of Robotics Research*, 19(3):225–237, March 2000.

[15] M. S. Triantafyllou and G. S. Triantafyllou. An efficient swimming machine. *Scientific American*, pages 64–70, March 1995.

[16] M. Zefran, J. P. Desai, and V. Kumar. Continuous motion plans for robotic systems with changing dynamic behavior. In *Workshop on Algorithmic Foundations of Robotics (WAFR) '96*, Toulouse, France, July 1996.

[17] H. Zhang and J. P. Ostrowski. Visual servoing with dynamics: Control of an unmanned blimp. In *Proc. Int. Conf. on Robotics and Automation*, pages 618–623, Detroit, MI, May 1999.

Quadruped Robot Running With a Bounding Gait

S. Talebi[1]   I. Poulakakis[1]   E. Papadopoulos[2]   M. Buehler[1]

[1]Ambulatory Robotics Laboratory, http://www.mcgill.cim.ca/~arlweb
Centre for Intelligent Machines, McGill University, Montreal, CANADA
[2]Department of Mechanical Engineering, NTU Athens, GREECE

Abstract: Scout II, an autonomous four-legged robot with only one actuator per compliant leg is described. We demonstrate the need to model the actuators and the power source of the robot system carefully in order to obtain experimentally valid models for simulation and analysis. We describe a new, simple running controller that requires minimal task level feedback, yet achieves reliable and fast running up to 1.2 m/s. These results contribute to the increasing evidence that apparently complex dynamically dexterous tasks may be controlled via simple control laws. An energetics analysis reveals a highly efficient system with a specific resistance of 0.32 when based on mechanical power dissipation and of 1.0 when based on total electrical power dissipation.

1. Introduction

Most existing four- or eight-legged robots are designed for statically stable operation - stability is assured by keeping the machine's center of mass above the polygon formed by the supporting feet. While this is the safest mode of locomotion, it comes at the cost of mobility and speed. Furthermore it requires a high mechanical complexity of three degrees of freedom per leg to provide continuous body support.

In contrast, we have pursued an agenda of low mechanical complexity in our Scout I and II robots, in order to decrease cost and increase reliability. We have shown in [1, 2] that dynamic walking, turning and step climbing can be achieved with a quadruped with stiff legs and only one hip actuator per leg. In this paper we show that Scout II with an additional compliant prismatic joint per leg is able to bound (Fig. 1). Dynamic running is possible with a very simple control strategy. Open loop control, simply positioning the legs at a fixed angle during flight, and commanding a fixed leg sweep angular velocity during stance results in a stable bounding gait. To our knowledge, Scout II is the first autonomous quadruped that achieves compliant running, features the simplest running control algorithm, and the simplest mechanical design to date.

This paper also addresses two subjects that have not yet received the attention they require in order to advance the state of the art in autonomous, dynamically stable legged locomotion - experimentally validated models and

D. Rus and S. Singh (Eds.): Experimental Robotics VII, LNCIS 271, pp. 281–289, 2001.

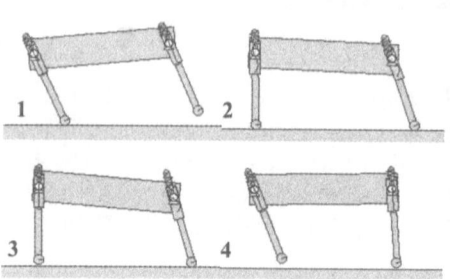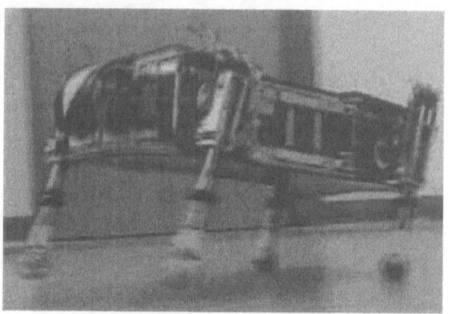

Figure 1. Illustration of a bound gait (left) and Scout II bounding (right)

energetics. Autonomous legged robots operate at the limits of their actuators, and require a model of the actuator dynamics and their interaction with the power source. We show that for Scout II, and likely for most other robots in its class, ignoring these issues results in inaccurate models. In addition, energy efficiency and autonomy are essential for mobile robots. In order to characterize the energetics of Scout II, we document the running efficiency as a function of speed, based on both the mechanical actuator output power, and the total electrical input power.

Ongoing research addresses compliant walking, rough terrain locomotion and dynamic stair climbing with Scout II, while another paper [3] demonstrated a trotting (walking) gait, based on additional passive, but lockable knee joints and non-compliant legs. The approach of using only one actuated degree of freedom per leg, compliant legs, and task-space open loop controllers has recently also been applied successfully to a dynamic hexaped, RHex [4]. This biologically inspired robot has the added advantage of a low center of mass and sprawled posture and is able to negotiate rough terrain at one body length per second.

Only few cases of quadruped running robots have been reported in the literature. About 15 years ago, Raibert [5] set the stage with his groundbreaking work on a dynamically stable quadruped, which implemented his three-part controller, via generalizations of the virtual leg idea. The robot featured three hydraulically actuated and one passive prismatic DOF per leg. The robot was able to trot, pace and bound, with smooth transitions between these gaits. Furusho et al [6] implemented a bounding gait on the Scamper robot. Even though the robot's legs were not designed with explicit mechanical compliance, the compliance of the feet, legs, belt transmissions, and the PD joint servo loops were likely significant. Akiyama and Kimura [7] implemented a bounding gait in the Patrush robot. Each three DOF leg featured an actuated hip and knee, and an unactuated, compliant foot joint. Their neural oscillator based controller was motivated by Matsuoka [8], which also underlies the control of the simulated planar biped of Taga et al [9]. An additional reflex network was added to the neural oscillator to achieve the stability and robustness necessary for experimental success.

2. Mechanical Structure and Modeling

The mechanical design of Scout II (Fig. 2) is an exercise in simplicity. Besides its modular design, the most striking feature is the fact that it uses a single actuator per leg - the hip joint provides leg rotation in the sagittal plane. Each leg assembly consists of a lower and an upper leg, connected via a spring to form a compliant prismatic joint. Thus each leg has two degrees of freedom, one actuated hip and one unactuated linear spring. All components for autonomous operation are integrated: The two hip assemblies contain the actuators and batteries, and the body houses all computing, interfacing and power distribution.

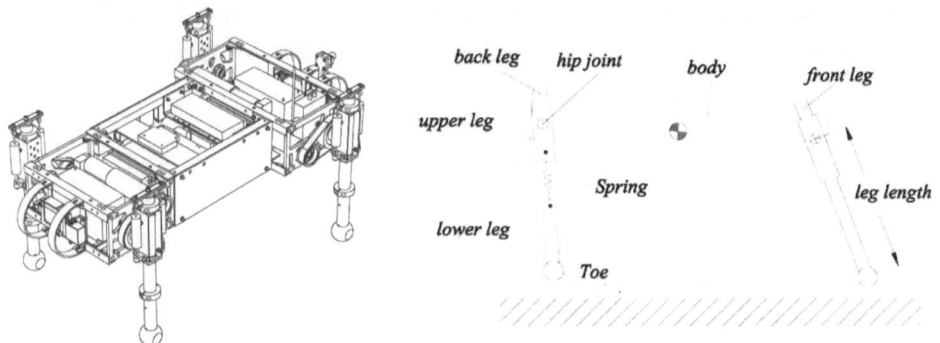

Figure 2. Scout II

The Scout II in planar motion is modeled in WorkingModel 2D [10] as a five-body kinematic chain, shown in Fig. 2. A linear spring and damper system models the leg compliance during stance phase. Since each of the two legs can be in stance or flight, there are four robot states.

3. Actuator and Power Source Modeling

It is well known that dynamically stable legged robots are complex dynamical systems with intermittent variable structure dynamics, fewer actuators than motion degrees of freedom, impacts, unilateral toe-ground constraints, and limited ability to apply tangential ground forces due to slip. These qualities greatly complicate modeling and usually prevent the application of classical control synthesis. In this section we demonstrate two additional modeling components which are dominant on our Scout II robot, and which are likely to be significant in dynamically stable legged robots in general - actuator and power source modeling.

Designing an autonomous dynamically stable robot is a formidable system design challenge. For example, the robot weight should be kept to a minimum, yet the actuators have to be capable not only to support the robot weight, but also to impart significant accelerations to the body, and support large dynamic loads. As a result, the actuators will typically operate at their limits, characterized by their torque-speed curve. While this fact is well known, it is typically not taken into account in robot modeling and control. As we will see below, ignoring this constraint will result in large differences between commanded and actually achieved torques.

The torque speed limitation of an electrical actuator can be characterized in the first quadrant by

$$\tau = min(\frac{K}{R}(V_T - K\omega), \tau_{max}) \tag{1}$$

where K is the motor torque constant, R is the motor armature resistance, ω is the motor speed, V_T is the motor terminal voltage, and τ_{max} is the fixed torque limit imposed by the motor amplifiers' current constraint. Figure 3 below shows the large difference between desired torques (top plots), and actually achievable torques (lower plots), for a fixed power supply or battery voltage.

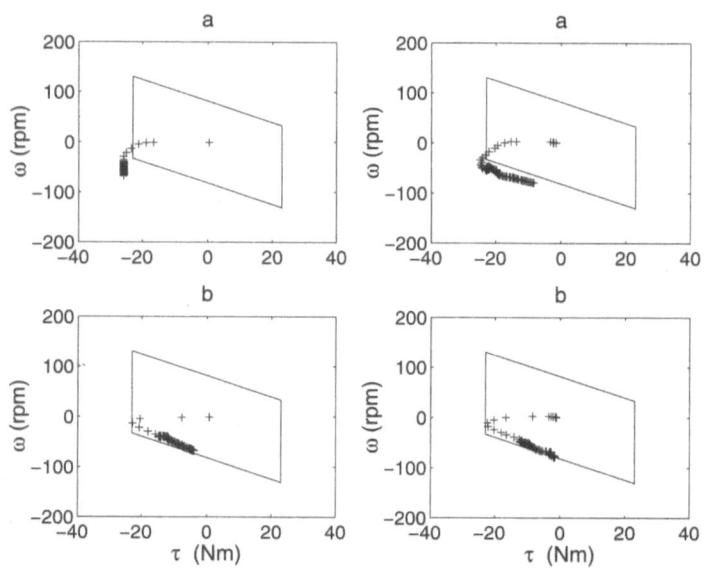

Figure 3. Experimental results. Torque speed plot for the back (left), and front legs (right) actuators. Top plot shows commanded torques and bottom plots show actually achievable torques, based on a fixed 24 V battery voltage (solid polygon).

Since electrically actuated autonomous robots can draw significant peak power and operate from non-ideal voltage sources, the variation of the supply voltage as a function of the total load current must be considered. Fig. 4 (left) shows the drastic supply voltage fluctuations, and that a simple battery model, consisting of a fixed internal voltage source of 24 V in series with an internal resistance of 0.15 Ω results in a very good match between the measured and modeled supply voltage.

Fig. 4 (right) demonstrates both the large discrepancy between desired (upper solid line) and achievable motor torques (lower solid line) and the accuracy of the combined actuator/power model. It is interesting to point out that, due to the multitude of dynamic, actuation, and power constraints, it is nearly impossible to control either torque or leg angular velocity during stance arbitrarily. The controller can only affect the system dynamics during stance in a limited fashion. For this reason it is important that the robot's passive (unforced) dynamics be as close as possible to the desired motion. Indeed, this is likely one of the reasons for the successful operation of Scout II. In addition, the actuation constraints during stance suggest the use of the leg touchdown angle (which is easily controlled during flight) as a dominant control input. As shown in the following section, this is one of the main control parameter in our bounding controller.

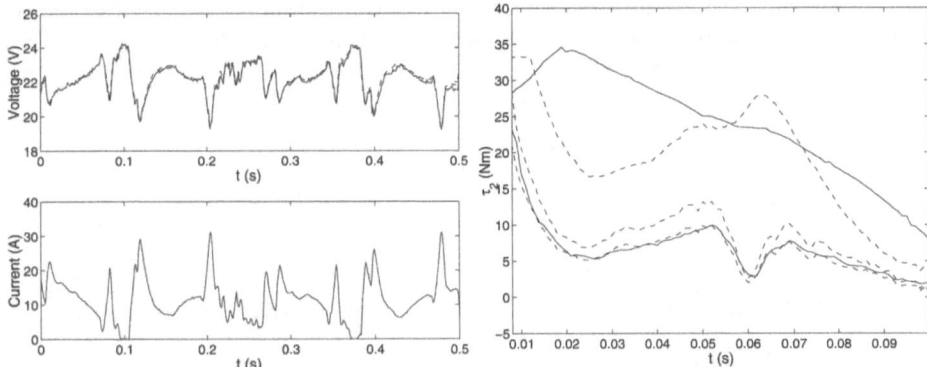

Figure 4. (left) Battery voltage fluctuation (top, solid) as a function of load current (bottom). The top graph shows both the measured battery voltage (solid) and the battery voltage estimation (dash). The exceptionally good match between experimental and model data validates the simple internal resistance model (24V nom. battery voltage with 0.15 Ω internal resistance). Figure 4. (right) Torque profiles during one stance phase from touchdown until the sweeplimit is reached. Data traces from top to bottom: Desired torque from controller (solid), maximum achievable torque based on torque speed curve with fixed 24 V supply voltage (dashed), max. achievable torque based on battery voltage model; additional loop gain fix due to amplifier gain modeling error; measured motor torque (solid).

4. Bounding Controller

Even though Scout II is an under-actuated, highly nonlinear, intermittent dynamical system, we found that simple a control laws can stabilize periodic motions, resulting in robust and fast running. Surprisingly, the controllers do not require task level feedback like forward velocity, or body angle. What is more, there seem to exist many such simple stabilizing controllers - in [11] three variations are introduced. It is remarkable that the significant controller differences have relatively minor effects on bounding performance! For this reason and for brevity we shall describe one of these controllers here.

The controller is based on two individual, independent leg controllers,

without a notion of overall body state. The front and back legs each detect
two leg states - stance (touching ground) and flight (otherwise), which are
separated by touchdown and lift-off events.

There is no actively controlled coupling between the fore and hind legs -
the resulting bounding motion is purely the result of the controller interaction
through the multi-body dynamic system. During flight, the controller servos
the flight leg to a desired touchdown angle ϕ_{td}, then sweeps the leg during
stance with a desired angular velocity $\dot{\phi}_d$ until a sweep limit ϕ_{sl} is reached.
Table 1 lists these controller parameters and the stance and flight PD gains.
Even though we show only the results for one of several controllers implemented,
experimental performance for all of them is very similar - resulting in stable
and robust bounding, at top speeds between 0.9 and 1.2 m/s.

	front and back legs
$\phi_{td}(^{o})$	20
$\phi_{sl}(^{o})$	0
$\dot{\phi}_d(^{o}/s)$	-200
$k_{p,s}$ & $k_{p,f}$ (Nm/ ^{o})	35
$k_{d,s}$ & $k_{d,f}$ (Nm/ ^{o})	0.15

Table 1. Controller Parameters and PD gains (average speed 0.7m/s).

Figure 5 compares the body angle trajectories and torque profiles between
simulations and experiment. The stride frequency as well as the body oscillation amplitude matches well. The torque traces are qualitatively similar, but
there are still many details which differ. Some of these are likely due to in-
accurately modeled ground-toe friction, and unmodeled compliance in the leg.
These differences are still the subject of ongoing work.

5. Energetics

For mobile robots to be of practical utility, they need to be energy efficient and
able to operate in a power-autonomous fashion for extended periods of time.
Thus, energy efficiency is an important performance measure of mobile robots.
An increasingly accepted measure of energy efficiency is the 'specific resistance'
- a measure proposed originally by Gabrielli and von Kármán [12] in 1950,

$$\varepsilon(\nu) = \frac{P(\nu)}{mg\nu} \qquad (2)$$

where P is the power expenditure, m is the mass of the vehicle, g is the gravitational acceleration, and ν is the vehicle speed. Since many vehicle specific
resistances quoted in the literature are based on the average mechanical output
power of the actuators, we have calculated this figure as a function of speed
(Fig. 6). Even though energy efficiency has so far not been optimized, Scout II
at top speed already achieves a low specific resistance of $\varepsilon = 0.32$. This value
places Scout II among the most energy efficient running robots, only slightly
higher than the (lowest published running robot efficiency) $\varepsilon = 0.22$ value for
the ARL Monopod II [13, 14], but still lower than any other running robot.

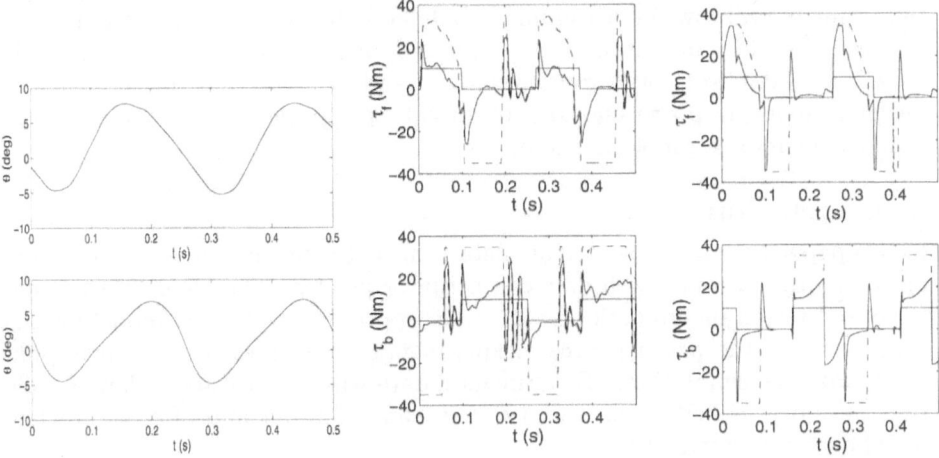

Figure 5. (left) Body angle . Experiment (top), simulation (bottom). Figure 5 (right) Front (top) and back (bottom) actuator torques. Commanded torques(dash) vs. measured torques (solid). Experiments (left) vs. simulation(right). The solid square wave denotes the leg state: stance (high)and flight (low).

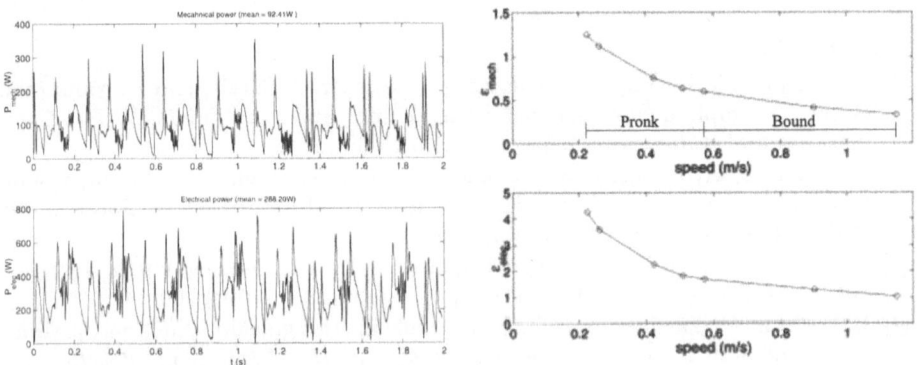

Figure 6. (left) Mechanical (top) and total electrical (bottom) power consumption at 1.15 m/s. Figure 6. (right) Specific resistance as a function of forward speed based on mechanical (top) and total electrical (bottom) power consumption.

The specific resistance based on mechanical output power has drawbacks, since it does not take the actuator efficiency or the power consumption of the entire system into account. Both of these effects can have a dramatic negative influence on runtime. Therefore, a more useful measure of energy efficiency, is the specific resistance based on total power consumption. For a system with a battery as the main power source, this is the total average product of battery current and voltage. For Scout II, this value is approximately $\varepsilon = 1.0$, three times the specific resistance based on mechanical power. We suspect that this value is still low for a running robot, and that even the large difference between electrical and mechanical power is normal; however, little comparable data is available from other robots to date, and we hope that the reporting of mechanical output power and total electrical input power will become standard practice for mobile robots in the future.

6. Conclusion

In this paper we have presented an algorithm that controls compliant bounding for a quadruped robot with only one actuator per leg. The algorithm was derived and tested in simulations, which incorporated a validated model for the actuators and the power source. Experimental runs showed good correspondence with the simulations. Experimental data was used to show a low specific resistance of $\varepsilon = 0.32$ when based on mechanical power and of $\varepsilon = 1.0$ when based on total electrical power.

7. Acknowledgements

This project was supported in part by IRIS, a Federal Network of Centers of Excellence and the National Science and Engineering Research Council of Canada (NSERC). We also acknowledge the generous and talented help of G. Hawker, M. de Lasa, D. Campbell, D. McMordie, E. Moore, and S. Obaid.

References

[1] M. Buehler, R. Battaglia, A. Cocosco, G. Hawker, J. Sarkis, and K. Yamazaki. Scout: A simple quadruped that walks, climbs and runs. In *IEEE Int. Conf. Robotics and Automation*, pages 1707–1712, May 1998.

[2] M. Buehler, A. Cocosco, K. Yamazaki, and R. Battaglia. Stable open loop walking in quadruped robots with stick legs. In *IEEE Int. Conf. Robotics and Automation*, pages 2348–2353, May 1999.

[3] G. Hawker and M. Buehler. Quadruped trotting with passive knees. In *IEEE Int. Conf. Robotics and Automation*, pages 3046–3051, April 2000.

[4] U. Saranli, M. Buehler, and D. E. Koditchek. Design, modelling and preliminary control of a compliant hexapod robot. In *IEEE Int. Conf. Robotics and Automation*, pages 2589–2596, April 2000.

[5] M. H. Raibert. *Legged Robots that Balance*. MIT Press, Cambridge, MA, 1986.

[6] J. Furusho, A. Sano, M. Sakaguchi, and E. Koizumi. Realization of bounce gait in a quadruped robot with articular-joint-type legs. In *IEEE Int. Conf. Robotics and Automation*, pages 697–702, May 1995.

[7] S. Akiyama and H. Kimura. Dynamic quadruped walk using neural oscillators - realization of pace and trot. In *13. Annual Conf. RSJ*, pages 227–228, 1995.

[8] K. Matsuoka. Sustained oscillations generated by mutually inhibiting neurons with adaptation. *Biol. Cybernetics*, 52:367–376, 1985.

[9] G. Taga, Y. Yamaguchi, and H. Shimizu. Self-organized control of bipedal locomotion by neural oscillators in unpredictable environment. *Biol. Cybernetics*, 65:147–159, 1991.

[10] Knowledge Revolution. *Working Model 2D User's Guide*. CA, 1996.

[11] S. Talebi. Compliant running and step climbing of the Scout II platform. Master's thesis, McGill University, 2000.

[12] G. Gabrielli and Th. von Kármán. What price speed? *Mechanical Engineering*, 72(10):775–781, 1950.

[13] M. Ahmadi and M. Buehler. The ARL Monopod II running robot: Control and energetics. In *IEEE Int. Conf. Robotics and Automation*, pages 1689–1694, May 1999.

[14] M. Ahmadi and M. Buehler. Stable control of a simulated one-legged running robot with hip and leg compliance. *IEEE Trans. Robotics and Automation*, pages 96–104, Feb 1997.

Evidence for Spring Loaded Inverted Pendulum Running in a Hexapod Robot

Richard Altendorfer, Uluç Saranli, Haldun Komsuoğlu, Daniel Koditschek
Artificial Intelligence Laboratory, University of Michigan
Ann Arbor, MI 48109
{altendor, ulucs, hkomsuog, kod}@eecs.umich.edu

H. Benjamin Brown Jr.
The Robotics Institute, Carnegie Mellon University
Pittsburgh, PA 15213
hbb+@cs.cmu.edu

Martin Buehler, Ned Moore, Dave McMordie
Ambulatory Robotics Laboratory, Dept. of Mech. Engineering, McGill University
Montréal, Québec, Canada H2A 2A7
{buehler, ned, mcmordie}@cim.mcgill.ca

Robert Full
Dept. of Integrative Biology, University of California at Berkeley
Berkeley, CA 94720
rjfull@socrates.berkeley.edu

Abstract:

This paper presents the first evidence that the Spring Loaded Inverted Pendulum (SLIP) may be "anchored" in our recently designed compliant leg hexapod robot, RHex. Experimentally measured RHex center of mass trajectories are fit to the SLIP model and an analysis of the fitting error is performed. The fitting results are corroborated by numerical simulations. The "anchoring" of SLIP dynamics in RHex offers exciting possibilities for hierarchical control of hexapod robots.

1. Introduction

We have recently reported on a prototype robot that breaks new ground in artificial legged locomotion [1]. Our shoe-box sized, compliant leg hexapod, RHex, travels at speeds better than one body length per second over terrain that few other robots can negotiate at all. RHex origins and construction are grounded in the interplay between biomechanics, controls, and engineering design that we have come to call "functional biomimesis." We aim to articulate broad principles with mathematically precise formulations of biomechanically observed fact and then translate these into specific design practices. This paper presents the first empirical evidence that our strategy to use a low degree

D. Rus and S. Singh (Eds.): Experimental Robotics VII, LNCIS 271, pp. 291–302, 2001.

of freedom mechanism as a "template" for a high degree of freedom task may be relevant and productive. Biomechanics research suggests that the Spring Loaded Inverted Pendulum (SLIP) functions as a sagittal plane template for all animal running [2]. Motivated by the success of Raibert's hoppers [3] that explicitly incorporate a physical SLIP in the working mechanism, we had previously begun to develop a theory to inform SLIP tuning [4]. We had also reported simulation evidence describing how the two degree of freedom SLIP template might be anchored in a four degree of freedom (all revolute) bipedal running model [5]. Adapting well-characterized methods developed at the UC Berkeley Polypedal lab to explore gait stabilization in animals and a model introduced in [1], we now offer a preliminary characterization of RHex center of mass (COM) trajectories respecting which the presumed relevance of the SLIP model can be empirically tested.

2. The SLIP Template for Legged Runners

A template [6] is a low dimensional model of a robot operating within a specified environment that is capable of expressing a specific task as the limit set of a suitably tuned dynamical system involving some controlled (robot) and uncontrolled (environment) degrees of freedom. To "anchor" this low dimensional model in a more physically realistic higher degree of freedom representation of the robot and its environment, we seek controllers whose closed loops result in a low dimensional attracting invariant submanifold on which the restriction dynamics is a copy of the template. Examples of this idea at work in functioning robots include a series of batting machines that anchored a "Raibert vertical" template [7] in a one degree of freedom paddle robot (operating into a two degree of freedom environment) [8] and a three degree of freedom paddle robot (operating into a three degree of freedom environment) [9]. This same idea is used to control a recently reported brachiating robot [10]. In this section we review the manner in which a hierarchical controller can be devised to shape and then exploit the appearance of the SLIP template in morphologically distinct legged machines.

2.1. Hierarchical Control of a Virtual SLIP Monopod

Biomechanical evidence for the existence of a SLIP template in human runners [11, 12] naturally leads to the possibility of template based controller designs for legged locomotion. Toward this end, previous work [5] demonstrated an approximate embedding of a SLIP template in a planar 4 DOF leg with ankle, knee and hip joints (AKH), similar in morphology to a human leg.

The hierarchical control of AKH involves defining a virtual leg between the toe and the COM of the system. The joint control torques are then computed using a SLIP template prescribing the ground reaction force (and hence the acceleration of the COM) together with an approximate, virtual work based embedding. In consequence of this hierarchical decomposition, a high level SLIP controller can be used to regulate the speed and hopping height of the overall system.

This approach to hierarchical design bears useful comparison to the notion of impedance control advanced by Hogan [13] and more recently introduced into the locomotion literature in the more specific form of "Virtual Model Control" by Pratt and colleagues [14]. This framework allows a user to program the robot's task in terms of a reference compliance imposed on a targeted part of the body. It is different from our approach in that the allowable reference models operate in the quasi-static regime, so, for example, running could not lie within the formal scope of the method. In contrast, the SLIP template provides an explicitly (hybrid) dynamical specification of the exchange between kinetic and potential energy that accomplishes the task at hand after transients in the many degrees of freedom unrelated to the task have died out. In this paper, we are concerned to find a means of effecting this "collapse of dimension" in the RHex mechanics.

2.2. Hierarchical Control of a Virtual SLIP Hexapod

We now describe two alternative approaches to hierarchical control for our hexapod. Both appeal to the SLIP template for the prescription of COM forces, but incorporate different anchoring mechanisms.

The hexapod model we consider is a rigid body with six massless legs [1]. Two of the spherical leg freedoms — the radial length and one of the angles — are driven by passive springs and dampers, whereas the hip angle is torque actuated. Consequently, there are only six actuated joints, and the overall system has six degrees of freedom, all due to the rigid body.

2.2.1. Active Control

In principle, the force and torque acting on the hexapod rigid body can be determined using the equations of motion of the model [1]. Sufficient conditions for exact embedding of an arbitrary dynamical template can be developed from the invertibility of the dynamics. However, complete input invertibility generally cannot prevail in our system. The morphology of the system, the hybrid nature of the problem and the structure and number of the actuators (especially when not all legs are on the ground) do not yield full control over the six body degrees of freedom.

A simpler planar model, on the other hand, provides an exactly invertible plant, except for co-dimension one and two singularities. The model consists of a three degree of freedom planar rigid body, with six torque actuated massless legs, with the assumption that three or more legs are in contact with the ground during stance.

Preliminary numerical experience with this model suggests that choosing a "reasonable" stance posture affords inverse dynamics controllers that pass transversally through the kinematic singularities and give good SLIP trajectories. Moreover, the planar model is structurally very close to the spatial model. As a consequence, it seems likely that the inverse dynamics anchoring in the planar model can be readily extended to yield an approximate embedding of the SLIP template in the spatial hexapod model.

Nevertheless, realizing the active template through inverse dynamics control suffers the traditional problems of all such approaches based on exact cancellations: the presumption of a perfect model; known parameters; and noise-free high bandwidth state information. It is not clear how effectively this exact embedding can be implemented in a physical platform in the face of the inevitable actuator, computational, and sensory limitations.

2.2.2. Passive Control

An alternative to active control relies on the passive dynamics of the system combined with low-bandwidth controllers to anchor the SLIP template. Demonstrating that this may be possible represents the chief concern of the paper as established in §3. Even with a very simple open-loop control strategy, our study reveals the presence of certain "sweet spots" in the RHex parameter space, wherein the SLIP emerges naturally. It is still unclear whether this respects the formal "anchored template" paradigm wherein the lower dimensional dynamics actually appears as an attracting invariant dynamical submanifold. However, experimental evidence revealing the template behavior in steady state from various different initial conditions suggests there are, indeed, operating regimes where the system trajectories are attracted to the low dimensional SLIP template dynamics. Further evidence for the SLIP template comes from numerical studies using SimSect — a simulation package developed by Saranli [15]. SimSect was devised to approximate the behavior of RHex by numerically integrating a set of simplified equations of motion which are expected to govern RHex's hybrid mechanical system. Currently, the same low-level controller as in RHex is implemented in SimSect. This makes SimSect an ideal test-bed for new control designs for RHex. Although an exact correspondence between RHex's and SimSect's parameter space has not yet been established, the simulation results in §3.4 lend credence to SimSect being a representative numerical approximation to RHex's dynamics.

3. Finding the SLIP in RHex's Motion

The central observations about cockroach locomotion that inform the design of the RHex prototype include: (i) that it operates via compliant legs; (ii) that its limb motions appear to be characterized by a strongly stereotypical "clock"; (iii) that it has a sprawled posture to enhance stability; and (iv) that the stabilizing controller must somehow be embedded in the very morphology itself. The impact of these observations for RHex are, indeed, directly apparent in the morphology and control approach that we have already reported. However, it is not obvious that we will find a SLIP in such a machine.

3.1. Data Collection, Experimental Setup and Procedures

In order to determine whether RHex passively anchors a SLIP, the ground reaction forces produced by RHex during locomotion were measured during 92 trials using two six-component force plates[1]. The force and torque signals were amplified[2] and each channel was recorded at 1000Hz by an analog to digital converter[3]. Each trial was also recorded by a high speed video camera[4].

---

[1] Biomechanics Force Platform, Advanced Mechanical Technology, Inc., Newton, MA.
[2] Model SGA, Advanced Mechanical Technology, Inc., Newton, MA.
[3] PCI board, National Instruments, Austin, TX.
[4] MotionScope PCI 1000, Redlake Imaging, Morgan Hill, CA.

In our experiments, the robot started walking approximately two meters away from the force plates in order to allow the robot to settle into an approximate steady state motion upon encountering the plates. While the robot was in contact with the force plates no directional adjustment was made since this would otherwise have broken the open loop symmetry between the right and left leg motion profiles.

During the trials, four parameters were varied: leg type; ground material; robot mass; and forward speed. For a fixed set of parameters, the experiments were repeated between 4 and 8 times. The experiments started off with the slowest forward speed on the bare force plates with Delrin legs (stiffness $\kappa \approx 4300 N/m$). Then the speed was increased in three steps by choosing different cycle times[5] ($t_c \in \{1.2s, 0.8s, 0.53s, 0.5s\}$) without changing the physical structure. To reduce bounce and slippage, which was observed especially at high speeds, the surface of the force plates was then covered with an elastic foam mat, and the same speed sweep with the Delrin legs was performed. In the second round of the experiments, a new 4-bar linkage composite leg design ($\kappa \approx 3100 N/m$) was used in conjunction with a similar speed sweep on both the bare force plates and the plates covered with the foam mat. Our preliminary observations during the first two trial runs suggested that the COM of the body behaves more like an inverted pendulum (IP) rather than a spring loaded inverted pendulum (SLIP). We reasoned that the leg-body system, which defines an overall lumped spring-mass system, has a much higher natural frequency than the stride frequency achievable by the hip actuators. In order to test this hypothesis, in the third round of the experiments, the body mass was increased, effectively decreasing the natural frequency of the spring-mass system. We ran the robot with composite legs at the highest speed setting on the elastic mat. Its mass was increased incrementally from $7.83 kg$ to $9.47 kg$ to $11.12 kg$ to $11.94 kg$. In the highest mass regime we observed the transition from IP to SLIP reported below.

3.2. Data Extraction

The data plotted in Fig. 1 arise from the summed leg or COM ground reaction forces imparted to the legs by the ground plate while the robot performs an alternating tripod gait. To remove noise from the recorded data, the forces were filtered using a second order Butterworth filter with a cutoff frequency of 50Hz. The minima of the vertical force data were used to isolate single strides. Since a stride is a complete cycle for all the legs, it contains two steps for each tripod and therefore two minima. Only strides from the middle section of the data for one force platform — where the force data exhibited oscillatory behavior of one predominant frequency and roughly constant amplitude — were selected. Only those trials where the maximum of the power spectrum $P(f)$ occurs at twice the cycle frequency[6] $f_c = 1/T_c$ were used, in rough accord with the

---

[5]The motion profile utilized by RHex is parametrized by cycle time, sweep angle, leg offset and flight time (for a detailed description of these parameters see [15]). For each forward speed setting a different set of values is assigned to these four parameters.

[6]During one cycle, two steps are taken.

criterion established in [16] to distinguish walking from running.[7] These criteria reduced the number of available trials to be used for SLIP fitting from 92 to 14.

Since the force platform is very narrow with respect to the width of RHex and only those trials for which RHex stayed on the track were recorded, the lateral force data are not used in this investigation.[8] This restriction respects the intended limitations on our scope of analysis in this initial study to the saggital plane only.

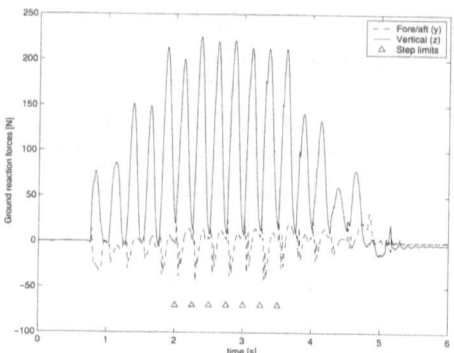

Figure 1.   Ground reaction forces for trial 6 (comp. legs, foam mat, $t_c = 0.5s$, $m = 11.94kg$). The triangles denote the beginning and the end points of steps selected for fitting.

Unfortunately, the camera's resolution was not good enough to provide integration constants for vertical and fore/aft speeds and positions at the beginning of each stride, required for our fitting study. Instead, the initial vertical speed is indirectly obtained from the assumption of periodicity – namely, that after one stride, the robot returns to its initial height. Similarly, the fore/aft initial speed is calculated by matching the average velocity over one stride to the average velocity over both force plates. The initial height is assumed to be at RHex's static equilibrium with all legs vertical to the ground (0.164m).

3.3. Data Analysis

The SLIP template imposes a very particular set of relationships — those specified by the Lagrangian mechanics of a single point mass prismatic-revolute (i.e., polar coordinate) kinematic chain between the ground reaction forces, motion of the COM, and system energies. Ruina has pointed out [19] that any convex curve supports in a neighborhood of its vertical minimum at least one time varying trajectory generated by some SLIP. RHex's COM inevitably rides along a convex curve: we wish to understand whether its actual time trajectory along this curve can be readily generated by some SLIP model.

3.3.1. A Protocol for Fitting SLIP to RHex's Running Data

The 14 remaining trials that satisfy the criteria in §3.2 are now used to test the presence of the SLIP template in RHex according to an adaptation of the methodology introduced in [11]. For each of the 136 steps (68 strides) in the 14 trials, a SLIP model is fit using ordinary least squares regression. Fitting a SLIP model to these data is not entirely straightforward: since RHex did not

---

[7]Specifically, the ratio of the integrated power spectrum around the maximum to the total integrated power spectrum was required to satisfy $(\int_{2f_c - \epsilon}^{2f_c + \epsilon} P(f)df)/(\int_0^\infty P(f)df) > 0.8$, where ϵ was appropriately chosen to include the global maximum alone.

[8]The magnitude of the lateral forces is comparable to the one for the fore/aft forces. A complete template should incorporate lateral forces, see e. g. [17] or [18].

exhibit any flight phase, the touchdown and lift-off points are not defined. In contrast, if the anchoring hypothesis has any validity, then the region around a bottom should be well described by a SLIP stance phase. Absent a specific model for determining the limits of this region, we adopt the ad hoc condition determined by the point of zero crossing of the vertical COM force.[9]

Fitting data to this central force model requires knowledge of the center of pressure (the pivot point of the virtual SLIP), which could not be determined with our experimental equipment. As a reasonable work-around, we assume that the center of pressure lies directly below the vertical minimum. This in turn, implies that the SLIP model operates at equilibrium on a "neutral orbit" [3, 11] characterized by this symmetry. However, the measured force data are not perfectly periodic, hence the integrations to yield velocity and position are necessarily not periodic, either. Notwithstanding this slight conceptual conflict, we see no better method for selecting the nominal center. These assumptions in force, the data population for SLIP fitting can be restricted to range only from a vertical minimum to the next zero crossing of the vertical force.

Given a COM trajectory fragment, $\{\mathbf{b}(t), \dot{\mathbf{b}}(t), \ddot{\mathbf{b}}(t)\}|_{t_0=T_{bottom}}^{t_N=T_{liftoff}}$, the COM position[10] $\mathbf{b} = (y \ z)^\top$ and acceleration $\ddot{\mathbf{b}}$ are fitted to a Hooke spring law with unknown spring length[11] q_{r0} and spring stiffness κ:

$$\frac{\mathbf{b}}{||\mathbf{b}||}(\kappa_1 - \kappa_2||\mathbf{b}||) = m(\ddot{\mathbf{b}} - \mathbf{g}) \ ,$$

where $\kappa = \kappa_2$ and $q_{r0} = \frac{\kappa_1}{\kappa_2}$. Note that this model is linear in parameters so that ordinary least squares applies directly.

The assessment of the quality of the fit proceeds in two steps. First, a SLIP simulation over the same period of time as the data trajectory is run with the values of κ and q_{r0} obtained in the first step. The initial conditions are taken to be the positions and velocities of the data trajectory at the minimum. Second, the resulting SLIP trajectories z^{SLIP}, $\dot{z}^{\text{SLIP}}$, y^{SLIP}, $\dot{y}^{\text{SLIP}}$ are compared to the data trajectories by L_1 and L_2 percent errors:

$$\Delta X_{L_1} = 100\frac{||X - X^{\text{SLIP}}||_1}{\text{Range}(X)} \ , \qquad \Delta X_{L_2} = 100\frac{||X - X^{\text{SLIP}}||_2}{||X||_2} \ ,$$

where $\text{Range}(X) = |\max(X) - \min(X)|$. Here, $||X||_p = (\int_{t_0}^{t_1}|X(t)|^p dt)^{\frac{1}{p}}$ and $X \in \{z, y, \dot{z}, \dot{y}\}$.

In an effort to simplify the assessment of the fitting error, the quality of the fit is reported as a single number — the average L_p percent error $\Delta_{L_p} = (\Delta z_{L_p} + \Delta y_{L_p} + \Delta \dot{z}_{L_p} + \Delta \dot{y}_{L_p})/4$.

[9]Presumably, the phase interval corresponding to flight in the simple SLIP template must be replaced with an appropriately more complex (but still low dimensional) model. Since we have not yet developed this model, we rely on the ad hoc termination criterion.

[10]In this notation, z gives the co-ordinates in the vertical direction and y gives the co-ordinates in the fore/aft direction relative to an inertial frame located at the center of pressure.

[11]Fitting the spring length q_{r0} alleviates the arbitrariness of selecting the equivalent lift-off point for RHex data, because now the zero crossing of the vertical COM force need not correspond to the lift-off point of the fitted SLIP model.

As an illustration of the fitting results, the worst and the best SLIP fits amongst the 136 steps are presented in the next figure. The data trajectories

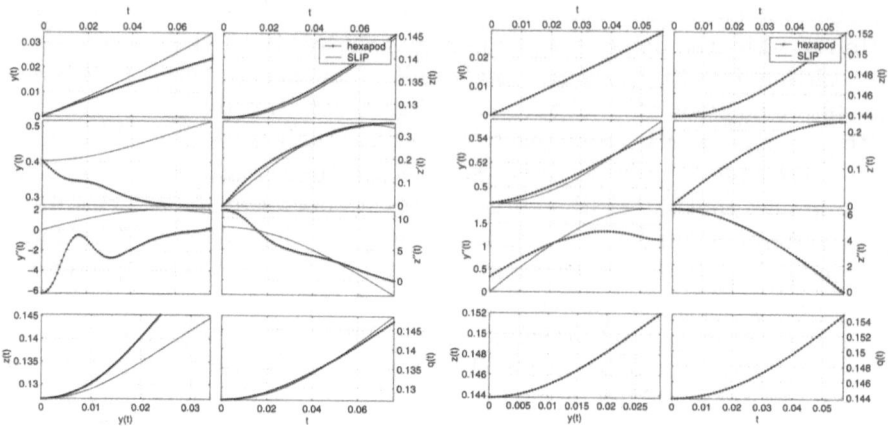

(a) RHex trial no. 2, 4th stride, 2nd minimum (composite legs, foam mat, $t_c = 0.5s$, $m = 11.12kg$) incurs the largest SLIP fitting error of $\Delta_{L_2} = 21.6\%$.

(b) RHex trial no. 10, 5th stride, 1st minimum (composite legs, foam mat, $t_c = 0.5s$, $m = 11.94kg$) incurs the smallest SLIP fitting error of $\Delta_{L_2} = 0.3\%$.

Figure 2. Worst and best SLIP fits: Dotted lines represent experimental data; solid lines represent SLIP trajectories with fitted values κ and q_{r0}.

of $y(t)$, $\dot{y}(t)$, $\ddot{y}(t)$, $z(y)$, $z(t)$, $\dot{z}(t)$, $\ddot{z}(t)$, and $q(t) = \sqrt{y^2(t) + z^2(t)}$ (dotted lines) are plotted together with the SLIP predictions computed with the fitted stiffness κ and spring length q_{r0}. The worst SLIP fit with $\Delta_{L_2} = 21.6\%$ is shown in Fig. 2(a), in contrast to $\Delta_{L_2} = 0.3\%$ in Fig. 2(b).

As an internal consistency check, we introduce a simple form of cross validation: The available COM trajectory components $X \in \{z, y, \dot{z}, \dot{y}\}$ each comprising a time sampled fragment of length N are partitioned by $r = \lfloor N/20 \rfloor$, sub-sampling them into a fitting population X^i_{fit}, $i \in \{1, \ldots, r\}$ of length $N_{fit} = \lfloor N/r \rfloor$ and its complement $X^i_{cross} = X \backslash X^i_{fit}$ — the cross validation populations. The fitting procedure is applied to a fitting population X^i_{fit} to yield κ^i and q^i_{r0}. The quality of fit is assessed not only on X^i_{fit} but also on the corresponding cross-validation population X^i_{cross} with κ^i and q^i_{r0} obtained from X^i_{fit}. In Table 1 the fitting errors $\Delta^i_{L_p,fit}$ and cross validation errors $\Delta^i_{L_p,cross}$ are subsumed under $\Delta_{L_p,fit} = \text{mean}\{\Delta^i_{L_p,fit}\}^r_{i=1}$ and $\Delta_{L_p,cross} = \text{mean}\{\Delta^i_{L_p,cross}\}^r_{i=1}$, and the standard deviations $\text{std}\{\Delta^i_{L_p,fit}\}^r_{i=1}$ and $\text{std}\{\Delta^i_{L_p,cross}\}^r_{i=1}$.

In addition to the L_1 and L_2 errors that compare the experimental data to fitted SLIP predictions, experimental measurement errors introduce noise into the data. To evaluate the quality of fit, one must compare the relative noise floor, $\delta X/\text{Range}(X)$ to the L_1 fitting errors. The noise floor comes from two sources: the measurement error of the ground reaction forces, and the uncertainties of the integration constants for the velocity and position trajectories. In Table 1 the estimated relative noise floors for the $z, y, \dot{z}, \dot{y}$ trajectories are averaged over each stride.

3.3.2. Evidence for SLIP in RHex Running

The fitting protocol outlined in the previous section is now applied to all 136 steps[12]. For the sake of brevity, we refrain from listing the fitted parameters and the fitting errors for each step. Instead, average values over all steps are calculated together with the standard deviation. The mean fitted stiffness is $\kappa = (6100 \pm 940)N/m$, and the mean fitted relaxed spring length is $q_{r0} = (0.171 \pm 0.007)m$. The results of the error analysis are listed in Table 1.

(%)	Δ_{L_p}	$\Delta_{L_p,fit}/\Delta_{L_p,cross}$	$\Delta_{L_p,cross}$	$std(\Delta_{L_p,cross})$
L_2	0.069 ± 0.054	0.985 ± 0.013	0.069 ± 0.054	0.001 ± 0.001
L_1	0.258 ± 0.197	1.002 ± 0.012	0.259 ± 0.199	0.004 ± 0.006

(%)	Δy_{L_1}	Δz_{L_1}	$\Delta \dot{y}_{L_1}$	$\Delta \dot{z}_{L_1}$
L_1	0.041 ± 0.036	0.010 ± 0.010	0.96 ± 0.78	0.020 ± 0.018
Noise floor	0.009	0.003	0.38	0.009

Table 1. Error analysis of SLIP fitting to 136 steps.

The average Δ_{L_2} error of $\approx 7\%$ seems to be remarkable for a mechanical device that a priori bears to resemblance to a SLIP model. The L_1 average percent errors are all similarly low, except for the $\dot{y}$ fits, which are corrupted in part by the high noise floor (see discussion at the end of §3.3.1) and in part by real discrepancies with the putative model, for example, as depicted in Fig. 2. Since the component-wise L_1 errors are considerably above the noise floor, and since the L_p fitting and cross validation errors are comparable, there is little concern that we are merely fitting to noise.

The question naturally arises why we only present fitting results for the decompression phase interval of the COM data instead of the whole compression-decompression phase interval between the zero-crossings of the vertical force. After all, this alternative would serve as a test of our neutral orbit assumption (see §3.3.1). Indeed, the fitting errors for the compression phase interval are of comparable magnitude to the ones for decompression above. However, the fitted spring stiffness is higher and its standard deviation over all 136 steps is twice as large as compared to the decompression phase interval study reported here. An explanation might be the more frequent occurrence of "double stance" (more than 3 legs on the ground) during compression than decompression. This is in fact observed in sample SimSect simulations. As double stance events alter the dynamics of the robot, we single out the decompression phase interval of the COM data as the likeliest candidate of the trajectories for the validation of the SLIP model.

---

[12]In all the selected experiments, RHex operated in the highest speed setting with the composite leg design and the elastic foam ground. Moreover, they were mainly experiments with high body masses: 8 runs with $m = 11.94kg$, 5 runs with $m = 11.12kg$ and 1 run with $m = 9.47kg$.

3.4. Supporting Numerical Study

3.4.1. SLIP Fitting in SimSect

The previous sections suggest that the SLIP model provides a good low dimensional approximation to the "stance" dynamics of RHex. In this section, we describe a parallel numerical investigation of SimSect simulations to determine the "sweet spots" wherein the hexapod might actually be presumed to anchor the SLIP. Specifically, we show that SLIP-like behavior of the mechanical system modeled by SimSect occurs in specific ranges of SimSect's parameter space.

In order to compare the SLIP fitting results from SimSect to those from RHex experiments, the same decompression phase interval as in §3.3.1 is used for fitting.[13] Only those runs where exactly 3 legs of the same tripod are on the ground at the vertical minimum were considered to be acceptable; this was inspired by the assumption that the jointly controlled three legs, which constitute a tripod, can be thought of as a virtual SLIP leg and reflects our experimental experience. As an additional filter, only those simulations which exhibit periodic behavior after a certain amount of time are used for fitting to the SLIP model.

The simulations are run for a fixed set of physical parameters (e.g. total mass, moments of inertia, etc.) and initial conditions, whereas the control parameters sweep angle, cycle time, and leg offset[14] are varied in certain ranges described below. The flight time is chosen to be ≈ 0.4 the cycle time in order to match the highest speed setting for RHex. With force, velocity, and position data from SimSect simulations, the fitting procedure is carried out as in §3.3.1.

3.4.2. Fitting Results

Although SimSect is modeled with many simplifications respecting the actual robot, RHex's main dynamical features are believed to be incorporated in SimSect. Hence the SimSect simulations were run in parameter regimes that include the high mass, high speed regimes of RHex where good SLIP fits could be obtained. In particular, with total mass $m = 11.9kg$, sweep angle= $0.44 \ldots 0.76rad$, cycle time=$0.42 \ldots 0.54s$, leg offset=$0 \ldots - 0.15rad$, and individual leg stiffnesses $\kappa \approx 2700N/m$, SLIP like behavior could be found in SimSect, too. This is demonstrated in Fig. 3, which shows — on the left side — a histogram of the number of simulations from the above listed parameter range with respect to the average L_2 error, Δ_{L_2}. The total number of simulations is reported above the histogram. For comparison, a two-parameter exponential distribution is fit to the histogram bins. On the right side a graph shows the average L_2 error Δ_{L_2} as a function of two of the four control parameters: cycle time and sweep angle. The flight time is kept proportional to the cycle time and only a small dependence of Δ_{L_2} on the leg offset was observed. Instead of a scatter plot, a quadratic surface is fit to the data, and the spread

---

[13]In SimSect, information about which legs are on the ground and which are not is, of course, available. A detailed investigation of the quality of the SLIP fitting as a function of footfall patterns will be the subject of a future report.

[14]For a detailed description of SimSect's parameter space, see [15].

of the data is characterized by its second moment around the surface, which is represented by vertical bars at the corners of the surface. For all SimSect

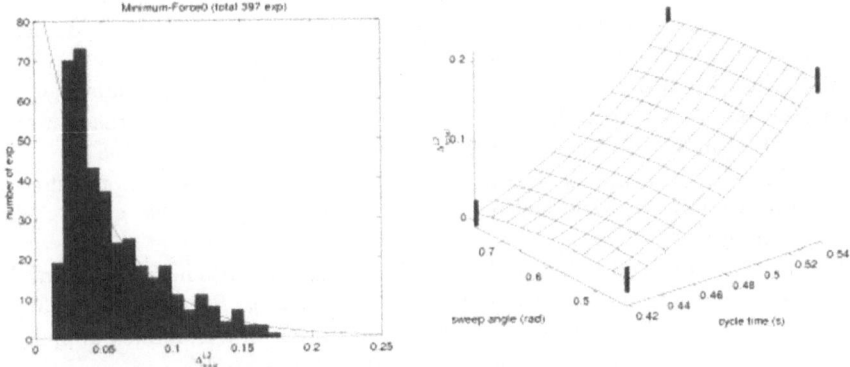

Figure 3. Average errors Δ_{L_2} for SimSect SLIP fits. The left side shows a histogram of the number of simulations; the right side shows a map of Δ_{L_2} as a function of the (reduced) control parameter space.

simulations the ratio of the average fitting to the average cross validation error $\Delta_{L_2,fit}/\Delta_{L_2,cross}$ did not deviate by more than 10% from unity, thus providing a successful self-consistency check for our fitting procedure. The results in this section, in particular the low average fitting errors of $\approx$ 6% in Fig. 3 lend strong support to the presumption that the SLIP template may be anchored in SimSect's dynamics.

4. Conclusion: Implications for More Autonomous Control of RHex

Hierarchy promotes the use of few parameters to control complex systems with many degrees of freedom. In this light, as we understand matters, the emergence of an anchored SLIP in RHex is most fortunate. The pogo-stick can function as a useful control guide in developing more complex autonomous locomotion behaviors such as registration via visual servoing, local exploration via visual odometry, obstacle avoidance, and, eventually, global mapping and localization. In the longer term, we propose to work with the anchored SLIP in RHex in analogy to the manner in which the simple two-bead template has been exploited in juggling. Namely, as we shape behavior via manipulation of gains-in-the-loop [20], we hope to develop a formal programming language with semantics in the world of dynamical attractors [21].

Acknowledgements

This work is supported by DARPA/SPAWAR under contract N66001-00-C-8026. We thank Rodger Kram and Claire Farley for the use of the force platform, Irv Scher for collaboration at an early stage of this project and William Schwind for the insight and the analytical tools he provided.

References

[1] Saranli U, Buehler M and Koditschek D E 2000 Design, Modeling and Preliminary Control of a Compliant Hexapod Robot. *Proc. IEEE Int. Conf. Rob. Aut.* 3:2589-2596.

[2] Blickhan R and Full R 1993 Similarity in multilegged locomotion: Bouncing like a monopode. *J. J. Comp. Physiol. A* 173, 509-517.

[3] Raibert M 1986 *Dynamic Robots that Balance*, MIT Press, Cambridge.

[4] Schwind W J and Koditschek D E 2000 Approximating the Stance Map of a 2 DOF Monoped Runner. *Journal of Nonlinear Science* 10:533-568.

[5] Saranli U, Schwind W J, and Koditschek D E May 1998 Toward the Control of Multi-Jointed, Monoped Runner. *IEEE Int. Conf. on Rob. and Aut.* Leuven, Belgium pp 2676-2682.

[6] Full R J and Koditschek D E 1999 Templates and Anchors: Neuromechanical Hypotheses of Legged Locomotion on Land. *J. Exp. Bio.* 202:3325-3332.

[7] Koditschek D E and Bühler M Dec 1991 Analysis of a simplified hopping robot. *International Journal of Robotics Research* 10(6):587-605.

[8] Bühler M, Koditschek D E, and Kindlmann P J 1990 A Family of Robot Control Strategies for Intermittent Dynamical Environments. *IEEE Control Systems Magazine* 10(2):16-22.

[9] Rizzi A A, Whitcomb L L, and Koditschek D E 1992 Distributed Real-Time Control of a Spatial Robot Juggler. *IEEE Computer* 25(5):12-24.

[10] Nakanishi J, Fukuda T, and Koditschek D E 2000 A Brachiating Robot Controller. *IEEE Trans. Rob. Aut.* 16(2):109-123.

[11] Schwind W J 1998 Spring Loaded Inverted Pendulum Running: A Plant Model. PhD thesis, University of Michigan.

[12] Full R J and Farley C T 2000 Musculoskeletal Dynamics in Rhythmic Systems: A Comparative Approach to Legged Locomotion. In: Winter, Crago (eds) *Biomechanics & Neural Control of Posture & Movement* Springer Verlag, New York, pp 192-205.

[13] Hogan N Mar 1985 Impedance Control: An Approach to Manipulation. *ASME Journal of Dynamic Systems, Measurement, and Control* 107:1-7.

[14] Pratt J and Pratt G May 1998 Intuitive Control of a Planar Bipedal Walking Robot *ICRA* Leuven, Belgium pp 2014-2021.

[15] Saranli U 2000 SimSect Hybrid Dynamical Simulation Environment. *University of Michigan Technical Report* CSE-TR-437-00.

[16] Alexander R McN 1992 A Model of Bipedal Locomotion on Compliant Legs *Phil. Trans.: Biol. Sc.* 338(1284):189-198.

[17] Carver S 2000 The Limits of Deadbeat Control for the Spatial SLIP. in preparation.

[18] Schmitt J and Holmes P 2000 Mechanical models for insect locomotion: Dynamics and stability in the horizontal plane I: Theory; II: Application. *Biological Cybernetics* 83(6):501-515 and 517-527.

[19] Ruina A, personal communication.

[20] Burridge R R, Rizzi A A, and Koditschek D E 1999 Sequential Composition of Dynamically Dexterous Robot Behaviors. *Int. J. Rob. Res.* 18(6):534 - 555.

[21] Klavins E and Koditschek D E 2000 A formalism for the composition of concurrent robot behaviors. *Proc. IEEE Conf. Rob. and Aut.* 4:3395 - 3402.

A Framework and Architecture for Multirobot Coordination

R. Alur, A. Das, J. Esposito, R. Fierro, G. Grudic, Y. Hur
V. Kumar, I. Lee, J. P. Ostrowski, G. Pappas
B. Southall, J. Spletzer, and C. J. Taylor
GRASP Laboratory and SDRL Laboratory
University of Pennsylvania, Philadelphia PA 19104, USA

Abstract: In this paper, we present a framework and the software architecture for the deployment of multiple autonomous robots in an unstructured and unknown environment with applications ranging from scouting and reconnaissance, to search and rescue and manipulation tasks. Our software framework provides the methodology and the tools that enable robots to exhibit deliberative and reactive behaviors in autonomous operation, to be reprogrammed by a human operator at run-time, and to learn and adapt to unstructured, dynamic environments and new tasks, while providing performance guarantees. We demonstrate the algorithms and software on an experimental testbed that involves a team of car-like robots using a single omnidirectional camera as a sensor without explicit use of odometry.

1. Introduction

It has long been recognized that there are several tasks that can be performed more efficiently and robustly using multiple robots [1]-[4]. In fact, there is extensive literature on robot control and the coordination of multiple robots. Our goal, in this paper, is to describe a set of software tools that allows the development of controllers and estimators for multirobot coordination. The tools consist of a framework for developing software components, architecture for control and estimation modules, and a set of decentralized control, planning and sensing algorithms. Our software framework divides the overall multi-robot control task into a set of modes or behaviors, which may be executed either sequentially or in parallel. Modes can consist of high-level behaviors such as planning a path to a goal position, as well as low-level tasks such as obstacle avoidance. We use a high-level language to formally describe how and when transitions between these modes are to take place in order to achieve a set of global objectives. Finally, because it is difficult to predict exactly under what conditions switching between modes should occur, we parameterize mode boundary transitions within each robot's information space and use reinforcement reward to obtain locally optimal mode boundary locations. Thus the multirobot system can learn to continually improve overall performance through interaction with the environment, without human intervention.

D. Rus and S. Singh (Eds.): Experimental Robotics VII, LNCIS 271, pp. 303–312, 2001.

2. Motivation

There is extensive literature on the control of robot manipulators or mobile robots in structured environments, and robot control is a well understood problem area. However, traditional control theory mostly enables the design of controllers in a single mode of operation, in which the task and the model of the system are fixed. When operating in unstructured or dynamic environments with many different sources of uncertainty, it is very difficult if not impossible to design controllers that will guarantee performance even in a local sense. A similar problem exists in developing estimators in the context of sensing. If one views planning as an extension of control, and mapping as an extension of estimation, similar problems exist at higher levels of control and coordination. In contrast, we also know that it is relatively easy to design reactive controllers or behaviors that react to simple stimuli or commands from the environment. This is the basis for the subsumption architecture [5] and the paradigm for behavior-based robotics [6]. While control and estimation theory allows us to model each behavior as a dynamical system, it does not give us the tools to model switches in behavior or the hierarchy that might be inherent in the switching behavior, or to predict the global performance of the system. Our goal in this paper is to present the software tools that are at the core of the development of intelligent robotic systems. Specifically, we describe an architecture and a high-level language, CHARON, with formal semantics, that can be used to describe multiagent, networked robotic systems with multiple control and estimation modes, and discrete communication protocols in a principled way. The architecture allows the development of complex multirobot behavior via hierarchical and sequential composition of control and estimation *modes*, and parallel composition of *agents*. We present our ongoing work to automatically generate control and simulation code from the high-level language description. We also illustrate the application of these ideas to the development of an experimental platform of multiple mobile robots that cooperate in tasks that require sensing, mapping, navigation and manipulation using vision as a sensing modality. Experimental results illustrate the benefits and the limitations of mode switching and the methodology underlying the implementation of robot formation control.

3. Software Architecture

We have developed CHARON, an acronym for Coordinated Control, Hierarchical Design, Analysis, and Run-Time Monitoring of Hybrid Systems, a high-level language to facilitate the programming of multiple, interacting hybrid systems [7]. The language is designed with the goal of being able to control multiple mobile, autonomous robots for mission-critical applications and stringent requirements on safety. A hybrid system here refers to a collection of digital programs that interact with each other in a physical world that is analog in nature. A hybrid system has multiple modes or behaviors of operation. Each mode is a reactive, sensor-based, control law that generates a behavior in a robot, and indirectly in a group of robots, see Figure 1. More details about the language, the semantics and the formal description are presented in [8].

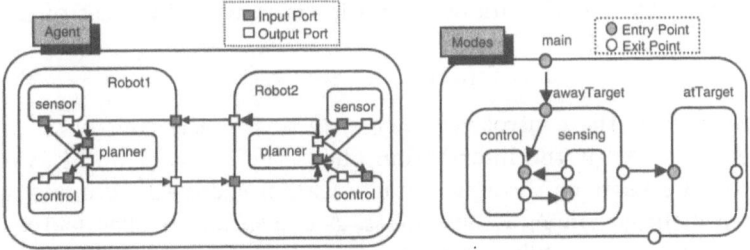

Figure 1. Hierarchy in CHARON.

The architecture proposed here allows the development of complex multirobot behavior via hierarchical and sequential composition of control and estimation modes, and parallel composition of agents. This is schematically illustrated in Figure 2. All software components are called agents. For example, all robots are modeled as agents. Agents can communicate with each other and the human operator can interact with the agents. Agent definitions can have parameters that can be used to create different agents of the same type. Variables, in ad-

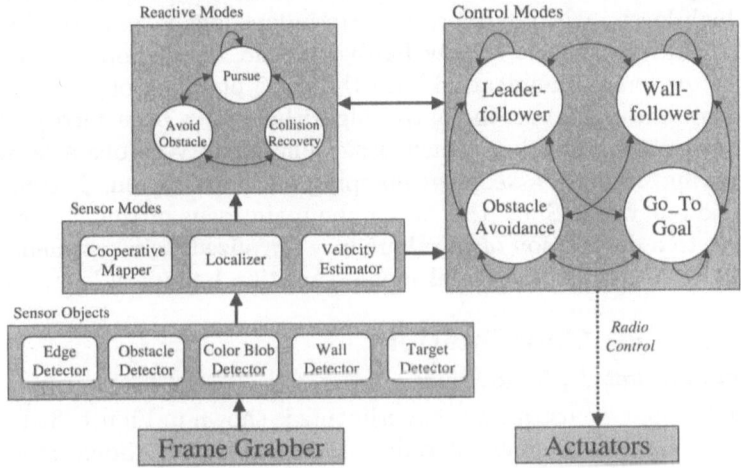

Figure 2. Architecture for multirobot coordination.

dition to being typed, can be discrete or analog. Analog variables are updated continuously, while discrete variables are updated only upon initialization and mode switches. The variables of an agent are partitioned into *read*, *write*, and *private* to allow modular specifications. For example, the robot can receive estimates of the obstacles from other robots, and commands and specifications from the human operator on input channels, and it can send its own information to other robots or to the human operator on the output channels. While physical variables such as the position and velocity of the robot are public, the sensory or control information that is internal to a robot is designated as private. The agent definition contains modes describing behaviors that are available to the robot. Modes specify evolution of control. If the state of an agent is given by $x \in \Re^n$, its evolution is determined by a set of differential equations:

$$\dot{x} = f_q(x, u), \qquad u = k_q(x, z), \tag{1}$$

where $u \in \Re^m$ is the control vector, $q \in Q \subset Z$ is the control mode for the agent, and $z \in \Re^p$ is the information about the external world available either through sensors or through communication channels. A mode definition includes *transitions* among its *submodes*. A transition specifies source and destination modes, the enabling condition, and the associated discrete update of variables. Each mode can have submodes, and there is a hierarchy of modes that is typical in most robot software. Our low-level implementation in C++ uses *Live Objects*. *Live Objects* have been developed as part of the software architecture for implementation on the hardware platforms. A live object encapsulates algorithms and data in the usual object-oriented manner together with control of a thread within which the algorithms will execute, and a number of events that allow communication with other live objects. At the top of the hierarchy, the algorithms associated with the objects are likely to be planners, whilst at bottom they will be interfaces to control and sensing hardware. The planner objects are able to control the execution of the lower level objects to service high-level goals. To offer platform independence, only the lowest level objects should be specific to any hardware, and these should have a consistent interface for communication with the more planning objects that control their execution. Visual servo control algorithms have been incorporated into the live object framework for such basic functionality as obstacle avoidance, wall-following, formation keeping, mapping and localization. Learning is also relevant to our work. Since this is not the main focus of this paper, we point the reader to a description of the Boundary Localized Reinforcement Learning (BLRL) to obtain locally optimal mode transition boundary locations [9].

4. Multirobot Coordination

4.1. Experimental platform

The mobile robot we use for our experiments is shown in Figure 3. It has been constructed from a commercial radio-control truck kit. Some modifications have been made to improve shock absorption and to house an omnidirectional vision system, a 2.4 GHz wireless video transmitter, and a battery pack.

The robot has a servo controller on board for steering and a digital proportional speed controller for forward/backward motion. A parallel port interface, also designed in our lab, allows driving up to 8 mobile robot platforms from a single Windows NT workstation. The receiver, located at the host computer, feeds the signal to a frame grabber that is able to capture video at full frame rate (30 Hz.) for image processing. This yields a video signal in a format for viewing and recording, as well as image processing.

4.2. Sensors

Color feature extraction and target tracking Pixels corresponding to a target can be identified in the image using a YUV based color extractor which provides robustness to variations in illumination. Three-dimensional color models

 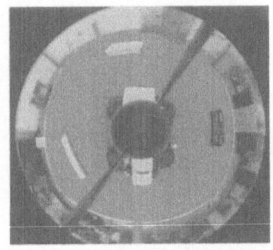 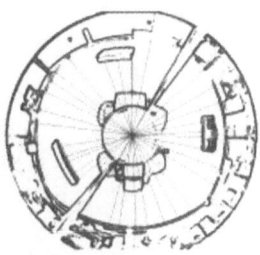

Figure 3. The mobile robot platform with Omnicam (left), range mapping (right).

are generated a priori from images of the target at numerous distances, orientations, and illumination levels. These data are stored in a pair of look-up-tables (LUTs) to speed image processing. During operation, the target detection algorithm - the *blobExtractor* sensor, is initially applied to the entire image and can run at frame rate (30 Hz). Once the target is acquired, the sensor switches to target tracking mode.

The target tracking scheme is simple yet robust. To increase the speed of color feature extraction, a region of interest is dynamically constructed surrounding the target in the current image based on its location in the previous image. By constraining image processing operations to this region of interest, we are able to run multiple target trackers at frame rate. This allows us to assume little motion of the targets between consecutive image captures. Such small inter-frame movement thus permits the straightforward tracking process whereby the position of the region of interest (which is centered upon the target) is moved to coincide with the centroid of the target extracted from each frame.

4.2.1. Range mapping

A Sobel gradient was applied to the original omnidirectional image. The resulting edges in the image were assumed to be features of interest, see Figure 3. By assuming a ground plane constraint, the distance to the nearest feature in the sector of interest was determined from the its relative elevation angle to the mirror. This provides a range map to all obstacles at frame rate.

4.2.2. Localizer

We have implemented a localization algorithm for our mobile robots. The algorithm employs an extended Kalman filter (EKF) to match landmark observations to an a priori map of landmark locations. The *Localizer* object uses the *blobExtractor* sensor to determine the range and the bearing of an observed landmark. If the observed landmark is successfully matched, it will be used to update the vehicle position and orientation. Figure 4 depicts a typical image used for localization.

The kinematic model of the mobile robot is given by

$$\dot{x} = u_1 \cos\theta, \quad \dot{y} = u_1 \sin\theta, \quad \dot{\theta} = \frac{u_1}{l}\tan\phi, \quad \dot{\phi} = \lambda(u_2 - \phi), \qquad (2)$$

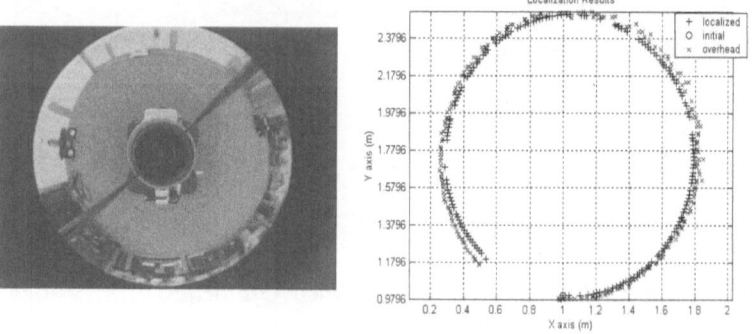

Figure 4. Image used for localization (left), experimental results (right).

Figure 5. Cooperative mapping.

where l is the body length, u_2 is the steering command, $|\phi| < 70°$ is the steering angle, and $\lambda \approx 4s^{-1}$ is a parameter that depends on the steering servo time constant and wheel-ground friction. The control vector is given by $u = [u_1 \ u_2]^T$.

4.2.3. Velocity estimator

The *leader-following* control object described in the next section, requires reliable estimation of the linear velocity and angular velocity of a leader mobile robot. The velocity *estimator* algorithm is also based on an extended Kalman filter. It uses the *blobExtractor* sensor to determine the range ρ and the bearing β of the observed leader. In addition, the filter requires a sensor model, and the relative kinematic equations of the leader and follower robots.

4.2.4. Mapper

We have implemented a cooperative mapping using three nonholonomic platforms. A simulated room 4m×4m was constructed. The positions of two robots are held fixed, while the third robot, called mapper, is driven around the test area. A global map updates is accomplished at 3-5 Hz. The experimental setup and results are displayed in Figure 5.

4.3. Controllers

4.3.1. Obstacle avoidance and wall following

The wall follower works by using inputs from two live object sensors - a *wall detector* and an *obstacle detector*. Both take as input the image from an it

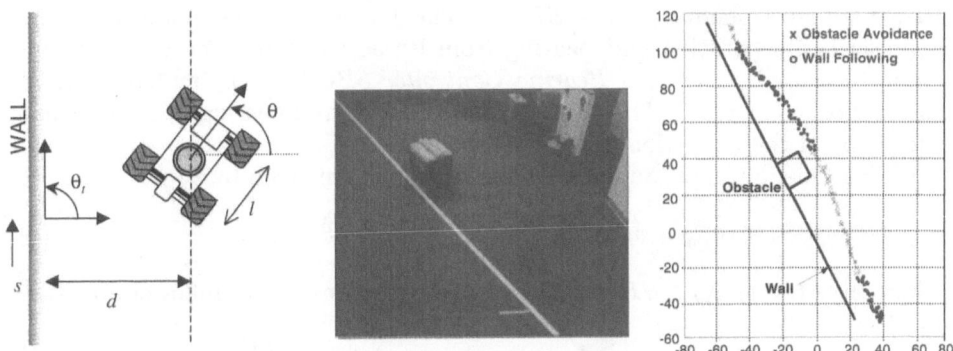

Figure 6. The wall-follower, sample wall-following configuration, and corresponding mode vs. position results.

edge detector, and use range map data to find the relative position of the wall/obstacle. The wall detector has a 40° field-of-view from 160 to 200 degrees. A line is fit to these points using RANSAC (random sampled consensus), which gives us a line fit robust to outliers. From this we are able to extract the relative position and orientation of the robot to the wall. We use I/O feedback linearization techniques to design a PD controller to regulate the distance of the vehicle to the wall, Figure 6 (left). Wall following can be considered as a particular case of path following. Thus, the kinematics in terms of the path variables become

$$\dot{s} = v_1 \cos \theta_p, \quad \dot{d} = v_1 \sin \theta_p, \quad \dot{\theta}_p = \frac{v_1}{l} \tan \phi, \quad \dot{\phi} = v_2. \tag{3}$$

In this case $\theta_t = \frac{\pi}{2}$ and $\theta_p = \theta - \theta_t$. Assuming the robot is to follow the wall with a piecewise constant velocity $v_1(t)$, the controller is given by

$$u = \tan^{-1}\left[\frac{l}{v_1^2 \cos \theta_p}(k_p(d_0 - d) - k_v v_1 \sin \theta_p)\right], \tag{4}$$

where $u(t)$ is the steering command, $v_1(t)$ is the linear velocity, and k_p, k_v are positive design controller gains. Usually, we may want a critically damping behavior i.e., $k_v = 2\sqrt{k_p}$.

The obstacle detector picks up objects in its 80° forward-staring field-of-view. Since the position and orientation relative to the wall are known, the detector is able to discriminate which *obstacles* are actually the wall, and which are truly obstacles that must be avoided. Mode switching between wall following and obstacle avoidance is accomplished by giving priority to the latter. Experimental results are depicted in Figure 6 (axes units are inches).

4.3.2. Leader-Following Control

We consider a team of n nonholonomic mobile robots that are required to follow a prescribed trajectory while maintaining a desired formation. The desired formation may change based on environmental conditions or higher-level commands. A robot designated as the *lead* robot follows a trajectory generated by a high-level planner $g(t) \in SE(2)$. The *follower* robots should maintain a prescribed separation and bearing from its adjacent neighbors. This controller (denoted *Separation Bearing Controller SBC* here) is implemented on each robot in the team. The desired separations l_{ij}^d and bearings ψ_{ij}^d will define the shape of the formation, see Figure 7 (left).

The kinematics of the nonholonomic *i*-robot are given by

$$\dot{x}_i = v_i \cos \theta_i, \quad \dot{y}_i = v_i \sin \theta_i, \quad \dot{\theta} = \omega_i, \tag{5}$$

where $x_i \equiv (x_i, y_i, \theta_i) \in SE(2)$. The control velocities for the follower are given by [10]

$$v_j = s_{ij} \cos \gamma_{ij} - l_{ij} \sin \gamma_{ij} (b_{ij} + \omega_i) + v_i \cos(\theta_i - \theta_j), \tag{6}$$

$$\omega_j = \frac{1}{d_j} [s_{ij} \sin \gamma_{ij} + l_{ij} \cos \gamma_{ij} (b_{ij} + \omega_i) + v_i \sin(\theta_i - \theta_j)], \tag{7}$$

where

$$\gamma_{ij} = \theta_i + \psi_{ij} - \theta j, \tag{8}$$

$$s_{ij} = k_1(l_{ij}^d - l_{ij}), \tag{9}$$

$$b_{ij} = k_2(\psi_{ij}^d - \psi_{ij}), \quad k_1, k_2 > 0. \tag{10}$$

The closed-loop linearized system becomes

$$\dot{l}_{ij} = k_1(l_{ij}^d - l_{ij}), \quad \dot{\psi}_{ij} = k_2(\psi_{ij}^d - \psi_{ij}), \quad \dot{\theta}_j = \omega_j. \tag{11}$$

In the following theorem, we provide a stability result for the *SBC* [11].

Theorem 4.1 *Assume that the reference trajectory $g(t)$ is smooth, the reference linear velocity is large enough and bounded i.e., $\beta_{max} > v_i > \beta_{min} > 0$, the reference angular velocity is small enough i.e., $\|\omega_i\| < W_{max}$ and the initial relative orientation is bounded i.e., $\|\theta_i - \theta_j\| < \varepsilon_\theta < \pi$. If the control velocities (6)-(7) are applied to R_j, then system (11) is stable and the output system error of the linearized system converges to zero exponentially.*

□

While the two output variables in (11) converge to the desired values arbitrarily fast (depending on k_1 and k_2), the behavior of the follower's internal dynamics, θ_j, depends on the controlled angular velocity ω_j. In our analysis we have considered the internal dynamics which is required for a complete study of the stability of the system. Let the orientation error be expressed as

$$\dot{e}_\theta = \omega_i - \omega_j \tag{12}$$

After some work, we have

$$\dot{e}_\theta = -\frac{v_i}{d_j} \sin e_\theta + f_\theta(u, e_\theta) \tag{13}$$

where u is a vector that depends on the output system error and reference angular velocity ω_i. $f_\theta(\cdot)$ is a nonvanishing perturbation for the nominal system

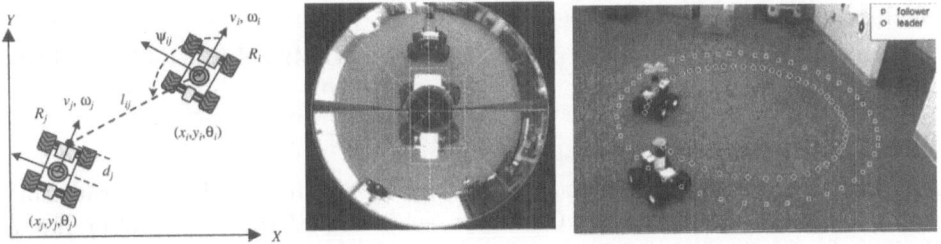

Figure 7. The *Separation Bearing Control* (*SBC*), and formation control experimental setup.

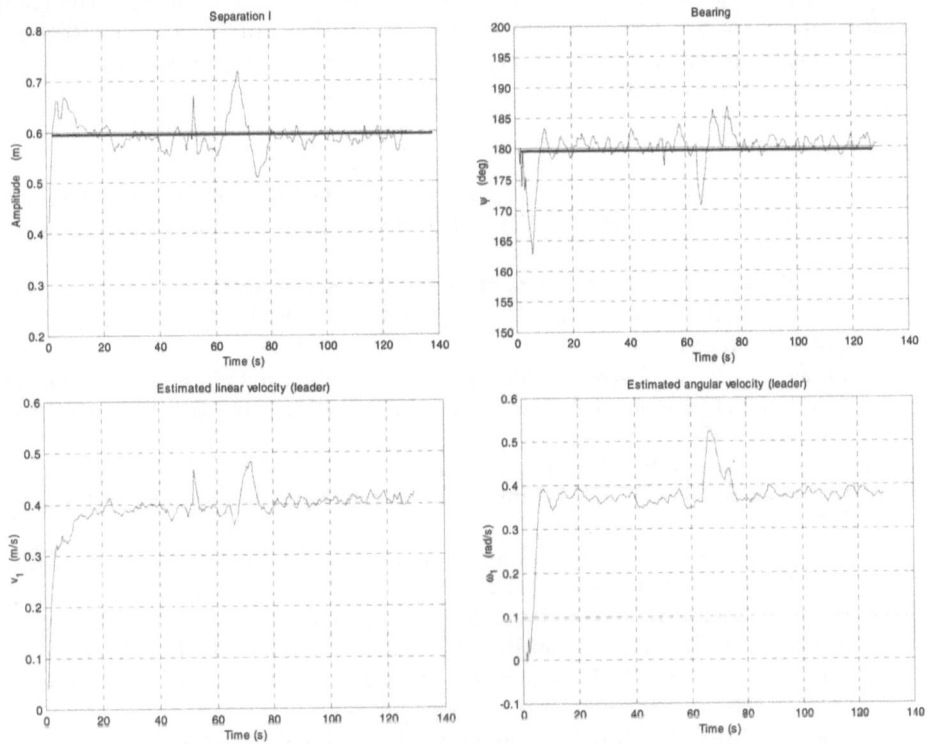

Figure 8. Leader following experimental results.

(13) which is (locally) exponentially stable. By using stability of perturbed systems [12], it can be shown that system (13) is stable, thus the stability result in Theorems 1 follows.

Figure 7 (right) shows a view of leader-following experiment. These are actual data points collected from an overhead camera installed in our lab for ground truth purposes. Figure 8 depicts the estimated linear and angular velocity of the leader robot, and the measured separation and bearing. We choose $l_d = 0.6$ m and $\psi_d = 180°$. The robustness of the system is verified when we manually hold the follower for a few seconds at $t \approx 65$ s.

Acknowledgements

This research was supported in part by DARPA ITO MARS 130-1303-4-534328-xxxx-2000-0000.

5. Concluding Remarks

We describe a formal architecture and high-level language for programming multiple cooperative robots. Our approach assumes that each robot has a finite set of behaviors or modes that it can execute, and the programming language is used to formally specify a set of conditions under which mode transitions take place. Thus the tasks performed by the multirobot system are uniquely specified as mode transition boundaries that are defined in the robots information space. Experiments have been carried out in complex scenarios where robots need to exhibit a variety of behaviors such as localization, target acquisition, collaborative mapping and formation keeping.

References

[1] Donald B., Gariepy L., Rus D. (2000) Distributed manipulation of multiple objects using ropes. Proc. IEEE Int. Conf. on Robotics and Automation, 450-457.

[2] Khatib O., Yokoi K., Chang K. et al. (1996) Vehicle/arm coordination and mobile manipulator decentralized cooperation. IEEE/RSJ Int. Conf. on Intelligent Robots and Systems, 546-553.

[3] Parker L. (2000) Current state of the art in distributed robot systems, Distributed Autonomous Robotic Systems 4. Parker L., Bekey G., Barhen J. (Eds.). Springer, 3-12.

[4] Rus D., Donald B., Jennings J. (1995) Moving furniture with teams of autonomous robots. IEEE/RSJ Int. Conf. on Intelligent Robots and Systems, 235-242.

[5] Brooks R. (1986) A robust layered control system for a mobile robot. IEEE J. Robotics and Automation, 2(1):14-23.

[6] Balch T., Arkin R. (1998) Behavior-based formation control for multi-robotic teams. IEEE Transactions on Robotics and Automation, 14(6):926-934.

[7] Alur R., Henzinger T., Lafferriere G., Pappas G. (2000) Discrete abstractions of hybrid systems. Proceedings IEEE, 88(2):971-984.

[8] Alur R., Grosu R., Hur Y., Kumar V., Lee I. (2000) Modular specification of hybrid systems in CHARON. LNCS 1790, Lynch N. A., Krogh B. H. (Eds.). Springer, 6-19.

[9] Grudic G., Ungar L. (2000) Localizing search in reinforcement learning. National Conference on Artificial Intelligence (AAAI 2000), 590-595.

[10] Desai J., Ostrowski J., Kumar V. (1998) Controlling formations of multiple mobile robots. Proc. IEEE Int. Conf. on Robotics and Automation, 2864-2869.

[11] Fierro R., Das A., Kumar V., Ostrowski J. (2001) Hybrid control of formation of robots. IEEE Int. Conf. on Robotics and Automation, ICRA01, Seoul, Korea, May, 2001. To appear.

[12] Khalil H. (1996) Nonlinear Systems. Prentice Hall.

Motion Control of Distributed Robot Helpers Transporting a Single Object in Cooperation with a Human

Yasuhisa Hirata, Kazuhiro Kosuge
Department of Machine Intelligence and Systems Engineering, Tohoku University
Aoba-yama01, Sendai 980-8579, JAPAN
hirata@irs.mech.tohoku.ac.jp

Hajime Asama, Hayato Kaetsu, Kuniaki Kawabata
Biochemical Systems Laboratory,
The Institute of Physical and Chemical Research, RIKEN
Hirosawa 2-1, Wako, Saitama 351-0198, JAPAN

Abstract: In this paper, we propose a concept of distributed robot helpers referred to as DR Helpers, and a decentralized control algorithm for them to transport a single object in cooperation with a human/humans. The proposed control algorithm could be applied to DR Helpers more than three, even if each DR Helper has a slippage between its wheels and the ground. The proposed decentralized control algorithm is experimentally applied to three DR Helpers for illustrating the validity of the proposed algorithm.

1. Introduction

Most of robots have been used as industrial robots in factories to replace humans doing tasks, which humans do not want to do or could not do. Most of these robots have been isolated from humans. If robots could do tasks together with a human/humans, robots could be applied to other fields, such as applications in a house, in a hospital, in a shopping center, in a construction site, for elderly care, etc. A robot could not be applied to these applications without any interactions with humans.

Much researches have been done for the human-robot cooperation by several researchers[1]-[4]etc. However, the working space of these cooperation systems is restricted, because the control algorithms were designed for a manipulator/manipulators. Mobility is the important function to cover a large working space in an environment.

Colgate and Peshkin have proposed a control algorithm of transporting an object by a mobile robot referred to as a Cobot in cooperation with a human [5]. However, this type human-robot cooperation is not suitable for transporting a large or a heavy object, because there is a limitation with respect to the size and the weight of an object handled by a single robot. The system has no actuator to generate the motion of the robot.

D. Rus and S. Singh (Eds.): Experimental Robotics VII, LNCIS 271, pp. 313–322, 2001.

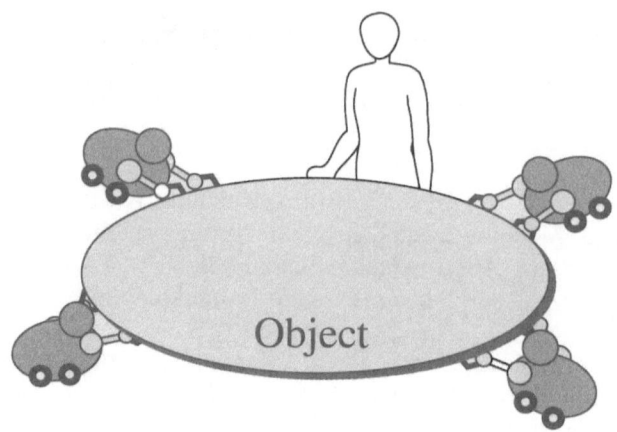

Figure 1. Handling an Object with Robot Helpers

To overcome these problems, we consider a human-robots cooperation system using multiple mobile robots as shown in Figure 1. Khatib has proposed the control system by multiple mobile manipulators in cooperation with a human [6]. However, this control system could not realize effective human-robots cooperation without integration of the human arm inertial properties and a description of the human grasp.

We have proposed a concept of a human-robots cooperation system referred to as distributed robot helpers and a decentralized control algorithm for them to transport an object together with a human/humans with unknown dynamics [7]. In this algorithm, each robot is controlled as if it has a dynamics of caster wheel as shown in Figure 2 and transports an object in cooperation with a human. However, this algorithm could not be applied to mobile robots more than three, when a slippage between the wheels of each robot and the ground could not be negligible. The motion of an object could be constrained completely, when each robot has a slippage between its wheels and the ground during a transportation of an object.

In this paper, we extend the caster-like dynamics proposed in [7] and propose a decentralized motion control algorithm for mobile robots more than three, even if each robot has a slippage between its wheels and the ground during a transportation of an object. The proposed control algorithm is experimentally applied to three DR Helpers and experimental results illustrate the validity of the system.

2. Distributed Robot Helpers

We briefly explain the concept of distributed robot helpers referred to as DR Helpers proposed in [7]. When we would like to move a large or a heavy object, which could not be handled by a human, we move it with other people or helpers. If a robot/robots could play a role of human helpers, we could move it without any help of humans. A robot helper is a robot, which plays a role of the human helpers as shown in Figure 1.

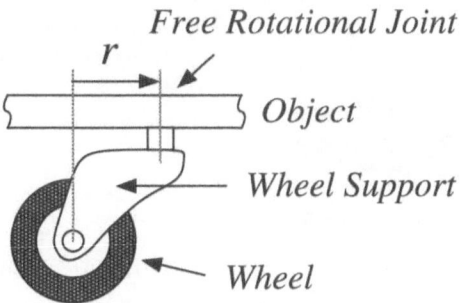

Figure 2. Real Caster

The robot helper is expected to do tasks in an ordinary environment with humans. Mobility is an important function to cover a working space in the environment. Multiple small robots are more appropriate for such a system than a large and heavy robot from a safety point of view. Because each small robot has less kinetic energy than the large and heavy one, when they are moving with the same speed, and less harmful to a human when it collides with a human/humans. The distributed robot helper is a small mobile robot, and helps humans to carry an object together with other robot helpers as shown in Figure 1.

3. Motion of Object

In this section, we consider how the distributed robot helpers are controlled so that a human/humans can transport a single object in cooperation with them. We design a passivity-based control system to guarantee the stable realization of the human-robots interaction through a manipulated object, under the assumption that the passivity conditions for human/humans are satisfied.

The passivity-based controller design is well known in the area of tele-operation [8]etc., and has been applied to DR Helpers transporting an object in cooperation with a human/humans [7]. In this algorithm, each DR Helper is controlled as if it has a dynamics of a caster wheel as shown in Figure 2. However, we could not apply the same control principle proposed in [7] to DR Helpers more than three, when each DR Helper has a slippage between its wheels and the ground during a transportation of an object.

When two DR Helpers handle an object, the motion of the object supported by them is characterized by two kinds of motion based on the heading direction of the caster wheel. One is the translational motion of the object along the heading direction of the caster wheel as shown in Figure 3(a), and the other is the rotational motion of the object around a point defined by an angle of the caster wheel as shown in Figure 3(b).

When DR Helpers more than three handle an object, the motion of the object supported by them is characterized by three kinds of motion based on the heading direction of caster wheel. One is the translational motion of the object as shown in Figure 3(a), another is the rotational motion of the object around

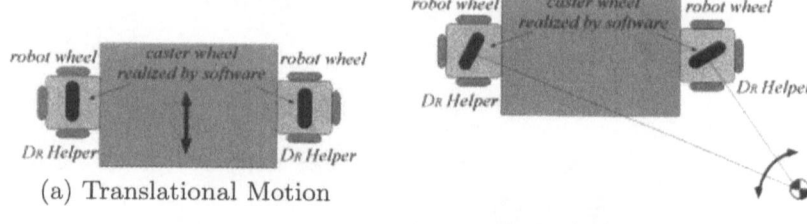

(a) Translational Motion

(b) Rotational Motion

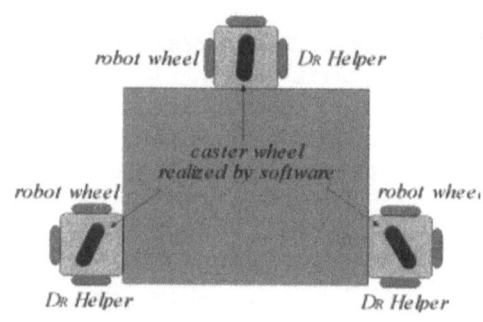

(c) Constrained Motion

Figure 3. Motion of Object

a point as shown in Figure 3(b), and the other is the completely constrained motion as shown in Figure 3(c) because of the misalignment of the caster wheel.

Under the assumption that the caster wheel of each DR Helper could rotate to the direction of the force applied by a human/humans using the adaptive dual caster action proposed in [7] and each DR Helper does not have a slippage between its wheels and the ground, the motion of the object supported by multiple DR Helpers is the translational one or the rotational one around a point, and each DR Helper could transport a single object in cooperation with a human/humans.

However, if each DR Helper has a slippage between its wheels and the ground during a transportation of an object, the motion of the object supported by multiple DR Helpers is constrained completely because of the misalignment of the caster wheel. In this case, each DR Helper could not move along the direction of the force applied by a human/humans, and a human could not transport the object in cooperation with multiple DR Helpers.

To overcome this problem, we extend the caster-like dynamics proposed in [7]. In the algorithm proposed in [7], each DR Helper is controlled as if it has a dynamics of a caster wheel. Therefore, each DR Helper could not move along the direction of the caster wheel axis and the motion of an object could be constrained completely because of the misalignment of the caster wheel. In this paper, we consider that each robot is controlled as if it has a dynamics of a free rotational joint of the caster.

In this case, the motion of an object supported by multiple DR Helpers is not constrained completely, because each DR Helper could generate the velocity in all directions. Therefore, a human could transport a single object in cooperation with multiple DR Helpers, even if each DR Helper has a slippage between its wheels and the ground.

4. Velocity-based Caster-like Motion

To realize a dynamics of the free rotational joint, we define three coordinate systems of i-th DR Helper as shown in Figure 4; a base coordinate system $^b\Sigma_i$, a robot coordinate system $^r\Sigma_i$ and a caster coordinate system $^c\Sigma_i$. The origins of these coordinate systems are located at the center of the force/torque sensor.

The base coordinate system is fixed to the mobile robot. The orientation of the coordinate system is not changed even if the orientation of the mobile robot changes. The mobile robot coordinate system is fixed to the mobile robot and moves together with the robot. The force/moment applied to the robot is measured in this coordinate system. Let $^r\theta_i$ be the rotational angle of the robot coordinate system with respect to the base coordinate system as shown in Figure 4.

The caster coordinate system rotates around its origin to mimic the free rotational motion of the caster support as shown in Fiuger 2. The direction of $^c x_i$-axis of the caster coordinate system is defined as the heading direction of the caster wheel. Let $^c\theta_i$ be the rotational angle of the caster coordinate system with respect to the base coordinate system as shown in Figure 4.

A velocity-controlled servomotor drives each wheel of the DR Helper which realizes the omni-directional motion, and we assume that each wheel rotates with a specified angular velocity. We could generate the translational motion

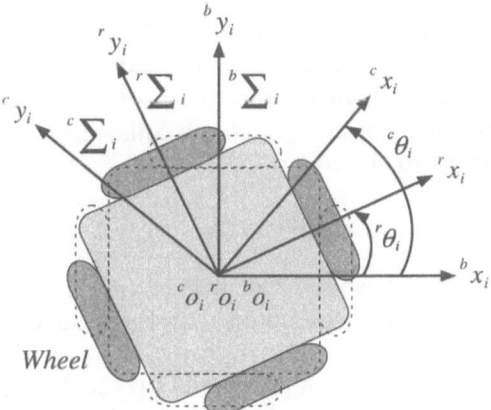

Omni-directional Mobile Robot

Figure 4. Coordinate System of i-th DR Helper (top view)

of the caster wheel based on the force applied to the robot as follows;

$$^{tran}D_i{}^c\dot{x}_i = {}^cf_{xi}$$
$$= ({}^rf_{xi} - {}^rf_{xi}^{in})\cos({}^c\theta_i - {}^r\theta_i) + ({}^rf_{yi} - {}^rf_{yi}^{in})\sin({}^c\theta_i - {}^r\theta_i) \quad (1)$$

where $^{tran}D_i \in R$ is a positive damping coefficient and $^c\dot{x}_i \in R$ is the velocity of the robot along ^cx_i-axis of the caster coordinate system. $^rf_{xi}$, $^rf_{yi} \in R$ are the forces applied to the robot with respect to the $i-th$ robot coordinate system and $^rf_{xi}^{in}$, $^rf_{yi}^{in} \in R$ are the specified internal forces applied to the object by the robot with respect to the $i-th$ robot coordinate system.

To mimic the motion of the wheel support as shown in Figure 2, we obtain the angular velocity of the caster coordinate system $^c\dot{\theta}_i$ using the following equation.

$$^{cast}D_i{}^c\dot{\theta}_i = \frac{1}{r_i}{}^cf_{yi}$$
$$= \frac{1}{r_i}\{-({}^rf_{xi} - {}^rf_{xi}^{in})\sin({}^c\theta_i - {}^r\theta_i) + ({}^rf_{yi} - {}^rf_{yi}^{in})\cos({}^c\theta_i - {}^r\theta_i)\}$$
$$(2)$$

where $^{cast}D_i \in R$ is a positive damping coefficient, r_i is the caster offset as shown in Figure 2, and $^cf_{yi}$ is the force applied to the robot along ^cy_i-axis of the caster coordinate system. To realize the caster-like motion around the free rotational joint, we generate the motion of the robot based on the angular velocity of the caster coordinate system using the following equation and make the caster coordinate system rotate around its origin based on $^c\dot{\theta}_i$.

$$^c\dot{y}_i = r_i{}^c\dot{\theta}_i \quad (3)$$

where $^c\dot{y}_i \in R$ is the velocity of the robot along ^cy_i-axis of the caster coordinate system.

When the robot holds the object rigidity, the kinematic relation between the robot and the object is kept unchanged. Each robot has to generate the motion of the free rotational joint. For this purpose, the rotational motion of each robot is controlled so as to have the following dynamics based on a moment $^rn_i \in R$ applied to the robot.

$$^{rot}D_i{}^r\dot{\theta}_i = {}^rn_i \quad (4)$$

where $^{rot}D_i \in R$ is a positive damping coefficient and $^r\dot{\theta}_i \in R$ is the real angular velocity of the robot. It should be noted that the relative angle between $^r\theta_i$ and $^c\theta_i$ is not changed.

The adaptive dual caster action proposed in [7] is also implemented in this system, so that a human could manipulate the object easily together with the DR Helpers. The adaptive dual caster action adjusts the position of the caster offset and changes the apparent dynamics of an object supported by multiple DR Helpers according to the tasks.

5. Experiments

We did experiments of transporting a single object using three Dʀ Helpers in cooperation with a human as shown in Figure 5 to illustrate the validity of the proposed algorithm. We did two types of experiments to compare the conventional algorithm proposed in [7] with the proposed algorithm explained in this paper. The control algorithms were implemented using VxWorks. The sampling rate was 1024[Hz].

In each experiment, we applied the force to the object along y-axis as shown in Figure 5(b) and transported the object along y-axis in cooperation with the Dʀ Helpers. It should be noted that each robot is controlled using the same control parameters in each experiment. Experimental results are shown in Figure 6. These figures show the motion of a Dʀ Helper and the force/moment applied to the Dʀ Helper during the transportation of the object.

In the experiment using the conventional algorithm as shown in Figure 6(I), the force applied to the robot by a human along y-axis of the object coordinate system is larger than the force in the experiment using the proposed algorithm, and the force/moment applied to the robot had the vibratory part. These results means that a constrained motion of the object as shown in Figure 3(c) was occurred slightly during the transportation of the object because of a slippage between the wheels of each robot and the ground.

As shown in Figure 6(II), a human could transport a single object easily in cooperation with multiple Dʀ Helpers using the proposed control algorithm, even if each robot has a slippage between its wheels and the ground. An example of tasks by three Dʀ Helpers is shown in Figure 7. In this experiment, a human transported a single object in cooperation with Dʀ Helpers successfully.

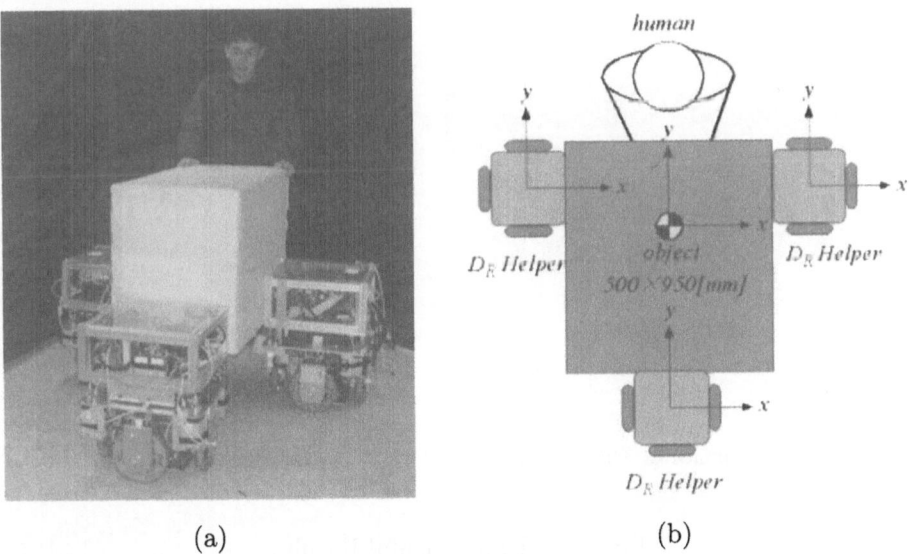

(a)   (b)

Figure 5. Experimental System

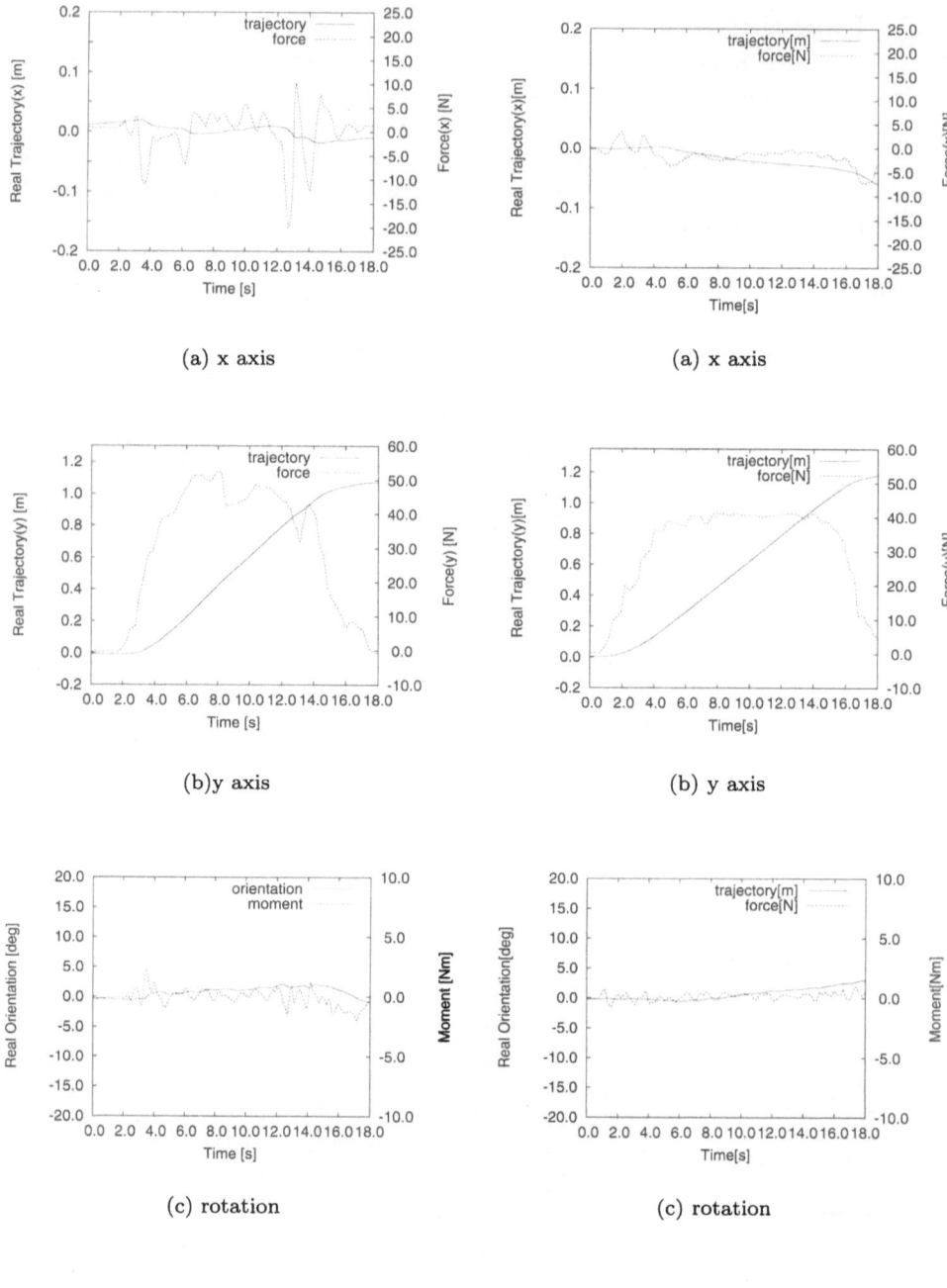

(I) Conventional Algorithm   (II) Proposed Algorithm

Figure 6. Experimental Results

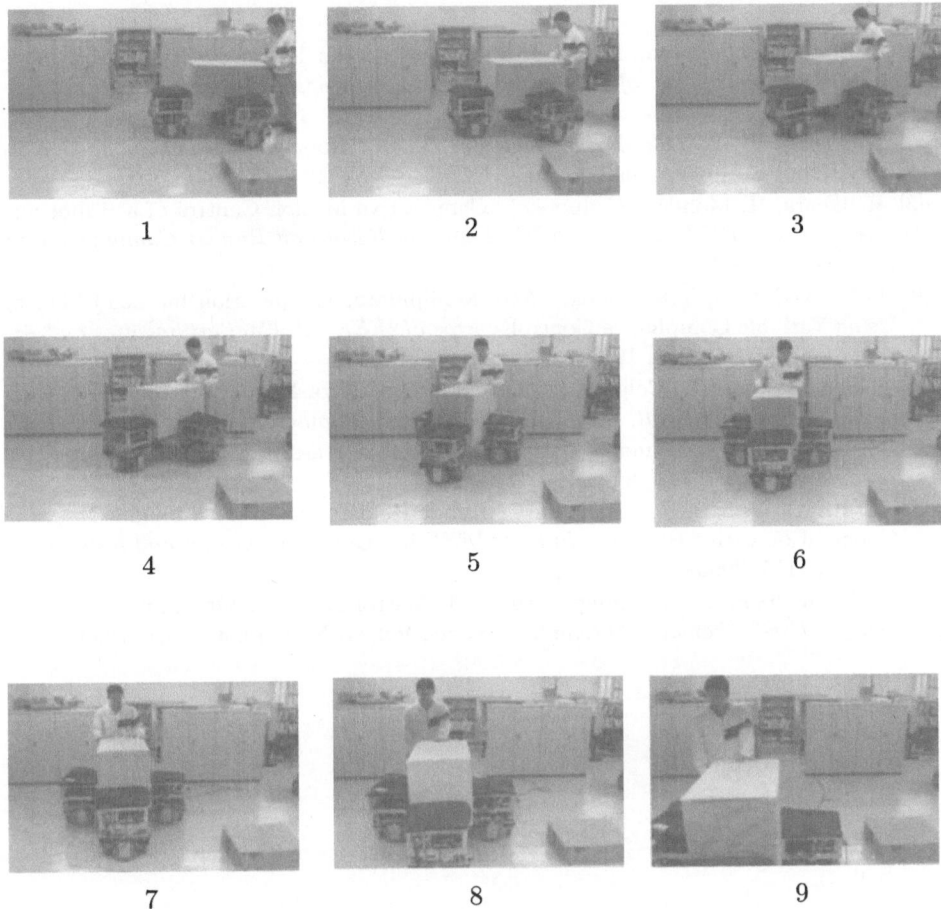

Figure 7. Task by Dʀ Helpers

6. Conclusions

In this paper, we proposed a concept of distributed robot helpers referred to as
Dʀ Helpers, and a decentralized control algorithm for them to transport a single
object in cooperation with a human/humans. The proposed control algorithm
could be applied to Dʀ Helpers more that three, even if each Dʀ Helper has
a slippage between its wheels and the ground. The proposed decentralized
control algorithm is experimentally applied to three Dʀ Helpers for illustrating
the validity of its algorithm.

Acknowledgments

This project has been partially supported by Grant-in-Aid for Scientific Research of Japan Society for the Promotion of Science (No. A-2-12305028).

References

[1] H. Kazerooni, "Human Machine Interaction via the Transfer of Power and Information Signals", *Proc. of IEEE Int. Conf. on Robotics and Automation*, pp.1632–1642, 1989.

[2] K. Kosuge, H. Yoshida, D. Taguchi, T. Fukuda, "Robot-Human Collaboration for New Robotic Application", *Proc. of IEEE IECON'94*, pp.713–718, 1994.

[3] R. Ikeura, H. Monden, H. Inooka, "Cooperative Motion Control of a Robot and a Human", *IEEE International Workshop on Robot and Human Communication*, pp.112–117, 1994.

[4] O.M. Al-Jarrah, Y.F. Zheng, "Arm Manipulator Cooperation for Load Sharing Using Variable Compliance Control", *Proc. of IEEE Int. Conf. on Robotics and Automation* , pp.895–900, 1997.

[5] R.B. Gillespie, J.E. Colgate, M. Peshkin, "A Genera Framework for Cobot Control", *Proc. of IEEE Int. Conf. on Robotics and Automation* , pp.1824–1830, 1999.

[6] O. Khatib, "Mobile manipulation : The robotic assistant", *Robotics and Autonomous Systems*, No.26, pp.175–183, 1999.

[7] Y. Hirata, K.Kosuge, "Distributed Robot Helpers Handling a Single Object in Cooperation with a Human", *Proc. of IEEE Int. Conf. on Robotics and Automation* , pp.458–463, 2000.

[8] R.J. Anderson, M.W. Spong, "Bilateral Control of Telemanipulators with Time Delay", *IEEE Trans.on Automatic Control*, Vol.34, No.5, pp.494-501, 1989.

First Results in the Coordination of Heterogeneous Robots for Large-Scale Assembly

Reid Simmons, Sanjiv Singh, David Hershberger, Josue Ramos, Trey Smith
Robotics Institute
Carnegie Mellon University
Pittsburgh, PA 15217
{reids, ssingh, hersh, josue, trey}@ri.cmu.edu

Abstract: While many multi-robot systems rely on fortuitous cooperation between agents, some tasks, such as the assembly of large structures, require tighter coordination. We present a general software architecture for coordinating heterogeneous robots that allows for both autonomy of the individual agents as well as explicit coordination. This paper presents recent results with three robots with very different configurations. Working as a team, these robots are able to perform a high-precision docking task that none could achieve individually.

1. Introduction

As robots become more autonomous and sophisticated, they are increasingly being used for more complex and demanding tasks. Often, single robots are insufficient to perform the tasks. For some types of tasks, such as exploration or demining, multiple robots can be used to increase efficiency and reliability. For many other tasks, however, not only are multiple robots necessary, but explicit coordination amongst the robots is imperative. Our research focus is on the latter class of problems, particularly those in which the individual robots have vastly different capabilities. For many tasks, the use of heterogeneous robots is indicated because of the difficulties of constructing a single robot that has the needed size, strength, dexterity, etc.

One such application domain is assembly of large-scale structures, such as terrestrial buildings, planetary habitats, or space solar power structures. Such domains need both heavy lifting capabilities, as well as precise, dexterous manipulation to connect parts together. A motivating scenario is that of assembling the steel structure of a large building. In such cases, a large crane is used to lift beams and move them near their destinations; a worker near the destination uses hand signals to guide the crane operator; when the beam is close enough, the worker grabs the end and moves it into place.

Our short-term research goal is to accomplish that scenario using a team of three autonomous robots. Our initial assembly scenario is to emplace a long heavy beam precisely. This task needs both strength and dexterity. Our approach is to coordinate three robots— an overhead crane, a mobile manipulator, and a roving eye. The crane provides heavy lifting capability and has a large workspace, but is not precise; the manipulator provides dexterity and precise control, but is weaker and has a relatively smaller workspace from a fixed position of the base; the roving eye provides accurate

D. Rus and S. Singh (Eds.): Experimental Robotics VII, LNCIS 271, pp. 323–332, 2001.

views of the workspace, which are used to guide the other two robots.

This task has been chosen to highlight issues with heterogeneous robots. Research issues include techniques for explicit coordination between the robots, distributed visual servoing, planning and execution techniques that take advantage of the heterogeneous nature of the robot team, and robust monitoring and exception handling within teams. Longer-term issues include dynamic team formation with large numbers of robots and high-level, distributed planning for building complex structures with many parts. In this paper we present our approach and compare it with work done by other researchers. We discuss two topics, distributed coordination and distributed visual servoing, in the context of a beam placement task accomplished by a team of three robots.

2. Approach

Our approach to coordinating multiple, heterogeneous robots is based on the layered architectures that are becoming increasingly popular for single-agent autonomous systems [3], [12], [17]. In our architecture, each robot is an autonomous agent, consisting of a *planning* layer that decides how to achieve high-level goals, an *executive* layer that synchronizes agents, sequences tasks and monitors task execution, and a *behavioral* layer that interfaces to the robot's sensors and effectors (Figure 1). As is customary with single-agent tiered architectures, each layer interacts with those above and below it. In addition, in our multi-robot architecture, agents can interact with one another through direct connections at each of the three layers. This type of layer-specific interaction provides for increased flexibility and efficiency in the way the robots can coordinate.

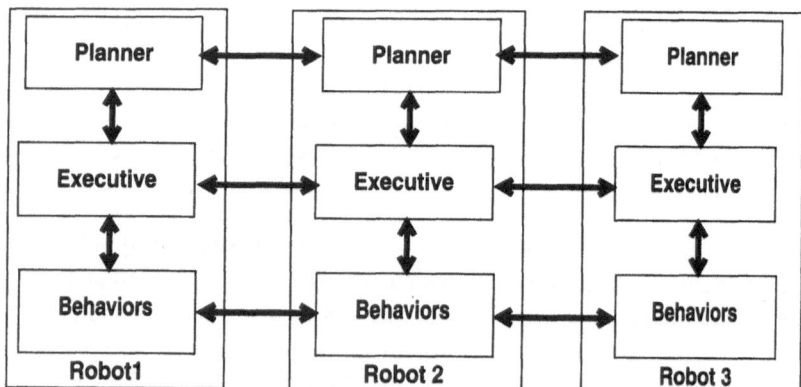

Figure 1.   Layered multi-robot architecture. Each robot has three layers that can directly interact with one another and with the appropriate layers of the other robots.

For instance, the behavioral level typically consists of sensor/effector feedback loops. By allowing connections between the sensor behaviors of one robot and the effector behaviors of another, we create efficient distributed servo loops, such as the visual servoing described in Section 6. Similarly, by enabling the robot executives to interact with one another (see Section 5), we can easily synchronize tasks performed by multiple robots, have robots monitor each other's progress, and even have one robot handle exceptions raised by another robot. This is particularly important when the robots must coordinate explicitly to perform complex tasks.

Finally, by having the planning layers coordinate, we can flexibly construct multi-agent plans that try to optimize overall resource utilization. Our approach allows for this to be done either in a totally distributed fashion, using distributed negotiation between agents [14] to decide which agents will perform which roles and how the agents will cooperate, or else in a more global fashion, where agents bid on becoming "foremen" for particular subtasks. In the latter approach, the foreman agent (which may itself be one of the robots, and may change depending on the subtask) dynamically negotiates with other agents to form teams and assigns them tasks.

The teams form "commitment groups" with joint intentions that provide the basis for their coordinated actions [8], [22]. The individual agents can also negotiate with one another, if necessary, to carry out their assigned tasks. For instance, if two robots are jointly holding a workpiece, one may request the other to move in order to obtain increased freedom of motion. In addition to task negotiation, the foreman monitors progress, adding or replacing team members if problems arise.

3. Related Work

Our approach stands in contrast to much of the current work in multi-robot systems. Most current approaches can be categorized as either "group behavior" or "highly centralized". In the "group behavior" approach [2],[5],[6],[11],[13] each agent is autonomous, but there is usually no explicit coordination among the robots: coordination (or, more accurately, *cooperation*) is an emergent property of the way the behaviors of the robots interact with the environment. For instance, in Parker's ALLIANCE architecture [13], robots decide which tasks to perform in a behavior-based fashion: They have "motivations" that rise and fall as they notice that tasks are available or not. While ALLIANCE can handle heterogeneous robots (robots can have different motivations for different tasks), it does not deal with the problem of explicit coordination. In particular, it has not been demonstrated on tasks that *require* multiple robots.

At the other end of the spectrum, in the "highly centralized" approach a centralized planner plans out detailed actions for each robot. For example, a planner might treat two 6 DOF arms as a single 12 DOF system for the purpose of planning detailed trajectories that enable the arms to work together in moving some object, without bumping into each other [10]. While this approach provides for tight coordination, it does so at the expense of local robot autonomy. In particular, this approach usually employs centralized monitoring and, if anything goes wrong, the planner is invoked to replan everything. This approach also suffers from single point failure.

Under our scheme, individual robots can autonomously solve many problems themselves or by negotiating with each other, without having to invoke a high-level planner. These characteristics reduce the need for inter-robot communication and improve overall reliability. As such, our approach is similar to some work in which coordination strategies are explicitly represented and reasoned about [8],[21],[22].

Our architecture also supports dynamic team formation. Coordination occurs between agents filling specific roles in the structure of the team, and roles can be dynamically assigned to agents, in a manner similar to [9]. We also plan to use distributed methods to optimize the assignment of roles to agents, as in [4],[15],[20].

4. Testbed

Our experimental testbed is comprised of three robots— a crane, a roving eye and a mobile manipulator (Figure 2). The crane, called Robocrane, is a 20-foot high, inverted Stewart platform built by the National Institute of Standards and Technology (NIST) [1]. Robocrane consists of a large triangular platform supported by six cables attached to winch motors. This enables Robocrane to move freely with six degrees of freedom in a roughly 10 foot cubed workspace. We have added a winch motor on the platform, which pays out a cable to which an 8-foot long beam is attached. The roving eye is the robot Xavier, a 4-foot tall, 2-foot diameter synchro-drive robot with stereo cameras mounted on a pan-tilt head [17]. The mobile manipulator is built on top of a four wheeled robot testbed, called Bullwinkle, which can drive and avoid obstacles using stereo vision [19]. The manipulator itself, which mounts to the front of Bullwinkle, is a 5 DOF arm designed and built at NASA Johnson Space Center. The end effector is an electromagnet mounted on springs at the end of the wrist and is used to attach to the underside of the hanging beam. The three robots communicate with each other and an off-board workstation using Wavelan radio Ethernet.

Figure 2.   Experimental testbed consisting of 6 DOF crane, mobile manipulator, and roving eye robots.

5. Distributed Coordination

To perform large-scale assembly tasks, the robots must coordinate their actions. For instance, the crane and the mobile manipulator must coordinate so the manipulator has enough freedom to move the beam without having to support much of the beam's weight. Similarly, the roving eye and crane must coordinate so that the position of the beam can be well estimated.

Our approach to the problem of distributed coordination extends work we have done in single robot task-level control [16],[18]. The basic idea is that agents execute plans by dynamically constructing *task trees*. Nodes in a task tree represent commands (which are primitive behaviors executed by the robot), goals (which are further decomposed into subgoals and/or commands), or monitors (which are periodically executed). Tasks within the tree are partially ordered, with temporal constraints between them. For instance, one can constrain goal **B** to start after goal **A** ends, which implies that no subtask of **B** can start until all the subtasks of **A** have completed. Tasks can also raise exceptions and terminate other tasks. Temporal constraints and goal decomposition strategies are encoded using the Task Description Language (TDL), a superset of C++ that has explicit syntax to support creating task-level control programs [18].

For this work, we are extending TDL to deal with synchronization of multiple agents. The idea is to distribute the task tree representation so that each of the robots maintains only a part of the complete tree (that portion dealing with their own goals and actions). Temporal constraints can be associated between nodes on different robots. For instance, one can encode that task **A** on robot 1 must start 10 seconds after task **B** on robot 2 starts. The extensions made to TDL also enable robots to monitor each other's execution, handle exceptions raised by others, and terminate tasks of other agents.

The multi-agent version of TDL forms an infrastructure for coordination— it allows expression of the necessary synchronization constraints. However, it does not address what coordination needs to take place to do the task. This is the responsibility of the planning layers. Consider, for instance, the following scenario for the task of connecting a beam at a given location: A call is put out for a foreman to manage this task, which could be filled by an agent that has sufficient knowledge and available computational resources. The chosen foreman would put out a request for an available crane, a roving eye or two and, possibly, a mobile manipulator, depending on the precision needed for the particular task at hand. Agents can participate in more than one task— for instance, a roving eye with a pan-tilt head could conceivably assist in two different assembly subtasks, if they are within proximity. Once a team is chosen and roles assigned, the agents coordinate amongst themselves. For instance, the roving eye and the crane coordinate to exchange information, and the crane and mobile manipulator coordinate to decide which will move when, and by how much.

While the scenario described above illustrates our longer-term goals for multi-agent coordination, our current implementation uses a fixed set of three robots, a fixed "foreman" agent, and fixed task assignments (Figure 3). The foreman agent decides which robot should be moving the beam at which times. It initially tasks the crane to move the beam to the vicinity of the emplacement point, which the crane does based solely on encoder feedback. This gets the fiducials on the beam within the roving eye's field of view. The foreman then sets up a behavioral loop between the roving eye and crane robots to servo the beam to near the emplacement point (Section 6). The foreman monitors the progress and, when the difference between the desired and observed poses of the fiducials is within the resolution of the crane's motion, it tasks the roving eye and the mobile manipulator to servo the arm to grasp the beam. When the arm indicates that it is in contact with the beam, the foreman initiates the task of having the roving eye and mobile manipulator coordinate to servo the beam to the emplacement point, which completes the task. The foreman also handles some

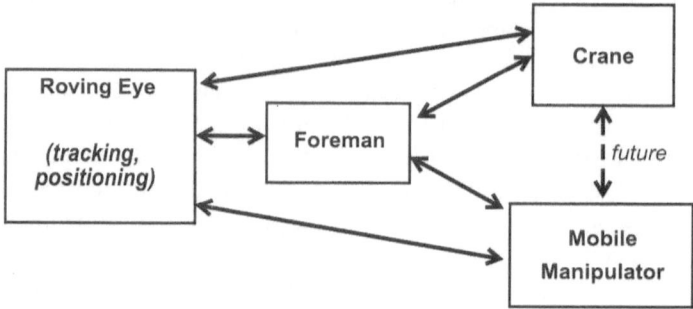

Figure 3.   Robot agents for the assembly task. Agents can communicate directly with each other, or through a foreman agent. In the future, the crane and mobile manipulator will be able to negotiate directly with each other.

simple task failures. For instance, if the arm loses contact with the beam, the foreman restarts the arm-grasp-beam task. In the near future, if the mobile manipulator finds itself at the limits of its workspace, it will negotiate directly with the crane robot to provide it more slack on the beam.

6. Distributed Visual Servoing

An important step in our research has been to develop a technique for distributed visual servoing. The roving eye uses a pair of cameras to track fiducials that are placed on the beam, the mobile manipulator arm, and the destination site (Figure 4).

Figure 4.   Tracking fiducials by the roving eye robot (left). Fiducials (right) are mounted on the fixed structure, on the beam being emplaced, and on the mobile manipulator.

The roving eye moves in order to maintain the best view of the fiducials. It pans and tilts the cameras and drives around the workspace to keep the targets in sight and centered in the image, and it moves back and forth to ensure that the targets fill most of the cameras' fields of view. Stereo is used to compute the 6 DOF pose of each object marked with fiducials, and the differences between the poses of the objects are used to compute manipulator motion commands. Figure 5 illustrates the data flows between modules. The visual servoing runs as a set of distributed behaviors, implemented using the Skill Manager from the 3T architecture [3]. Information flow between modules is implemented using message passing.

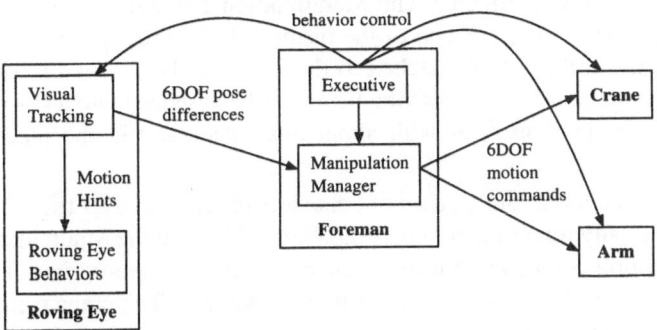

Figure 5.   Data flow between the distributed visual servoing modules.

6.1. Servo Control

The roving eye needs to communicate to the crane and mobile manipulator how to move the beam in order to position it properly. Since none of the robots knows anything about the others' positions, they must communicate information solely in terms of the task space (i.e., the relationship between the beams, or the beam and the arm). The Manipulation Manager module (a component of the "foreman" agent) is used to perform geometric transformations between task space and manipulator space. It uses the pose differences between the objects, combined with knowledge of goal positions, to compute the end-effector motion needed. This motion transform is sent to the appropriate robot agent, which uses its own kinematic model to determine how to move to achieve the desired transform.

The servoing process starts with the roving eye providing information to the crane. It continues until the crane is close enough to the destination, where "close enough" is based on how accurately the crane can be expected to move the beam and how close the beam must be before the arm can grasp it for the final positioning. After the crane moves the beam close to the goal position, the arm is visually servoed to grasp the beam. The grasping motion of the arm works similarly to the crane motion. The roving eye tracks the end-effector of the arm and the horizontal beam held by the crane. The Manipulation Manager computes desired end-effector motion from this combined with knowledge of the desired grasp point on the beam. The arm motion uses a dynamic look-then-move scheme, in which position commands are given to the arm, but they can be interrupted by new ones before they complete. This allows for smooth arm motion combined with the safety that the arm will stop if it does not receive new motion commands for some reason. This servo loop stops when the gripper (an electromagnet) contacts the beam and sticks to it.

In the final phase of the task, the arm moves the beam to dock with the stationary beam. The grasped beam hangs from the crane by a cable, which provides compliance but also complicates motion since the arm does not have full control of the beam. It can effectively control only the position of the end of the beam, but not its orientation. In addition, if the angle between the beam and the arm's end-effector is too large, the magnet will not be able to hold the beam. Therefore for this subtask the roving eye tracks three objects: the positions of the stationary beam, the moving beam, and the arm end-effector. The Manipulation Manager computes end-effector motions which will move the end of the beam to the correct position while keeping the angle of the end-effector matched to the angle of the grasped beam. This servo loop stops when the end of the moving beam has been placed into a slot atop the fixed beam. This currently operates with about 5mm accuracy in the placement of the hanging beam.

The visual servoing runs as a set of behaviors distributed over the robots. The roving eye continually tracks the fiducials (at about 3 Hz) and the Manipulation Manager calculates motion commands for the crane or the arm, as appropriate. The crane and the arm each obey these commands as often as they can. The crane can not yet move continuously, so it makes discrete moves one at a time, ignoring new commands until the previous one finishes. The arm can move continuously, so it adjusts its motion with every new command. An executive module implemented using TDL (a part of the Foreman), manages the process by starting, monitoring, and stopping these behaviors.

The bar coded fiducials used on the beams allow unique identification of several fiducials (8 in the current scheme), and is quite robust to background noise. The tracking starts with an adaptive threshold, to correctly separate black and white even if some fiducials are in shadow and some in strong light. Next, connected components are found which have the same centroids, thus picking out the "bullseyes" on each end of each fiducial. Each pair of bullseyes are used as the endpoints for a bar code scan line, and the pairs with valid bar codes between them are kept as fiducials. 3D data is found by triangulating the positions of the corners of the bullseyes and fitting a model of the object's fiducials to this sensed data.

6.2. Roving Eye Motion

Control of the roving eye motion is accomplished with three behaviors: panning to keep the fiducials centered in the images, moving forward or backward to keep the cameras as close as possible to the fiducials, and lateral motion to move to face the fiducials as directly head-on as possible. Running concurrently, these behaviors keep the roving eye directly in front of the fiducials and close enough to see them well, but not so close that they are in danger of moving outside the field of view of the cameras. The behaviors are diagrammed in Figure 6 (a) and the resulting motion is depicted in Figure 6 (b).

The roving eye behaviors receive information from the vision system in the form of "eye motion hints". These consist of the bounding box of the fiducials in the images and the average angle of the surface normals relative to the camera pointing angle. The bounding box of the fiducials is used by the panning behavior to keep the edges of the fiducials as far as possible from the edges of both fields of view simultaneously. This bounding box is also used by the forward motion behavior that drives the roving eye towards or away from the fiducials. If any side of the bounding box is too close to the edge of the frame, the roving eye backs away. If all sides are too far from the edges of the frame, it drives forward. The lateral motion behavior uses the average of the fiducial surface normal angles projected onto the ground plane. It moves the robot left or right to be most directly in front of the fiducials. This is

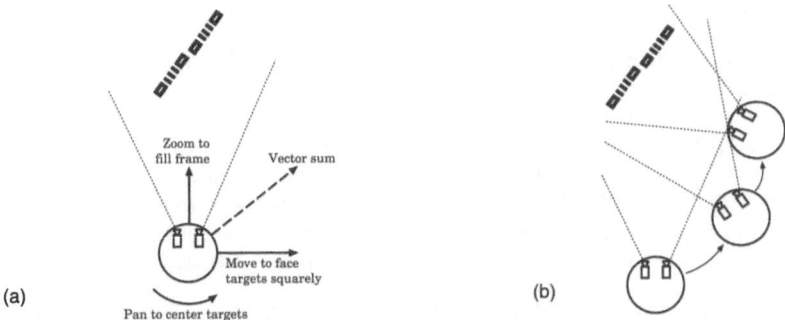

Figure 6.   (a) The three motion behaviors of the roving eye robot. (b) The resulting motion of the roving eye.

important since the fiducials are planar: when viewed from an angle that is too steep, tracking will fail.

The three roving eye behaviors combine to produce smooth motion when the vision updates are fast enough relative to the driving speed of the roving eye. Figure 6 (a) shows how the lateral motion and forward motion behaviors' outputs are combined in a vector sum. These vectors are defined relative to the orientation of the cameras so that when the panning behavior turns the cameras, the directions of the vectors from the other behaviors change accordingly. Lateral robot motion moves the fiducials off-center in the images, triggering the panning behavior. Together these two effects generate smooth motion in a spiral arc.

7. Conclusions

We have demonstrated preliminary results for the coordination of a team of heterogeneous robots performing an assembly task. The roving eye provides higher servoing accuracy more consistently for a larger workspace than a fixed camera system. The roving eye has greater robustness to tracking failures because of its ability to stay aligned with the fiducials. Accurate camera calibration has not been necessary because visual servoing provides relative positions of the fiducials— errors due to calibration affect all measurements roughly equally. Currently, our implementation with a Manipulation manager falls short of the ideal distributed three layer architecture (Figure 1). In the near future, we will distribute the functionality of the Manipulation Manager amongst the behavioral layers of the multiple agents to remove a bottleneck from the high bandwidth behavior-level communication. In this scenario, each robot will be more autonomous as well, since each will calculate its own motion rather than being commanded by the Manipulation Manager.

Acknowledgements

This research was sponsored in part by a grant from NASA (NAG9-1226). The robot arm used on the mobile manipulator was developed by Metrica under a NASA grant. Josue Ramos was funded by a grant from CNPq-Brazil (200501/86-0).

References

[1]   J. Albus, R. Bostelman, N. Dagalakis, 1992 The NIST ROBOCRANE, Journal of Robotics System, 10:5.

[2]   R. Arkin, 1992 Cooperation without Communication: Multiagent Schema-Based Robot Navigation, Journal of Robotic Systems, 9:3, pp. 351-364.

[3]   P. Bonasso, D. Kortenkamp, D. Miller, M. Slack, 1997 Experiences with an Architecture for Intelligent, Reactive Agents, Journal of Artificial Intelligence Research, 9:1.

[4]   M.B. Dias, A. Stentz, 2000 A free market architecture for distributed control of a multi-robot system, In Proc. 6th International Conference on Intelligent Autonomous Systems (IAS-6), pp 115-122.

[5]   L. Chaimowicz, T. Sugar, V. Kumar, and M. Campos 2001 An Architecture for Tightly Coupled Multi-Robot Cooperation , in Proceedings International Conference on Robotics and Automation, Seoul, Korea, May 21-26.

[6]   B. Donald, 1995 Distributed robotic manipulation: experiments in minimalism, In: Proc. International Symposium on Experimental Robotics (ISER), Springer-Verlag, pp 11-25.

[7]   D. Hershberger, R. Burridge, D. Kortenkamp, R. Simmons, 2000 Distributed Visual Servoing with a Roving Eye, In Proc. Conference on Intelligent Robots and Systems (IROS), Takamatsu Japan, October.

[8]   N.R. Jennings. 1993 Specification and Implementation of a Belief-Desire-Joint-Intention Architecture for Collaborative Problem Solving, International Journal of Intelligent and Cooperative Information Systems, 2(3), pp. 289-318.

[9]   J. Jennings, C. Kirkwood-Watts, 1998 Distributed mobile robotics by the method of dynamic teams, In Proc. Conference on Distributed Autonomous Robot Systems (DARS).

[10]   O. Khatib, 1995 Force Strategies for Cooperative Tasks in Multiple Mobile Manipulation Systems, In Proc. International Symposium of Robotics Research, Munich, October.

[11]   M. Mataric, 1992 Distributed Approaches to Behavior Control, In Proc. SPIE Sensor Fusion V, pp. 373-382.

[12]   N. Muscettola, P. P. Nayak, B. Pell, and B. Williams, 1998 Remote Agent: To Boldly Go Where No AI System Has Gone Before, Artificial Intelligence 103(1-2), pp. 5-48, August.

[13]   L. Parker, 1998 ALLIANCE: An Architecture for Fault Tolerant Multirobot Cooperation, IEEE Transactions on Robotics and Automation, 14:2, pp. 220-240, April.

[14]   T. Sandholm, O. Shehory, M. Andersson, K. Larson, and F. Tohm , 1998 Anytime Coalition Structure Generation with Worst Case Guarantees, In Proc. Fifteenth National Conference on Artificial Intelligence (AAAI), pp. 46-53, Madison WI, July.

[15]   O. Shehory O, S. Kraus,1998 Methods for task allocation via agent coalition formation, Artificial Intelligence Journal, 101:1-2, pp 165-200, May.

[16]   R. Simmons, 1994 Structured Control for Autonomous Robots, IEEE Transactions on Robotics and Automation, 10:1, pp 34-43, February.

[17]   R. Simmons, R. Goodwin, K. Haigh, S. Koenig, J. O Sullivan, 1997 A Layered Architecture for Office Delivery Robots, In Proc. First International Conference on Autonomous Agents, Marina del Rey, CA, February.

[18]   R. Simmons and D. Apfelbaum, 1998 A Task Description Language for Robot Control, In Proc. Conference on Intelligent Robotics and Systems, Vancouver Canada, October.

[19]   S. Singh, R. Simmons, T. Smith, A. Stentz, V. Verma, A. Yahja, and K. Schwehr, 2000 Recent Progress in Local and Global Traversability for Planetary Rovers In Proc. International Conference on Robotics and Autonomous, San Francisco CA, April.

[20]   R. Smith, 1980 The contract net protocol: high-level communication and control in a distributed problem solver, IEEE Transactions on Computers, C-29:12, pp 1104-1113,

[21]   K. Sycara and D. Zeng, 1996 Coordination of Multiple Intelligent Software Agents, International Journal of Cooperative Information Systems, 5:2-3.

[22]   M. Tambe, 1997 Towards Flexible Teamwork, Journal of Artificial Intelligence Research, No. 7, pp. 83-124.

Towards A Team of Robots with Repair Capabilities: A Visual Docking System

Curt Bererton
Robotics Institute, Carnegie Mellon University
Pittsburgh, Pennsylvania, USA
curt@cs.cmu.edu

Pradeep K. Khosla
Electrical and Computer Engineering, Carnegie Mellon University
Pittsburgh, Pennsylvania, USA
pkk@ece.cmu.edu

Abstract: In the future, we propose that there will be largely self-sufficient robot colonies operating on distant planets and in harsh environments here on earth. A highly desirable quality of such a colony would be the capability of the robots to repair each other. Towards the goal of autonomous repair, we designed a robot that can replace the modules composing a similar robot. In this paper we highlight a visual docking system for the repairable robot design that allows the robots to autonomously replace their teammate's modules. The primary contribution of this work lies in the application of known techniques to the more constraining platforms of very small robots. This forces the use of very simple hardware and algorithms that perform robustly. The results obtained consist of initial configurations from which the robots could successfully complete the docking operation and the average time required to dock.

1. Introduction

There are several applications for robots capable of cooperative repair. Our main vision is of a colony of robots on Mars (or another planet) preparing a station for human habitation (Figure). NASA's technology plan [12] lists self-diagnosing and self-repairing systems as one of the technologies necessary for self-sustained long duration human operations. Such a colony would need to operate without outside assistance for extended periods of time, although there is the possibility for some teleoperation of the robots if humans are present. Other applications include operation in any kind of harsh environment where human intervention is either costly or dangerous. A team with cooperative repair capabilities would be able to operate longer and more efficiently than a comparable team without such capabilities.

Repairable robots are also useful in non-automated repair tasks. If a robot can be repaired by another robot, then it is likely that the task will be trivial for a human technician. This can lead to smaller downtimes in factory settings as well as decreasing the maintenance cost of robotic systems in general.

D. Rus and S. Singh (Eds.): Experimental Robotics VII, LNCIS 271, pp. 333–342, 2001.

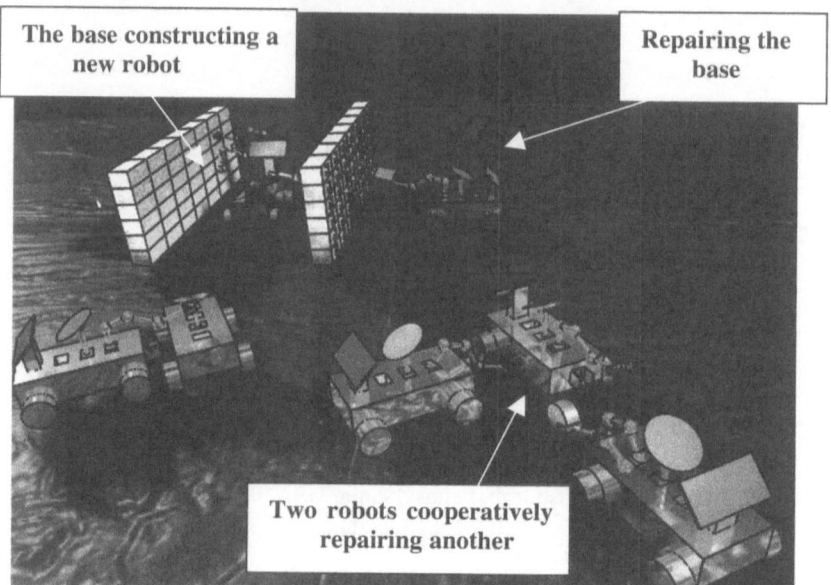

The base constructing a
new robot

Repairing the
base

Two robots cooperatively
repairing another

Figure 1. Possible tasks for a self-sufficient team of robots

The concept of modular redundancy leading to hardware availability is an extremely old concept [5]. Cooperative repair uses modular redundancy between robots so that one need not replicate individual components on a given robot to achieve dependability. The components of failed teammates provide the modular redundancy we require for a dependable robot team. In previous work [16], we have designed a small mobile robot that has a standard skid-steered base, and yet also has the capacity to replace modules on teammates as well as have its own modules be replaced. This type of module replacement is a key step towards *cooperative repair*. In our implementation, the robots also have the ability to perform *cooperative reconfiguration*, by which we mean that different module types can be placed on the robots in a wide variety of configurations.

The closest related work to the concept of systems capable of cooperative repair/reconfiguration is that of metamorphic or reconfigurable robots. The primary difficulties with these systems are due to the exponential increase in planning complexity for a large set of modules [6]. Some of these systems have the capability of self-repair, in that they can expel failed modules and continue to function [7][8]. The key difference is that the above systems must carry the redundant modules with them even if they don't serve any purpose. These modules can then be used to perform self-repair when a module breaks. In our design, the redundancy comes from the other teammates. All modules are in use until that robot fails. These modules can then be used to repair the next robot that fails. Although reconfigurable robots may one day be the solution to many problems, they will be useless in practical situations until an effective means to power them has been developed. All of the reconfigurable and metamorphic robots built to date either use tethered power or quickly drain any battery-powered supply due to the large number of actuators required.

It is generally the case that systems made for a specific task are usually cheaper and more efficient. However, in applications where the exact task and environment is not known before hand, a team of robots with repair and reconfiguration capabilities will yield a more versatile solution. The advantages of such a system are similar to those promised by metamorphic robots [6]: Versatility; in that robots can be autonomously reconfigured to optimally perform a task given a limited set of functional modules. Reliability; in that as the number of modules increases in any system, be it a team of robots or one very expensive robot with a large amount of redundancy, the probability of at least one component failing as the number of components increases goes to one. Low Expense; were such a team to be built, a large number of identical modules would lead to decreased production cost per module. The design of such a system also avoids some of the main difficulties with metamorphic systems in that planning complexity is not directly proportional with the number of modules. Placing the modules on separate robots gives a natural hierarchy that simplifies control and repair. Multi-robot research [6][9][10] can then be applied to the control of the team as a whole. Perhaps the most appealing feature of this type of system is that redundant components are actually in use prior to a failure. Therefore, no components are sitting idle waiting to take over when another component fails and no productivity or expense is wasted carrying redundant components that are never used.

Repair in dependability and reliability literature is known as *fault removal* [1][2]. In order for fault removal to occur in a self-sufficient robot colony, the robots must first determine that there is some fault present. Once a fault is known to exist, one must then determine the location and nature of such a fault. Action is then taken to remove the fault. Of these three steps: *verification*, *diagnosis*, and *correction*, we have concentrated on the correction aspect. In our case, correction corresponds to replacing a faulty module in a modular design. In order to correct any fault in this case, we must replace the faulty module. To replace the faulty module, we must be able to dock with the failed robot and replace the module. This paper concentrates on the docking system that was designed for this purpose.

There is a relatively small amount of literature that is directly applicable to docking mobile robots. Most of the literature about pose-estimation and navigation based on landmarks is highly relevant to docking [20][21]. The main difference between this type of work and the systems presented here has to do with complexity. The typical task in such work is to accurately determine the robot's cartesian location. This work could be used to perform docking tasks, but tends to be overly complex for a simple docking task and often does not yield the required accuracy. Other docking tasks specifically designed for mobile robots tend to be highly task-specific [22], or use complex sensors and algorithms unsuitable for use on small robots [23].

2. Task Description

In [4] we developed a localization system based on ultrasonic trilateration. The system could position a robot reliably within one or two centimeters. Thus we required a docking system to dock from a start position that was anywhere within that area. In the results section, we describe the initial positions from which the robot can begin the docking procedure and successfully complete the docking task. These initial positions consist of an X and Y offset from the goal position. We will henceforward refer to this as the initial configuration of the robot.

In this case, the task definition was very precise. This task was to dock a forklift robot (the repair robot) with a second (stationary) robot and remove a module. The sensor used to accomplish this was a black and white wireless camera. The robot was required to fully insert the forklift pins into the forklift receptacle on the other robot. Contact with the second robot caused a bump switch to be triggered, indicating that docking was complete. The two robots are shown in Figure . In this task, the visual target was identified and tracked autonomously in order to dock the robot.

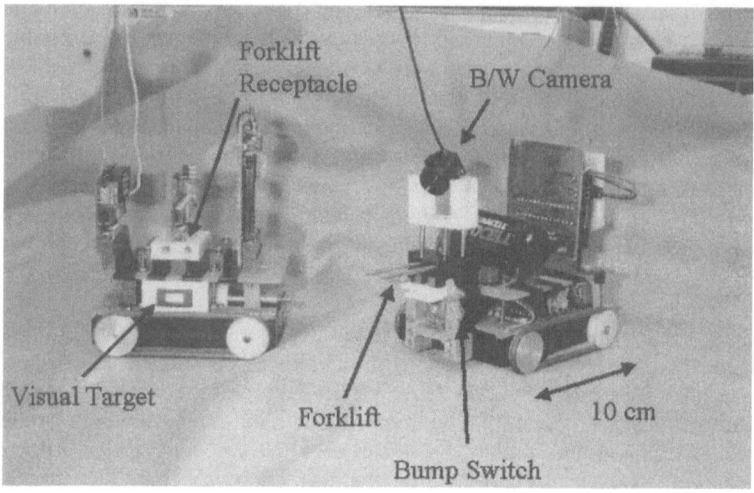

Figure 2. Repair robot and the robot to be repaired, including the visual target

The repair robot was placed in an arbitrary initial configuration and was required to complete the docking procedure, including the removal of the module, or it should indicate to the user that it could not complete the task. Success in this task occurred when the robot successfully removes the module on the second robot.

3.   Hardware and Algorithms

3.1.   Hardware Description

The basic platform for the repairable and repair capable is discussed in detail in [16]. To this platform we added a black-and-white wireless camera and a bump switch, as indicated in Figure . The camera transmitted a 320x240 pixel video stream to a Pentium II 400MHz PC, which then digitized the image using a Matrox Meteor II capture card. The PC performed the image processing and target tracking and then sent control commands to the robot to perform the docking task via an RF link. Most of the image processing was done using the Matrox imaging library.

One of the key features of the forklift and the forklift receptacle was that they were designed to have approximately a 30 degree allowable error angle between the line formed by the forklift pin and the axis of the hole in which the pin was to be inserted. This was partially achieved through rounding the pins of the forklift and shaping each of the two holes in the forklift receptacle as cones. The other main feature that allowed the 30 degree error is the fact that the robot is skid-steered. Thus, if the pins were partially inserted into the receptacle and the robot pushes straight ahead, then the wheels would slip on the ground and allow the robot to center the pins in the receptacle.

Although the image processing was done on the PC, our goal is to eventually perform all processing on the robot itself. To this end, the image processing, machine vision algorithms, and robot control strategies were deliberately kept as simple as possible.

3.2.   Image processing and machine vision algorithms

The location of the visual target in the image could be identified given only a single image. The basic idea for this type of visual target came from [18]. In that example they used it to control a 5-degree of freedom manipulator. In order to determine the pixel coordinates of the target within the image, the following steps were performed:

- Capture a single grayscale image
- Perform histogram equalization on the image (equalization of the image intensity distribution)
- Binarize the image using a constant threshold which is determined empirically
- Find all 8-connected white regions (blobs) in the image and calculate the center of gravity of each region (i.e. the center of the region)
- Find all 8-connected black regions (blobs) in the image and calculate their centers of gravity
- Exclude all regions that are less than 30 pixels in size (30 pixels is the area the visual target occupies when it can barely be detected by this algorithm). This step eliminates noise in the image and reduces the processing required by the following steps
- Exclude all black regions that do not have exactly one white hole inside them. Since the target only has one white hole then all other regions are not of interest.
- For all remaining black regions, compare their centers of gravity to the centers of gravities of the white regions
- If a white center of gravity matches a black center of gravity to within 2 pixels, then that is our visual target. The coordinates of the black region's center of gravity give the center of the visual target. The value of 2 pixels was determined empirically.

The image processing routine returned the x and y location of the target in the image. We only had one 1cm tall target in the image and it was tracked extremely reliably up to a distance of 50cm from the target. At distances of greater than 50cm, the connected regions composing the target became too small to discriminate from the background. For comparison, a human observer was unable to observe the connected regions in the image (shown on the screen) at approximately 70cm. Obviously, if we had a larger target or a higher resolution camera, the distance at which the target could be tracked would have been greater.

3.3.   Controlling the robot using the results from the machine vision system

The control of the robot followed a simple state machine:

- Acquire target state: This is the initial state in which the robot begins the docking procedure. Essentially, the robot turns counter-clockwise (clockwise would work equally as well) until the target is seen. This allows the robot to start from an arbitrary orientation. After the target is acquired, the robot proceeds to the approach state
- Approach state: First we determine the position of the visual target in the image. If the target is to the left of center, turn slightly to the left while moving forward. If it is to the right of center, turn slightly to the right while moving forward. Repeat until the target approaches the bottom of the image at which point we move to the centering state
- Centering state: In this state the robot turns left or right very slowly until the visual target is directly centered below the forklift pins and then proceeds to the insertion state
- Insertion state: The robot drives straight ahead until either the bump switch is triggered or a timeout occurs. If the bump switch is triggered, the docking is complete and the forklift can be raised to remove the module. This is considered a successful completion of the docking task. If a timeout occurs, the robot enters the retry state.
- Retry state: The robot drives directly backwards until the visual target is within view or a timeout occurs. If a timeout occurs during the retry state, then the docking procedure is considered a failure. If the visual target is seen again, then we proceed to the approach state. If this is the third retry then the docking procedure has failed.

There are several features to note about this state machine. Firstly, the robot was only allowed to retry 3 times before the docking procedure is considered to have failed. The justification here was that there is likely something blocking the forklift pins from entering the holes. Not mentioned in the above state machine is the fact that each state has an associated timeout. If the robot takes too long to perform any given state, the docking procedure is considered to have failed.

Perhaps the weakest feature of this control algorithm was that it relied on the bump switch for confirmation of a successful dock. Though we only checked the state of the bump switch in the Insertion state, it may have been possible for something to have accidentally triggered the switch. The hardware actually supported a superior method for determining docking success though it was unimplemented in these experiments. Just below the forklift pins on the repair robot, the white docking guide has connections for one power pin and four input pins. When properly positioned on the failed robot, these four input pins generate a four-bit address for each module, thus we would know if we had successfully docked with the failed robot as well as exactly which module on the robot we had docked with. This will be implemented in future work.

4. Docking System Results

Our primary concern was to determine the initial configurations from which the robot could successfully dock with and remove the module from the second robot The second robot is stationary at the goal position is at 0,0. Note in Figure that we do not show the initial orientation of the robot, as the acquire target state mentioned above will find the target if it is within the recognition range of the machine vision algorithm or it will timeout and indicate a failure.

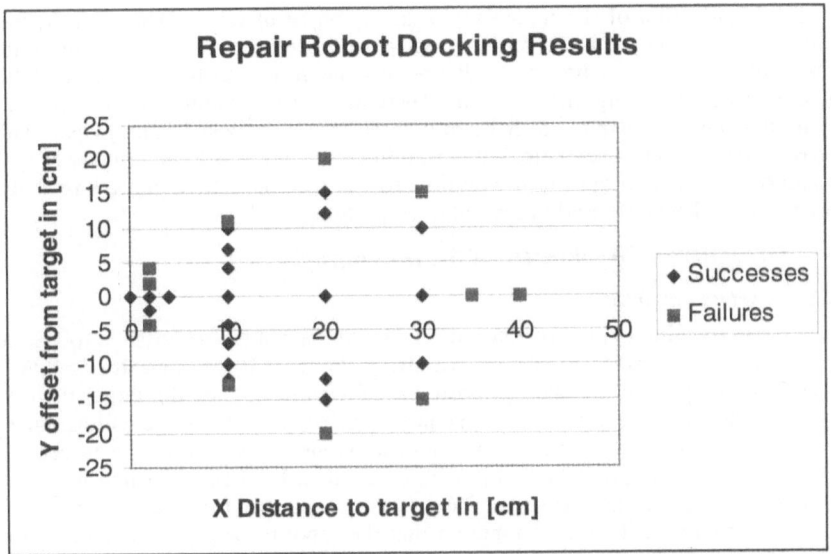

Figure 3. Docking Results

In Figure we performed 31 trials to determine the initial configurations from which the repair robot could successfully dock with the failed robot. Note that the area in which the robot succeeded in docking forms a cone-like shape. Thus, if we could position the robot anywhere within this cone we could successfully dock with the failed robot and remove the module. In the trials shown in Figure we measured the average time of a successful docking procedure to be 80 seconds, and the average time to notify the user of a failed docking procedure to be 75 seconds. From the furthest point that the robot could successfully dock without retries, the robot required 80 seconds if it performed no retries. A retry on average took 45 seconds, and retries were only required if the robot was on the edge of the area in which a docking procedure could be successfully performed. The system is quite robust; all the trials that were not on the border of the cone were successful. Considering that the robot operates for approximately one and a half to two hours on a single battery charge, the docking time was a minor portion of the operational time of the robot.

5.   Discussion

5.1.   Docking System Failure Modes
The failure modes of this system were due almost entirely to the visual system and the machine vision algorithms being used. Firstly, one should note that the camera was positioned high on the robot and looking towards the ground at an angle. This configuration allowed us to use the y position of the target in the image as a rough estimation of the distance to the target. The disadvantage was that we were unable to see the target at a distance of further than 40cm. The only reason for placing the

camera there (as opposed to mounting the camera lower on the chassis) was that we wanted to have the camera be a replaceable module.

The binarization of the image was a large source of error. Though a human could clearly distinguish the target in the image at 55 cm, the binarization caused the outlines of the target to become indistinguishable from the background and the remaining steps in the algorithm to fail. There are several solutions to this. In [17], they used a self-similar pattern that could be easily recognized in the image. This approach could also be used here, but is significantly more processing intensive than our approach. Another approach would be to use a color camera and a target of a unique color as has been used in several other systems.

5.2.   Strengths and Weaknesses of the Docking System

5.2.1.   System advantages:

Visual target tracking explicitly determines whether or not it was possible to see the target from the current position, thus we always knew if it was indeed possible to successfully dock from the starting position. If we did not see the target then we could alert the user or a higher-level planner of the error. If the target is lost during the docking procedure, the robot can undo previous steps to find the target and reattempt the docking procedure. In many cases the robot being repaired will have lost power, if this is the case the visual system will still function correctly. In general one cannot make the assumption that the robot to be repaired can assist in the repair procedure in any way.

5.2.2.   System disadvantages:

The hardware required for the docking system is expensive compared to the cost of the robot: Wireless B/W Camera $250, Receiver $100, Capture Card $100-$600. The robot itself costs approximately $500, most of which is in the cost for the motors. Though in this particular case we performed the vision processing off-board, a more realistic scenario would be to have the vision processing done on-board. Vision is a computing-intensive application, and thus we would need a fair amount of processing power on the robot, which is difficult for a robot of this size.

5.3.   Docking Systems for Repairable Robots

The above system was a second attempt at a docking system. The first system used infrared light emitting diodes (LEDs) to perform the docking procedure. This docking system was very inexpensive and computationally simple, but had several properties that were unacceptable in a docking system for cooperative repair. In order to find a damaged robot, the damaged robot had to turn on two LEDs in order for the repair robot to dock with it. Obviously, if the damaged robot does not have power, this is impossible. As a general rule, one cannot assume that a robot in need of repair can assist in the repair procedure in any way. Thus any docking system used in a repair procedure should be passive with respect to the damaged robot.

For our specific class of repairable robots, where repair consists of replacing failed modules, the docking system should allow the repair robot to dock with any replaceable module. Though not included in this system, the extension is simple. The addition of a visual bar code around the border of the visual target will allow us to have a visual target for every replaceable module. A more general approach might be to have one or more reference markers on the robot from which the relative position of all the replaceable modules could be determined.

Another crucial feature of docking systems for repair robots is that there must be a retry feature. There are an infinite number of possible failures that may occur during the docking phase. Many of these failures are only transient, and thus can be overcome if the docking procedure is repeated. Also, if the retries fail, it should be possible to re-plan at a higher level. In our case, this means using our ultrasonic localization system to move the robot to a configuration from which it can attempt the docking procedure again.

6. Conclusion

We have presented the concept of *cooperative repair*. Cooperative repair requires three distinct steps: *verification, diagnosis,* and *correction*. The docking system in conjunction with the mechanical design of the robot will allow us to complete the correction aspect of these three steps. Our current goal is to integrate the docking system with the localization system developed in [4]. We must then further develop the system to be able to distinguish and dock with multiple modules and various teammates.

The overall objective is to build a platform with which to develop the remaining two aspects required for cooperative repair, namely fault detection and fault diagnosis. Once the platform is complete, we will have a real platform to test the algorithms and procedures being developed for fault detection and diagnosis.

References

[1]   J. -C. Laprie et al.. Dependability – It's Attributes, Impairments and Means. In Predictably Dependable Computing Systems, pp.3-24, ISBN: 3-540-59334-9, 1995.

[2]   B. Randell. Facing Up to Faults. Turing Lecture, 2000.

[3]   C. Bererton, L.E. Navarro-Serment, R. Grabowski, C. J.J. Paredis and P. K. Khosla. Millibots: Small Distributed Robots for Surveillance and Mapping. Government Microcircuit Applications Conference, 2000.

[4]   L.E. Navarro-Serment, C.J.J. Paredis and P.K. Khosla. A Beacon System for the Localization of Distributed Robotic Teams. In Proceedings of the International Conference on Field and Service Robotics, 1999.

[5]   J. Gray. Why Do Computers Stop and What Can Be Done About It?. Technical Report 85.7, PN87614, 1985.

[6]   M. Yim, D. Duff and K. Roufas. PolyBot: a Modular Reconfigurable Robot. In proceedings of the IEEE Conference on Robotics and Automation, 2000.

[7]   S. Murata et al. Self-Repairing Mechanical System. In proceedings of SPIE Conference on Sensor Fusion and Decentralized Control in Robotic Systems II, 1999.

[8]   E. Yoshida et al. Experiment of Self-repairing Modular Machine. In proceedings of SPIE Conference on Sensor Fusion and Decentralized Control in Robotic Systems II, 2000.

[9]   P. Stone and M. Veloso. Multiagent systems: A survey from a machine learning perspective. In Autonomous Robots, Volume 8, number 3, July 2000.

[10]   T. Balch and R. Arkin. Motor schema-based formation control for multiagent robot teams. In proceedings of the First International Conference on Multi-Agent Systems, pp.17-24, 1995.

[11]   S. Thrun, D. Fox, and W. Burgard. A Probabilistic Approach to Concurrent Mapping and Localization for Mobile Robots. Machine Learning 31, 29--53 and Autonomous Robots 5, 253—271.

[12]   NASA Technology plan. Available at http://technologyplan.nasa.gov/, pp. 100-118, 2000.

[13]   S. Hutchinson and G. Hager. A Tutorial on Visual Servo Control. In IEEE Transactions on Robotics and Automation, Vol. 12, No. 5, 1996.

[14]   J. Shi and C. Tomasi. Good Features to Track. In IEEE Conference on Computer Vision and Pattern Recognition, 1994.

[15]   L. Navarro, R. Grabowski, C. Paredis, and P. Khosla, 1999. Modularity in Small Distributed Robots. In proceedings of the SPIE conference on Sensor Fusion and Decentralized Control in robotic Systems II.

[16]   C. Bererton and P. Khosla. Towards A Team of Robots with Reconfiguration and Repair Capabilities. In proceedings International Conference on Robotics and Automation, 2000.

[17]   A. Briggs, D. Scharstein and S. Abbott. Reliable Mobile Robot Navigation From Unreliable Visual Cues. Workshop on the Algorithmic Foundations of Robotics (WAFR 2000).

[18]   D. Hershberger, Burridge, D. Kortenkamp and R. Simmons,. Distributed Visual Servoing with a Roving Eye. In proceedings, IROS 2000.

[19]   C. Colombo, B. Allotta and P. Dario. Affine Visual Servoing for Robot Relative Positioning and Landmark-Based Docking, AdvRob(9), No. 4.

[20]   S. Hutchinson, G. Hager and P. Corke. A Tutorial on Visual Servo Control, RA(12)

[21]   W. Wilson, C. Hulls and G. Bell. Relative End-Effector Control Using Cartesian Position Based Visual Servoing, RA(12), No. 5 1996.

[22]   S. Mascaro and H. Asada. Docking control of holonomic omnidirectional vehicles with applications to a hybrid wheelchair/bed system. In proceedings IEEE International Conference on Robotics and Automation, Volume: 1, 1998.

[23]   H. Roth and K. Schilling. Navigation and docking maneuvers of mobile robots in industrial environments. Industrial Electronics Society. IECON '98. In proceedings of the 24th Annual Conference of the IEEE, Volume: 4 , 1998.

Merging Gaussian Distributions for Object Localization in Multi-Robot Systems

Ashley W. Stroupe, Martin C. Martin, and Tucker Balch
Robotics Institute, Carnegie Mellon University
Pittsburgh, Pennsylvania, USA
{ashley, mcm, tcb}@ri.cmu.edu

Abstract: We present a method for representing, communicating, and fusing distributed, noisy, and uncertain observations of an object by multiple robots. The approach relies on re-parameterization the two-dimensional Gaussian distribution to correspond more naturally to a robots' observation space. The approach enables two or more observers to achieve greater effective sensor coverage of the environment and improved accuracy in object position estimation. We demonstrate empirically that, using our approach, more observers achieve more accurate object position estimates. The method is tested in three application areas: object location, object tracking, and ball position estimation for robot soccer. We provide quantitative evaluation of the technique on mobile robots.

1. Introduction

Typically, individual robots can only observe part of their environment at any moment in time. In dynamic environments, information previously collected about currently unobservable parts of the environment grows stale and becomes inaccurate. Sharing information among robots increases the effective instantaneous visibility of the environment, allowing for more accurate modeling and more appropriate response. If processed effectively, information collected from multiple points of view can provide reduced uncertainty, improved accuracy, and increased tolerance to single point failures in estimating the location of observed objects.

In order to meet the time demands of a highly dynamic environment (e.g. robotic soccer), both the information transmitted between robots and the computational demands to combine observations must be minimal. Our approach makes use of a few easily obtainable parameters describing an observation and simple computations to meet these needs. We use two-dimensional statistical representations of target location observations generated by individual robots. Each robot independently combines the multiple observations provided it in order to produce improved estimates of target locations. The local computation allows for robust use of all available information without failure due to communication loss.

2. Background and Related Work

One common method of object position estimation, including robot location, uses Kalman filters to track objects [17, 21, 22, 23] or robots [13, 15, 19] through time, updating estimates based on previously known positions. Data fusion of

D. Rus and S. Singh (Eds.): Experimental Robotics VII, LNCIS 271, pp. 343–352, 2001.
© Springer-Verlag Berlin Heidelberg 2001

multiple sensors (such as odometry, sonar, cameras, and laser range scanners) on a single robot, using Kalman filters, has been used for robot localization [14, 15, 18, 19]. The technique is especially useful in dynamic applications such as robotic soccer. The CS Freiburg RoboCup team of Germany, for example, estimates position using odometry and by finding the field borders in laser range scans [10]. These two estimates are fused using a Kalman filter to localize the robots. Recent developments in localization in unknown environments by simultaneously mapping and localizing rather than relying on previously generated maps [3, 8]. The specific issues of multi-robot localization in a mapped environment are also investigated [9].

The ability to rapidly share distributed observations is critical in distributed dynamic tasks like robotic soccer. Most robot soccer team approaches use vision and/or sonar to localize and vision to locate objects in the environment. Some share information for planning and dynamic role assignment (ART [16]). Others fill-in blank areas in the world model with shared data (CS Freiburg [10, 11], RMIT [4], 5dpo [7]). Other distributed sensing approaches include merging independent grid cell occupancy probabilities measured by multiple robots (possibly distributed in time) [5, 6], and curve fitting of models and observations by multiple robots [12].

The task we address is distinct from the others described above. We focus on fusing multiple simultaneous observations of the same object from distributed vantage points (as opposed to observations from the same vantage point over multiple instants in time). Our objective is to provide more accurate instantaneous estimations of the location of dynamic objects that are simultaneously visible by multiple robots without relying on historical data. Additionally, most probabilistic methods rely on decomposing the space into discrete cells [5, 6, 9, 14, etc]. Our approach does not require discretization, working in the continuous spatial domain.

3.   Fusing Gaussian Distributions

3.1.   Overview

We represent a single observation of an object as a two-dimensional Gaussian distribution (Figure 1). The center, or mean, of the distribution is the estimated location of the object and the standard deviations along the major and minor axes of the distribution correspond to estimates of the uncertainty (or noise) in the observation along each axis. The distribution corresponds to the conditional probability that the object is in that location, given the observation.

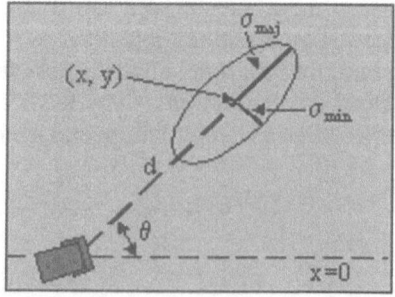

Figure 1. Distribution parameter definitions: mean (x, y), angle of major axis (θ), major and minor axis standard deviations $(\sigma_{maj}, \sigma_{min})$, distance to mean (d).

Provided two observations are independent and drawn from normal distributions, the observations can be merged into an improved estimate by multiplying the distributions. To meet cycle time requirements of a highly reactive system, an efficient method of multiplying distributions is necessary. We use a two-dimensional statistical approach based on Bayes' Rule and Kalman filters, first introduced by Duffin [1]. In this approach, multi-dimensional Gaussian distributions can be combined using simple matrix operations. Since multiplying Gaussian distributions results in a Gaussian distribution, the operation is symmetric, associative, and can combine any number of distributions in any order.

Our approach, illustrated in Figure 2, is to collect observations of multiple robots, and then merge the corresponding Gaussian distributions to yield a better estimate of the location and uncertainty of the observed object.

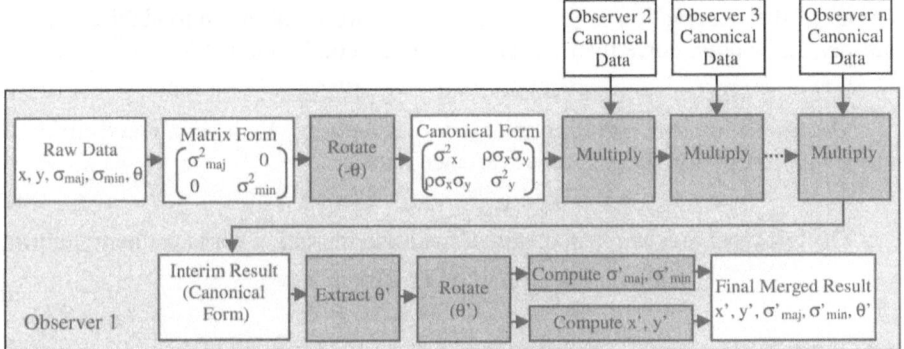

Figure 2. Block diagram of the multi-distribution merging process. Multiplication is conducted using the mathematical formulation described above. Each subsequent distribution is merged with the previous result, and the final parameters are extracted.

The canonical form of the two-dimensional Gaussian distribution depends on standard deviations, σ_x and σ_y, a covariance matrix, C, and the mean, as shown [20]:

$$p(X) = \frac{1}{2\pi\sqrt{|C|}}\exp\left(-\frac{1}{2}(X - \bar{X})^T C^{-1}(X - \bar{X})\right), \text{ where } C = \begin{bmatrix} \sigma_x^2 & \rho\sigma_x\sigma_y \\ \rho\sigma_x\sigma_y & \sigma_y^2 \end{bmatrix} \quad (1)$$

The parameterization of the Gaussian distribution in this representation does not correspond the parameters of our observations (Figure 1). We address the problem through a transformation of parameters from observation form to canonical form. In this form, the distributions can be merged using matrix operations. After all observations are merged, we extract the mean and standard deviations from the merged result (these correspond to estimated location and uncertainty of the object).

3.2. Mathematical Details

We wish to determine the mean, standard deviations, and angle of the combined distribution to estimate object position and characterize the quality of the estimate. We can compute these parameters from sensor readings and models of sensor error (deviations). Thus, we require a method of determining combined parameters from those of individual distributions. The formulation we use is adopted from Smith and Cheeseman [20] and derivations are provided in more detail in a technical report [21]. Since the mean, standard deviations, and orientation of the major axis are independent of scaling, they can be extracted from the resulting merged covariance matrices without considering absolute probability values.

The covariance matrix, C, of an observation relative to coordinates aligned with the major and minor distribution axes is initially determined from the major and minor axis standard deviations in the local coordinate frame (designated L).

$$C_L = \begin{bmatrix} \sigma^2_{maj} & 0 \\ 0 & \sigma^2_{min} \end{bmatrix} \tag{2}$$

Since observations may be oriented arbitrarily with respect to the global coordinate frame (angle θ relative to global x-axis), they must be transformed to this frame. Rotation of X in equation 1 by leads to the following relationship.

$$C^{-1} = R(-\theta)^T C_L^{-1} R(-\theta) \quad \Rightarrow \quad C = R(-\theta)^T C_L R(-\theta) \tag{3}$$

Once the observation is in canonical form, we combine individual covariance matrices into a covariance matrix representing the combined distribution.

$$C' = C_1 - C_1 [C_1 + C_2]^{-1} C_1 \tag{4}$$

The mean of the resulting merged distribution, X, is computed from the individual distribution means and covariance matrices.

$$\hat{X}' = \hat{X}_1 + C_1 [C_1 + C_2]^{-1} (\hat{X}_2 - \hat{X}_1) \tag{5}$$

The principal axis angle is obtained from the merged covariance matrix entries:

$$\theta' = \frac{1}{2} \tan^{-1} \left(\frac{2B}{A-D} \right) \tag{6}$$

A, B, and D are top left, top right/lower left, and lower right entries, respectively.

Lastly, the resulting major and minor axis standard deviations are extracted by rotating the covariance matrix to align with those axes and reversing Equation 2.

$$C' = R(\theta')^T C' R(\theta') \tag{7}$$

3.3.   Simulated Example

Two robots observe a target object (Figure 3).   Each observation produces a Gaussian distribution of possible locations for the object; typically, each distribution provides greater accuracy along a different direction than the other distributions.

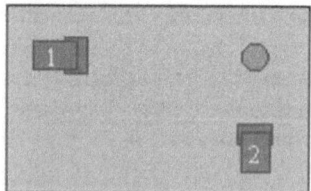 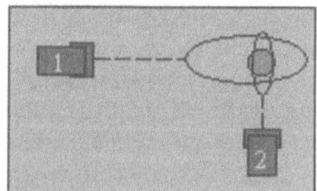

Figure 3.   Left: Two (distributed) robots see a target.   Right: Observations generate Gaussians; uncertainty (1-σ ovals shown) increases with distance.

For this example, the robots are positioned with relative headings 90 degrees apart and looking directly at the target. The target is located at (10,10). The two simulated robot observations were drawn from a random normal distribution centered at the object's true position. The major and minor axis standard deviations of these distributions were (5,3) for robot 1 and (3,1) for robot 2. Robot 1 reports a mean of (12.34, 9.02) and robot 2 reports a mean of (9.90, 11.69). In Figure 4, the distribution resulting from the single measurements by robot 1 and robot 2 are shown at left. The resulting merged distribution is shown in at right. The narrowing of the distribution indicates that implied uncertainty (standard deviations) is reduced, and the mean is more accurate relative to the actual target position. The merged mean is (9.97, 9.57), with major and minor axis standard deviations (0.89, 0.49).

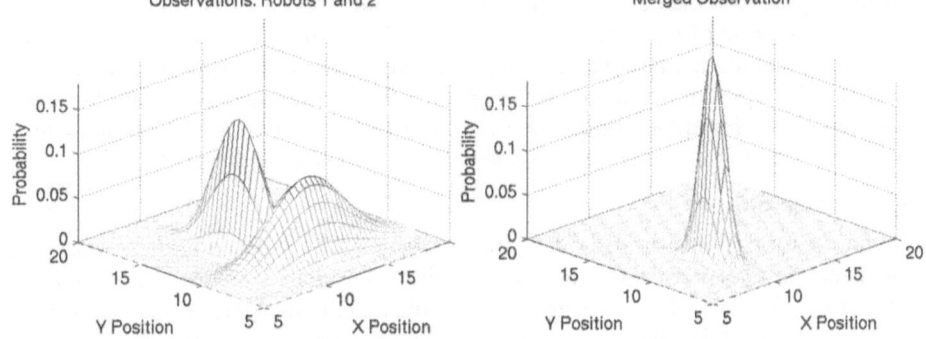

Figure 4.   Left: The individual distributions to be merged.   Right: Resulting merged distribution, with reduced error and higher accuracy in the mean.

4.   Validation On Robots

4.1.   Hardware Platform

The hardware platform used is a modified Cye robot, an inexpensive and commercially available platform (Figure 5). This platform consists of a drive section and a trailer. The drive section uses differential drive with two wheels and is equipped with a front bump sensor; the trailer is passive. On board the Cye is a motor controller processor. High level commands and image processing algorithms are implemented in C and Java on a Pentium 266 (running linux) using *TeamBots* [2]. A wide field-of-view NTSC video camera provides sensory input. *CMVision* performs color analysis and color blob detection and merging [2]. Robots communicate with wireless ethernet.

Camera calibration was conducted at two levels. First, *Flatfish* [14] determined the parameters describing the aberrations of the lens. These parameters enable mapping from pixel location to points in three-space.

Figure 5.   Enhanced Cye Robot platform.

A second calibration step characterizes systemic errors. Targets are placed at a set of fixed distances and angles relative to the robot and the distance and angle calculated by the vision system is recorded. Comparing measured distance versus actual distance provides a mean bias as a function of measurement distance. After correcting measurements for this bias, proportional errors are determined.

A histogram of these sensing errors determined that the corrected distances are distributed about actual distance approximately normally. From these errors, a standard deviation in percent distance can be directly determined. A similar process was completed for angle, though no bias correction was conducted. These deviation functions are used to compute parameters of the observation distributions.

4.2. Assumptions

Several assumptions are implicit in this approach. In addition to assuming independent, Gaussian sensor errors, the robot coordinate frames are assumed to be coincident. Without this, data are incompatible and the merging is meaningless.

We do not take into account robot positional uncertainty in the generation of target location distributions; our localization is not fully implemented at this time. Once determined, robot positional uncertainty can be incorporated by encoding it in object positional uncertainty. Merging the measurement uncertainty distribution with the robot's position distribution translated to the same mean spreads the object's position to accommodate both measurement and localization uncertainty.

Several additional assumptions were introduced for simplicity in experimentation and camera calibration. First, the camera parameter calibration assumes that objects are at a known height from the ground plane; unknown objects are therefore assumed to be on the ground plane. This reduces the transformation from three dimensions to two. This is not a highly restrictive limitation, as common obstacles, agents, landmarks, (etc) in environments are generally on the ground plane. Second, objects of interest are assumed to be unique in order to avoid the necessity of solving the association problem.

4.3. Experimental Setup

An experiment was devised to directly test this approach to distributed sensing and merged distributions without complications due to motion and positional uncertainty. Three stationary robots sequentially locate a stationary object at several pre-determined points (Figure 6). Robots share observations and compute resulting merged position estimates for each point. Computations were separately conducted on pairs of data points. Thus, accuracy of single-robot measurements can be directly compared to the accuracy obtained by combining data from two and three robots.

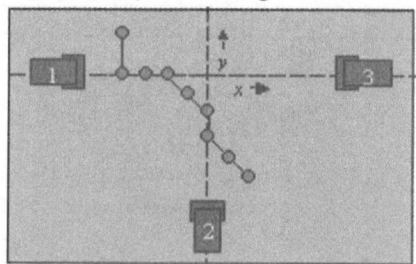

Figure 6. Three robots observe a ball at several fixed locations (shown as circles on line) and combine observations into single position estimates.

4.4.   Experimental Results

Example experimental results are shown graphically. Figure 7 shows the results of successive merging compared to the actual trajectory. The top left compares robot 3's observations (largest errors at the greatest distances from the robot). The top right similarly compares the result of merging two robots' observations. The estimate resulting from merging three robots' measurements is shown at the bottom.

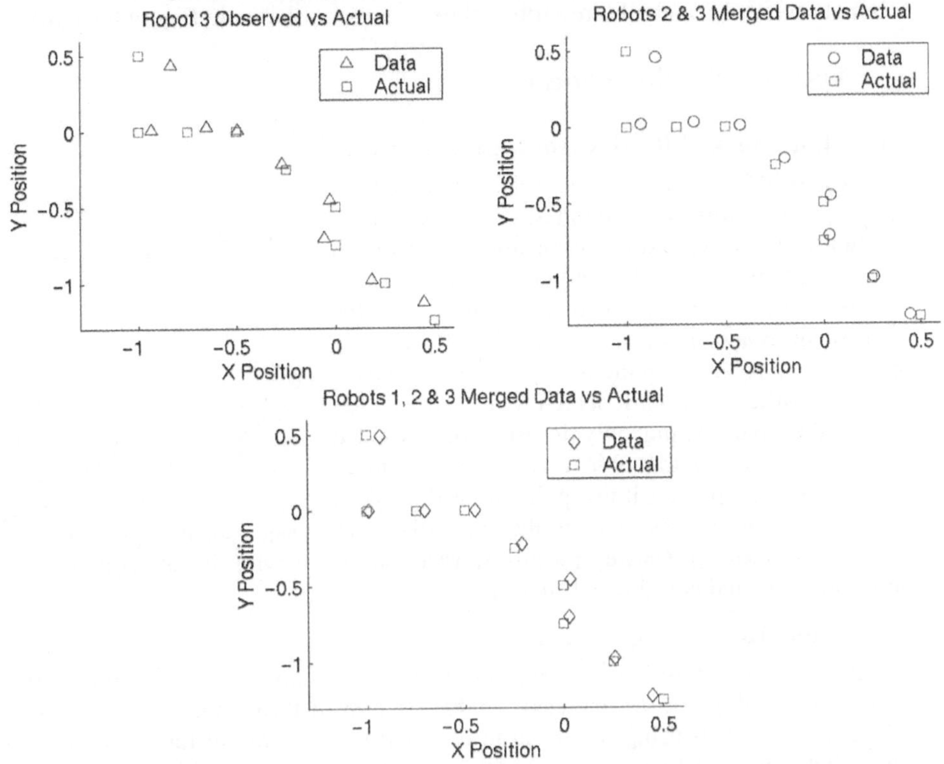

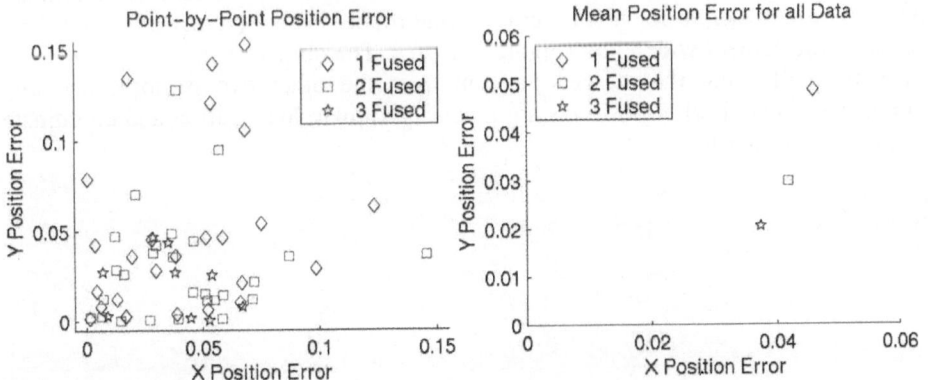

Figure 7.   Reported data versus actual ball position for robot 3 (top left), robots 2 and 3 merged estimates (top right), and robots 1, 2, and 3 merged estimates (bottom).

Figure 8. Left: Position error in x and y for each measurement. Each merging lowers position error bounds and reduces outlier frequency. Right: Mean position error in x and y over all single observations, all 2-observation merges, and all 3-observation merges.

Point-by-point trajectory errors are compared for all single-robot, two-robot, and three-robot measurements in Figure 8 (left). While individual trajectories are sometimes accurate at single points (in fact, occasionally more accurate than combined data), the consistency of accuracy in combined results is absent in the single-robot trajectories. Additionally, outlier frequency is reduced. This is best characterized by plotting the mean error of all single-robot observations, all two-robot observations, and all three-robot observations, as shown in Figure 8 (right).

5.   Test Applications and Results

5.1.   Location and Retrieval of Unseen Targets

This test exhibits the agents' increased effective field of view and ability to function with merged target positions. Initially, a robot is positioned so that it can see, but not reach, the target. Another cannot see the target, even with camera panning, but the path to the object is clear (Figure 9). By sharing information, robot 1 immediately obtains a target position without random search and successfully locates the object using only information provided by robot 2. Once the object is located, robot 1 reaches and manipulates it using the merged position provided by both robots. Due to the small distances traveled from known starting positions, assumptions on localization and coordinate frames hold.

Figure 9.   Robot 1 initially cannot see the target. Robot 2 provides initial target location.

5.2.   Blind Robot Target Tracking

In this experiment, three robots are positioned around a target area (for convenience, at relative headings of 90 degrees). A ball is moved throughout the area, and all robots track the ball using the position obtained by merging all three observations. The robots are able to track the target in most cases, even at higher speeds, and always quickly recover lost objects. Even when the target travels along the line of sight of a single robot (diminished accuracy in the depth dimension), the additional point of view make up for this accuracy. One robot is subsequently blindfolded by covering the camera with a box (Figure 10, left). The ability of the blinded robot to track the ball using the merged position from the other two is not observably diminished. Fixed robot positions are precisely known; localization and coordinate frame assumptions hold.

Figure 10. Left: A blind robot (left) can track targets by merging data provided by other robots. Right: A soccer test. Attacker (right) attempts to score on blue defender (left).

5.3. Robot Soccer

This approach to distributed sensing was applied to the CMU Hammerheads middle-sized robot soccer team (Figure 10, right). Robots transmit the position of the ball, if visible, so that it can be combined with all other current observations. This provides robots far from the ball with more accurate positions and allows robots to quickly locate an unseen ball. Conflicting observations (means differing by more than 2 standard deviations) were not merged to prevent false (or multiple) targets and data collected in incompatible coordinate frames from resulting in confusion.

The 2000 Hammerheads robots were localized entirely by odometry. Differences in coordinate frames arise from odometry drifts over time. As a result, the primary impact of distributed sensing was to provide a starting point for robots to locate lost balls. Despite coordinate frame discrepancy, frames were generally coherent enough that the camera's wide field of view allowed searching robots to immediately locate an unobstructed ball by looking at a provided target position.

6. Conclusions

We present a method to improve target position estimates by fusing data from two or more robot agents. This approach, based on Bayes' Rule and Kalman filter theory, implements real-time sensor data fusion on a reactive multi-robot system for many different applications. The successful ability to fuse these statistical measurements and the ability to receive position estimates of targets not visible allows our robots to quickly acquire targets and to more accurately estimate object position. While this work uses only vision for sensing, the approach can be applied to any sensor or suite of sensors which can be modeled by approximately Gaussian distributions.

This approach to distributed sensing and information sharing is very promising based on the applications presented here: unseen target location, accurate target acquisition and manipulation, and robot soccer. However, several extensions of this work are necessary for practical implementation. Even in well-localized systems, disparity between coordinate frames can arise and must be accommodated. Autonomously determining the relative transformation between coordinate frames using sensors will be investigated. Additionally, the accommodation of robot positional uncertainty will be incorporated into the target position distributions, as described previously. Lastly, it may be possible to implement a pixel-to-world coordinate transformation that does not assume that objects are at a known elevation, but this would require the development of a new method of camera calibration.

Acknowledgements

The authors wish to acknowledge the contributions and support of Jim Bruce, Scott Lenser, Kevin Sikorski, Hans Moravec, Manuela Veloso, and DARPA's Programs in Mobile Autonomous Robot Software and Control of Agent Based Systems.

References

[1] W. Anderson and R. Duffin. *J Mathematical Analysis Applications*, 26:576, 1969.
[2] T. Balch. TeamBots. http://www.teambots.org, 2000.

[3] R.G. Brown and B.R. Donald. Mobile robot self-localization without explicit landmarks. *Algorithmica*, 26(3/4):515-559, 2000.

[4] J. Brusey et al. RMIT Raiders. In: Veloso, Pagello, and Kitano (eds) *RoboCup-99: Robot Soccer World Cup III.* Springer-Verlag, Berlin, pages 741-744, 2000.

[5] A. Cai, T. Fukuda, and F. Arai. Information Sharing among Multiple Robots for Cooperation in Cellular Robotic System. In: *Proc IROS 97*, pages 1768-1773, 1997.

[6] A. Cai, T. Fukuda, F. Arai, and H. Ishihara. Cooperative Path Planning and Navigation Based on Distributed Sensing. In: *Proc 1996 IEEE Int Conf Robotics and Automation*, pages 2079-2084, 1996.

[7] P. Costa et al. 5dpo-2000 Team Description. In: Veloso, Pagello, and Kitano (eds) *RoboCup-99: Robot Soccer World Cup III*, Springer-Verlag, Berlin, pages 754-757, 2000.

[8] G. Dissanayake, H. Durant-Whyte, and T. Bailey. A Computationally Efficient Solution to the Simultaneous Localisation and Map Building (SLAM) Problem. In: *Proc 2000 IEEE Int Conf on Robotics & Automation*, pages 1009-1014, 2000.

[9] D. Fox, W. Burgard, H. Kruppa, and S. Thrun. Collaborative Multi-Robot Localization. In: *Autonomous Robots on Heterogeneous Multi-Robot Systems, Special Issue*, 8(3), 2000.

[10] J.S. Gutmann et al. Reliable self-localization, multirobot sensor integration, accurate path-planning and basic soccer skills: playing an effective game of robotic soccer. In: *Proc Ninth Int Conference on Advanced Robotics*, pages 289-296, 1999.

[11] J.S. Gutmann et al. The CS Freiburg Robotic Soccer Team: Reliable Self-Localization, Multirobot Sensor Integration, and Basic Soccer Skills. In: Asada and Kitano (eds) *RoboCup-98: Robot Soccer World Cup II*, Springer-Verlag, Berlin, pages 93-108, 1999.

[12] R. Hanek, T. Schmitt, M. Klupsch, and S. Buck. From Multiple Images to a Consistent View. To appear in: Stone, Balch, and Kraetzschmar (eds) *RoboCup-2000: Robot Soccer World Cup IV*, Springer-Verlag, Berlin, 2001.

[13] T.D. Larsen, M. Bak, N.A. Andersen, and O. Raven. Design of Kalman filters for mobile robots; evaluation of the kinematic and odometric approach. In: *Proc 1999 IEEE Int Conf Control Applications, Vol 2*, pages 1021-1026, 1999.

[14] H. Moravec. Robust Navigation by Probabilistic Volumetric Sensing. http://www.ri.cmu.edu/~hpm/project.archive/robot.papers/2000/ARPA.MARS.reports.00/Report.0001.html, 2000.

[15] L. Moreno, J.M. Armingol, A. de la Escalera, and M.A. Salichs. Global integration of ultrasonic sensors information in mobile robot localization. In: *Proc Ninth Int Conf Advanced Robotics*, pages 283-288, 1999.

[16] D. Nardi et al. ART99–Azzurra Robot Team. In: Veloso M, Pagello E, Kitano H (eds) *RoboCup-99: Robot Soccer World Cup III*, Springer-Verlag, Berlin, pages 695-698, 2000.

[17] G. Petryk, and M. Buehler. Robust estimation of pre-contact object trajectories. *Robot Control 1997, Vol 2*, pages 793-799, 1997.

[18] D. Rembold, U. Zimmermann, T. Langle, and H. Worn. Detection and handling of moving objects. In: *Proc 24th Annual Conf IEEE Ind Electron Soc, IECON '98, Vol 3*, pages 1332-1337, 1998.

[19] J.Z. Sasiadek and P. Hartana. Sensor data fusion using Kalman filter. In: *Proc Third Int Conf Information Fusion, Vol 2*, pages 19-25, 2000.

[20] R. C. Smith and P. Cheeseman. "On the Representation and Estimation of Spatial Uncertainty." In: *The International Journal of Robotics Research*, 5(4):56-68, 1986.

[21] A.W. Stroupe, M.C. Martin, and T. Balch. Merging Probabilistic Observations for Mobile Distributed Sensing. Carnegie Mellon University Techical Report CMU-RI-TR-00-30, 2000.

[22] C.Y. Tang, Y.P. Hung, S.W. Shih, and Z. Chen Z. feature-based tracker for multiple object tracking. In: *Proc National Science Council, Republic of China, Part A*, 23(1):151-168, 1999.

[23] H. Wang, C.S. Chua, and C.T. Sim. Real-time object tracking from corners. *Robotica*, 16(1):109-116, 1999.

Principled Communication for Dynamic Multi-Robot Task Allocation

Brian P. Gerkey and Maja J Matarić
Robotics Research Labs
University of Southern California
Los Angeles, CA, USA
{bgerkey|mataric}@cs.usc.edu

Abstract: In the pursuit of an efficient cooperative multi-robot system, the researcher must eventually answer the question "how should robots communicate?"; a natural way to attack this question is to decompose it into three simpler corollaries: "what should robots communicate?", "when should they communicate?" and "with whom should they communicate?". In this paper, we propose answers to these questions in the form of a general framework for inter-robot communication and, more specifically, advocate its use in dynamic task allocation for teams of cooperative mobile robots. We base our communication model on *publish/subscribe messaging* and validate our system by using it in a tightly-coupled multi-robot manipulation task and a loosely-coupled long-term experiment involving many robots concurrently executing different tasks.

1. Introduction

We posit that the long-term goal for collective robotics is to have a decentralized collection of cooperative robots that are anonymously taskable. That is, irrespective of which and how many robots are available and what if any differences exist among them, the group should be viewed as a pool of resources to whom work assignments, or tasks, can be delegated. For a variety of practical reasons, the robots must be individually autonomous, and so the delegation of tasks must be accomplished in a distributed manner such that this autonomy is ensured. Further, we are striving toward *long-term* autonomy, in which, for long periods of time (days or weeks or even longer), each robot is a self-sufficient entity either executing a task or awaiting a new assignment; human intervention should be required only in severe failure situations.

In our approach, each robot has a set of individual resources, or capabilities, which it can use to achieve tasks. These capabilities, which could be physical, like having a gripper, or ethereal, like having knowledge of the topological layout of a building, can and do vary over time. Thus, even if two robots are initially identically physically configured, they will, through experience and interaction with the world, become heterogeneous with respect to both their capabilities and their fitness for any given task. Each robot might have many different resources, and thus be capable of a variety of tasks. Also, any robot might fail at any time in a variety of ways. The problem we address here is how to intelligently allocate tasks, both simple and complex, to such a

D. Rus and S. Singh (Eds.): Experimental Robotics VII, LNCIS 271, pp. 353–362, 2001.

dynamic, heterogeneous group of physically-embodied individuals working in an unpredictable environment.

Toward this end, we have implemented and tested a novel task allocation system which we call MURDOCH. This system is based on a principled, resource-centric, *publish/subscribe* communication model and makes extensive (but efficient) use of explicit inter-robot communication. Since our goal is not to exploit resources in a globally optimal fashion, but rather to investigate practical methods for allocating tasks to groups of autonomous and heterogeneous physical robots, we have built MURDOCH as a completely distributed system. As such, it offers a distributed approximation to a global optimum of resource usage which is equivalent to an instantaneous greedy scheduler.

We have tested this system in two very different domains: a tightly-coupled multi-robot physical manipulation task (see Section 6.1) and a loosely-coupled many-robot experiment in long-term autonomy (see Section 6.2).

2. Related Work

There has been a great deal of work on architectures for automated cooperation among agents. The Open Agent Architecture (OAA) [1] and RETSINA [2] are two such architectures. In both, the focus is on providing a maximally general environment in which very different agents, such as user interfaces and legacy database management systems, can interact and coordinate. Since we are concerned specifically with task allocation for physical mobile robots, we do not require the overhead (such as ontology specifications) that allows these systems to be so general. There is a component of task allocation among their agents, although the tasks are purely informational, rather than physical, in nature. Both architectures accomplish task allocation through a broker (called *facilitator* and *matchmaker*, respectively) which matches new task requests with agents that have previously advertised relevant capabilities. In MURDOCH, we completely distribute the matchmaking process and thus do away with the centralized broker. A different approach is taken in STEAM [3]; this model-based system relies on each agent's explicitly tracking the actions of both itself and its team and then applying *joint intention* principles to the resulting world model in order to make decisions. While this approach has been validated on software agents, it is not clear how, in a physically embodied world, an autonomous agent could gather and interpret the necessary data to perform the requisite tracking and modeling.

In stark contrast to these very general architectures are ALLIANCE [4] and BLE [5]. These two special-purpose systems provide coordination among multiple robots by having the execution of a behavior on one robot directly inhibit the execution of the same behavior on another. Although it is not strictly necessary, in practice robots in these systems share an identical internal (behavior-based) structure and the inhibition relationships between the relevant behaviors are hand-coded. Further, while a system built in either ALLIANCE or BLE may be well-suited to performing a single task, neither architecture has been demonstrated as being easily retaskable.

Whereas most of the architectures described above fundamentally communicate by point-to-point message passing, an alternate method is the broadcast-oriented blackboard model [6]. The standard blackboard system has many similarities to the publish/subscribe system used in MURDOCH. In fact, one might say that a publish/subscribe system is an instantaneous blackboard in which no central state need ever be kept, with the tradeoff that the past history of the blackboard is lost. This is an acceptable tradeoff in our domain, since old information quickly becomes irrelevant in physical, real-time systems.

In order to allocate a given task, we use a simple auction in which each capable agent evaluates its own fitness for the task. A similar approach for deriving supply chains in a producer/consumer problem is described in [7].

3. The Communication Model

In implementing distributed control systems for teams of robots, researchers typically resort to *ad hoc* communication strategies. These specialized strategies are often implemented as hand-crafted, task-specific communication graphs. For example, in [4] and [5], communication channels are explicitly created among individual behaviors on the different robots. While this model is well-suited to the design of some special-purpose, single-task systems, it has not been demonstrated for the general case of controlling a large population, in which the members dynamically form teams to accomplish different tasks as they are presented to the system.

As an alternative to such special-purpose systems, we propose a principled communication model based on *publish/subscribe messaging*. Also known as *dissemination-oriented communication* and the *announce/listen metaphor* [8], publish/subscribe messaging is a commercially viable ([9]) message delivery paradigm that has been studied in a distributed systems context ([10]).

The unifying concept of publish/subscribe systems is that *messages are addressed by content rather than by destination*. This idea, often called *subject-based addressing*, is used to divide the network into a loosely-coupled association of anonymous data producers and data consumers. A data producer simply tags a message with a subject describing its content, and "publishes" it onto the network; any data consumers who have registered interest in that subject by "subscribing" will automatically receive the message. Data producers need not have any knowledge of which consumers, if any, are receiving their messages, and vice versa. This kind of communication represents a fundamental departure from the traditional communication model, in which each message is unicast from the sender to a single receiver at a specific known destination. As an aside, we have tailored this idea slightly so that when a message is published, it is addressed to a set of subjects, rather than just one. A data consumer will receive a message if the subjects in the message comprise any subset of the consumer's current subscription list. Although we have not optimized the subset matching algorithm in our implementation, others have investigated the topic [11].

4. Murdoch

Several important but hopefully admissible assumptions were made in the design of MURDOCH. First, the robots have a reasonable communication system. By "reasonable", we mean that, although the communication paths among the robots need not be reliable, they should work most of the time and they should easily provide the meager bandwidth required by our system. At a higher level, we assume that the robots share a common vocabulary. Thus, when one robot communicates about a task called **push-box-on-the-right-end**, the others know what that means. Now, this assumption does not imply homogeneity, for the mapping from the task name **push-box-on-the-right-end** to an internal control system for actually achieving it could be different on each robot, either out of necessity (e.g., one robot has legs and the other wheels), or by design (e.g., we are comparing different pushing algorithms simultaneously). We also expect the robots to be fundamentally cooperative and honest. For example, when evaluating a fitness metric for a potential task, each robot faithfully reports its score and gracefully withdraws when beaten. Of course, since the fitness metric is likely based on the robot's noisy and imperfect sensory input, the reported score may not actually be "correct" (see Section 4.3).

4.1. Subject Namespace

The first step in creating a publish/subscribe system is designating the semantics of the subject namespace. Analogous to deciding the layout of a database, the interpretation of subjects will heavily influence the rest of the system. In MURDOCH, since we are allocating tasks among a group of potentially heterogeneous robots, we use subjects to represent their "resources". Resources can be physical devices (e.g., **camera**, **gripper**, **sonar**), higher-level capabilities (e.g., **mobile**, **door-opener**) or abstracted notions of current state (e.g., **idle**, **have-puck**, **currently-pushing-box**). Thus, if we have a task that involves going to some physical location and observing it, we can reach the appropriate robots by addressing a message to the set {**mobile camera idle**}. Since messages are addressed to subjects and subjects represent resources, all interrobot communication will necessarily be resource-centric, which we believe to be fundamental in achieving our goals. The robots never interact with each other by name and in fact have no explicit knowledge of each others' existence; rather they only communicate about tasks and all messages are addressed in terms of the resources required to do the tasks.

4.2. Task Structure

Since we are concerned with task allocation, we must of course choose a representational structure in which we can describe (and the robots can understand) a given task. Although we do not currently use the Hierarchical Task Network (HTN) formalism [12] (we will soon transition to an HTN representation), the flexible hierarchical scheme that we do use is quite similar. Much like HTN's *compound tasks*, we have *high-level tasks*; these tasks are sufficiently high-level that a human user can readily understand and reason about them. High-level tasks can be (and often are) composed of what we simply call *tasks*; a task is an atomic unit of computation and control that a single robot will execute. Our tasks are akin sometimes to HTN *compound tasks* and sometimes to HTN *atomic tasks*, since the execution of our tasks can involve allocating yet more tasks.

Since in this work we are investigating task allocation, not task decomposition, we delegate to the task designer the work of defining a high-level task in the form accepted by MURDOCH[1]. However, it is important to note that in practice the tasks we want the robots to achieve are structurally simple enough (typically just a single layer of decomposition even for a tightly-coupled task) that the designer is not mired in exploration of a vast space of combinations.

We simplify the job of writing the single-robot tasks, like **push-right-end-of-box** through the use of basis behaviors [14]. In fact, we follow a behavior-based model ([15], [14]) for controlling all the sensors and actuators on our robots. However, we do not require that the robots be controlled in a behavior-based manner; any control architecture that allows for the starting and stopping of task execution will fit within our model. Our behaviors are low-level controllers, such as **avoid-obstacles-with-sonar** and **visual-servo-to-color** and are each implemented as a separate operating system thread. A task, then, is defined in terms of a concurrent instantiation of a collection of properly parameterized behaviors. For example, we can define **push-right-end-of-box** as particular instances of **avoid-obstacles-with-sonar** and **visual-servo-to-color** (our box is brightly colored) and we can define **goto-goal** as a differently parameterized instance of **avoid-obstacles-with-sonar** paired with **servo-to-location**. While writing these individual behaviors is not a trivial matter, we believe that having a library of robust and parameterizable basis behaviors is well worth the work involved as the result is a natural way of specifying tasks.

4.3. Negotiation

At the heart of MURDOCH lies a simple distributed negotiation protocol that allocates tasks one by one via a sequence of one-round auctions. The process is triggered by the introduction of a task to the system. The task could be introduced in many ways, including a human user, a cron-style alarm, or an already ongoing task. In every case, the first step is for an agent (which is working on behalf of a user, alarm, or task) to publish an announcement message for the task. This message contains details about the task, such as its name, length, and a new subject on which to negotiate it. The announcement message is addressed to the set of subjects which represent the resources required to execute the task; thus only those robots currently capable of the task will receive the message. Of course, there may be more than one capable robot available for a single task and we need a method for deciding among them; in fact, this decision process is the very basis of achieving sensible task allocation. For this purpose, we employ metric functions, or simply *metrics*.

Metrics can take many forms, with the restriction that each one, when evaluated in the context of a specific robot, should return a scalar "score" representing that robot's fitness for the task. A metric is usually defined as some function of the robot's current state, although in general, a metric could perform any arbitrary computation, including inter-robot communication. As an example, if the task under consideration is to go to a certain location and pick up an object, one possible metric is to compute the Cartesian distance from the robot's current position to the goal position, with a shorter distance being

---

[1]Alternatively, an off-line planner, such as the one described in [13], endowed with knowledge of the preconditions, postconditions and interdependencies of the tasks involved could be employed here. Note that the planner would not be deriving specific motion sequences, but rather overall task structure; the behaviors themselves generate *in situ* trajectories.

better. Multiple metrics can be defined for a single task, with the final score being some combination of the individual scores; we have experimented with combining metrics through both simple sums and weighted sums, the latter of which provides for a prioritization of metrics. It is important to note that metrics, being functions of each robot's own sensor data, may not accurately represent the current state of the robots, possibly resulting in a non-optimal allocation of the task. Since finding an optimal allocation would require gathering global data, guaranteeing its accuracy, and centralizing control, we find our metric-based distributed approximation to be a parsimonious alternative.

Upon receipt of a task announcement message, each capable robot participates in a one-round auction, determining its fitness by evaluating the indicated metrics and broadcasting its score back to the others. Everyone immediately knows who is best suited (i.e., who won the auction) and so the losers can go back to listening for new tasks while the winner begins the task. Since each task will always be claimed by the most capable robot at the time, MURDOCH acts as an instantaneous greedy task scheduler. Thus we suffer from the well-known problems of greedy algorithms; they are manifested in our domain as situations in which, although sufficient resources exist to achieve a given set of tasks, the order in which they are presented causes resources to be exploited in a non-optimal manner such that not all the tasks are actually achieved. Centralized broker and matchmaker systems avoid this pitfall by analyzing concurrent tasks before allocating them; of course, this kind of planning will not help in what we view as the common case in which single tasks are input stochasticly from some outside source, such as a human user. As a distributed alternative, we are investigating the inclusion of simple optimizations in the metrics themselves; for example, a specialist who has few resources to offer could always increase its own score by some amount such that it will defeat a generalist who has more resources and might be put to better use on a later task.

5. Experimental Platform

MURDOCH is implemented on our group of seven ActivMedia Pioneer 2-DX mobile robots. The robots are equipped with a variety of sensors and each is configured differently. The sensors include: ultrasonic rangefinders, laser rangefinders, compasses, color cameras, and tactile bumpers. With regard to actuators, each is equipped with two drive wheels and a passive caster (steering is differential), some have grippers, and some have pan-tilt-zoom camera units.

Internally, each robot houses a Pentium-based computer running Linux; MURDOCH runs on this platform and communicates with the robot via a server[2] developed at the USC Robotics Research Labs. Inter-robot communication is provided by way of wireless Ethernet; the topology is such that every machine (robot or not) on the network can communicate freely with every other machine.

---

[2]The server, Player [16], runs on the Pioneer's computer and offers a unified TCP socket-based interface to all the sensors and actuators attached to the robot. Player is freely available under the GNU Public License. Consult http://fnord.usc.edu/player for more details.

 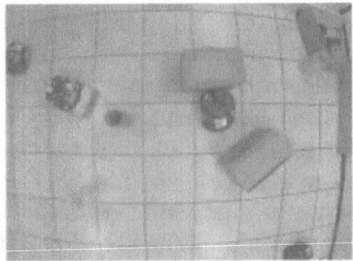

Figure 1. *Our two experimental task domains. In the cooperative box-pushing task (left), the robot on the left (the watcher) is coordinating the actions of the other two robots (pushers). In the other experiment (right), a group of robots concurrently execute a variety of single-agent tasks, including box-pushing and target-tracking.*

6. Experimental Task Domains

We validated our approach in two different task domains, one featuring tightly-coupled coordination of multiple robots engaged in a cooperative box-pushing task, and the other loosely-coupled coordination of many robots executing a collection of independent single-robot tasks. For more information about these tasks, and videos of experiments, consult: http://robotics.usc.edu/~agents/projects/pub-sub.html.

6.1. Tightly-Coupled Task Allocation

Our experimental setup (see Figure 1) is as follows: one large box must be moved from a start location to a goal location by the collective pushing efforts of a group of robots. The robots do not have manipulators with which they could grip the box, and the box is large enough and characterized by sufficiently complex physical properties that a single robot cannot predictably manipulate the box by pushing in the middle of a side. We chose a cooperative pushing algorithm inspired by the distributed manipulation work of [17] and [18]. We have two pushers and a single watcher, equipped with a laser rangefinder and color camera. The watcher drives in front of the box and toward the goal, which it finds with the camera, all the while monitoring the orientation of the box with the laser. Depending on the current orientation of the box, the watcher generates the appropriate pushing tasks to rotate and translate the box as necessary. It is important to note that the pushers are not inherently "handed"; they have no preference for pushing a particular end of the box and are in fact both executing the same control program. When a new pushing task is announced, each available pusher executes a vision-based metric that reflects how well-positioned it is to achieve the particular task.

After verifying that the simple case of continuous straight-line pushing was successful, we stressed the system in various ways. First, we moved the goal far enough off to one side that simultaneous action by both pushers would have been insufficient to move the box to the goal; the box had to be rotated as well. In this situation, the watcher senses that the box is out of alignment with respect to the goal and consequently allocates only the appropriate single pushing task which would correct the alignment; they then return to the simple case of straight pushing. We also conducted experiments with a moving goal and even changing the watcher task dynamically to a different goal. Since the watcher algorithm involves no path planning, but rather constantly reacts to current sensor data, moving and changing goals do not constitute a special case in the algorithm. We were able to reliably lead the whole group (robots and box) along many sufficiently widely curved paths around the lab.

In order to test the fault-tolerance capability of our box-pushing system, we purposefully introduced failure by turning off one pusher while the robots were in the middle of a coordinated box movement. With the removal of one potential pusher, nothing changes with respect to the algorithm that the remaining robots follow; only the observed behavior changes. Since the remaining pusher is the only one capable of new pushing tasks, it continues pushing but switches sides many times as new corrective pushing tasks are allocated by the watcher, and eventually is able to see-saw the box to the goal. Thus, after a robot failure, the system exploits the other resources and degrades naturally to a less efficient pushing method. Of course, when the remaining resources are insufficient for the task, the task is deemed unachievable and an informative error message is returned to the user. However, if resources again become available, they are immediately put to use. We verified this behavior by turning off one robot, letting the other continue for a while, then reintroducing the failed robot on the opposite side from which it had started; the new pusher was automatically brought into the team, and the pushers had swapped roles. In summary, the system properly handled all failure classes for which we tested.

6.2. Loosely-Coupled Task Allocation

In the pursuit of long-term autonomy, we also experimented with MURDOCH in a very different task domain. In this domain (see Figure 1), we ran a large group, consisting of 6 heterogeneous robots and 1 desktop computer connected to an overhead camera, over a long period of time (approximately 3 hours), all the while randomly injecting new tasks of random lengths into the system. Each machine was controlled by a copy of the same program; this program simply queried its host for the list of currently available devices, then made the proper resource subscriptions. For example, the robots equipped with both cameras and lasers subscribed to {camera laser mobile}, while the desktop computer only subscribed to {overhead-camera}. We used four different tasks. The first, **object-tracking**, requires camera and mobile. The task is to find and track from a safe distance a certain colored object. The second, **sentry-duty**, requires camera, laser, and mobile. The task is to find a certain colored object, then turn about and remain still, watching for any motion with the laser and setting off an intruder alarm if motion is detected. Our third task, **cleanup**, requires camera, bumpers, and mobile. The task for the robot is to find each small box of a certain color and use its tactile bumpers to push the box to the edge of the room, thereby cleaning the room. The final task, **monitor-object**, requires only overhead-camera. The task is to monitor the positions of various colored objects, such as boxes and robots, from the overhead view, and log the information for later review.

Each robot also runs a battery-monitoring behavior that checks the current charge whenever the robot is idle, in between tasks. At that time, if the battery is low enough, the robot will unsubscribe from all subjects (thereby removing itself from consideration for future tasks) and go to a clearly marked charging station. After charging for some time, the robot is freed to reenter the experiment.

We observed the following system behavior. Over the course of 3 hours, the 7 agents (6 robots and 1 computer) successfully executed 49 tasks and returned to charge their batteries 12 times. The same control program executed on each robot for the length of the experiment, with robots periodically idle (only executing passive collision avoidance), executing some task, or charging. Some of the randomly generated tasks were unachievable due to a lack of resources, because all the capable robots were either charging or otherwise engaged. In these situations, an error was returned, suggesting that the task be reintroduced later. We are currently implementing a smarter user-agent that will persistently reintroduce unachievable tasks on the user's behalf (following a randomized exponential backoff algorithm) until they are allocated. We also purposefully induced failures in this domain. Since the tasks are all single-robot jobs, a robot failure meant that the task had only to be automatically reallocated to another robot. When resources permitted this reallocation, it was done as expected; otherwise the task was deemed unachievable.

7. Conclusions

In this paper, we presented the details of a dynamic task allocation system based on a principled publish/subscribe messaging model that requires all inter-robot communication to be resource-centric. This system was implemented and tested on physical robots in both a short-term tightly-coupled task domain and a long-term loosely-coupled task domain. We demonstrated how the system is extremely reactive to changes in the environment, including abrupt failures of robots and random introduction of new tasks. MURDOCH is completely distributed, with no single point of congestion or failure, with the tradeoff that its task allocation solutions are always greedy.

As part of continuing work in the box-pushing domain, we will increase the granularity of our control of the box by dividing the box's alignments into more categories and parameterizing each pushing task by both time and velocity. Our hope is to successfully transport the box in a tightly-constrained corridor environment. As for the loosely-coupled task domain, we are currently designing a more ambitious experiment in which we will run the robots for much longer, possibly days, and record relevant data for an objective performance analysis of the system.

8. Acknowledgments

The research reported here was conducted at the Interaction Lab, part of the Robotics Research Lab at the University of Southern California Computer Science Department. The work is supported by Office of Naval Research Grants N00014-00-1-0140 and N0014-99-1-0162, Jet Propulsion Laboratory Contract No. 1216961, and DARPA Grant DABT63-99-1-0015. We thank Richard Vaughan and Andrew Howard for important insights concerning this work.

References

[1] David L. Martin, Adam J. Cheyer, and Douglas B. Moran. The open agent architecture: A framework for building distributed software system. *Applied Artificial Intelligence*, 13(1):91–128, Jan–Mar 1999.

[2] Katia Sycara, Keith Decker, Anandeep Pannu, Mike Williamson, and Dajun Zeng. Distributed intelligent agents. *IEEE Expert*, 11(6):36–46, December 1996.

[3] Milind Tambe. Agent architectures for flexible, practical teamwork. In *Proceedings of the Natl. Conf. on Artificial Intelligence (AAAI)*, Providence, Rhode Island, July 1997.

[4] Lynne E. Parker. Alliance: An architecture for fault-tolerant multi-robot cooperation. *IEEE Transactions on Robotics and Automation*, 14(2), 1998.

[5] Barry Brian Werger and Maja J Matarić. Broadcast of local eligibilty for multi-target observation. In *Proceedings of the Intl. Symp. on Distributed Autonomous Robotic Systems (DARS)*, Knoxville, Tennessee, October 2000.

[6] Daniel D. Corkill. Blackboard systems. *AI Expert*, 6(9):40–47, September 1991.

[7] William E. Walsh and Michael P. Wellman. A market protocol for decentralized task allocation. In *Proceedings of the Intl. Conf. on Multi Agent Systems (ICMAS)*, Paris, France, July 1998.

[8] Steven McCanne. Scalable multimedia communication with internet multicast, light-weight sessions, and the mbone. Technical Report CSD 981002, UC Berkeley, March 1998.

[9] Arvola Chan. Transactional publish/subscribe: The proactive multicast of database changes. In *Proceedings of ACM SIGMOD Conf. of Management of Data*, Seattle, WA, June 1998.

[10] Guruduth Banavar et al. An efficient multicast protocol for content-based publish-subscribe systems. In *Proceedings of the Intl. Conf. on Distributed Computing Systems*, Austin, Texas, June 1999.

[11] Marcos K. Aguilera et al. Matching events in a content-based subscription system. In *Proceedings of the ACM Symposium on Principles of Distributed Computing*, Atlanta, Georgia, May 1999.

[12] Kutluhan Erol, James Hendler, and Dana S. Nau. HTN planning: Complexity and expressivity. In *Proceedings of the Natl. Conf. on Artificial Intelligence (AAAI)*, Seattle, WA, July 1994.

[13] Kutluhan Erol, James Hendler, and Dana S. Nau. UCMP: A sound and complete procedure for hierarchical task-network planning. In *Proceedings of the Intl. Conf. on Artificial Intelligence Planning Systems*, Chicago, IL, June 1994.

[14] Maja J Matarić. Behavior-based control: Examples from navigation, learning, and group behavior. *Journal of Experimental and Theoretical Artifical Intelligence*, 9(2–3):323–336, 1997.

[15] Ronald C. Arkin. *Behavior-Based Robotics*. MIT Press, Cambridge, MA, 1998.

[16] Brian P. Gerkey, Kasper Støy, and Richard T. Vaughan. Player robot server. Technical Report IRIS-00-392, Institute for Robotics and Intelligent Systems, School of Engineering, University of Southern California, November 2000.

[17] Bruce Donald, Jim Jennings, and Daniela Rus. Information invariants for distributed manipulation. *The Intl. Journal of Robotics Research*, 16(5):673–702, October 1997.

[18] Bruce Donald, Jim Jennings, and Daniela Rus. Minimalism + distribution = supermodularity. *Journal of Experimental and Theoretical Artifical Intelligence*, 9(2–3):293–321, 1997.

Progress in RoboCup Soccer Research in 2000

M. Asada[1], A. Birk[2], E. Pagello[3], M. Fujita[4], I. Noda[5],
S. Tadokoro[6] D. Duhaut[7], P. Stone[8], M. Veloso[9], T. Balch[9],
H. Kitano[10], B. Thomas[11]

[1]Adaptive Machine Systems, Osaka University, Suita, Osaka 565-0871, Japan
[2]AI Lab., Vrije Universiteit Brussel, 1050 Brussels, Belgium
[3]DFI, The University of Padua, I-35131 Padova, Italy
[4]DCL, Sony Corp., Tokyo 141-0001, Japan
[5]ETL, AIST, MITI, 305 Japan
[6]Computer & Systems Engineering, Kobe University, Kobe 657-8501 Japan
[7]LRP, 78140 VELIZY, France
[8]AT&T Labs - Research, Florham Park, NJ 07932, USA
[9]School of Computer Science, CMU, Pittsburgh, PA 15213, USA
[10]Computer Science Lab, Sony Corp., Tokyo 141-0022, Japan
[11]Bellarine Secondary College, Victoria, Australia

Abstract: In addition to researchers in AI and robotics, RoboCup attracts ordinary people, especially kids, high school and university students. Over 3000 people from 35 nations around the world have participated in RoboCup since the great success of the First Robot World Cup Soccer Games and Conferences, RoboCup-97 [1] held in conjunction with the Fifteenth International Joint Conference on Artificial Intelligence (IJCAI-97). Every year, the number of participating teams is increasing about 50%, that is, 35 teams in RoboCup-97, 64 teams in RoboCup-98 [2], and 90 teams in RoboCup-99 [3], and almost same number of teams in RoboCup-2000. Attendance in 2000 was impacted by the application of a new qualification process, difficulties for some of the European teams to travel to Australia, and by the addition of a European RoboCup competition, EURO-2000. This paper focuses on a discussion of the challenging research problems present in RoboCup and how they have been concretely addressed in RoboCup competitions in 2000.

1 Introduction

Robocup is a research challenge encourages investigation in the fields of robotics and artificial intelligence, with a particular focus on developing cooperation between autonomous agents in a dynamic multiagent environment [4]. By provid-

D. Rus and S. Singh (Eds.): Experimental Robotics VII, LNCIS 271, pp. 363–372, 2001.
© Springer-Verlag Berlin Heidelberg 2001

ing a well defined problem, RoboCup enables many research groups to cooperate and compete with each other over many years in the pursuit of a grand challenge.

RoboCup-2000 was held between August 27th and September 3rd, 2000, at the Melbourne Exhibition Center in Melbourne, Australia. 83 teams, including a total of about 500 people attended RoboCup-2000. In RoboCup-2000, much improvement in all completion leagues (F180, F2000, Legged, Simulation) were obtained, and new activities (RoboCup Jr. and RoboCup Rescue) were demostrated with many attendants. In the RoboCup workshop, four papers were nominated for a Challenge Award; one was award the RoboCup Engineering Challenge Award, another the RoboCup Scientific Challenge Award.

In the remainder of this article, we summarize each league and their activities.

2   F180

In the F-180 league, 16 teams from nine different nations competed this year. The usage of on-board vision by three teams is among the most interesting achievements of this year. The ViperRoos from the University of Queensland, Australia, the CIIPS Glory team from the University of Western Australia, and the 4 Stooges from the University of Auckland, New Zealand managed to integrate an on-board camera and on-board vision into the severe size-limits of the F-180 league, namely maximum 18 cm diameter and maximum 180 cm^2 floor area for a robot. The ViperRoos even managed to win 2:0 against a team with global vision.

The Big Red team from Cornel University, USA, repeated their last year's success and won the tournament. Instead of well controlled power, which was their road to success last year, they focused on skillful play. On the hardware side, they came up with two mechanical novelties in the F-180 league: omnidirectional drive and a dribbling-device. Building an omni-directional drive with in the size limitation of F-180 is challenging and Big Red seems to have come up with a novel mechanical solution to it. Especially the goalie could profit from this feature as it can defend the sidelines of the goal as well as the space ahead of the goal equally well. The second mechanical novelty from Big Red, the dribbling device, consists of a rotating bar in front of each robot. It imparts a backspin on the ball which therefore tends to "stick" to the robot. This device was impressively used by individual robots for dribbling as well as for team-behaviors by receiving passes. The second place team, FU Fighters from the Freie Universität Berlin, Germany, also were finalists in the previous year. They mainly impressed by a strong kicking mechanism which allows very fast shots towards the goal. The third place went to LuckStar II from Ngee Ann Polytechnic, Singapore.

Figure 1: Players from Big Red and Fu-fighters after the final

3   F2000

F-2000 League Competition this year included 15 teams from eight different countries all around the world. The games showed a great improvement both in the technologies applied, and in the scientific achievements from all teams. The winner was CS Freiburg, from Germany, who earned an F-2000 First Place for the second time. The runner-up the Golem Team, a surprising new entry from a group of young students, supported by Elpro Innotek, an Italian small industrial automation company. The third place was obtained by Sharif CE, from Iran, the 1999 winner. Both the finals for the first place, and for the third place, ended with penalty kicks, since the difference in skill between the teams was quite small. This added great excitement to the games. Several other teams, in addition to the winners of the 3 cups, showed very good playing capabilities based on excellent research, like for example, among all, the Trackies (the fourth place), from Japan, that had a very flexible control capabilities of the ball without any kicking devices while top other three teams have them, and RMIT, from Australia, that were very fast and flexible in their motions.

The games affirmed the superiority of the planning system implemented by the University of Freiburg, that uses a positioning system based on laser range finders. They successfully demostrated ball passing, opponents dribbling, etc., through a good estimation of the position and velocity of the other players. With the simple improvements given by introducing new motors, and adding a powerful kicker, CS Freiburg succeeded to overcome some mechanical limitations of the original commercial platforms used for their players. However, the most interesting players were the new mobile robot designed by the Golem Team. This omnidirectional platform proved to be extremely powerful. The 360 degree vision system complemented very well the mechanical design, allowing to exploit the maximum of flexibility during the play (see Figure 2(a)). Golem Team's planning system was mainly based on a reactive approach, since their robots do not localize globally, like CS Freiburg, but use only relative positioning with respect to the opponent goal, and rely on a set of very well designed basic behaviors able to show emergent cooperative abilities.

Figure 2: Distinguished Players from the F-2000 and Legged Leagues. Left: An F-2000 player of Golem Team, Italy. Right: Legged player of UNSW United approaches a goal.

4 Legged

The selected twelve teams came to Melbourne to participate in the Legged Robot League. In this league, all participants use the same robot platform, the Sony four legged robot. Therefore, this is a software competition in the real world. We had an elimination round with 4 groups (3 teams/group), the top two teams of each group went to the quarter finals. This year, UNSW United from University of New South Wales, Australia was the champion of the league. The second place was awarded to Les 3 Mousquesaires from Laboratoire de Robotique de Paris, France, and the 3rd place was CM-Pack00 from Carnegie Mellon University, USA.

In the Legged Robot League, we have an additional RoboCup Challenge competition. Three different technical challenges were devised: Challenge-1 (Striker Challenge), Challenge-2 (Collaboration Challenge), and Challenge-3 (Collision Avoidance Challenge). These are not game type competitions, but qualification of basic behaviors such as finding a ball and goal, self-localization, passing, detection of other robots and so on. The 1st place of these 3 challenges is UNSW United, the 2nd is Baby Tigers from Osaka University, Japan, and the 3rd is CM-Pack00.

One of the most significant progress in this year is UNSW's behavior of creeping flat on the ground by two front legs (See Figure 2(b)), which enabled more stable walking and much more accurate ball control such as passing or shooting. In addition, they invented fast self-localization and collision avoidance with the team mates by exchanging audio signals. Besides them, almost all teams implemented "heading" (kicking the ball with the head) so that ball passing can be a better strategy than ball dribbling by a single player.

5   Simulation

In the simulation league, 40 teams from 15 countries all around the world joined the competition. We can summarize this year's competition as the game of real team strategies. The final winner was FC Portugal, which was also the champion of Euro-2000 and a new comer of this year. The runner-up was Karlsruhe Brainstormers from Germany. The third place was ATT-CMUnited2000 from USA, whose previous program, CMUnited-99, was the champion of RoboCup-99 and finished at the fourth place in this year. Final games were very sophisticated and exciting. Most of teams in final tournaments have good tactics to handle the ball and strategies to pass the ball smoothly. Some of games needed extended time. Especially, a game of Magma Freiburg and Sharif Arvand ended up with draw and the winner was decided by a coin toss. These happened because most of teams has the similar level of basic skills.

Offensive teamwork was an important factor this year. Because goal blocking techniques of goalies have been improved very much, it becomes very difficult for a single attacker to dribble and get a goal against such a skillful goalie. As a result, dynamic combination plays in the offending situation is necessary to get goals. For example, FC Portugal is very good at side attack strategy; A side player carries the ball and kick to the goal area, and an attacker rushes into the goal area to shoot the ball. These strategies require methods to assign roles and to synchronize actions among multiple players. In order to realize these methods, FC Portugal used situation based strategic positioning and dynamic positioning and role exchange, which enables dynamic role assignment without communication in the offending situation. On the other hand, defensive strategies have not improved so much in this year. Some teams uses man-to-man marking in the defensive strategy, which was used the last year's champion team, CMUnited-99. Although this was effective in the last year, it can be broken by dynamic combination plays like the centering strategy of FC Portugal. To oppose such offense, defending strategies, especially expectation of opponents strategies will become an important factor in the next few years.

6   RoboCup Jr.

RoboCup Jr. is one of the activities putting much more emphasis on educational and entertainment aspects of RoboCup. Education researchers from the world's leading universities met at Robocup 2000 to discuss methods of using Robocup Junior as part of a school's curriculum. Robofesta 2001 (www.robofesta.com) to be held in 4 major cities in Japan, is using the Australian Robocup Jr as a pilot project for their competition. Plans are already afoot for each state in Australia to conduct eliminations, followed by an Australian Championship to find a representative for World Robocup in Seattle next year.

The soccer competition along with a Line Tracking-Sumo and a Parade/Dance competition attracted entries from interstate and overseas. With the number of entries over 40, it was one of the larger leagues of Robocup 2000.

Students from 55 visiting schools had a chance to perfect their skills in virtual robot soccer competitions, not only planned to be fun, but having educational elements integrated into them. Who would have thought that evolution could be taught with a robot soccer game. These simulations were presented by their designers who have travelled from U.S.A. and Europe, in the interest of promoting the concept that "education can be fun."

7   RoboCup-Rescue

The RoboCup-Rescue Project was newly launched by the RoboCup Federation in 1999. Its objective is as follows.

1. Development and application of advanced technologies of intelligent robotics and artificial intelligence for emergency response and disaster mitigation for the safer social system.

2. New practical problems with social importance are introduced as a challenge of robotics and AI indicating a valuable direction of research.

3. Proposal of future infrastructural systems based on advanced robotics and AI.

4. Acceleration of rescue research and development by the RoboCup competition mechanism.

A simulation project is running at present, and a robotics and infrastructure project will soon start.

In Melbourne, a simulator prototype targeting earthquake disasters was open to the public to start international cooperative research. Distributed simulation technology combines the following heterogeneous systems to make a virtual disaster field.

1. Disaster Simulators:
   Collapse of buildings, blockage of streets, fire spread, and traffic flow are simulated considering mutual effects.

2. Autonomous Agents:
   Civilians, fire brigades, policemen, and rescue parties make autonomous action in the virtual disaster.

3. Simulation Kernel:
   Manages state values and networking of/between the systems.

4. Geographical Information System:
   Gives a spatial information to the whole system.

5. Simulation Viewers:
   Show 2D/3D image of simulation results in real time as shown in Fig. 3.

RoboCup-Rescue competition will begin in 2001. Details are described in papers and a book [7, 8, 9, 10, 11, 14]. The simulator prototype can be downloaded from the links at http://robomec.cs.kobe-u.ac.jp/robocup-rescue/.

Burning   Extinguished   Fire Brigade
Agents

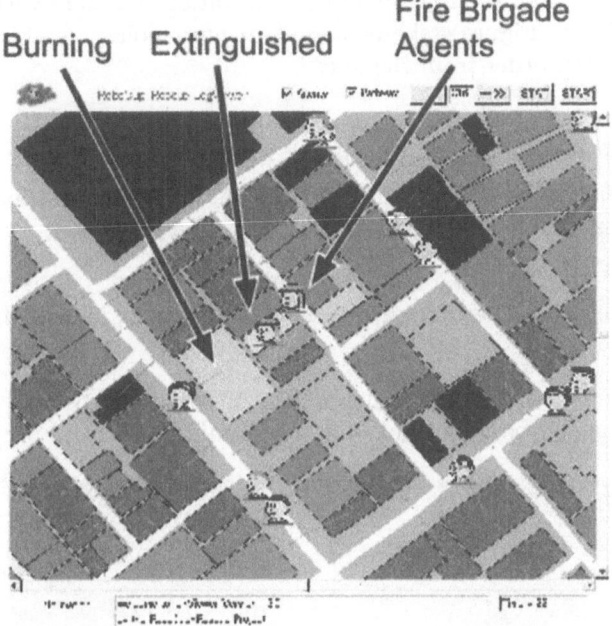

Figure 3: 2D viewer image of RoboCup-Rescue prototype simulator.

8 Humanoid Demonstration

The humanoid league will start from 2002 towards the final goal of RoboCup, that is, to beat the human world cup champion team by a team of eleven humanoid robots by 2050. This league is much much more challenging than the existing ones because dynamic stability of robot walking and running must be addressed.

The main steps of such development will be:

1. building an autonomous biped able to walk alone on the field,

2. locomotion of this biped on straight line, curve, turn on itself,

3. identification of the ball, the teammates, and the opponents,

4. kicking, passing, shooting, intercepting, and throwing the ball,

5. acquisition of cooperative behavior (coordination of basic behaviors such as passing and shooting), and

6. acquisition of team strategy.

Although the items 3, 4, 5, and 6 have been already attacked in the existing leagues, the humanoid league has its own difficulty to handle the ball by foot and hands.

At RoboCup-2000, the humanoid demonstration was held with four characteristic humanoids. Figure 4 shows these four humanoids, in which the number corresponds to the order from the left.

1. Mark-V: Prof Tomiyama group, Aoyama Gauin University, Japan. Mark-V showed his walking and kicking a ball into the goal.

2. PINO: Kitano Symbiotic Project, Japan. PINO showed his walking and waving his hand to say "Good Bye!"

3. Adam: LRP, France.

4. Jack Daniel: Western Australia University. Jack showed his walking in the air.

These humanoids are still under development, and we expect to see more humanoids with higher capability of walking, running and more in next year.

Figure 4: Four humanoids brought in RoboCup-2000

9   Workshop and Challenge Awards

The RoboCup-2000 workshop was co-chaired by Peter Stone, Tucker Balch, and Gerhard Kraetzschmar. From among more than 60 submissions, 20 papers were selected for oral presentation and an additional 20 were selected for poster presentation. Topics covered included research advances in all of the leagues as well as RoboCup rescue and RoboCup education initiatives.

Four workshop papers were selected as finalists for the RoboCup-2000 Scientific and Engineering Challenge Awards, are distinctions given annually to the RoboCup-related research that shows the most potential to advance their respective fields.

The RoboCup-2000 Scientific Challenge Award went to [5] for its contribution of a method for using Self-Organizing Maps to cluster spatio-temporal data. A novel vector representation is used as the inputs to the Self-Organizing Maps to capture relative player motions and player-ball interactions in the RoboCup simulator.

The RoboCup-2000 Engineering Challenge Award went to [6] for its novel localization method employed by robots that compete in the middle-size league. This method works with images captured from an omni-directional vision system consisting of a specially engineered mirror designed to obtain the bird's eye-view of the soccer field, with no need for software image transformations after image acquisition. It matches landmarks in the image, such as goals and field lines, using a priori knowledge regarding the geometry of the playing field.

10   Conclusion

Several RoboCup competitions and workshops have been already held, namely in 1997 in Nagoya, Japan, in 1998 in Paris, France, in 1999 in Stockholm, Sweden, and in 2000 in Melbourne, Australia. The participation and attendance of these events has been increasing every year, with about 500 participants and more than 5,000 visitors at RoboCup-2000.

RoboCup-2001 is going to be held in the United States for the first time, in Seattle, colocated with the International Joint Conference of Artificial Intelligence. RoboCup-2001 will include a 2-day research forum with presentations of technical papers, all competition leagues: soccer simulator competition, RoboCup simulation rescue competition (for the first time - details TBA), the small-size robot competition (F180), the middle-size robot competition (F2000), Four-Legged competition, the RoboCup robot rescue competition in conjunction with the AAAI robot competition (for the first time - details TBA), the RoboCup Jr. symposium including soccer 1 on 1 robot and robot dancing competitions, and other educational events for middle-school and high-school children, and an exhibition of humanoid robots.

For more information, please visit:
http://www.robocup.org

References

[1] Hiroaki Kitano, editor. *RoboCup-97: Robot Soccer World Cup I*. Springer, Lecture Note in Artificail Intelligence 1395, 1998.

[2] Minoru Asada and Hiroaki Kitano, editors. *RoboCup-98: Robot Soccer World Cup II*. Springer, Lecture Note in Artificail Intelligence 1604, 1999.

[3] Manuela Veloso, Enrico Pagello, and Hiroaki Kitano, editors. *RoboCup-99: Robot Soccer World Cup III*. Springer, Lecture Note in Artificail Intelligence (to appear), 2000.

[4] Hiroaki Kitano, Minoru Asada, Yasuo Kuniyoshi, Itsuki Noda, Eiichi Os-awa, and Hitoshi Matsubara. Robocup: A challeng problem for ai and robotics. In Hiroaki Kitano, editor, *RoboCup-97: Robot Soccer World Cup I*, pages 1–19. Springer, Lecture Note in Artificail Intelligence 1395, 1998.

[5] Michael W˙ Behavior classification with self-organizing maps.

[6] Carlos Marques and Pedro Lima. A localization method for a soccer robot using a vision-based omni directional sensor. In Peter Stone, Tucker Balch, and Gerhard Kraetzschmar, editors, *RoboCup-2000: Robot Soccer World Cup IV*. Springer Verlag, Berlin, 2001. To appear.

[7] Satoshi Tadokoro, Hiroaki Kitano, Tomoichi Takahashi, Itsuki Noda, Hitoshi Matsubara, Atsuhi Shinjoh, Tetsuya Koto, Ikuo Takeuchi, Hironao Takahashi, Fumitoshi Matsuno, Mitsuo Hatayama, Jun Nobe, Susumu Shimada, The RoboCup-Rescue Project: A Robotic Approach to the Disaster Mitigation Problem, Proc. IEEE International Conference on Robotics and Automation, 2000.

[8] Yoshitaka Kuwata, Atsushi Shinjoh, Proc. 4th International Workshop on RoboCup, 2000.

[9] Tomoichi Takahashi, Ikuo Takeuchi, Tetsuhiko Koto, Satoshi Tadokoro, Itsuki Noda, RoboCup-Rescue disaster simulator architecture Proc. 4th International Workshop on RoboCup, 2000.

[10] Toshiyuki Kaneda, Fumitoshi Matsuno, Hironao Takahashi, Takeshi Matsui, Masayasu Atsumi, Michinori Hatayama, Kenji Tayama, Ryousuke Chiba, Kazunori Takeuchi, Simulator complex for RoboCup-Rescue Simulation Project – As test-bed for multi-agent organizational behavior in emergency case of large-scale disaster, Proc. 4th International Workshop on RoboCup, 2000.

[11] Masayuki Ohta, RoboCup-Rescue simulation: in case of fire fighting planning, Proc. 4th International Workshop on RoboCup, 2000.

[12] Masayuki Ohta, Tetsuhiko Koto, Ikuo Takeuchi, Tomoichi Takahashi, Design and implementation of the kernel and agents for the RoboCup-Rescue, Proc. 4th International Conference on MultiAgent Systems, 2000.

[13] Hiroaki Kitano, RoboCup-Rescue: A grand challenge for multiagent systems, Proc. 4th International Conference on MultiAgent Systems, 2000.

[14] Satoshi Tadokoro, Hiroaki Kitano, ed., The RoboCup-Rescue: A challenge for emergency search & rescue at large-scale disasters, Kyoritsu Publ., 2000 (in Japanese).

View Planning via C-space Entropy for Efficient Exploration with Eye-in-Hand Systems

Yong Yu

yongyu@cs.sfu.ca

School of Engineering Science

Simon Fraser University

Burnaby, B.C. V5A 1S6 Canada

Kamal K. Gupta

kamal@cs.sfu.ca

School of Engineering Science

Simon Fraser University

Burnaby, B.C. V5A 1S6 Canada

Abstract: We present an implemented sensor-based planner for motion planning and exploration for eye-in-hand systems. A model-based motion planner is used to plan paths within the known part of the environment to further sense the unknown part of the environment. Each sensing action is viewed as gaining information about the status of configuration space. We introduce the notion of C-space entropy as a measure of ignorance or lack of information of C-space. The next view is planned so as to maximize expected entropy reduction (MER), or equivalently, expected information increase. Experimental results demonstrate that MER criterion results in efficient exploration of unknown environments and that the planner can make a robot arm move around safely (without collisions) while carrying out exploratory and purposive tasks in unknown environments.

1. Introduction

In sensor-based motion planning, the task for a robot, equipped with a sensor, is to sense and explore its unknown environment (and reach a given goal) while avoiding collisions with any obstacles in the environment. A variety of sensors have been used in the past, mostly with mobile robots, with some exceptions as in [1]. This paper falls in the category of eye-sensor based motion planning for robots with non-trivial geometry/kinematics. An "eye" sensor is essentially a distance or range sensor that senses distances from a single vantage point (or a reference frame); non-trivial[1] geometry/kinematics implies that the physical space and C-space for the robot are distinctly different. This class of robots is broad and includes robots ranging from a simple polygonal mobile robot to complex articulated manipulators, and humanoid robots [2]. In particular, our concern here is motion planning and exploration for an eye-in-hand system – a manipulator arm (called robot from here on), equipped with an eye type sensor as shown in Figure 1. The robot is required to plan and execute collision-free motions in an environment initially unknown to it. The decision to mount the sensor (eye) on the robot end-effector (hand) was motivated by the additional

---

[1] Idealizations such as point or circle robots, made for most mobile robots, are considered to have trivial geometry/kinematics.

D. Rus and S. Singh (Eds.): Experimental Robotics VII, LNCIS 271, pp. 373–384, 2001.

maneuverability for the sensor. Our framework, however, is general and could be extended to other types of sensor+robot mechanisms, for instance humanoid robots, where the eye sensor(s) is located on the head [2]. See [3] for some interesting issues for sensor-based planning with eye sensors. Figure 2 shows

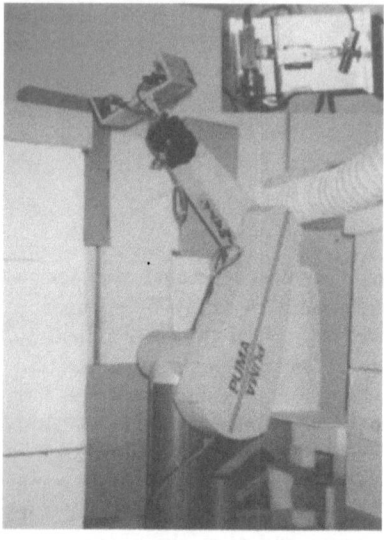

Figure 1. The experimental eye-in-hand system — a PUMA 560 with a wrist mounted area-scan laser range finder. Inset shows an enlarged view of the sensor with the camera on the left and the laser striper on the right.

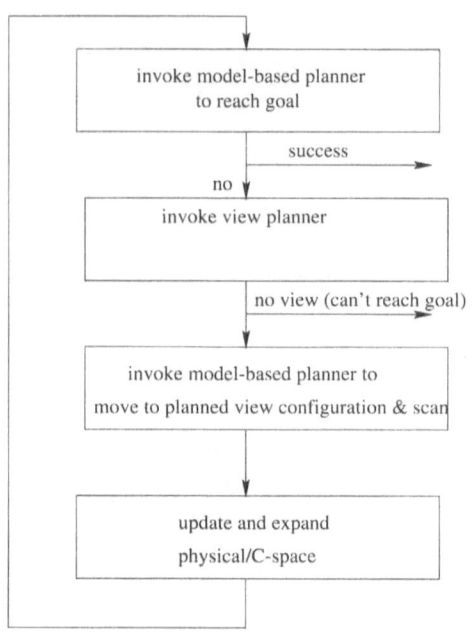

Figure 2. A general framework for eye-sensor based MP.

our "high-level" framework for eye-sensor based MP for robots with non-trivial geometry/kinematics. The sensing action is assumed to be discrete, i.e., it take place at discrete time instants. This framework assumes that a model-based[2] planner is available. This underlying model-based planner is used to plan paths within the known part of the environment to further sense the unknown part of the environment. We call this problem of determining the next sensing action the view planning problem. It has two component sub-problems: a) determine which region (in physical space) to scan and b) determine a reachable robot configuration from where the scan can be taken. The robot invokes the model-based planner to move to the planned view configuration and takes a scan. Having taken the new scan, the robot "updates" its internal representation (of physical and configuration space) and re-invokes the model-based planner to reach the goal. If it succeeds, it terminates with success. Otherwise, this process is repeated until the given goal is declared reachable by the model-

---

[2] We use the term model-based motion planning when the environment (model) is assumed known [4].

based planner, or the goal is declared unreachable[3]. The two key sub-problems that sensor-based MP therefore needs to solve are: (i) *view planning*, and (ii) *an "incrementalized" model-based planner* that repeatedly updates its physical and configuration space representation.

Within the framework described above, we have implemented a sensor-based planner for the real eye-in-hand test-bed system shown in Figure 1. The model-based planner component used in our overall algorithm is an incremental version of the probabilistic roadmap planner [5], adapted to deal with unknown environments. We call it Sensor-based Incremental Construction of Probabilistic Roadmaps (SBIC-PRM). Since a roadmap is constructed without explicitly representing the C-obstacles, it is particularly useful as a representation for high dimensional configuration spaces, our main motivation behind using this approach. While other roadmap type approaches could also be used, a key advantage of PRM is that it naturally handles disconnected components in the roadmap. See [6] for the case when ACA is used as the underlying model-based planner within our framework.

We have reported various aspects of SBIC-PRM in [7, 8]. This paper integrates a novel, systematic and more efficient information-theoretic approach to the view planning sub-problem within the overall SBIC-PRM planner. In this approach each sensing action is viewed as gaining information about the status of configuration space. It uses a novel notion of C-space entropy as a measure of ignorance or lack of information of C-space. The next view is planned so as to maximize expected entropy reduction (MER), or equivalently expected increase in information. The details of this information theoretic view planning approach are described in [9, 10]. The emphasis in this paper is on its integration within the overall SBIC-PRM planner and experimental results showing the efficacy of MER based view planning compared to other criteria.

Figure 3 shows SBIC-PRM running on a simulated 2-link eye-in-hand system. The robot has a range sensor on the end-effector. The sensor has a field of view of 40°, indicated by the triangular region. The sensor also has an additional rotational degree of freedom at the wrist. The left image in each column shows the physical space and the right image shows the corresponding C-space. In both physical space and C-space, white region is known free space, black region is known obstacles and gray region is unknown. In physical space, dark gray region is obstacles, but unknown to the robot. White region around the initial robot configuration (pointing vertically down and shown in gray) is known to be free at the start. The robot in the desired goal configuration is shown in black. SBIC-PRM distinguishes between three types of landmarks: *White* landmarks, similar to the nodes in [5] belong to $\mathcal{C}_{free}$, i.e., the robot in this configuration lies completely in $\mathcal{P}_{free}$; *Black* landmarks correspond to robot being in collision with a *(sensed or known)* obstacle at the corresponding configuration; and, *Gray* landmarks correspond to those configurations where the robot does not intersect with any sensed obstacles, and does not lie com-

---

[3]This termination condition would depend on whether it is a start-goal problem or exploration problem. This one is for start-goal problem.

pletely in $\mathcal{P}_{free}$ either. In Figure 3 the dots in gray (unknown) region[4] are the gray landmarks, the dots in the white (known) region are the white landmarks, and the dots in the black region (can't see them) are black landmarks. The nodes of the roadmap (a graph) are the white landmarks and two nodes are connected if a local planner — that simply tries to execute a discretized straight line in configuration space — returns a collision-free path between the nodes. Edges between nodes are not shown to avoid clutter. There are several connected components in the evolving roadmap (recall that the C-space is a torus, shown as a rectangle). The start node belongs to the one on lower left corner.

The robot carries out its motion within this evolving roadmap to further sense the physical environment. Each sensing action, as mentioned earlier, is chosen to maximize the expected entropy decrease (while satisfying reachability constraints), resulting in efficient and fast exploration of the environment. With each sensing action, as new free space (and obstacles) is sensed in the physical world, the roadmap is expanded by placing new landmarks and checking the status of all gray landmarks (old and new ones). The status of some may change to white (or black), thereby expanding the roadmap. This process is repeated until either the final goal is reachable from one of the nodes in the graph (as in this case), or the goal is declared unreachable.

The core computation in this process is that of collision detection with known obstacles/unknown region. The geometry of the robot and the sensor is modeled by a collection of polyhedra and the environment is modelled as spatial occupancy octrees. Each scan gives a range image which is used to update the octree(s) representing the environment. Details of these rather straightforward geometric computations are in [11].

For brevity, we omit a comparison with other similar systems [12, 13], except to note that our approach is more systematic, effective and efficient for both view planning and model-based planning components. See [11] for details. For example, the next view in [12] is chosen to maximize the unknown physical space volume. We show (see Section 3.5) that the MER criterion based view planning results in more efficient exploration. We also note there is vast view planning literature in machine vision [14], however, these works assume that the sensor (camera) can move arbitrarily and do not consider the geometric and kinematic constraints — critical for eye-in-hand systems — on where the sensor can be positioned.

2. Notation

Let $\mathcal{A}$ denote the robot. We assume that the sensor body is "absorbed" into the robot $\mathcal{A}$. Hence, we can treat the physical sensor as an abstract reference frame without any physical body. Let $\mathcal{P}$ denote the physical space (R^p, $p = 2$ or 3) and $\mathcal{C}$ denote the configuration space. $\mathcal{A}(q) \subset \mathcal{P}$ denotes the physical space occupied by the robot at configuration $q \in \mathcal{C}$. Let $\mathcal{V}_{\mathcal{S}}(q) \subset \mathcal{P}$ denote the sensed region when robot scans at configuration q. Subscripts $free$, obs and

---

[4]Note that the regions are not computed by the planner, only the corresponding landmarks are computed. Regions are shown for visualization only.

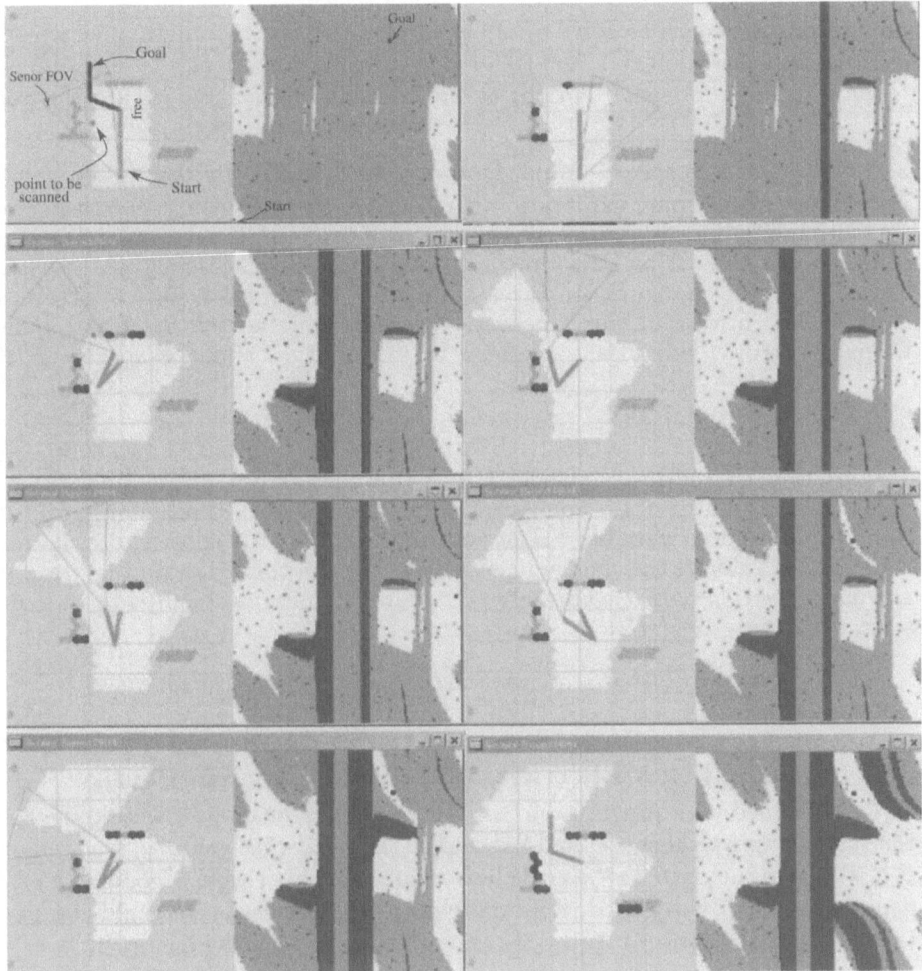

Figure 3. SBIC-PRM illustrated for a 2-link eye-in-hand system — a 2-link robot arm with an eye sensor (a range camera) mounted at the wrist. Sensor field of view is indicated by the triangular region. See text for further explanation.

unk are used to indicate the free, obstacle and unknown portions of physical or configuration space. For instance, $\mathcal{A}_{unk}(q)$ describes the part of the robot that lies in unknown physical space at configuration q. $\mathcal{C}_{free}$ is characterized by a *roadmap* $\mathcal{R}$. We use $\mathcal{R}(l_0)$ to represent the connected component of $\mathcal{R}$ that contains l_0, the landmark corresponding to the start configuration q_0. $\mathcal{C}_{free}(q_0)$ denotes the connected component of $\mathcal{C}_{free}$ containing q_0.

3. View Planning with Kinematic/Geometric Constraints

Recall that there are two sub-problems within the view planning problem: a) to determine where in the physical space should the sensor scan, and b) determine a robot configuration from where the scan should be taken. We first present our approach to solve a). In general, there are two main objectives in choosing which region to scan: (i) to gain information about the environment (exploratory component) and (ii) to preferentially "expand" toward the goal (goal component).

3.1. Where to Scan: C-space Entropy and MER Criterion

Recall that from a pure exploration perspective, we may view each sensing action as gaining information about the status of configuration space. The status of C-space is viewed as an n-dimensional discrete stochastic signal, in which each discretized configuration takes a value in a binary set {0(free), 1(obstacle)}. The entropy of this stochastic process, denoted by $H(C)$ and called C-space entropy, provides a measure of information about C-space. $H(C)$ can be written as

$$H(C) \sum_{i=0}^{N-1} H(C(q_i)) - \sum_{i \neq j} I(C(q_i), C(q_j)) + \sum_{i,j,k \text{ different}}$$

$I(C(q_i), C(q_j), C(q_k)) - \cdots$, where $H(C(q_i))$ is the entropy of the random variable $C(q_i)$, the status of the configuration q_i, and $I(\cdot)$ is the mutual information of multiple random variables. The next sensing action is chosen to maximize the *expected* value of the change in entropy[5]. We call this criterion the *maximal entropy reduction (MER)*. Formally, the view planning problem is to choose a scan, such that $-E(\Delta H(C))$ is maximized, i.e.

$$\max_{\substack{\text{all possible} \\ \text{scans}}} -E(\Delta H(C)) = \max_{\substack{\text{all possible} \\ \text{scans}}} \left(H(C_{\text{before scan}}) - E(H(C_{\text{after scan}})) \right)$$

3.2. Implementing MER criterion: Information Gain Density

$H(C)$ is defined as a function of C-space. But it is ultimately determined by the status of physical space, and indeed the the probability that a given region, $B \subset P$ is free, i.e., $p(B \subset P_{free})$. When a small neighborhood of a point $x \in P$, denoted by $B(x)$ is scanned, it would change the entropy. We could compute this (expected) change in entropy per unit volume around every point $x \in P$, essentially a histogram of $G_c(x) = \lim\limits_{vol(B(x)) \to 0} \dfrac{-E(\Delta H)}{vol(B(x))}$. The best point to be scanned is the one that maximizes $G_c(x)$, the *information gain density function* (IGDF). It[6] can be expressed as $\sum_q g_q(x)$ where

$$g_q(x) = \lim_{vol(B(x)) \to 0} \frac{-E\Delta H(C(q))}{vol(B(x))}$$

An interpretation of the above expression is as follows. When $B(x) \subset P$ is scanned, it affects the C-space entropy via each configuration q. $g_q(x)$ is the expected contribution of q to the resulting C-space entropy reduction (per unit volume). $G_c(x)$ is the summation of $g_q(x)$ over all the configurations in the

---

[5]For simplicity, we discuss the case of a single scanning action. This is easily generalized to a sequence of sensing actions.

[6]This limit may not always exist. In practice, one could always compute an approximation for a small region $B(x)$. Also, two simplifying assumptions are made here. First, mutual entropy terms are neglected. Second, we assume that the status (free or obstacle) of $B(x)$ can be determined after the scan, i.e. no obstacle occludes the view.

entire C-space. However, note that only certain configurations contribute to $G_c(x)$. $g_q(x)$ equals 0 when $\mathcal{B}(x) \cap \mathcal{A}(q) = \emptyset$, because changing the status of $\mathcal{B}(x)$ will not change the contribution to C-space entropy from such a configuration q. The set of configurations that do contribute belong to what we call the C-zone of x, denoted by $\mathcal{X}(x)$. Formally, $\mathcal{X}(x) = \{q : x \in \mathcal{A}(q)\}$. Points outside the C-zone of x do not contribute to $g_q(x)$. Therefore, we explicitly write $G_c(x) = \sum\limits_{q \in \mathcal{X}(x)} g_q(x)$

We have derived an expression for $G_c(x)$ for Poisson distribution of (point) obstacles. Recall that a stationary Poisson point process Φ is characterized by uniformly distributed points in space [15]. From a motion planning point of view, these points are obstacles in the physical space of the robot. The density parameter, λ, gives us the ability to set the density of obstacles in the physical space. For a more dense distribution, λ is higher. We simply state the result here, see [11] for details.

$$p(\mathcal{B} \subset \mathcal{P}_{free}) = e^{-\lambda \, vol(\mathcal{B})}$$

$$p(q) = p(\mathcal{A}_{unk}(q) \subset \mathcal{P}_{free}) = e^{-\lambda \, vol(\mathcal{A}_{unk}(q))}$$

$$G_c(x) = \sum_{q \in \mathcal{X}(x)} \lambda \left(\log \frac{p(q)}{1 - p(q)} - p(q) \log p(q) - (1 - p(q)) \log(1 - p(q)) \right)$$

These expressions completely determine the IGDF for a physical space with a Poisson point distribution.

3.2.1. IGDF: Practical Implementation

The calculation of IGDF can be approximated as follows. Eq. 1 tells us that computing $IGDF(x)$ requires a summation over C-zone $\mathcal{X}(x)$. The computational cost of finding the C-zone $\mathcal{X}(x)$ is high for a high dimensional C-space. One way out would be to approximate the summation over randomly selected configurations $q \in \mathcal{C}_{unk}$ (unknown part of C-space) when $x \in \mathcal{A}(q)$.

3.3. Overall Objective Function

The exploratory component is achieved by maximizing the IGDF, as we just discussed above. For the goal-directed component, a reasonable choice is to find the gray landmark closest to the goal (call it l_G) subject to the condition that $\mathcal{A}(q_{l_G})$ has at least a certain fraction of its volume in $\mathcal{P}_{free}$. $\mathcal{A}_{unk}(q_{l_G})$ is the region that should be sensed. Formally,

$$G_g(x) = \delta(\mathcal{A}_{unk}(q_{l_G})) = \begin{cases} 1, & x \in \mathcal{A}_{unk}(q_{l_G}) \\ 0, & otherwise \end{cases} \tag{1}$$

The combined objective function is then a weighted sum of G_c and G_g, i.e. $G(x) = w_e G_c(x) + w_g G_g(x)$. $G(x)$ can be considered as a generalized IGDF with a larger weight placed on the points (in physical space) belonging to configuration q_{l_G}. The point x to be scanned is the one that maximizes $G(x)$, i.e. $x_{max} = \{x : \max\limits_{x \in \mathcal{P}_{unk}} \{G(x)\}\}$. In Figure 3, for each snapshot, the small dot in the centre of the sensor FOV (triangle) corresponds to x_{max}. w_e and w_g, the weights for exploratory and goal components, can be chosen to emphasize the exploratory or goal-directed behaviour as needed by the task at hand.

3.4. Determine View Configuration To Scan From:

Given a point x_{max} to be scanned, the next step is to determine a reachable configuration from which to scan it, i.e. to determine a view configuration, say q_v. Obviously, $q_v \in C_{free}(q_0)$, since the robot needs to reach q_v in order to scan. We search $\mathcal{R}(l_0)$, the connected component of the roadmap to find the view node[7]. From all the white landmarks which satisfy the collision (between robot and known obstacles) and visibility (of x_{max} from the sensor) constraints, we choose the one whose view node configuration results in the center of the sensing region $V_S(q_v)$ being closest to x. This is the the view node. Note that a q_v may not exist for a given x_{max}. In this case, the planner chooses the next best x_{max} to scan. If the view planner has exhausted all the observable physical space, it will stop and report a failure. In Figure 3, the robot is at the chosen view node in each snapshot.

3.5. Effectiveness of MER criterion

We conducted a series of experiments on the simulated two link eye-in-hand system of Figure 3 with three different criteria for view planning: (i) the next view (position and direction) is randomly chosen, (ii) the next view is chosen so that it maximizes the unknown physical space volume, and (iii) the next view is chosen based on MER criterion. The task was to reach the goal configuration (shown in dark); the start configuration being vertically downwards in each of the three cases. Fig. 4 shows the explored physical space and C-space after 14 scans with each of these three different sensing strategies. The MER criterion results in the highest explored C-space (70%), and the maximal unknown physical space volume criterion results in the least explored C-space (33%), with the random scan strategy somewhere in between (47%). In fact the robot was able to reach the goal configuration only with the MER based view planning; it failed to make the goal configuration free with the other two strategies. It is interesting to note that even random strategy does better than the one based on maximizing the unknown physical space volume. One factor is that the overlap between viewing regions tends to be small in the latter strategy, and hence known physical space may tend to consist of several small islands scattered around. On the other hand, only large contiguous regions may make portions of C-space known. In addition, much of the physical space may even lie outside the workspace of the robot, hence contributing nothing to the C-space exploration. Maximizing unknown physical space volume may, therefore be the least efficient view planning strategy from C-space exploration perspective.

4. Experiments

We now present experimental results with SBIC-PRM running on the real test-bed shown in Figure 1 – a PUMA 560 with a triangulation-based area-scan laser

---

[7]If the sensor has internal degrees of freedom, one can further search over this sub-space to find the "best" viewing position.

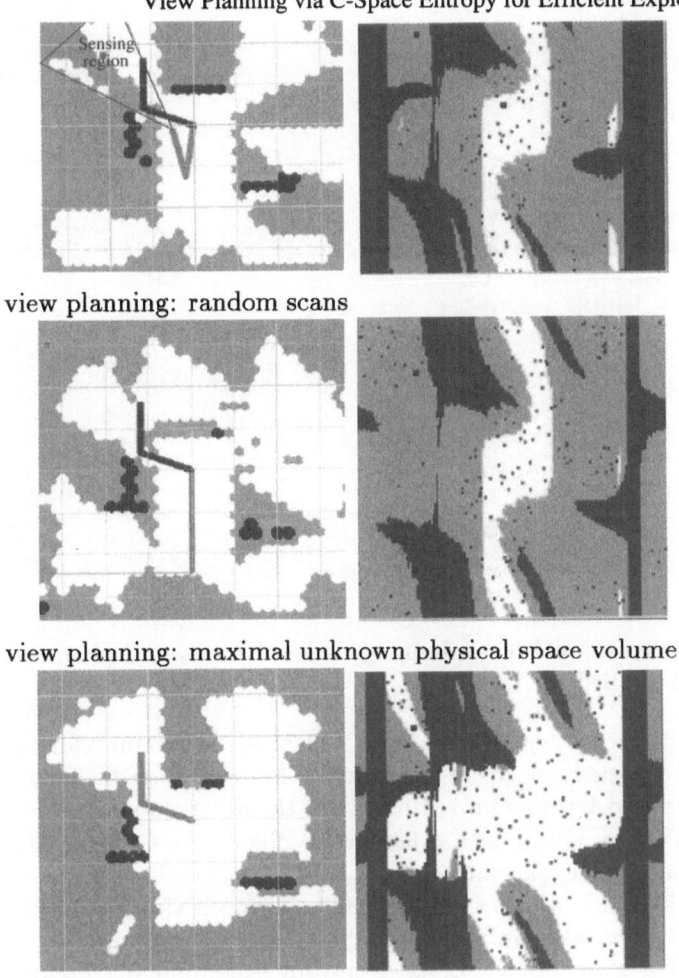

view planning: random scans

view planning: maximal unknown physical space volume

view planning: MER criterion

Figure 4. Comparison of C-space exploration efficiency for three different view planning criteria. White region shows the explored C-space. MER criterion results in significantly more C-space explored.

range finder (the eye) mounted on its wrist. The sensor field of view is a conical region, approximately one metre long with an apex (solid) angle of about 40 degrees. The sensor accuracy is about 1 cm in all three directions. For system and software details, see [7]. Figure 5 shows a real planning task solved by the planner. In this experiment, the robot has its forearm and upper arm inside a hole in a wall at the goal configuration (snapshot goal). The start configuration is robot pointing vertically upwards (snapshot 1). A tight small cuboid region around the initial configuration is assumed to be free at the start. Several boxes were scattered in the robot workspace that are unknown to the robot in the beginning. The robot reached the goal configuration in about 13 scans. Figure shows some of the intermediate viewing configurations that the robot moved to on its way to the goal configuration.

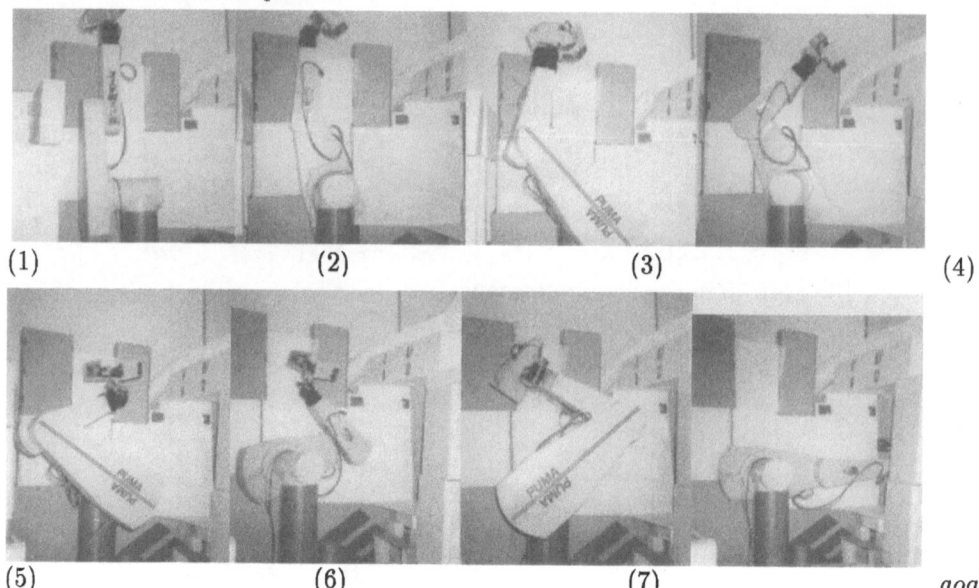

(1)   (2)   (3)   (4)

(5)   (6)   (7)   *goal*

Figure 5. An example of a planning task solved by the planner on the real
eye-in-hand testbed.

In the following experiments, SBIC-PRM is run 10 times for each case and
the results reported are average results. This was done to reduce statistical
variations caused by randomized nature of the algorithm.

SBIC-PRM was run with different for different weightings of exploratory
and goal components (w_e and w_g). $w_g = 0$ corresponds to a pure exploration
task. As a baseline, we also implemented a random view planning strategy in
which scans were taken in a random direction from a randomly chosen white
landmark. For each run, the C-space entropy is computed[8]. Figure 6(a) shows
the resulting plots. The horizontal axis is the number of scans (iterations).
The vertical axis is the C-space entropy. The random scan curve is shown with
diamond markers. Clearly, larger w_e (exploration weight) results in faster C-
space entropy reduction, or equivalently faster exploration. Furthermore, MER
based view planning explores the environment significantly faster than random
scans. Because of the random approach in generating landmarks, the number of
landmarks of different colors (white or black) should be roughly proportional to
the volume of C-space of the corresponding status (free or obstacle). Therefore,
the number of black and white landmarks essentially provides a measure of the
volume of explored (or known) C-space. Figure 6(b) shows the increase in
black and white landmarks vs. the number of iterations for different values of
exploration weight, w_e. The number of black and white landmarks increases at
a faster rate for greater values of w_e. The faster speed of C-space exploration
is, on average, at a cost of slightly increasing the number of scans to reach the
goal. For instance, with $w_e = .25$, about 14 scans were needed to reach the

---

[8] For computational reasons, entropy computation was carried out over a set of randomly
selected (unknown) C-space configurations.

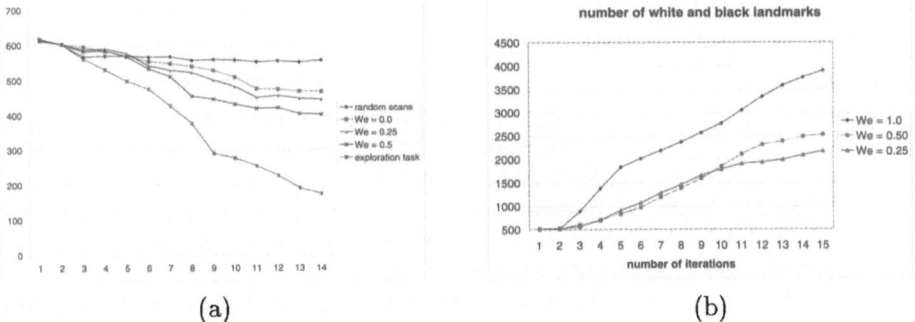

(a)   (b)

Figure 6. Higher exploration weight (w_e) results in faster decrease in entropy (a)and faster increase in number of black+white landmarks (b), thereby indicating faster rate of exploration.

goal; with $w_e = 1$, about 18 scans were needed. In some situations, however, a larger w_e may lead to fewer scans for reaching a goal. For instance, a larger w_e may help the planner to escape dead ends which were initially believed to be a free path towards the goal.

Typically, the planner is able to reach (planning as it senses) the goal configuration in about 7 - 25 scans (depending on the scene complexity), while avoiding collisions with the obstacles throughout. The run time (on a Pentium II 450) for one iteration varies from 1 minute to 2 minutes depending on the number of landmarks generated and the complexity of the scene. More details on run times are reported in [7].

5. Conclusions

We demonstrated that (i) our information theoretic approach to view planning, in particular C-space entropy and the resulting MER criterion provide an efficient way of exploring C-space; (ii) parameters w_g and w_e change the behavior of the algorithm, i.e., the C-space exploration rate increases with the weight of the exploratory component w_e, and (iii) SBIC-PRM is an effective eye-sensor based planner making a robot arm move around safely (without collisions) while carrying out exploratory and purposive tasks in unknown environments.

We have made certain simplifying assumptions in deriving the expression for MER criterion. In particular, (i) we assume Poisson point process, (ii) we ignore occlusion (visibility) constraint, i.e., we assume that a point to be sensed will either become free or obstacle, and (iii) we compute MER for a single point. Our immediate future work is to extend the MER formulation beyond these assumptions. Along another line, several fundamental yet novel issues arise for sensor-based motion planning and exploration for robots with non-trivial geometry/kinematics. We intend to explore them in future. See [3] for some initial thoughts.

Acknowledgements

This research has been jointly funded by the National Science and Engineering Research Council, Canada and National Research Council, Canada. Many thanks to Mehran Mehrandezh for the 2-dimensional implementation of the planar simulated eye-in-hand system. Thanks to Juan Manuel Ahuactzin for providing the initial simulation code.

References

[1] V.J. Lumelsky and E Cheung. Real-time collision avoidance in teleoperated whole-sensitive robot arm manipulators. *IEEE Transactions on Systems, Man and Cybernetics*, 23(1):194–203, Jan.-Feb 1993.

[2] K. Hirai, M. Hirose, M. Haikawa, and T. Takenaka. The development of honda humanoid robot. In *Proceedings of IEEE International Conference on Robotics and Automation*, pages 1321–1326, 1998.

[3] Kamal Gupta and Yong Yu. On eye-sensor based path planning for robots with non-trivial geometry/kinematics. accepted for *IEEE International Conference on Robotics and Automation*, 2001.

[4] Kamal K. Gupta and Angel del Pobil, editors. *Practical Motion Planning in Robotics: Current Approaches and Future Directions*. John Wiley, 1998.

[5] L. Kavraki, P. Svestka, J. Latombe, and M. Overmars. Probabilistic roadmaps for path planning in high-dimensional configuration spaces. *IEEE Transactions on Robotics and Automation*, 12(4):556 – 580, Aug 1996.

[6] J. Ahuactzin and A. Portilla. A basic algorithm and data structure for sensor-based path planning in unknown environment. In *Proceedings of IEEE International Conference on Intelligent Robots and Systems*, 2000.

[7] Y. Yu and K. Gupta. Sensor-based roadmaps probabilistic roadmaps: Experiments with an eye-in-hand system. *Advanced Robotics*, 14(8), August 2000. A version also appeared In *Proceedings of IEEE/RSJ International Conference on Intelligent Robot and System*, pages 1707–1714, 1999.

[8] Y. Yu and K. Gupta. Sensor-based motion planning for manipulator arms: An eye-in-hand system. In *IEEE International Conference on Robotics and Automation video session*, 2000.

[9] Y. Yu and K. Gupta. An information theoretic approach to view point planning for motion planning of eye-in-hand systems. In *Proceedings of 31st International Symposium on Robotics*, pages 306– 311, 2000.

[10] Yong Yu and Kamal Gupta. An information theoretical approach to view planning with kinematic and geometric constraints. accepted for *IEEE International Conference on Robotics and Automation*, 2001.

[11] Yong Yu. An information theoretical incremental approach to sensor-based motion planning for eye-in-hand systems. Ph.D. Thesis. School of Engineering Science, Simon Fraser University. Canada. 2000.

[12] E. Kruse, R. Gutsche, and F. Wahl. Effective, iterative, sensor based 3-d map building using rating functions in configuration space. In *Proceedings of IEEE International Conference on Robotics and Automation*, pages 1067 – 1072, 1996.

[13] P. Renton, M. Greenspan, H. Elmaraghy, and H. Zghal. Plan-n-scan: A robotic system for collision free autonomous exploration and workspace mapping. *Journal of Intelligent and Robotic System*, 24:207 – 234, 1999.

[14] S. Hutchinson and A. Kak. Planning sensing strategies in a robot work cell with multi-sensor capabilities. *IEEE Transaction on Robotics and Automation*, 5(6):765 – 783, December 1989.

[15] D. Stoyan and W.S. Kendall. *Stochastic geometry and its applications*. J. Wiley, 1995.

Motion Planning for a Self-Reconfigurable Modular Robot

Eiichi Yoshida, Satoshi Murata, Akiya Kamimura,
Kohji Tomita, Haruhisa Kurokawa and Shigeru Kokaji
Mechanical Engineering Laboratory,
1-2 Namiki, Tsukuba-shi, Ibaraki 305-8564 Japan
eiichi@mel.go.jp

Abstract: This paper addresses motion planning of a homogeneous modular robotic system. The modules have self-reconfiguration capability so that a group of the modules can construct a robotic structure generating dynamic motion. Motion planning for self-reconfiguration is not straightforward because the modular structure allows many combinatorial configurations, and also the proposed module has only two degrees of freedom. We will show a motion planning method for a particular class of multi-module structure, based on global planning and local motion scheme selection. The fundamental module motion will also be demonstrated through hardware experiments.

1. Introduction

Reconfigurable robotic systems have been attracting more interest, as their feasibility has been examined through hardware and software experiments in recent years [1]–[7]. This paper focuses on self-reconfigurable and homogeneous modular robotic systems that can adapt themselves to the external environment by changing their configuration. They can also repair themselves by using spare modules without external help owing to homogeneous modular structure. They have various potential applications, especially for structures or robots that should operate in extreme environments inaccessible to humans, for instance, in space or deep sea, or in nuclear plants.

The reconfigurable modular robotic system are classified into two types, lattice type [1, 2, 3] and thread type [5, 6, 7]. The former corresponds to a system where each module has several fixed connection directions, and a group of them can construct various static structures like jungle-gym. However, it is difficult for such a system to generate some dynamic robotic motions. On the other hand, the latter has snake-like shape that can generate various dynamic motions, nevertheless self-reconfiguration is difficult.

We have therefore developed a new type of modular robotic system that can realize both static structure and dynamic robotic motion. This has been realized by simplified design of a module and connecting mechanism. Fundamental reconfiguration motions are demonstrated in the experiment section of the paper.

This paper addresses some aspects of motion planning of a modular robotic system. There have been a number of studies on reconfiguration method for lattice-type modular robots. We have developed a series of distributed self-reconfiguration methods for two-dimensional and three-dimensional homogeneous modular robots [8, 9].

D. Rus and S. Singh (Eds.): Experimental Robotics VII, LNCIS 271, pp. 385–394, 2001.

These methods enabled homogeneous modular robots to self-assemble and self-repair in a distributed manner based on local inter-module interactions. In contrast, most of other methods are based on centralized planning. Kotay et. al [10] developed robotic modules called "Molecules," and described a global motion synthesis method for a class of module group to move in arbitrary directions. Ünsal et. al [11] reported two-level motion planners for a bipartite module "I-Cube" composed of cubes and links, based on heuristic graph search between module configurations. These methods are dedicated to modules that have sufficient degrees of freedom to move to desired neighboring lattice position. Since our module has incomplete spatial symmetry, the formerly developed motion planning method cannot be directly applied.

We propose a two-layered motion planning method, a global and local planners to transfer a class of module cluster, which is classified into a centralized method. The former part of the planner provides the *flow* of the cluster, which corresponds to a global movement. The latter generates local coordinated motions called *motion schemes* in consideration with incomplete spatial symmetry of modules. The planned motion have not been fully implemented yet in the hardware, however we will discuss issues which arise when the planned motions are executed.

2. Hardware Design

The developed module consists of two semi-cylindrical parts connected by a link (Fig. 1). Servomotors are embedded in the link so that each of parts can rotate by 180°. Figure 2 shows a hardware prototype of the module. Each module has six connecting surfaces (three for each part) that can attach and detach other modules by using magnets and shape memory alloy (SMA) actuator [12]. The connecting surfaces have also electrodes for power supply and serial communication. All the connected modules can be supplied power from one module connecting to the power source. This eliminates the tether entanglement that becomes significant in three-dimensional configuration.

Each module is equipped with a PIC microprocessor that drives servomotors and SMA actuators. In the current development, all the modules are controlled from a host PC that provides motion commands through serial communication lines. The size of one semi-cylindrical part is 6cm cube and a module weighs approximately 400g.

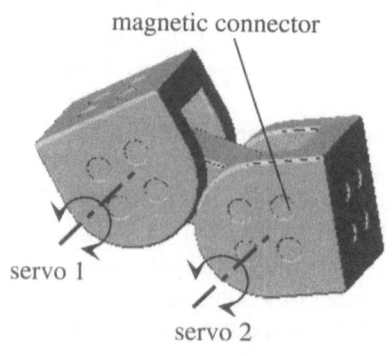

Figure 1. A robotic module.

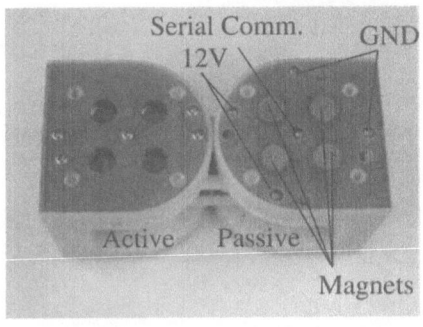
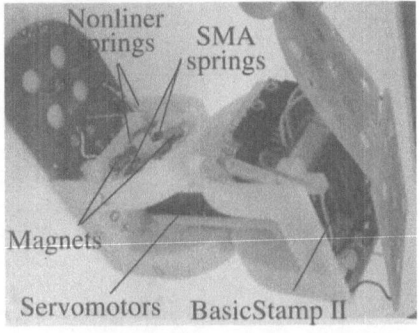

Overview of a module   Inside structure of a module

Figure 2. A hardware module.

3. Reconfiguration Motion Planning

We have shown the proposed module can generate various shapes such as legged walking robot or slime-like locomotion machine [12]. However, the motion planning from one configuration to another is not straightforward because the proposed modules' mobility strongly depends on its local configuration. Therefore, we deal with the motion planning to a particular class of module cluster as a basic case. We take a two-layered approach, namely global planner and local motion scheme selector. The global cluster movement, called *flow*, is planned in the former part as a set of paths of modules from initial and goal position. The latter outputs a sequence of *atomic motions* for local reconfiguration, called *motion schemes* based on a rule database to realize the generated paths. The *atomic motions* is the simplest module motion by one or two modules which will be described in section 3.1. The incomplete spatial symmetry is considered in the motion scheme selector by means of a rule database of feasible local coordinated motion associated with corresponding local configuration.

In the following, we suppose that only one motion scheme is allowed during the cluster flow for simplicity. Another assumption is that one module can lift only one other module in the planning, which comes from the limited torque capacity of the hardware.

3.1. Atomic Motion

There are mainly three types of atomic motion, *pivot mode*, *forward-roll mode* and *mode conversion*. Figures 3 and 4 show two different atomic motions on a plane, forward-roll and pivot, whose orientation of rotational axes are in different direction. Mode conversion is a two-module motion to convert from one mode to the other, where a helper module is required as illustrated in Fig. 5.

To construct three-dimensional structures, a cluster of module should include at least one module in forward-roll mode and pivot mode.

3.2. Cluster Flow and Global Planner

As a basic case of motion planning, we consider a class of module cluster mainly composed of two layer of pivot mode modules and a couple of forward-roll mode modules (Fig. 6). As this cluster includes both modes of atomic motion, it can change there configuration in three dimensions. The connectivity condition of the whole cluster is

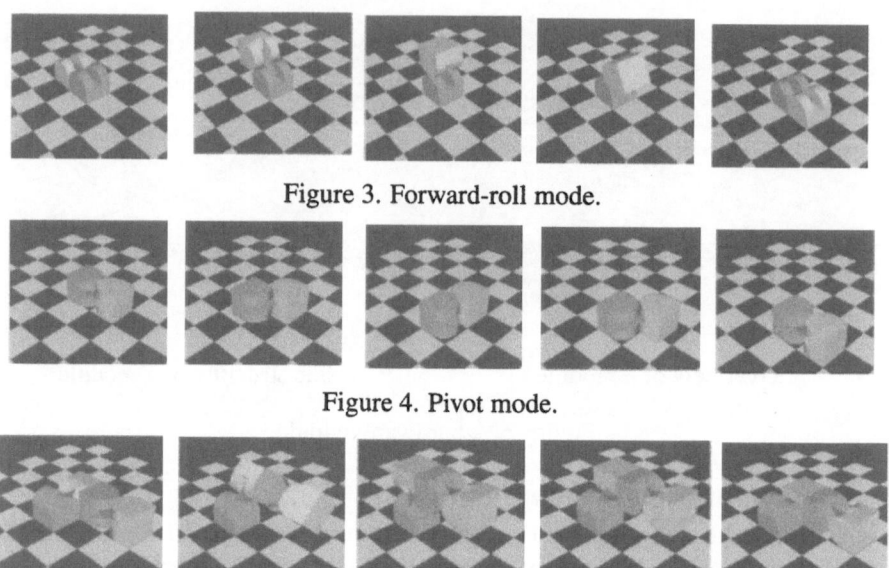

Figure 3. Forward-roll mode.

Figure 4. Pivot mode.

Figure 5. Mode conversion from pivot to forward-roll.

satisfied by placing the modules so that the direction of pivot mode (i.e. the direction of links between semi-cylindrical parts) is different in each layer.

We define the cluster *flow* as *paths* of the *tail* modules towards the *head* (from Fig. 7a to 7d). A *path* denotes a rough routing of a module to move from the head position to the tail. The *global planner* is in charge of generating these paths of modules to realize the desired flow. While there are several ways of generating reconfiguration motion of this kind of cluster, we adopt a simple conveyer-like motion to realize the desired flow. The tail modules move toward the heads by using forward-roll and some coordinated atomic motions on the side of the cluster (Fig. 7b and 7c). They become new heads when they reach the other end of the cluster. The next tails will be sent to

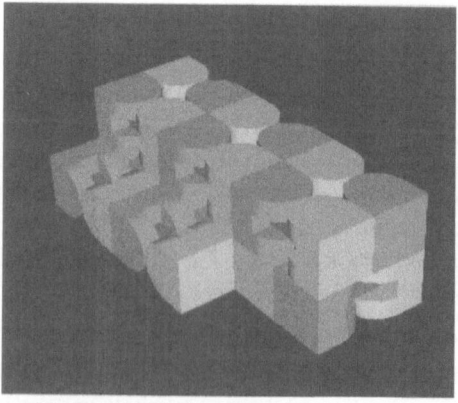

Figure 6. A cluster composed of two layers of pivot mode modules with two converter modules.

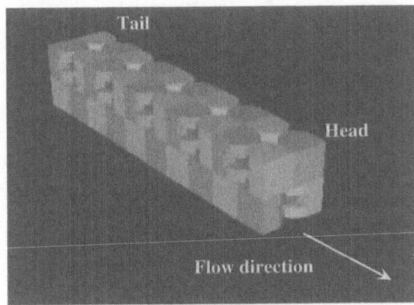
(a) Initial configuration.

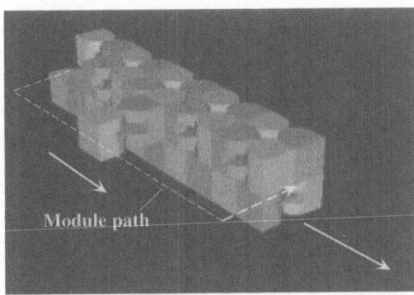
(b) Module moving on the side of cluster along the path.

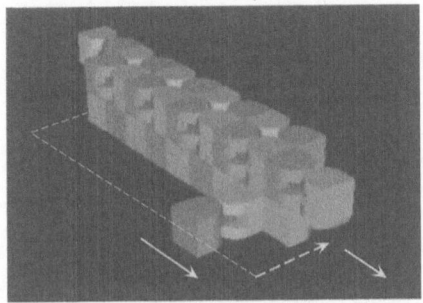
(c) Another module moving.

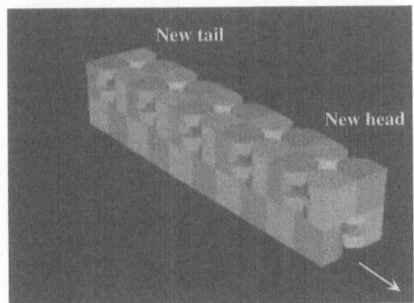
(d) Cluster moved in the direction.

Figure 7. Cluster flow based on "conveyer-like" module transfer.

the heads, and so forth. After these global paths are generated, the detailed sequence of motions for local configuration, named *motion scheme*, will next be determined by the *motion scheme selector*.

3.3. Motion Scheme Selector

After the global planner outputs the module path, appropriate *motion schemes* should be selected to achieve the paths, considering connectivity condition and collision avoidance. The *motion scheme selector* does this job based on a database of rules for local coordinated motion. Each rule includes a motion scheme associated with an initial configuration of involved modules that is described as a graph tree (Fig. 8). If a rule matches the current configuration, the associated motion scheme is added to the candidates to be applied. Then the best motion scheme for desired path is selected to update the motion plan.

In order to implement the motion scheme selector, we extracted several fundamental motion schemes as follows.

1. rolling on a side of the cluster (Fig. 8)
2. conveying a module by right-angle on a plane (Fig. 9)
3. converting the rotational axis of a module (Fig. 10)

Figure 8 shows a rule corresponding to a simple motion scheme of the rolling on the side of the cluster. Figures 9 and 10 illustrate how module configuration changes

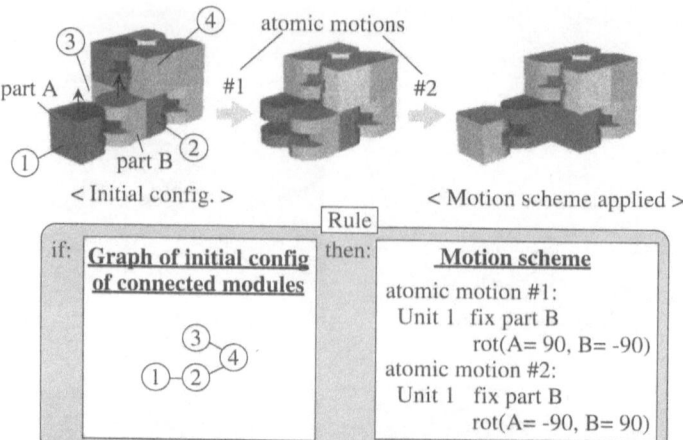

Figure 8. Example of a rule for a rolling motion scheme

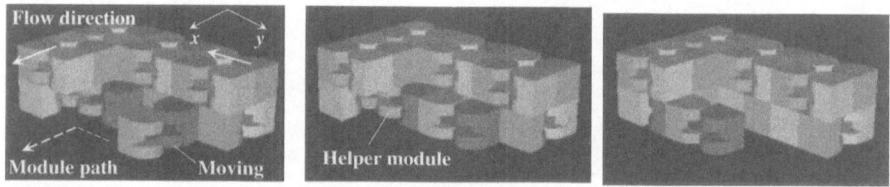

Figure 9. Direction change of cluster on a plane.

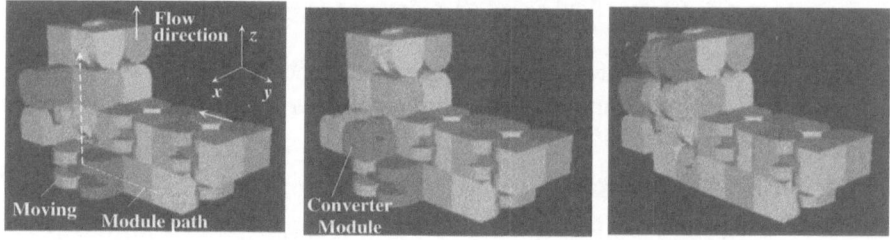

Figure 10. Direction change to vertical direction on a plane.

in the latter two motion schemes. Supposing that the initial module cluster are located on $x - y$ plane, the direction change between $x - y$ axis is done by alternating the layers (Fig. 9). The converter modules are used when the desired cluster flow requires change of rotational axis of the module (Fig. 10). The number of converter modules can be augmented if necessary.

3.4. Planning Results

The motion planning method can generate simple three-dimensional path and is scalable, i.e. applicable to variable number of modules. If modules are equipped with some external sensors, the planning method will allow the module cluster to move around in unknown environments with bumps or walls, adapting its shape to the outside world.

Figure 11 show some snapshots taken from planned motion of a cluster of twenty-

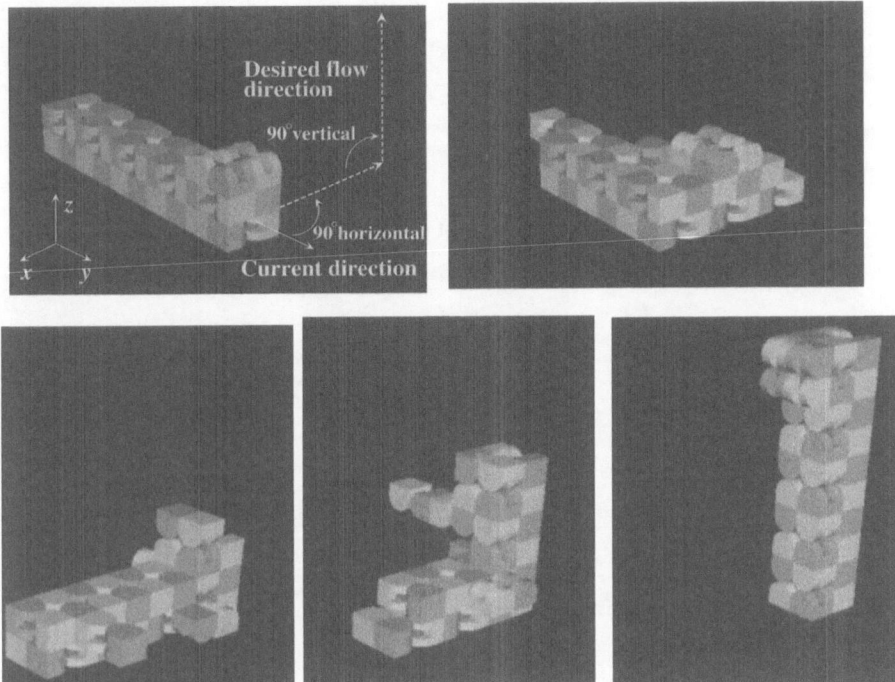

Figure 11. Simulated plan of motions in different flow directions from initial configuration on a plane.

two modules starting from a configuration on a plane. The cluster first changes its flow direction on the horizontal plane, then moves in vertical direction. To reduce the time required for reconfiguration, the simultaneous motions of several modules should be implemented by merging multiple module paths.

In the current implementation, the rule database does not cover all the situations that may occur. Thus the planner is not yet fully automated and human operator intervenes if the planner does not found applicable rules during planning.

4. Hardware Experiments

We are building hardware prototype of robotic modules. In this section, we will verify the motion capacity of hardware module and discuss some issues to be fixed in the future development.

Figure 12 and 13 show the experiment of forward-roll motion (Fig. 3) and mode conversion (Fig. 5). In these experiments, the connecting mechanism showed reliable performance; it has enough strength to hold the module against gravity and the smooth detachment is realized as well. We can also verify the module has sufficient torque to conduct certain two-module motions from Fig. 13. By combining these basic module motions, various motions are possible. Figure 14 is an example of combined motion using three modules.

Although we have verified the basic capacity of motion, the hardware still need

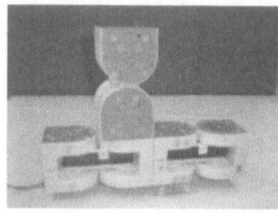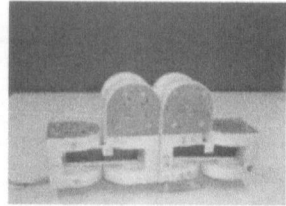

Figure 12. Experiment of forward-roll motion.

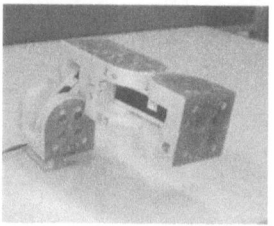

Figure 13. Experiment of mode conversion.

Figure 14. An example of motion sequence for three modules.

to be improved in several aspects to carry out the planned motion in section 3 using the hardware modules. The future improvements includes:

Redesigning electrodes. During the pivot motion (Fig. 4), it happens that the electrodes of power supply, ground and serial line of two modules contact, which causes short circuit. In the next prototype model, the electrodes will be allowed to slide with connecting mechanism to prevent the unnecessary contact.

Improving motor torque. Large torque is required for two-module motion in severest conditions. The actuator mechanism will be improved to provide enough torque by increasing gear reduction ratio.

Equipping modules with some sensors. By detecting the external environment, the group of modules are enabled to adapt to the various situations.

5. Conclusions

This paper discussed motion planning of a self-reconfigurable modular robot. The module was designed to generate both static structure and dynamic robotic motions. This module can form various three-dimensional shapes in spite of simple module design. On the other hand, it has incomplete spatial symmetry due to its limited degree of freedom, which imposes a stronger constraint on motion planning compared to ordinary lattice-type modules. The motion planning should consider this constraint, and we adopt a planning method based on global planning and motion scheme selector. The former outputs the global path to realize overall cluster flow, and the latter selects feasible motion schemes as a combined atomic motions based on a rule database associating a local configuration with a motion scheme. The constraints against module motion is considered in the latter part of the planner. The future work concerning the motion planner includes building a general global path-finding algorithm applicable to wider classes of configuration and investigating more compact and robust motion rule description. We also verified the basic functions of hardware modules through experiments. Some issues on hardware design were clarified to carry out the planned motions. After these refinement, we intend to demonstrate the capacity of self-reconfiguration and dynamic robotic motion, by using more than ten modules.

References

[1] Murata S, Kurokawa H, Yoshida E, Tomita K, Kokaji S 1998 A 3-D Self-Reconfigurable Structure. *Proc. 1998 IEEE Int. Conf. on Robotics and Automation*, 432–439

[2] Kotay K, Rus D, Vona M, McGray C 1998 The Self-Reconfiguring Robotic Molecule. *Proc. 1998 IEEE Int. Conf. on Robotics and Automation*, 424–431.

[3] Ünsal C, Kılıççöte H, Khosla P K 1999 I(CDS)-cubes: a Modular Self-Reconfigurable Bipartite Robotic System. *Proc. SPIE, Sensor Fusion and Decentralized Control in Robotic Systems II*, 246–257.

[4] Yoshida E, Kokaji S, Murata S, Tomita K, Kurokawa H 2000 Miniaturization of Self-Reconfigurable Robotic System using Shape Memory Alloy. *J. of Robotics and Mechatronics*, **12**-2: 1579–1585.

[5] Hamlin G, Sanderson A 1998 *A Modular Approach to Reconfigurable Parallel Robotics*. Kluwer Academic Publishers, Boston.

[6] Yim M and Casal A 1999 Self-Reconfiguration Planning for a Class of Modular Robots. *Proc. SPIE, Sensor Fusion and Decentralized Control in Robotic Systems II*, 246–257.

[7] Castano A, Chokkalingam R, Will P 2000 Autonomous and Self-Sufficient CONRO Modules for Reconfigurable Robots. *Distributed Autonomous Robotics 4*, Parker L E, Bekey G,
Barhen J eds., Springer, 155–164.

[8] Yoshida E, Murata S, Kurokawa H, Tomita K, Kokaji S 1999 A Distributed Method for
Reconfiguration of 3-D homogeneous structure. *Advanced Robotics*, **13**-4:363–380.

[9] Tomita K, Murata S, Yoshida E, Kurokawa H, Kokaji S 1999 Self-assembly and Self-Repair
Method for Distributed Mechanical System. *IEEE Trans. on Robotics and Automation*, **15**-
6:1035–1045.

[10] Kotay K and Rus D 1998 Motion Synthesis for the Self-Reconfigurable Molecule. *Proc.
1998 IEEE/RSJ Int. Conf. on Intelligent Robots and Systems*, 843–851.

[11] Ünsal C, Kılıççöte H, Patton M E, Khosla P K 2000 Motion Planning for a Modular Self-
Reconfiguring Robotic System. *Distributed Autonomous Robotics 4*, Parker L E, Bekey G,
Barhen J eds., Springer, 165–175.

[12] Murata S, Yoshida E, Tomita K, et al. 2000 Hardware Design of Modular Robotic System.
Proc. 2000 IEEE/RSJ Int. Conf. on Intelligent Robots and Systems, CD-ROM, F-AIII-3-5.

Experimental comparison of techniques for localization and mapping using a bearing-only sensor

Matthew Deans and Martial Hebert
Carnegie Mellon University
Pittsburgh, PA, USA
{mdeans,hebert}@cmu.edu

Abstract: We present a comparison of an extended Kalman filter and an adaptation of bundle adjustment from computer vision for mobile robot localization and mapping using a bearing-only sensor. We show results on synthetic and real examples and discuss some advantages and disadvantages of the techniques. The comparison leads to a novel combination of the two techniques which results in computational complexity near Kalman filters and performance near bundle adjustment on the examples shown.

1. Introduction

In this paper we will present a comparison and experimental evaluation of techniques for localization and mapping for a mobile robot equipped with a bearing-only sensor. The robot has no *a priori* knowledge of the environment (no map) and can only observe its egomotion through odometry and the bearings to landmarks in the environment with an omnidirectional camera. The robot has no external position or heading reference, and cannot measure the range to landmarks.

2. Related work

Much of the work published in Simultaneous Localization and Mapping (SLAM), also referred to as Concurrent Localization and Mapping (CLM), makes use of active range sensors. Approaches generally rely on recursive filtering using a Kalman filter or some variant [1]. There have also been attempts to find optimal estimates using batch techniques [2] or to deal with large environments using a hierarchical approach [3, 4].

The Computer vision literature contains a large body of work on multiframe Structure From Motion (SFM). Bundle adjustment is a batch technique that is widely used for SFM under perspective projection [5, 6] and is applicable to SLAM. Some attempts have also been made to recursively compute structure and motion using Kalman filters [7, 8].

The SLAM and SFM problems contain features relevant to localization and mapping with a bearing-only sensor. The goal of this paper is to present some comparisons on how techniques from SLAM and SFM perform on the bearing-only SLAM problem including results from experiments on simulated

D. Rus and S. Singh (Eds.): Experimental Robotics VII, LNCIS 271, pp. 395–404, 2001.
© Springer-Verlag Berlin Heidelberg 2001

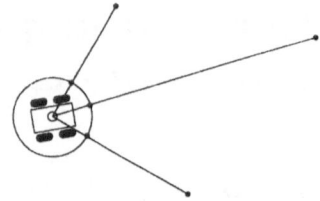

Figure 1. Bearing-only sensor projects landmark position onto unit circle

and real data. To date, little has been published in terms of direct experimental comparison of existing techniques for solving this problem.

3. Localization and Mapping

For simplicity we will only consider a 2D planar environment. The robot position at time i is $\mathbf{m_i} = (x_i, y_i, \theta_i)^T$ and the feature locations are $\mathbf{x_j} = (x_j, y_j)^T$. The odometry is modelled as a stochastic measurement of motion between two consecutive robot poses such that

$$\mathbf{d_i} = f(\mathbf{m_{i-1}}, \mathbf{m_i}) + \nu_i \tag{1}$$

where ν is $\mathcal{N}(0, R_d)$ noise. The bearings are modelled as stochastic measurements of the bearing towards each visible landmark in the rover coordinates,

$$z_{ij} = h(\mathbf{m}_i, \mathbf{x}_j) + \omega_{ij} \tag{2}$$

where ω is $\mathcal{N}(0, \sigma_\omega)$ noise.

3.1. SLAM as 2D Structure from Motion

The Structure from Motion (SFM) work in computer vision provides insight into what we can expect from bearing-only SLAM. In SFM the 3D structure of an environment is reconstructed using 2D projections (images) from unknown camera locations. The perspective camera projection model is very similar to a bearing-only sensor; depth information is lost in the projection. The only knowledge of the scene from a single image is that scene points lie somewhere along a ray originating at the camera center of projection and passing through the image plane at the point where the feature appears. In bearing-only SLAM, landmarks are projected onto a unit circle around the robot (Figure 1), and a similar constraint applies for the ray intersecting this projection surface.

If only the bearing measurements are available, and not the odometry, then a simple counting argument gives us a necessary condition for when the 2D structure from motion problem can be solved. For each robot position there are 3 unknowns, and for each landmark 2 unknowns. If there are M robot positions and N landmarks, then there are $3M + 2N$ unknowns. Assuming that all landmark features are observable from all robot positions, there will be MN measurements, and in general these will be linearly independent. There are 4 gauge freedoms[9] in the estimation, i.e. we can translate, rotate, and scale the solution and the measurements will be unchanged. We must set an absolute coordinate frame and absolute scale with 4 additional constraints. We can find a solution to the problem when $MN + 4 > 3M + 2N$, or

$$(M - 2)(N - 3) > 2 \qquad (3)$$

This constraint on the solution to the problem shows that there is no solution to the two frame SFM problem in 2D, regardless of methodology. Only when there are three robot positions and at least five landmarks or more than three robot positions and at least four landmarks can we get a solution, and this solution is only defined up to the gauge freedoms. In contrast, aligning just two scans from an accurate range-bearing sensor can give an estimate of vehicle motion and make up for errors in odometry.

In most SFM algorithms, little or no a priori information about the camera motion is used for reconstruction, camera motion is recovered in the estimation along with scene structure using only the image measurements. Odometry can be seen as providing a prior for the camera motion which is absent in general SFM problems. If we add odometry then we can immediately disambiguate the scale. There is even a unique solution to the two frame problem if odometry is available since it will provide a direct estimate of the relative pose of the second view. However until the third view is incorporated, bearing measurements cannot begin to correct odometry errors within the estimation algorithm. This can cause initialization problems with Kalman filters which are observed in practice, particularly when odometry is poor.

A final note is that bias in odometry cannot always be corrected in bearings-only SLAM. If the measurements of distance travelled are biased by a scale factor then the estimate of robot motion *and* landmark positions will be scaled by the same factor and no estimator can recover the bias using the bearing measurements. This is an inherent ambiguity. In contrast, SLAM with range-bearing sensors does not suffer from this problem since the range measurements provide direct information about the scale of the solution.

4. Estimating parameters

Finding an estimate for the model parameters $\mathbf{m} = (\mathbf{m_0}^T, \mathbf{m_1}^T \cdots \mathbf{m_M}^T)^T$ and $\mathbf{x} = (\mathbf{x_0}^T, \mathbf{x_1}^T \cdots \mathbf{x_N}^T)^T$ can be done in many ways. In this paper we wish to make some comparisons between two of these methods in particular. The first method is the Extended Kalman filter and the second is bundle adjustment. Each of these methods has properties which make it attractive for use in SLAM, and the nature of the SLAM problem is important to the way in which these techniques are used and the results that can be expected.

4.1. Extended Kalman Filter

The Extended Kalman Filter (EKF) has been used in SLAM and SFM to recursively estimate a state vector which consists of the current rover pose or camera parameters and the positions of all landmark or scene features [1, 8]. As a new measurement is made, the filter goes through prediction and update steps which incorporate the new measurement and generates a best estimate (mean) and an uncertainty estimate (covariance) for the current rover pose and the map. The mean and covariance are kept as a sufficient statistic for the posterior probability over state space, and old observations are discarded after being incorporated. Nonlinearities in bearing-only SLAM cause difficulties to the EKF; in particular, getting started is tricky. The posterior over robot pose and landmark position is significantly non-Gaussian such that the mean and covariance in the filter are not a reasonable sufficient statistic for the data. For this reason, we use bundle adjustment (to be explained in the next section) to get an initial state estimate and initial covariance matrix using some portion of the data, then recursively process the rest of the data with a Kalman filter. The quality of the result depends on how much of the data is used to get the initial estimates.

Space considerations prohibit detailed explanation of the use of Kalman filters in SLAM but there exists a large body of literature on the topic, particularly for problems where range and bearing measurements are available [1, 10]. Although they should be included in a complete treatment on this topic, we will also omit discussion of modifications to the Kalman filtering method such as implicit Jacobian computations [11] and covariance intersection [12].

4.2. Bundle Adjustment

Bundle adjustment is a full nonlinear optimization which does not rely on a mean and covariance as a sufficient statistic for previous observations and state estimates. Instead it linearizes the estimation problem at every step using all available observations and the current best estimate. Bundle adjustment may optimize all motion and structure parameters at every step, so the state vector is the entire history of robot pose (trajectory) and the entire map.

We wish to minimize the cost function

$$E(\Theta) = \frac{1}{2}(\mathbf{d} - \mathbf{f}(\Theta))^T \mathbf{R_d}^{-1}(\mathbf{d} - \mathbf{f}(\Theta)) + \frac{1}{2}(\mathbf{z} - \mathbf{h}(\Theta))^T \mathbf{R_z}^{-1}(\mathbf{z} - \mathbf{h}(\Theta)) \quad (4)$$

where $\Theta = (\mathbf{m}^T, \mathbf{x}^T)^T$ is the vector of parameters to be estimated, $\mathbf{d}$ and $\mathbf{z}$ are all odometry and all bearing measurements stacked into vectors, and $\mathbf{f}()$ and $\mathbf{h}()$ are all predicted odometry and bearing measurements stacked into vectors. The first term in (4) penalizes robot motion that does not agree well with odometry measurements and the second term penalizes robot motion and landmark map combinations that do not agree well with bearing measurements.

Taking the first and second derivatives of (4) we find

$$\nabla_\Theta E = -J_f^T R_d^{-1} \epsilon_d - J_h^T R_z^{-1} \epsilon_z \quad (5)$$

$$\nabla_\Theta^2 E = -\nabla_\Theta^2 f R_d^{-1} \epsilon_d + J_f^T R_d^{-1} J_f - \nabla_\Theta^2 h R_z^{-1} \epsilon_z + J_h^T R_z^{-1} J_h \quad (6)$$

where $J_f = \nabla_\Theta f$ and $J_h = \nabla_\Theta h$ are the Jacobian matrices of the measurement equations for odometry and bearings, $\epsilon_d = (d_1 - f(x_0, x_1), d_2 - f(x_1, x_2)...)^T$ is the difference between measured odometry and odometry predicted by the model parameters, and ϵ_z is the difference between measured and predicted bearings. These last two quantities are analogous to innovations in a Kalman filtering context.

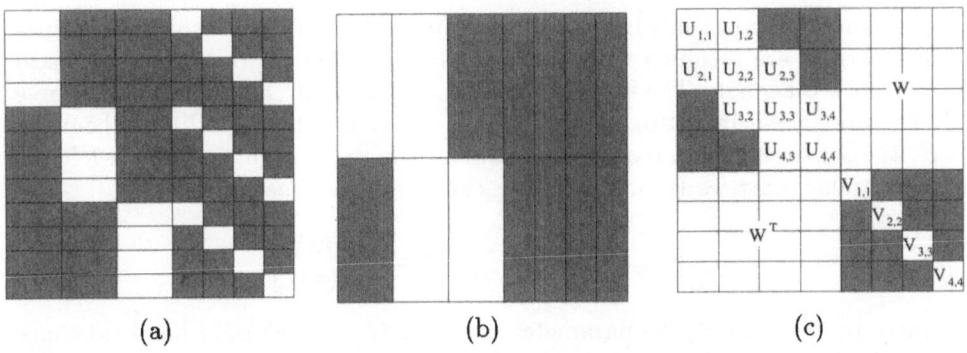

Figure 2. Sparse structure of derivatives. (a) Jacobian of bearing measurement equation. (b) Jacobian of odometry measurement equation. (c) Approximation to the Hessian of the cost function.

When the innovations are small the Jacobian inner products dominate expression (6). The Jacobian term will also dominate when the measurement equations are approximately linear (or linear), since $\nabla_\Theta^2 f$ or $\nabla_\Theta^2 h$ will be small (or zero). In our case, $\mathbf{f}$ is in fact linear, but $\mathbf{h}$ contains an arctangent. Bundle adjustment drops the second derivative terms and approximates the Hessian of the cost function using only the Jacobian inner products, i.e.

$$H = \nabla_\Theta^2 \approx J_f^T R_d^{-1} J_f + J_h^T R_z^{-1} J_h. \tag{7}$$

Note that the true H is the Fisher Information Matrix when Θ is the true parameter vector. We will use it as the inverse covariance matrix for estimating uncertainties. Newton iterations are used to minimize (4) by solving

$$H\Delta\Theta_k = -(J_f^T R_d^{-1}\epsilon_d + J_h^T R_z^{-1}\epsilon_z) \tag{8}$$

and then computing the update

$$\Theta_{k+1} = \Theta_k + \Delta\Theta_k \tag{9}$$

In general solving equation (8) is cubic in the number of model parameters ($3M + 2N$ here). However, photogrammetrists have known for decades how to speed up bundle adjustment using the sparse nature of the structure from motion problem. We can take a similar approach here by exploiting the nature of the bearing-only SLAM problem.

Each bearing measurement depends only on the landmark being measured and the robot pose at the time that the measurement is taken. Therefore, each row of the Jacobian of the bearing measurement equation, J_h, has nonzero entries only for the columns corresponding to the parameters which represent that landmark position and robot pose. The sparse structure is shown in Figure 2(a). The Jacobian of the odometry measurement equation, J_f, depends only on two consecutive robot poses. Each row of the Jacobian of the odometry measurement equation, then, has nonzero entries only for the robot poses before and after the motion. Odometry contains no information about landmark positions and corresponding columns of J_f are zero. The Jacobian has the structure shown in Figure 2(b). Because odometry and bearing measurement errors are assumed to be uncorrelated, R_d and R_z (and their inverses) are block diagonal. The Hessian $H = J_f^T R_d^{-1} J_f + J_h^T R_z^{-1} J_h$ has the sparse structure shown in Figure 2(c). The upper left is a block tridiagonal matrix U, the lower right is a block diagonal, and the upper right and lower left are rectangular matrices W and W^T. We can write the equation (8) as

$$\begin{bmatrix} U & W \\ W^T & V \end{bmatrix} \begin{bmatrix} \Delta \mathbf{m} \\ \Delta \mathbf{x} \end{bmatrix} = \begin{bmatrix} \epsilon(\mathbf{m}) \\ \epsilon(\mathbf{x}) \end{bmatrix} \tag{10}$$

where the Hessian H, the parameter update $\Delta \Theta$, and the right hand side have been partitioned. Now we premultiply both sides by

$$\begin{bmatrix} I & 0 \\ -W^T U^{-1} & I \end{bmatrix}$$

$$\begin{bmatrix} U & W \\ 0 & V - W^T U^{-1} W \end{bmatrix} \begin{bmatrix} \Delta \mathbf{m} \\ \Delta \mathbf{x} \end{bmatrix} = \begin{bmatrix} \epsilon(\mathbf{m}) \\ \epsilon(\mathbf{x}) - W^T U^{-1} \epsilon(\mathbf{m}) \end{bmatrix} \tag{11}$$

and solve equations (11) in two steps. We first solve the bottom equation

$$(V - W^T U^{-1} W) \Delta \mathbf{x} = \epsilon(\mathbf{x}) - W^T U^{-1} \epsilon(\mathbf{m}) \tag{12}$$

to find $\Delta \mathbf{x}$ and then substitute it into the top equation, rearranging to get

$$U \Delta \mathbf{m} = \epsilon(\mathbf{m}) - W \Delta \mathbf{x} \tag{13}$$

and solve for $\Delta \mathbf{m}$. Computation of $(V - W^T U^{-1} W)$ is $O(MN^2)$, and solving (12) is $O(N^3)$. Substituting $\Delta \mathbf{x}$ into (13) and solving only requires $O(MN)$ due to the block tridiagonal structure of U, so the overall complexity of bundle adjustment for bearing-only SLAM is $O(MN^2 + N^3)$, which scales linearly with the size of the trajectory. The complexity of bundle adjustment is larger than that of a Kalman filter, which has a complexity of $O(N^3)$ for N landmarks, but it is not complex as a general inverse Hessian approach which might require $O((N + M)^3)$.

5. Results

Results were generated for synthetic and real data. In the synthetic example, we simulated a robot driving in a large circle of radius 100 meters, with 50 landmarks scattered uniformly in a square area circumscribing the robot path. Odometry was simulated by computing the true robot motion and adding Gaussian noise to the estimates of along-track and cross-track motion to simulate slip as well as the rotation. Figure 3 shows estimates of robot motion and landmark map from a Kalman filter and from a bundle adjustment. In each figure, the robot trajectory is the circular path of dots and the landmarks are scattered inside and outside the path plotted with their corresponding uncertainty ellipses.

Both methods were also used to analyze real data from an RWI ATRV. Unfortunately there is no ground truth for this experiment. The robot drove

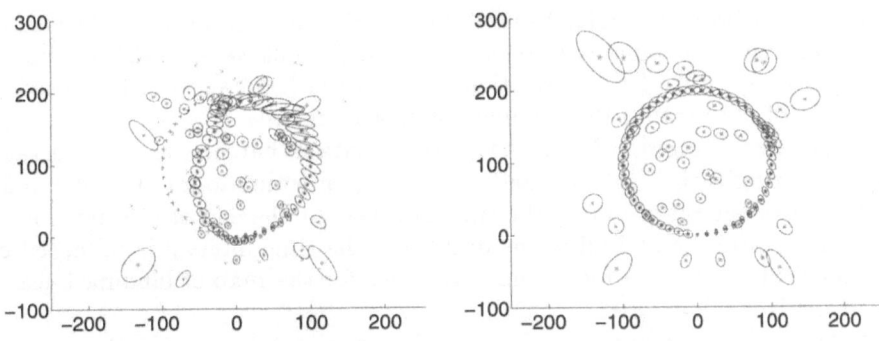

Figure 3. Synthetic data. (a) Kalman filter result (b) Bundle adjustment result

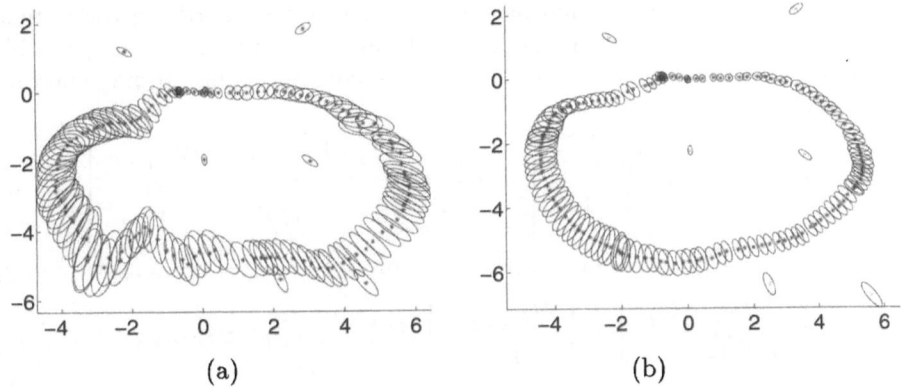

Figure 4. Results on real data (a) Kalman filter (b) Bundle adjustment.

in an approximately elliptical path between six different artificial landmarks. There was a problem with the drive mechanism which caused bias in the odometry measurements. Dead reckoning estimated that the robot drove in a spiral shaped path. Bundle adjustment results are shown in Figure 4(a). Without modelling the bias the bundle adjustment recovers the elliptical path of the robot.

The Kalman filter diverges in the presence of the odometry bias. The odometry was calibrated by fitting a linear function of the data to the bundle adjustment trajectory estimate. With the calibrated odometry and an initial model from bundle adjustment using the first 20 robot poses, the Kalman filter was able to produce the result in Figure 4(b). The lower left portion of the trajecory shows some reconstruction behavior which is most likely due to the introduction of the lower left feature. Since it processes one measurement at a time the Kalman filter does not initialize features well with bearing-only measurements.

5.1. Synthesis

The properties of the methods discussed above and the performance seen in experiments suggests a new approach to bearing-only localization and mapping.

We wish to have a recursive formulation so that the computational burden does not grow with the number of sensor scans made, but we also wish to avoid the bias introduced in the linearization phase in the Kalman filter by considering batches of data rather than individual measurements.

A way to accomplish this is to process data in batches, using observations made during a time interval and computing an estimate for the map and for the pose of the robot during the time interval. At the end of the time interval, the state estimate and all of the data from the time interval is collapsed onto a sufficient statistic, a mean and covariance for the map of landmark features and the last robot pose.

The very first batch of data can be processed in exactly the same way as was described for bundle adjustment. At the end of that estimation, the information matrix will have the same sparse structure as before and can be partitioned as it was in equation (10). We will further partition the U matrix into the block dealing with robot pose from time 1 through $k - 1$, the block dealing with robot pose at time k, and the corresponding off-diagonal blocks as follows

$$P^{-1} = \begin{bmatrix} U & W \\ W^T & V \end{bmatrix} = \begin{bmatrix} U_{1..k-1,1..k-1} & U_{1..k-1,k} & W_{1..k-1} \\ U_{1..k-1,k}^T & U_{kk} & W_k \\ W_{1..k-1}^T & W_k^T & V \end{bmatrix} \qquad (14)$$

In order to marginalize this matrix and remove the states $m_1..m_k$, we compute

$$P_k^{-1} = \begin{bmatrix} U_{kk} & W_k \\ W_k^T & V \end{bmatrix} - \begin{bmatrix} U_{1..k-1,k}^T \\ W_{1..k-1}^T \end{bmatrix} U_{1..k-1,1..k-1}^{-1} \begin{bmatrix} U_{1..k-1,k} & W_{1..k-1} \end{bmatrix}$$

$$(15)$$

The state (m_k, x) and the marginalized information matrix P_k^{-1} are used as a sufficient statistic in the next step of the filter. In general, P_k^{-1} will be a full matrix, it will no longer have the sparse structure of the Hessian from bundle adjustment due to the marginalization step.

As observations are made from new robot positions, the state vector grows to include $\{m_k, m_k + 1, m_k + 2, ..., x\}$ and the information from the bearing and odometry measurements are added into the information matrix, growing the inverse covariance matrix up and to the left. The upper left block will once again have a block tridiagonal form which can be used to provide fast optimization of the parameters. Since the blockwise elimination and backsubstitution done in equations (10) through (13) fill in the lower right block anyway, the computational complexity is not affected by filling in the block in the previous step. The update step of the new recursive/batch filter is $O(N^3 + kN^2)$ where k is the size of the time window used for the batch update. For small fixed k this is approximately the same computational complexity as the Kalman filter. The algorithm is similar to one reported recently for SFM [13].

Results of applying this filter with a batch size of 10 to the experiment discussed earlier are shown in Figure 5(a). The uncertainty ellipses on each robot pose are plotted as computed at the end of a batch update, so smoothing does not take place over the whole trajectory as it does with full bundle adjustment. Due to this the groups of 10 poses per batch are apparent from the

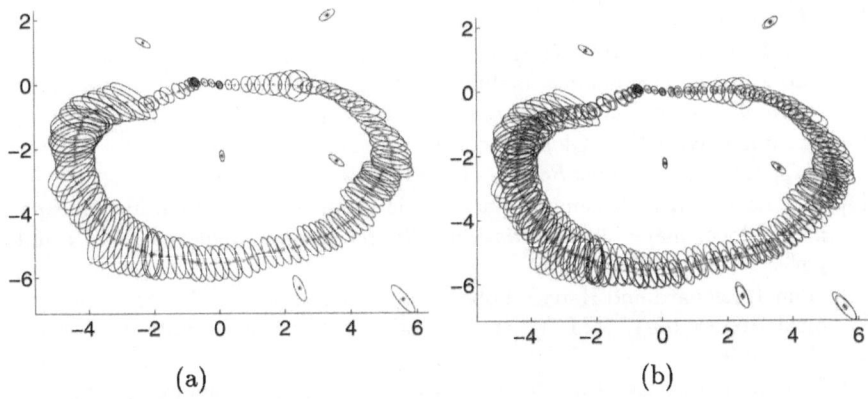

(a)   (b)

Figure 5. (a) map and trajectory estimated by recursive/batch filter using a batch size of 10. (b) result superimposed with result from bundle adjustment to show agreement.

uncertainty ellipses. Figure 5(b) shows the results from bundle adjustment and recursive/batch methods plotted against each other for comparison. Results for the example considered are very similar for the two methods.

6. Discussion

The examples here illustrate the advantages of bundle adjustment over Kalman filtering for the bearing-only SLAM problem. When Kalman filters are used, they need to be initialized with a good estimate of the robot state and landmark locations and bundle adjustment is an efficient way to compute optimal least squares estimates in order to initialize the filter. Bundle adjustment is empirically more robust to bias in the system model, which is to be expected since the linearizations are recomputed each time and are computed near the optimum. We did not investigate robustness to outliers since there exist both Kalman filters and bundle adjustment techniques specifically designed to deal with outliers[14, 6] but these were not implemented.

For computational and memory requirements, bundle adjustment is not practical for large problems. Because the entire observation history must be recorded, the memory requirement scales as $O(MN)$. By contrast the Kalman filter requires $O(N^2)$ memory. The Kalman filter requires $O(N^3)$ computation for each update. Computation required in bundle adjustment is $O(N^3 + MN^2)$ which grows linearly in the number of robot positions in the estimate. This is significantly less than a general second order least squares optimization but still not practical for a robot operating for long periods of time.

By combining Kalman filtering and bundle adjustment concepts, a new recursive/batch filter was introduced. The filter allows a flexibility in the batch size, trading off computational requirements and performance. If the batch size is set to 1, the extended Kalman filter results, and if the batch size is unlimited, full bundle adjustment results.

References

[1] J. J. Leonard and H. F Durrant-Whyte. Simultaneous map building and localization for an autonomous mobile robot. In *IEEE/RSJ International Workshop on Intelligent Robots and Systems IROS '91*, pages 1442–1447, 1991.

[2] F. Lu and E. Milios. Globally consistent range scan alignment for environment mapping. *Autonomous Robots*, 4(4):333–349, 1997.

[3] K. Chong and L. Kleeman. Large scale sonarray mapping using multiple connected local maps. In *International Conference on Field and Service Robotics*, pages 278–285, 1997.

[4] John J. Leonard and Hans Jabob S. Feder. Decoupled stochastic mapping. Technical Report 99-1, MIT Marine Robotics Laboratory, Cambridge, MA 02139, USA, 1999.

[5] Richard I. Hartley. Euclidean reconstruction from uncalibrated views. In Zisserman Mundy and Forsyth, editors, *Applications of Invariance in Computer Vision*, pages 237–256. Springer Verlag, 1994.

[6] B. Triggs, P. McLauchlan, R. Hartley, and A. Fitzgibbon. Bundle adjustment - a modern synthesis. In *To appear in Vision Algorithms: Theory & Practice*. Springer-Verlag, 2000.

[7] T. J. Broida, S. Chandrashekhar, and R. Chellappa. Recursive 3-d motion estimation from a monocular image sequence. *IEEE Trans. on Aerospace and Electronic Systems*, 26(4):639–656, 1990.

[8] Ali Azarbayejani and Alex P. Pentland. Recursive estimation of motion, structure and focal length. *IEEE Transactions on Pattern Analysis and Machine Intelligence*, 17(6):562–575, 1995.

[9] Philip F. McLauchlan. Gauge invariance in projective 3d reconstruction. In *Proceedings IEEE Workshop on Multi-View Modeling and Analysis of Visual Scenes (MVIEW'99)*, pages 37–44, 1999.

[10] M.W.M.G Dissanayake and et al. An experimental and theoretical investigation into simultaneous localization and map building (slam). In *Proc. 6th International Symposium on Experimental Robotics*, pages 171–180, 1999.

[11] S. Julier, J. Uhlmann, and H. Durrant-Whyte. A new approach for filtering nonlinear systems. In *Proceedings of the 1995 American Controls Conference*, pages 1628–1632, 1995.

[12] J. K. Uhlmann, S. J. Julier, and M. Csorba. Nondivergent simultaneous map-building and localization using covariance intersection. In *SPIE Proceedings: Navigation and Control Technologies for Unmanned Systems II*, volume 3087, pages 2–11, 1997.

[13] P. Mclauchlan. A batch/recursive algorithm for 3d scene reconstruction. In *Proceedings of CVPR 2000*, pages II:738–43, 2000.

[14] R. P. N. Rao. Robust kalman filters for prediction, recognition, and learning. Technical Report 645, Computer Science Department, University of Rochester, 1996.

Robot Navigation for Automatic Model Construction using Safe Regions

Héctor González-Baños   Jean-Claude Latombe

Department of Computer Science
Stanford University, Stanford, CA 94305, USA
e-mail:{hhg,latombe}@robotics.stanford.edu

Abstract: Automatic model construction is a core problem in mobile robotics. To solve this task efficiently, we need a motion strategy to guide a robot equipped with a range sensor through a sequence of "good" observations. Such a strategy is generated by an algorithm that repeatedly computes locations where the robot must perform the next sensing operation. This is called the next-best view problem. In practice, however, several other considerations must be taken into account. Of these, two stand out as decisive. One is the problem of safe navigation given incomplete knowledge about the robot surroundings. The second one is the issue of guaranteeing the alignment of multiple views, closely related to the problem of robot self-localization. The concept of *safe region* proposed in this paper makes it possible to simultaneously address both problems.

1. Introduction

Automatic model construction is a fundamental task in mobile robotics [1]. The basic problem is easy to formulate: After being introduced into an unknown environment, a robot, or a team of robots, must perform sensing operations at multiple locations and integrate the acquired data into a representation of the environment. Despite this simple formulation, the problem is difficult to solve in practice. First, there is the problem of choosing an adequate representation of the environment — e.g., topological maps [2], polygonal layouts [1], occupancy grids [4], 3-D models [12], or feature-based maps [6]. Second, the representation must be extracted from imperfect sensor readings — e.g., depth readings from range-sensors may fluctuate due to changes in surface textures [3], different sets of 3-D scans must be zippered [13], and captured images must be aligned and registered [11]. Finally, if the system is truly automatic, the robot must decide on its own the necessary motions to construct the model [5].

Past research in model construction has mainly focused on developing techniques for extracting relevant features (e.g., edges, corners) from raw sensor data, and on integrating these into a single and consistent model. There is also prior research on the computation of sensor motions, mostly on finding the *next-best view* (NBV) [3, 11]: Where should the sensor be placed for the next sensing operation? Typically, a model is first built by combining images taken from a few distributed viewpoints. The resulting model usually contains gaps. An NBV technique is then used to select additional viewpoints that will provide the data needed to fill the remaining gaps.

D. Rus and S. Singh (Eds.): Experimental Robotics VII, LNCIS 271, pp. 405–415, 2001.

Traditional NBV approaches are not suitable for mobile robotics. One reason is that most of the existing NBV techniques have been designed for systems that build a 3-D model of a relatively small object using a precise range sensor moving around the specimen. Collisions, however, are not a major issue for sensors that are mechanically constrained to operate outside the convex hull of the scene. In robotic applications, by contrast, the sensor navigates within the convex hull of the scene. Therefore, *safe navigation* considerations must always be taken into account when computing the next-best view for a robot map builder.

The second reason why most existing NBV techniques cannot be applied to mobile robots is that very few of the proposed approaches explicitly consider image-registration issues (one exception is the sensor-based technique presented in [11]). Localization problems particularly affect mobile sensors, and image registration becomes paramount when it is the means by which a mobile robot re-localizes itself (this is the so-called *simultaneous localization and map building* problem) [9, 7]. Although many image-registration techniques can be found in the literature, all require that each new image significantly overlaps with portions of the environment seen by the robot at previous sensing locations [9].

The system presented in [5] deals with the safe navigation and localization problems by applying the concept of *safe region* and the NBV algorithm introduced in this paper. With safe regions, it is possible to iteratively build a map by executing union operations over successive views, and use this map for motion planning. Moreover, safe regions can be used to estimate the overlap between future views and the current global map, and to compute locations that could potentially see unexplored areas.

The work in [5] is mainly about system integration and proof of concept. Instead, this paper focuses on the formal definition of a safe region (Section 2), and describes how to compute such region from sensor data (Section 3). An NBV algorithm based on safe regions is outlined in Section 5, and Section 6 describes an experimental run using our system.

2. Definition of Safe Regions

Suppose that the robot is equipped with a polar range sensor measuring the distance from the sensor's center to objects lying in a horizontal plane located at height h above the floor. Because all visual sensors are limited in range, we assume that objects can only be detected within a distance d_M. In addition, most range-finders cannot reliably detect surfaces oriented at grazing angles with respect to the sensor. Hence, we also assume that surface points that do not satisfy the sensor's incidence constraint cannot be reliably detected by the sensor. Formally, our visibility model is the following:

Definition 2.1 (Visibility under Incidence and Range Constraints) *Let the open subset* $W \subset \Re^2$ *describe the workspace layout. Let* ∂W *be the boundary* W. *A point* $w \in \partial W$ *is said to be* visible *from* $q \in W$ *if the following conditions are true:*

1. Line of sight constraint: *The segment from* q *to* w *doesn't intersect* ∂W.

2. Range constraint: $d(q, w) \leq d_M$, *where* $d(q, w)$ *is the Euclidean distance between* q *and* w, *and* $d_M > 0$ *is an input constant.*

3. Incidence constraint: $\angle(\boldsymbol{n}, \boldsymbol{v}) \leq \tau$, *where* $\boldsymbol{n}$ *is a vector perpendicular to* ∂W *at* w, $\boldsymbol{v}$ *is oriented from* w *to* q, *and* $\tau \in [0, \pi/2]$ *is an input constant.*

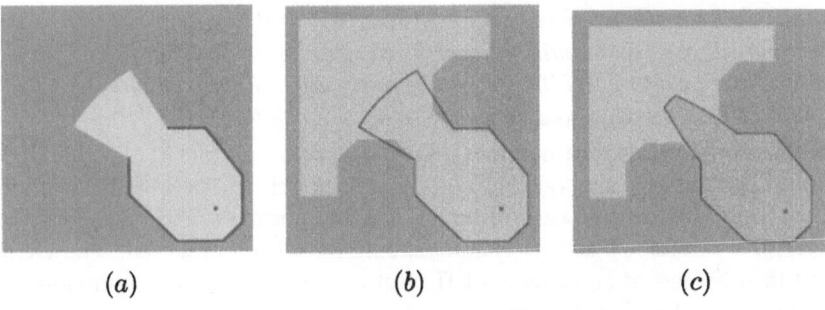

(a)   (b)   (c)

Figure 1. Effect of incidence on safe regions.

Without any loss of generality we assume the sensor is located at the origin (the workspace can always be re-mapped to a reference frame centered on the sensor). The sensor's output is assumed to be as follows:

Definition 2.2 (Range Sensor Output) *The output of a range sensor is an ordered list* Π, *representing the sections of* ∂W *visible from the origin under Definition 2.1. Every* $r(\theta; a, b) \in \Pi$ *is a polar function describing a section of* ∂W, *and such function is continuous* $\forall\, \theta \in (a, b)$ *and undefined elsewhere.* Π *contains at most one function defined for any* $\theta \in (-\pi, \pi]$ *(i.e., no two functions overlap), and the list is ordered counter-clockwise.*

Given an observation Π made by the robot at a location q, we define the *local safe region* s_l at q as the largest region guaranteed to be free of obstacles. While range restrictions have an obvious impact on s_l, the effect of incidence is more subtle. In Figure 1(a), a sensor detects the surface contour shown in black. A naive approach may construct the region in light color (yellow) by joining the detected surfaces with the perimeter limit of the sensor, and consider this region free from obstacles. Because the sensor is unable to detect surfaces oriented at grazing angles, this region may be not be safe, as shown in (b). A true safe region is shown in (c), for an incidence constraint of $\tau = 70$ deg.

3. Computing Safe Regions

The region s_l is bounded by solid and free curves. A *solid curve* represents an observed section of ∂W, and is contained in the list Π. Given two solid curves $\{r_1(\theta; a_1, b_1), r_2(\theta; a_2, b_2)\} \subseteq \Pi$, r_2 is said to *succeed* r_1 if no other element in Π is defined in the interval $[b_1, a_2]$. A curve $f(\theta; b_1, a_2)$ joining a pair (r_1, r_2) of successive sections is called a *free curve* if: (1) no undetected obstacle is contained in the polar region $b_1 < \theta < a_2$ bounded by f; and (2) this region is the largest possible.

The main goal of this section is to find the free curves that join each successive pair in Π in order to bound the region s_l. It turns out that the complexity of f is $O(1)$. In fact, a free curve f can be described using no more than 3 function primitives:

Theorem 3.1 (Free Curves) *Suppose* $r_2(\theta; a_2, b_2)$ *succeeds* $r_1(\theta; a_1, b_1)$ *in the output list* Π. *If* ∂W *is continuously differentiable, then the free curve* $f(\theta; b_1, a_2)$ *connecting* r_1 *to* r_2 *consists of at most three pieces. Each piece is either a line segment, a circular arc, or a section of a logarithmic spiral.*

The rest of this section proves this claim. But first we need the following lemma:

Lemma 3.2 *Let $r_2(\theta; a_1, b_1)$ succeed $r_2(\theta; a_2, b_2)$ in the list Π. Let C be some ob-stacle, and suppose that neither r_1 nor r_2 are part of the boundary of C (i.e., C is disjoint from r_1 and r_2). If ∂W is continuously differentiable, then no portion of C lies within a distance d_M from the origin in the polar interval $b_1 < \theta < a_2$.*

Proof: Suppose the lemma is not true — that is, there is a portion of C within d_M of the origin inside the polar interval (b_1, a_2). Let p be the closest point to the origin in the boundary of C. Because ∂W is differentiable, the normal of ∂W at p points toward the origin. Therefore, p and its vicinity should have been observed. The vicinity of p must then be part of an element of Π. But this contradicts our assumption that r_2 succeeds r_1 and that C is disjoint from r_1 and r_2. $\square$

From here on, let $\beta = a_2 - b_1$, $\rho_1 = r_1(b_1)$ and $\rho_2 = r_2(a_2)$; and let l_1 and l_2 denote the rays joining the origin with $p_1 = (\rho_1, b_1)$ and $p_2 = (\rho_2, a_2)$, respectively.

Each endpoint of a curve in Π represents one of the following events: the sensor line-of-sight was occluded (denoted as case $\{o\}$), the range constraint was exceeded (case $\{e\}$), or the incidence constraint was exceeded (case $\{v\}$). To join p_1 with p_2 there are a total of 6 distinct cases: $\{v,v\}$, $\{v,o\}$, $\{v,e\}$, $\{e,e\}$, $\{o,o\}$ and $\{e,o\}$. The cases $\{o,e\}$, $\{o,v\}$ and $\{e,v\}$ are mirror images of other cases.

Case $\{v,v\}$:   The incidence constraint was exceeded at $\theta = b_1$ and $\theta = a_2$. There-fore, the normal to ∂W immediately after r_1, and immediately before r_2, is oriented at a grazing angle with respect to the sensor. Suppose that ∂W continues after r_1 with its surface normal constantly oriented at exactly an angle τ with respect to the sensor's line-of-sight. This curve in polar coordinates satisfies the following relations:

$$n \;\dot{=}\; -r\,\delta\theta\,\hat{e}_r + \delta r\,\hat{e}_\theta, \tag{1}$$

$$n \cdot (-r\hat{e}_r) = r|n|\cos(\tau) \;\implies\; \frac{1}{r}\frac{\delta r}{\delta\theta} = \pm\lambda, \text{ with } \lambda \;\dot{=}\; \tan(\tau). \tag{2}$$

Hence, the curve's equation is $r = r_o \exp[\pm\lambda(\theta - \theta_o)]$, with $r_o = \rho_1$ and $\theta_o = b_1$. The equation now defines two spirals: a spiral s_1^+ growing counter-clockwise from p_1 (or shrinking clockwise), and a second spiral s_1^- shrinking counter-clockwise from p_1 (or growing clockwise). ∂W must continue counter-clockwise from p_1 either "above" s_1^+ or "below" s_1^-; otherwise, the incidence constraint would not have been violated.

Similarly, for the opposite end p_2, let $r_o = \rho_2$ and $\theta_o = a_2$. The solution to equation (2) now defines a spiral s_2^- growing clockwise from p_2 (or shrinking counter-clockwise), and a second spiral s_2^+ shrinking clockwise from p_2 (or growing counter-clockwise). ∂W must continue clockwise from p_2 either "above" s_2^- or "below" s_2^+.

Remark 1. ∂W cannot continue below s_1^- when $\rho_1 \exp(-\lambda\beta) < \rho_2$. In other words, ∂W cannot continue below s_1^- if this spiral curve cuts l_2 below the point p_2 (Figure 2(a)). To show this, suppose ∂W continues below s_1^-, which implies that ∂W bends toward the sensor immediately after r_1. We know that ∂W does not cross the origin, else nothing is visible under Definition 2.1 and Π would be empty. Hence, ∂W would have to bend outwards before cutting the ray l_2, otherwise r_2 will be occluded. Since ∂W is differentiable, there must then be a point p where the normal to ∂W points towards the origin. Because of Lemma 3.2, this point p is not occluded by any other section of ∂W that is disjointed from r_1 and r_2. Therefore, the vicinity of p is a visible portion of ∂W. This violates our assumption that r_2 succeeds r_1. Thus, when $\rho_1 \exp(-\lambda\beta) < \rho_2$, the first section of the curve f joining r_1 to r_2 coincides with s_1^+.

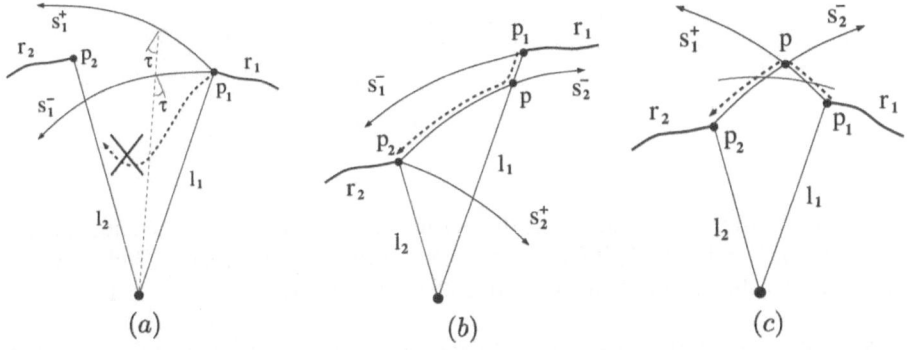

Figure 2. Example of a free-curve construction: (a) this situation is impossible; (b) in this case the free curve is composed of the segment joining p_1 with p and the spiral s_2^- joining p with p_2; (c) here the free curve is composed of the spiral s_1^+ joining p_1 with p and the spiral s_2^- joining p with p_2 (unless p is beyond range, in which case a circular arc of radius d_M is added).

Remark 2. By symmetry, when $\rho_2 \exp(-\lambda\beta) < \rho_1$ (i.e., s_2^+ cuts l_1 below p_1), the last section of the curve f coincides with s_2^- (which grows clockwise from p_2).

The point p_2 may lie below the intersection of s_1^- with l_2, above the intersection of s_1^+ with l_2, or between both intersections. Likewise, the point p_1 may lie below the intersection of s_2^+ with l_1, above the intersection of s_2^- with l_1, or between both intersections. There are total of 9 combinations of events for case $\{v,v\}$, but only 3 of them are independent:

(a) s_1^- cuts l_2 above p_2. Thus, $\rho_1 \exp(-\lambda\beta) > \rho_2$, and this is equivalent to $\rho_2 \exp(\lambda\beta) < \rho_1$. That is, s_2^- cuts l_1 below p_1.

(b) s_1^+ cuts l_2 below p_2. Thus, $\rho_1 \exp(\lambda\beta) < \rho_2$, and this is equivalent to $\rho_2 \exp(-\lambda\beta) > \rho_1$. That is, s_2^+ cuts l_1 above p_1.

(c) s_1^- cuts l_2 below p_2 and s_1^+ cuts l_2 above p_2. Thus, $\rho_1 \exp(-\lambda\beta) < \rho_2 < \rho_1 \exp(\lambda\beta)$, and this is equivalent to $\rho_2 \exp(-\lambda\beta) < \rho_1 < \rho_2 \exp(\lambda\beta)$. That is, s_2^+ cuts l_1 below p_1 and s_2^- cuts l_1 above p_1.

Let us analyze the first situation. $\rho_1 \exp(-\lambda\beta) > \rho_2$ is equivalent to $\rho_2 \exp(\lambda\beta) < \rho_1$, which in turn implies that $\rho_2 \exp(-\lambda\beta) < \rho_1$. In other words, both the clockwise-growing s_2^- and the clockwise-shrinking s_2^+ cut l_1 below p_1 (see Figure 2(b)). From Remark 2, the last section of the free curve f coincides with s_2^-. Let p be the intersection between s_2^- and l_1. The free curve f joining r_1 to r_2 is thus composed of the segment joining p_1 with p and the spiral s_2^- joining p with p_2.

A symmetric argument applies to the second situation, when $\rho_2 \exp(-\lambda\beta) > \rho_1$ (i.e., s_2^+ cuts l_1 above p_1), except that Remark 1 is used in this case.

The only remaining situation is (c): $\rho_1 \exp(-\lambda\beta) < \rho_2$ and $\rho_2 \exp(-\lambda\beta) < \rho_1$. From Remarks 1 and 2, these inequalities imply that the first section of f coincides with s_1^+ while the last section of f coincides with s_2^-. Let p be the intersection of s_1^+ and s_2^-. If p is within d_M, then f is composed of the spiral s_1^+ joining p_1 with p and the spiral s_2^- joining p with p_2 (Figure 2(c)). Otherwise p is beyond range, and f is composed of a section of s_1^+, a circular arc of radius d_M, and a section of s_2^-.

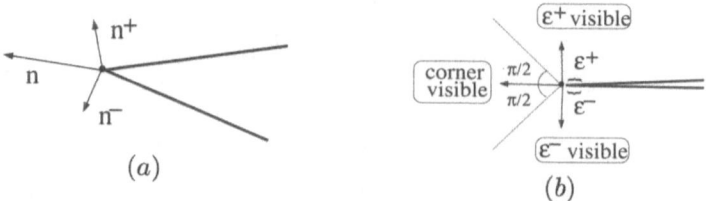

Figure 3. Dealing with corners: (a) the normal to $\partial\mathcal{W}$ at a corner is generalized as the average of n^+ and n^-; (b) if we assume that a corner has "thickness", and is therefore detectable by the sensor, then any wedge-shaped object is visible if $\tau \geq 45$ deg.

Case $\{v,o\}$:   As in the previous case, the curve r_1 was interrupted at $\theta = b_1$ because the incidence constraint was exceeded. The curve r_2, however, was interrupted at $\theta = a_2$ because a portion of $\partial\mathcal{W}$ blocked the sensor's line-of-sight. In order to produce the occlusion, $\partial\mathcal{W}$ must be tangent to l_2 at some point p_t below p_2. We know from Lemma 3.2 that the portion of $\partial\mathcal{W}$ producing the occlusion cannot be disjointed from r_1. Thus, p_t is part of the same curve as r_1.

$\partial\mathcal{W}$ cannot continue from r_1 below s_1^-. To show this, suppose $\partial\mathcal{W}$ continues below s_1^-. This implies that $\partial\mathcal{W}$ bends toward the sensor immediately after r_1. But to cause the occlusion, $\partial\mathcal{W}$ has to bend outwards before it reaches the tangent point p_t. Since $\partial\mathcal{W}$ is differentiable, there must be a point where the normal to $\partial\mathcal{W}$ points towards the origin. But we already know that this violates our assumption that r_2 succeeds r_1. Therefore, $\partial\mathcal{W}$ must continue above s_1^+.

For case $\{v,o\}$, it is always true that s_1^+ cuts the ray l_2 below p_2 at some point p. Otherwise, it will be impossible to produce the occlusion at p_t, because $\partial\mathcal{W}$ continues from r_1 above s_1^+. Thus, f is composed of the spiral s_1^+ joining p_1 with p, and the segment joining p with p_2.

Case $\{v,e\}$:   As before, the incidence constraint was exceeded at $\theta = b_1$, but r_2 was interrupted because the range constraint was exceeded at $\theta = a_2$. That is, $\rho_2 = d_M$.

The point p_1 is within range, hence $\rho_1 \exp(-\lambda\beta) < \rho_2$ because $\rho_2 = d_M$. This is exactly the situation described in Remark 1 of case $\{v,v\}$. Thus, $\partial\mathcal{W}$ cannot continue below s_1^-, and the first section of f coincides with s_1^+.

If s_1^+ cuts the ray l_2 below p_2 at some point p, then f is composed of the spiral s_1^+ joining p_1 with p, and the segment joining p with p_2. Otherwise p is beyond range, and f is composed of a section of s_1^+ and a circular arc of radius d_M.

Case $\{e,e\}$:   This case is trivial. The free curve is a circular arc joining p_1 to p_2.

Cases $\{o,o\}$ and $\{e,o\}$:   The reader may verify that these cases are impossible by following the same line of reasoning used throughout this proof. We skip the details for lack of space.

We have accounted all possible cases. This concludes our proof of Theorem 3.1.

4. Extracting Safe Regions from Real Sensor Data

The main practical problem with the theoretical results of the previous section is that the sensor output is usually a list of points, not a list of curves. Therefore, a pre-processing stage is needed to convert the raw data into the output list Π.

Let L be the list of points acquired by the sensor at q. L is transformed into a collection Π of polygonal lines called *polylines*. The polyline extraction algorithm operates in two steps: (1) group data into clusters, and (2) fit a polyline to each cluster. The goal of clustering is to group points that can be traced back to the same object in $\mathcal{W}$. Clustering is done using thresholds selected according to the sensor's accuracy.

The points in each cluster are fitted with a polyline so that every data point lies within a distance ϵ from a line segment, while minimizing the number of vertices in the polyline. The computation takes advantage of the fact that the data delivered by polar sensors satisfy an ordering constraint along the noise-free θ-coordinate axis. By applying the mapping $u = \cos\theta / \sin\theta, v = 1/(r\sin\theta)$, the problem is transformed into a linear fit of the form $v = a + bu$ (which maps to $bx + ay = 1$ in Cartesian (x, y)-space). Several algorithms exist to find polylines in (u, v)-space. We used a divide-and-conquer algorithm. Examples of our polyline-fit technique with real sensor data can be found in [5].

4.1. Corners

Corners pose a problem even under idealized conditions. Suppose the robot is surrounded by one or several wedge-shaped walls oriented toward the sensor. The sensor is then unable to see any of these wedges, and the safe region is empty. This is not a failure of our mathematical analysis, but a physical limitation of the sensor. This limitation was not taken into account by Definition 2.1, along with several others (e.g., that some surfaces could be perfect mirrors). We can only assume that the angle between any pair of incident walls is large enough such that at least one section at either side of the corner is visible to the sensor. Or that the corner itself is not sharp enough to remain undetected by the sensor (i.e., the corner has "thickness").

Under the above assumptions, we generalize the concept of a surface normal to include corners. The normal n to $\partial\mathcal{W}$ at a corner is the average of n^+ and n^-, where n^+ and n^- are the normals to $\partial\mathcal{W}$ immediately after and before the corner (Figure 3(a)). The corner is visible if the conditions of Definition 2.1 are satisfied for this generalized n. That is, a corner behaves like any other point in $\partial\mathcal{W}$, as long as our hypotheses about $\partial\mathcal{W}$ hold true.

The system described in [5] expects corners to have thickness, and therefore to be detectable by the sensor. Under this supposition, it is easy to verify that any wedge-shaped object within range is visible if $\tau \geq 45$ deg (Figure 3(b)).

5. A Next-Best View Algorithm

In a static environment, a safe region remains safe under the union operation. Hence, the layout model can be expanded iteratively. A first partial layout — a local safe region — is constructed from the data acquired by the range sensor at the robot's initial position q_0. At each iteration, the algorithm updates the layout model by computing the union of the safe region build so far with the local safe region generated at the new position q_k. The new safe region is then used to select the next sensing position q_{k+1}. To compute this next-best-view position, the procedure first generates a set of potential candidates. Next, it evaluates each candidate according to both the expected gain of information that will be sensed at this position, and the motion cost required to move there. These steps are described below.

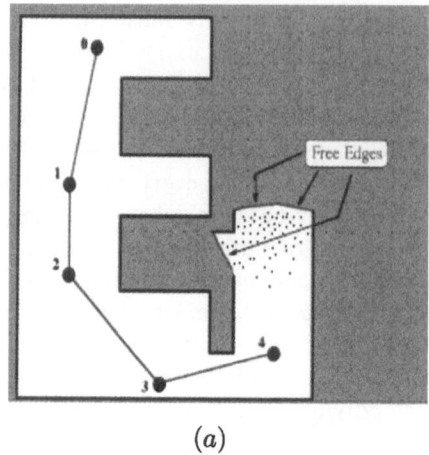 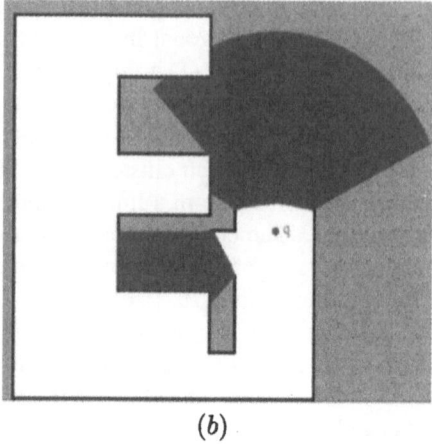

(a) (b)

Figure 4. Next-best view computation after 5 sensing operations have already taken place: (a) candidate generation; (b) evaluation of one candidate.

5.1. Model Alignment and Merging

Let $\langle \Pi_g(q_{k-1}), S_g(q_{k-1}) \rangle$ be the partial *global* model built at q_{k-1}. The term $S_g(q_{k-1})$ is the union of all local safe regions up to stage $k-1$. The boundary of $S_g(q_{k-1})$ is composed of free and solid curves, the latter representing physical sections of $\partial \mathcal{W}$. Let $\Pi_g(q_{k-1})$ be the list of solid curves in the boundary of $S_g(q_{k-1})$.

The robot performs a sensing operation once it moves into a new location q_k. From the local measurement $\Pi_l(q_k)$, we compute a local safe region $s_l(q_k)$ using the techniques from Section 3. Let $\langle \Pi_l(q_k), s_l(q_k) \rangle$ be the *local* model at q_k.

Suppose there exists an algorithm ALIGN that computes the transformation T aligning the line segments in $\Pi_l(q_k)$ with those in $\Pi_g(q_{k-1})$. We will *not* assume that this technique is perfect: ALIGN computes a correct T only when there is enough overlap between $\Pi_l(q_k)$ and $\Pi_g(q_{k-1})$.

Once T is calculated, the new global safe region $S_g(q_k)$ is computed as the union of $T(S_g(q_{k-1}))$ and $s_l(q_k)$. The new model $\langle \Pi_g(q_k), S_g(q_k) \rangle$ is represented in a coordinate frame centered over the robot at its current position q_k.

5.2. Candidate Generation

The next location q_{k+1} must be contained inside $S_g(q_k)$. Otherwise, q_{k+1} would not be reachable from the current position q_k.[1]

We generate at random a number of possible NBV candidates in $S_g(q_k)$ within the vicinity of the free curves bounding $S_g(q_k)$ (Figure 4(a)). For each possible candidate q, we compute the total length $\zeta(S_g(q_k), q)$ of the non-free curves bounding $S_g(q_k)$ that are visible from q under Definition 2.1 (this operation is done using a line-sweep technique [10]). ζ is the measure of the expected overlap between a new image $\Pi_l(q)$ and the current list of solid curves $\Pi_g(q_k)$. If $\zeta(S_g(q_k), q)$ is greater than some threshold, then q is actually selected as an NBV candidate. This filtering stage ensures that the function ALIGN will successfully find a transform T.

---

[1] Strictly speaking, $S_g(q_k)$ must first be shrunk by the radius of the robot before computing a safe route.

5.3. Evaluation of candidates

To decide whether a position q in $S_g(q_k)$ is a good candidate for q_{k+1} we must estimate how much new information about the workspace we expect to obtain at q — i.e., q should potentially see large unexplored areas *through* the free boundary of $S_g(q_k)$.

The score of every NBV candidate q is given by the function $g(q) = A(q) \exp(-\lambda L(q, q_k))$, where λ is a positive constant, $L(q, q_k)$ is the length of the shortest path connecting q_k with q, and $A(q)$ is a measure of the unexplored area of the environment that may be visible from q (see next paragraph). q_{k+1} is selected as the sample q that maximizes the function $g(q)$. The factor λ weights the relative cost of motion with respect to visibility gains. $\lambda = 0$ implies that the map builder incurs no cost while moving, and the NBV planner is allowed to select new locations exclusively in terms of their potential visibility gain. $\lambda \gg 0$ implies that motion is so costly that locations close to q_k are preferred over distant ones, as long as they produce a marginal gain in visibility.

Computation of $A(q)$ We measure the potential visibility gain of each candidate q as a function of the area $A(q)$ outside the current safe region that may be visible *through the free curves* bounding $S_g(q_k)$ (Figure 4(b)). For polygonal models, $A(q)$ can be computed by the same ray-sweep algorithm used to compute classic visibility regions [10], with the following modifications:

1. The sweeping ray may cross an arbitrary number of free edges before hitting a solid one. Therefore, the computation-time of the ray-sweep algorithm becomes $O(n \log(n) + n \, k_f)$, where k_f is the number of free edges bounding $S_g(q_k)$.

2. The resultant visible region is cropped to satisfy the range restrictions of the sensor. This operation can be done in $O(n \, k_f)$.

5.4. Termination Condition

If $S_g(q_k)$ contains no free curves, the 2-D layout is assumed to be complete; otherwise, $S_g(q_k)$ is passed to the next iteration of the mapping process. A weaker termination test is employed in practice: the length of any remaining free curve is smaller than a specified threshold.

5.5. Iterative Next-Best View Algorithm

The iterative NBV algorithm is summarized below:

> **Algorithm** *Iterative Next-Best View*
> **Input:** A new sensing position q_k and the local measurement $\Pi_l(q_k)$
> An image alignment function $T = \text{ALIGN}(\Pi_l(q_k), \Pi_g(q_{k-1}))$
> The number of samples m, and a weighting constant $\lambda > 0$
> **Output:** A next-best view position q_{k+1}
> 1. Compute the local safe region $s_l(qk)$. Set the list of samples $\mathcal{N}_{sam} = \emptyset$.
> 2. Compute $T = \text{ALIGN}(\Pi_l(q_k), \Pi_g(q_{k-1}))$, and the union $S_g(q_k) = s_l(q_k) \bigcup T(S_g(q_{k-1}))$.
> 3. Repeat until the size of $\mathcal{N}_{sam}$ is greater or equal than m:
>
>   (a) Randomly generate $q \in S_g(q_k)$ in the vicinity the free curves bounding $S_g(q_k)$.
>
>   (b) If $\zeta(S_g(q_k), q)$ is below the requirements of ALIGN, discard q and repeat Step 3.
>
>   (c) Compute $A(q)$ and $L(q, q_k)$. Add q to $\mathcal{N}_{sam}$ and repeat Step 3.
> 4. Select $q_{k+1} \in \mathcal{N}_{sam}$ maximizing $A(q) \exp(-\lambda L(q, q_k))$ as the next-best view.

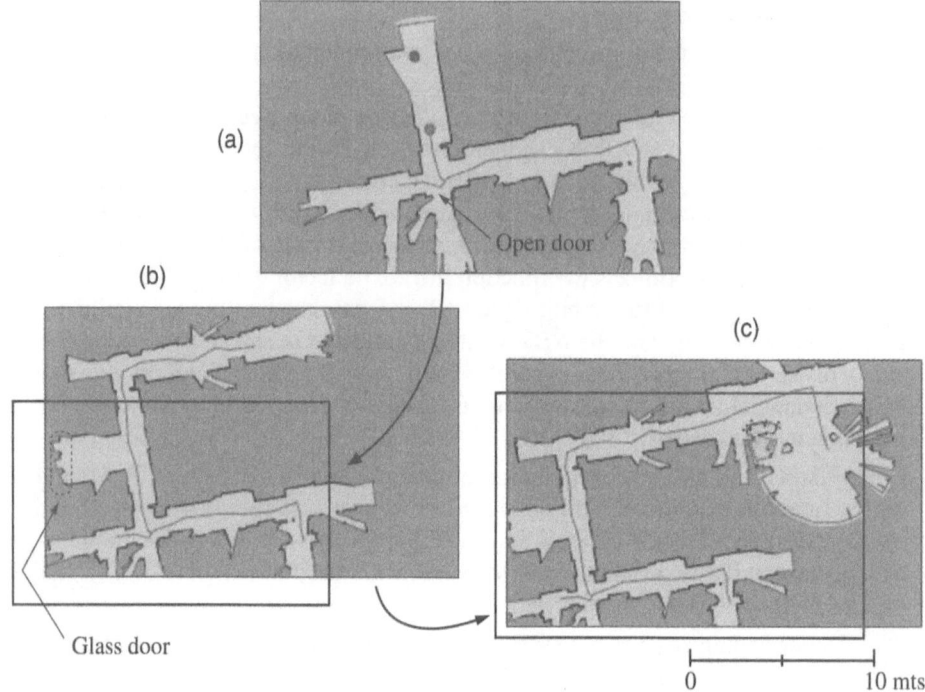

(a)

(b)

(c)

Open door

Glass door

0   10 mts

Figure 5. A run around the Robotics Lab. at Stanford University.

6. Experiments

The map-building system was implemented on a Nomadic SuperScout robot. The on-board computer is a Pentium 233 MMX, connected to the local-area network via 2 Mbs radio-Ethernet. The robot is equipped with a laser range sensor from Sick Optic Electronic which uses a time-of-flight technique to measure distances. The NBV planner runs off-board in a Pentium II 450 MHz Dell computer. The software was written in C++ and uses geometric functions from the LEDA library [8].

The sensor acquires 360 points in a single 180-deg scan request. A 360-deg view is obtained by taking 3 scans. The sensor readings where observed to be reliable within a range of 6.5 mts, at grazing angles not exceeding $\tau = 85$ deg. For the NBV planner, $\lambda = 20$ cm^{-1}, a value that prevents the robot from oscillating back and forth between regions with similar visibility gains.

An experimental run is shown in Figure 5. The robot mapped a section of the Robotics Lab. at Stanford U. The first 6 iterations are shown in (a). At the corridor intersection, the robot faces three choices, including going into an office. Nevertheless, the planner opted to continue moving along a corridor, all the way into the upper hall (b). Glass is transparent to the sensor's laser, so the robot failed to detect the glass door indicated in (b). At this point, the operator overrode the decision of the NBV planner, who interpreted the vicinity of the glass door as the threshold of an unexplored open area. Finally, in (c), the robot moved down the second hall until it reached the lab's lounge. The planner decided then to send the robot to explore this newly detected area.

7. Conclusion

Motion planning for model building applications has received little attention so far despite its potential to improve the efficiency of autonomous mapping. In this paper we introduced the concept of safe region, and described how it can be used to produce collision-free motions and next-best view locations under image-alignment considerations. Our research combines theoretical investigation of planning problems with simplified visibility models to produce algorithms that reach a compromise between algorithmic rigor and system practice. The result is a system able to construct models of realistic scenes.

Acknowledgments: This work was funded by DARPA/Army contract DAAE07-98-L027, ARO MURI grant DAAH04-96-1-007, and NSF grant IIS-9619625.

References

[1] R. Chatila and J.P. Laumond. Position referencing and consistent world modeling for mobile robots. In *Proc. IEEE Int. Conf. on Robotics and Automation*, pages 138–143, 1985.

[2] H. Choset and J. Burdick. Sensor based motion planning: The hierarchical generalized voronoi diagram. In J.-P. Laumond and M. Overmars, editors, *Proc. 2nd Workshop on Algorithmic Foundations of Robotics*. A.K. Peters, Wellesley, MA, 1996.

[3] B. Curless and M. Levoy. A volumetric method for building complex models from range images. In *Proc. ACM SIGGRAPH*, 1996.

[4] A. Elfes. Sonar-based real world mapping and navigation. *IEEE J. Robotics and Automation*, RA-3(3):249–265, 1987.

[5] H. González-Banos, A. Efrat, J.C. Latombe, E. Mao, and T.M. Murali. Planning robot motion strategies for efficient model construction. In *Robotics Research - The Eight Int. Symp.*, Salt Lake City, UT, 1999. Final proceedings to appear.

[6] B. Kuipers, R. Froom, W.K. Lee, and D. Pierce. The semantic hierarchy in robot learning. In J. Connell and S. Mahadevan, editors, *Robot Learning*. Kluwer Academic Publishers, Boston, MA, 1993.

[7] J.J. Leonard and H.F. Durrant-Whyte. Stochastic multisensory data fusion for mobile robot location and environment modeling. In *Proc. IEEE Int. Conf. on Intelligent Robot Syst.*, 1991.

[8] K. Mehlhorn and St. Nahër. *LEDA: A Platform of Combinatorial and Geometric Computing*. Cambridge University Press, Cambridge, UK, 1999.

[9] P. Moutarlier and R. Chatila. Stochastic multisensory data fusion for mobile robot location and environment modeling. In H. Miura and S. Arimoto, editors, *Robotics Research - The 5th Int. Symp.*, pages 85–94. MIT Press, Cambridge, MA, 1989.

[10] J. O'Rourke. Visibility. In J.E. Goodman and J. O'Rourke, editors, *Handbook of Discrete and Computational Geometry*, pages 467–479. CRC Press, Boca Raton, FL, 1997.

[11] R. Pito. A sensor based solution to the next best view problem. In *Proc. IEEE 13th Int. Conf. on Pattern Recognition*, volume 1, pages 941–5, 1996.

[12] S. Teller. Automated urban model acquisition: Project rationale and status. In *Proc. 1998 DARPA Image Understanding Workshop*. DARPA, 1998.

[13] G. Turk and M. Levoy. Zippered polygon meshes from range images. In *Proc. ACM SIGGRAPH*, pages 311–318, 1994.

Simulation and Experimental Evaluation of Complete Sensor-based Coverage in Rectilinear Environments

Zack J. Butler,* Alfred A. Rizzi and Ralph L. Hollis
Robotics Institute
Carnegie Mellon University
zackb@cs.dartmouth.edu, {arizzi,rhollis}@ri.cmu.edu

Abstract: Although sensor-based coverage is a skill which is applicable to a variety of robot tasks, its implementation has so far been limited, mostly by the physical limitations of traditional mobile robots. This paper presents sensor-based coverage algorithms both for a single robot and for a team of independent robots which have been designed to allow for easy integration on to real robots. The specific robots in question are planar robots called couriers, components of the minifactory, an automated assembly system. The couriers have excellent position sensing, which enables them to perform coverage, but have no explicit range or contact sensors to detect boundaries, which adds to the complexity of the coverage algorithm. A set of experiments from simulation is presented to show the overall efficiency of the single-robot and cooperative coverage processes in a variety of environments. A second set of experiments performed on a real robot demonstrates the ability to reliably perform sensor-based coverage and also illuminates the effects of specific choices in the type of control used.

1. Introduction

Sensor-based coverage is the problem of directing a robot operating in an initially unknown environment to explore each and every point of the environment. The applications are numerous, including demining, floor cleaning, and similar tasks. While several algorithms have been proposed, demonstrations of theoretical correctness and successful applications have led to different approaches. On the theoretical side, several works have shown how to reach every point in an unknown environment in a provably correct way [1, 2, 3, 4], but few of these have been applied to a real robot, and with limited success [4]. Much of this difficulty can be attributed to the challenge of mobile robot localization, compounded by the need to operate over a long time and large space in order to perform coverage of interesting environments. In contrast, one successful application of sensor-based coverage has been the development of autonomous lawn mowers that use pseudo-random motions. This approach requires no onboard mapping or odometry, but gives no theoretical guarantee of a complete traversal of the environment, and is typically quite inefficient [5].

*Currently with the Dept. of Computer Science, Dartmouth College, Hanover, NH

D. Rus and S. Singh (Eds.): Experimental Robotics VII, LNCIS 271, pp. 417–426, 2001.
© Springer-Verlag Berlin Heidelberg 2001

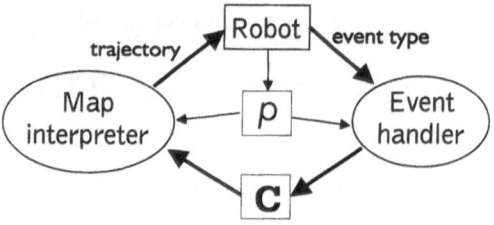

Figure 1. A schematic of the components of CC_R.

In the *minifactory*, an automated assembly system designed for rapid development and deployment [6], autonomous self-calibration is a critical step — once a factory has been assembled from its component modules (each of which is an independent robot or modular piece of infrastructure), the overall structure of the factory must be verified and the precise relative locations of all modules must be determined. Sensor-based coverage by the factory's *couriers* represents one way in which this task can be performed. Couriers are small robots based on planar motors that have reliable position sensing (due to a novel precision AC magnetic position sensor [7]) and motion capability in $I\!R^2$, although they can only detect the boundaries of their workspace by noting an inability to move in a particular direction. The couriers' position sensing abilities and the system's highly structured (rectilinear) environment provide a domain in which complete sensor-based coverage can be reliably performed.

This paper documents the successful application of new sensor-based coverage algorithms in simulation and on a minifactory courier. We discuss key features of (and additions to) the proven algorithm and present experimental results verifying that with appropriate attention to detail in terms of the algorithmic inputs and outputs as well as the incorporation of appropriate models for position error, a provably correct coverage algorithm can be successfully implemented on a real robot.

2. Coverage algorithms

Because of the unusual nature of the couriers' sensing and environment, a new sensor-based coverage algorithm was required. The algorithm developed, CC_R (*Contact-based Coverage of Rectilinear environments*), was also designed specifically to allow for the addition of cooperation as well as straightforward implementation on the couriers. CC_R, shown in schematic form in Fig. 1, achieves complete coverage of any rectilinear environment by incrementally building an exact cellular decomposition of the environment $\mathbf{C}$ in a reactive way. In each cycle of the algorithm, the *map interpreter* (half of CC_R) uses an ordered list of rules to evaluate $\mathbf{C}$ and the current position p and chooses a single straight-line trajectory with which to continue coverage. The robot executes the trajectory without interference from CC_R until a specified maximum distance has been traveled or a collision is experienced. At this point the *event handler* updates $\mathbf{C}$ based on the new event and the position at which it occurred. The basic behavior of CC_R is to cover each cell with a *seed-sowing* path, as

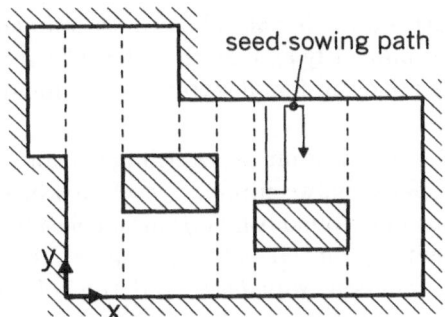

Figure 2. An example of a cellular decomposition that would be constructed by CC_R, along with a portion of a seed-sowing path used to perform coverage.

shown in Fig. 2. It starts with the assumption that the environment can be represented by a single rectangular cell, and commands the robot to perform seed-sowing until a corner is discovered in the environment. When this occurs, it will construct additional cells, localize the corner, and continue seed-sowing in one cell. This process continues until the boundary of **C** is known and closed and its interior is covered. A proof that CC_R will produce complete coverage of any finite rectilinear environment was presented in [8], although the proof assumes perfect position sensing.

In addition, an algorithm for distributed sensor-based coverage, DC_R, was developed under which each robot independently runs a slightly modified version of CC_R to perform coverage [9]. Cooperation is induced by adding an algorithmic component which alters **C** to reflect data obtained from other robots. This function, the *overseer*, operates in parallel with (and independent of) the event handler, so that the map interpreter can examine **C** and plan coverage without knowing anything about the cooperation process. A final additional function works to determine the robots' relative locations in their environment, as this information is assumed to be unknown when coverage begins. This algorithmic structure allows CC_R (and its proof) to be retained virtually without alteration while allowing for increased efficiency in terms of total time required by the team. Starting with the proof of CC_R, a proof was developed that shows that any number of cooperating robots running DC_R will collectively produce complete coverage of their environment [8].

3. Algorithm deployment

Although the coverage algorithms outlined above were developed with real-world application in mind (for example, the use of straight-line trajectory outputs simplifies deployment), there are necessary additions that are not fundamental to the proof of correctness. One important point to consider is localization ability — while the couriers have high precision position sensing, especially compared to most mobile robots, it is still not perfect, nor is the environment (in terms of being strictly rectilinear). Therefore the development of the cell decomposition and the way it is used to generate trajectories for the robot must allow for non-cumulative error in the robot's position. For example, the rules of the map interpreter implicitly assume that the cells of the decomposition do not overlap, and so as the event handler creates and updates cells this property must be maintained.

Under DC_R, these localization issues still apply for each robot, but in addition, when data is shared between robots, each of which have uncertainty in their maps, each robot must build a map that is consistent and representative of the underlying environment. In addition, since DC_R necessarily involves multiple robots in a common workspace, and the robots do not know their relative initial locations, inter-robot collisions are inevitable and must be handled in the context of coverage. The current simulations use a simple reactive avoidance strategy, wherein after two robots collide, they will decide which one should attempt to move out of the other's way. This works fairly well for two robots, but leads to frequent deadlock in large teams and in confined spaces.

Another practical issue is the ability to follow walls and detect gaps in them. Since the couriers have only contact sensing, CC_R was written (and proven) based on the ability of the robot to perform *sliding* motions, in which the robot maintains contact with a boundary while moving along it. The robot must be able to detect a loss of contact with the boundary as well as contact in the direction of motion. If sliding motions cannot be executed, an alternative is to approximate them by interleaving small free-space motions along the boundary with short motions to contact the boundary. These interleaved motions can be produced at the control level (from a sliding trajectory specification) or at the algorithmic level (by slightly altering the rules of the map interpreter that generate the sliding motions). In either case, a proof has been developed that shows that the robot will still produce complete coverage in virtually all rectilinear environments [8].

In addition to these algorithmic issues, when implementing coverage on the courier, there is a choice to be made about the type of control used. The courier's position sensor allows for micron-level precision and accuracy in the range of tens of microns throughout its workspace, enabling a variety of closed-loop control policies (with widely variable parameters) in addition to the open-loop microstepping commonly used to control planar motors. The most important choice is whether to attempt the sliding motions described above. The use of these motions inherently requires smooth boundaries and control based on good force sensing (or estimation). While the distance traveled by the robot to perform coverage is similar whether or not sliding is used, sliding motions are much more efficient in terms of time required, as shown in Sec. 5. A previously developed *dynamic force* controller [10] was used which allows for straight-line motion (both in contact and in free space). This controller uses an estimator to determine the disturbance force on the courier from position data, and attempts to apply a given desired force in each of x and y. Damping is then added in each direction to implicitly set a maximum velocity. By using the same control law for motion and maintaining contact, instability due to controller switching can be avoided. In addition, since the courier has no contact sensors, the controllers also use the force estimate to determine if contact

Environment size	$5w \times 5w$	$10w \times 10w$	$20w \times 20w$
Average cf	2.483	1.710	1.367
Std. deviation	0.1300	0.0636	0.0219

Table 1. Performance of CC_R in square environments of various sizes, where w is the width of the robot.

Environment	Random	Fig. 3a	Fig. 3b
Average cf	2.307	1.986	3.557
Std. deviation	0.329	0.086	0.144

Table 2. Performance of CC_R in various complex environments.

has occurred, and are able to immediately return this result to the higher level interface code.

4. Simulation experiments

Implementations of CC_R and DC_R were developed and operated in simulations that incorporated the position error models and optional sliding motions described above. The simulation was developed using a *world modeler* function in place of real physics, where the "Robot" block appears in Fig. 1. Since both CC_R and DC_R interact with the world through simple interfaces (straight-line trajectory output and a single position reading and one-bit contact sensor inputs), integrating such a function with the coverage algorithms was straightforward. These implementations were then run in a wide variety of simulated environments to determine typical efficiency for a single robot as well as the efficiency gain for robots in a cooperative team. To describe the efficiency of the algorithm as performed experimentally (both in simulation and on the courier), we define the *coverage factor* metric as the average number of times the robot passes over each point in the environment. This is easily calculated as:

$$cf = \frac{d \times w}{\mathrm{Area}(\mathbf{C})},$$

where d is the total distance traveled and w the robot width.

Ideally, the coverage factor for pure seed-sowing would be exactly 1, but two factors make this impossible to achieve in practice. First, unless the robot starts exactly an integer multiple of w away from the edge of the cell, it will finish seed-sowing with a pass that does not add a full robot-width of covered area (because it has only contact sensing, it cannot detect the edge of the cell until it has reached it). Also, in order to ensure detection of all gaps in the top and bottom of each cell, CC_R requires the robot to cover these edges twice. To empirically determine the magnitude of these effects, CC_R was run from 50 different starting locations in each of three empty square environments. The results of these experiments are given in Table 1. These experiments show that these inefficiencies have decreasing effect as the environment gets larger, which is as expected, since they add distance proportional to the perimeter of the cell, which is then divided by the area of the cell to calculate cf.

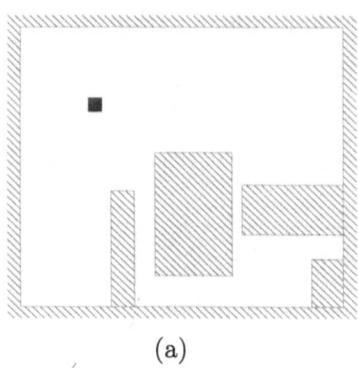   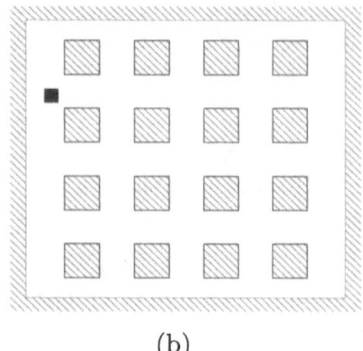

(a)   (b)

Figure 3. Environments used to test CC_R. The black square in each represents the size of the robot.

CC_R was then tested in a variety of randomly generated environments, with results reported in the first column of Table 2. These environments were generated by populating an open square of dimension $\sim 20w \times 20w$ with between three and eight rectangular obstacles, each with a random height and width between $w/20$ and $10w$. In addition, the two environments shown in Fig. 3 were chosen for repeated tests (50 runs in each), the results of which are also reported in Table 2. The coverage factor for CC_R ranged from under 2.0 for reasonably open environments, such as that of Fig. 3a, to nearly 4.0 for very cluttered spaces, such as Fig. 3b (which was manually designed to be adversarial to CC_R). It is worth noting that these data compare favorably to previously reported values for other (simulated) algorithms, and very favorably to the ratio of about 10 for the randomized technique used by the Friendly Robotics lawn mower[1].

To measure the efficiency gain of cooperating robots running DC_R, teams of sizes from two to ten robots were run in the environment of Fig. 3a. The coverage factor for each robot was then compared to that achieved by a single robot in the same environment. Under DC_R, each robot will develop a complete cell decomposition of the environment, so that the coverage factor for each will simply be proportional to the total distance it has traveled, which will hopefully be smaller for each robot than if it was working alone. This was in fact the case for these experiments, as shown in Table 3. The coverage factor decreased about 30% for each robot in a two-robot team, and almost 50% for each of three robots (this was also true for other similar environments). In addition, these experiments showed that the work was divided fairly evenly among the robots. This was measured by noting the largest coverage factor among the team for each run, and as can be seen in Table 3, in general no robot had to do more than 10% more work than the average robot.

For larger teams, in order to avoid frequent deadlock, the world modeler was set up such that collisions between the simulated robots would not occur (i.e. the robots simply traveled through each other). With this allowance, the

---

[1] Using data from the manufacturer's data sheet [11], $cf \approx \frac{0.5[\text{m/s}] \times 0.56[\text{m}]}{1000[\text{ft}^2/\text{hr}]} = 10.84$

	Two robots		Three robots		Five	Ten
	∥ orient.	⊥ orient.	w/ coll	w/o coll	robots	robots
# trials	15	15	10	10	10	10
Average cf	1.309	1.294	1.134	1.093	0.922	0.698
Avg. max. cf	1.408	1.365	1.268	1.205	1.096	0.790

Table 3. Performance of DC_R in the environment of Fig. 3a. Note that all runs with 2 robots include include inter-robot collisions while all runs with 5 and 10 robots do not.

	Single	Two robots		Three
	robot	∥ orient.	⊥ orient.	robots
Number of trials	50	10	11	10
Average cf	3.557	1.936	1.981	1.450
Avg. maximum cf	—	1.947	1.985	1.547

Table 4. Performance of DC_R in the environment of Fig. 3b.

efficiency continued to improve with increasing team size in teams up to ten robots. It is likely that with collisions in place (which would require a better collision avoidance strategy for success), this increase would not be achievable, as the robots would spend considerable time avoiding each other. However, experiments with and without collisions for three robots show only a very slight loss of efficiency with the addition of collisions.

A set of experiments in the adversarial environment of Fig. 3b was also performed, and shows that the constricting nature of this environment was actually beneficial to DC_R. Since this environment will always be decomposed into many small cells, the cooperation between the robots (which happens via transfer of completed cells) could be done more frequently. This allowed the robots to spend less time working in the same areas, leading to a decrease in coverage factor of approximately 45% for each of two robots and nearly 60% for three robots.

5. Courier experiments

As mentioned earlier, the use of straight-line trajectories as outputs from CC_R and the acceptance of small position errors allowed for a straightforward transition from simulation to the courier. The same algorithmic code was used as in simulation by simply replacing the world modeler with calls to the underlying courier motion control system, which in turn used the trajectory specification to create appropriately parameterized controllers. CC_R was tested in three environments: two were those shown in Fig. 4, and the third was a simple rectangle of similar size (approximately 70×100 cm, compared to a courier width of 15 cm). Ten runs were performed in each of two orientations and varying starting positions in each environment to determine the efficiency and qualitative capabilities of the algorithm. The results of these experiments are given in Table 5. These data confirm that the efficiency obtained in simulation is

(a)

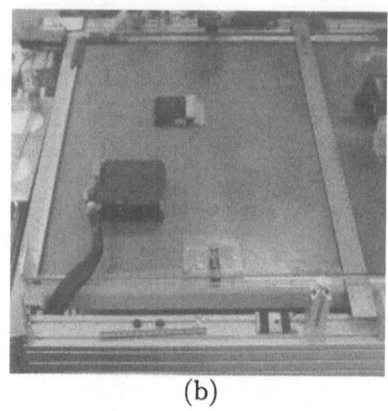

(b)

Figure 4. Environments used for CC_R testing, consisting of half of a commercial platen surface with additional obstacles. The tethered courier performing CC_R is included for scale.

Environment	Empty platen		Fig. 4a		Fig. 4b	
Orientation	std.	rot.	std.	rot.	std	rot.
Number of Runs	10	10	10	10	10	10
Average cf	1.69	1.72	2.99	2.68	2.92	3.05
Std. deviation	0.13	0.08	0.27	0.26	0.23	0.30

Table 5. Performance of CC_R on the courier in the environments of Fig. 4. "Standard" orientation is as shown in Fig. 4 (note overlaid coordinate axes) and "rotated" orientation is 90° counter-clockwise.

comparable to that seen in the real system — note that the environments in Fig. 4 are fairly constricted relative to that in Fig. 3a and most other simulated environments, accounting for the somewhat higher coverage factors.

The major failure modes resulted from the courier's tether producing an occasional large disturbance force and the boundaries not being amenable to sliding motions, both of which were overcome with careful engineering. One point of particular interest is the environment shown in Fig. 4b, in which the free space is not simply connected, requiring the robot to successfully attach the cells of the decomposition around the obstacle (in the presence of position uncertainty) to complete coverage. This was successfully done under a variety of conditions, although small additions to CC_R were required.

Finally, although the coverage factor indicates the distance traveled to perform coverage, when instantiated on a real system, the time taken for coverage is also of interest. Therefore, experiments were performed using different types of control and different maximum velocities v_{max} to determine empirically how the type of control affects the overall performance. Each set of control parameters was run from two starting locations in each of two environments, making four coverage "tasks". The simplest type of control used was open-loop trajectory following based on microstepping of the planar motor, using the position sensor only to detect collision (by noticing a significant difference

Control type	O.L.	C.L.	C.L.	C.L.	C.L.
Sliding?	No	No	No	Yes	Yes
v_{max} [mm/s]	70	70	250	70	250
Empty platen (p_1)	310	281	230	115	46
Empty platen (p_2)	318	295	234	140	53
Fig. 4a (p_1)	409	399	291	224	91
Fig. 4a (p_2)	365	341	250	201	79

Table 6. Elapsed time (in seconds) for CC_R under various open-loop (O.L.) and closed-loop (C.L.) control strategies.

between the commanded position and the actual position). While this is the easiest to implement, and is the standard mode of operation for planar motors, it is not capable of recovering from large disturbances such as those caused by strong impact. (This type of operation is fairly insensitive to small disturbances, however, making it suitable for certain types of operations.) Therefore, the maximum speed at which coverage could reliably be completed under this technique was approximately 70 mm/s. The amount of time required for the four coverage tasks is shown in the first column of Table 6. For some of these tasks, as well as some of the closed-loop experiments described below, the experiment was run several times, to confirm that variations from one run to the next due to random disturbances were slight (on the order of 1-3 s) compared to the overall time required.

To measure the effect of the sliding motions, since these operate under closed-loop control, it was first necessary to test the interleaved motions under closed-loop control. As can be seen in Table 6, this resulted in a slight improvement over the open loop technique for the same v_{max}, probably due to the collision detection being more responsive under closed-loop control. Then, since the controller added the capability for recovery from collisions at higher speeds (reliably up to about 250 mm/s), the interleaved motions were tested at a v_{max} of 250 mm/s, and the sliding control was run for the same four tasks at the same two velocities. It was found that without the sliding motions, increasing v_{max} had little effect, as the majority of the elapsed time was spent executing the short interleaved motions along boundaries during which v_{max} was not reached. However, with sliding enabled with small v_{max}, the elapsed time was about half as much as without sliding, since the exploration of the boundaries did not require stopping. In addition, for higher v_{max}, the sliding motions could realize even more advantage during the boundary explorations, and used as little as one quarter of the time required for the interleaved motions with the same maximum velocity.

6. Conclusions

This work has demonstrated that with an appropriately designed coverage algorithm along with a well-engineered robot system, it is possible to reliably execute provably complete sensor-based coverage in the real world. The algorithms interface to the robot through simple channels, allowing them to plug in to a simulation or a robot system in a straightforward way. Experiments both in simulation and on a robot have returned efficiencies that are comparable to previous theoretical results and much better than current commercial hardware implementations.

The major avenue for future work is to implement the cooperative algorithm DC_R on a set of couriers. The biggest challenge there is to implement a strategy for the tethered couriers to share a workspace (in which they initially have no knowledge of each other's location) without tangling their tethers while still making progress toward coverage.

Acknowledgements

The courier experiments would not have been possible without the work of Arthur Quaid, who developed the courier control software and interface code for it. This work was supported in part by NSF grants DMI-9523156 and DMI-9527190.

References

[1] A. Pirzadeh and W. Snyder, "A unified solution to coverage and search in explored and unexplored terrains using indirect control," in *Proc. of IEEE Int'l. Conf. on Robotics and Automation*, pp. 2113–2119, April 1990.

[2] S. Hert, S. Tiwari, and V. Lumelsky, "A terrain covering algorithm for an AUV," *Autonomous Robots*, vol. 3, pp. 91–119, 1996.

[3] I. A. Wagner, M. Lindenbaum, and A. M. Bruckstein, "MAC versus PC: Determinism and randomness as complementary approaches to robotic exploration of continuous domains," *Int'l Journal of Robotics Research*, vol. 19, pp. 12–31, January 2000.

[4] E. Acar and H. Choset, "Critical point sensing in unknown environments for mapping," in *Proc. of IEEE Int'l Conf. on Robotics and Automation*, April 2000.

[5] Friendly Robotics, "RoboSim: RL500 simulator." Available at http://www.friendlyrobotics.com/sim/RoboSim.exe.

[6] R. L. Hollis and J. Gowdy, "Miniature factories for precision assembly," in *Int'l Workshop on Microfactories*, (Tsukuba, Japan), pp. 9–14, 1998.

[7] Z. J. Butler, A. A. Rizzi, and R. L. Hollis, "Integrated precision 3-DOF position sensor for planar linear motors," in *Proc. of IEEE Int'l. Conf. on Robotics and Automation*, May 1998.

[8] Z. J. Butler, *Distributed Coverage of Rectilinear Environments*. PhD thesis, Carnegie Mellon, September 2000.

[9] Z. J. Butler, A. A. Rizzi, and R. L. Hollis, "Distributed coverage of rectilinear environments," in *Proc. of the Workshop on the Algorithmic Foundations of Robotics*, (Hanover, NH), March 2000.

[10] A. Quaid, *A Planar Robot for High-Performance Manipulation*. PhD thesis, Carnegie Mellon, July 2000.

[11] Friendly Robotics, *RL500 Owner Operating Manual*. Available at http://www.friendlyrobotics.com/um/RL500_manual.pdf.

An interactive model of the human liver

F. Boux de Casson   D. d'Aulignac
C. Laugier
GRAVIR/INRIA Rhône Alpes
38330 Montbonnot Saint-Martin, France
{Francois.Boux-de-Casson, Diego.d_Aulignac, Christian.Laugier}@inrialpes.fr

Abstract: In the aim of building a surgical simulator we have developed a model of the human liver. The model respects both the heterogeneous (different material properties depending on the tissue) and non-linear nature of the organ, using binary connectors. We validate that the local behavior of the connector is accurately reproduced on a global scale. Interaction, including collision detection and response, is possible in real-time using a haptic device. For smoother force feedback we introduce a local modeling technique that approximates forces at high frequency. Further we describe a fast method that allows real-time changes of the topology by avoiding subdivision. Finally we illustrate all these techniques by several experimental results.

1. Introduction

Objectives With increasingly complex surgical procedures doctors need to learn new skills. However, to practice during a real intervention may be dangerous for the patient. It is our long-term goal to build a surgical simulator for hepatic procedures to be used in medical training. A cornerstone of such an endeavor is the underlying physical model of the liver. Very little strain/stress experimental data on the liver is currently available, and thus we want our model to be as realistic as possible in a qualitative way. Finally, since the simulation is to be used interactively real-time performance is of paramount importance.

Relevant mechanical characteristics of a human liver The liver is a very malleable body, its exact shape strongly depends on the contact interactions with the other organs located in its vicinity. It is composed of tree major parts :

1. the Parenchyma, which presents a mechanical behavior near to those of a sponge full of liquid;

2. a complex vascular network, irrigating the liver;

3. an elastic skin called Capsule of Glisson, which is quite elastic and stiffer than the Parenchyma.

To simplify the problem, we do not modelise the vascular network behavior. So, intuitively, one could coarsely represent the liver by a sponge filled with a liquid and covered by an elastic skin.

D. Rus and S. Singh (Eds.): Experimental Robotics VII, LNCIS 271, pp. 427–436, 2001.
© Springer-Verlag Berlin Heidelberg 2001

2. Outline of the Model

Because the above mentioned anatomic and biomechanical properties, a heterogeneous model is required for modeling the mechanical behavior of the liver. In the sequel, we show how we have modeled the liver using two main components : a 2D component for modeling the Capsule of Glisson and a 3D one for modeling the Parenchyma. Each of this models include a geometrical and a physical component.

2.1. Geometrical Component of the Model

The geometric component of the model is used for performing the display operations and for detecting the interactions. This model is also used as a spatial frame for constructing the physical component.

We have chosen to make use of 2D mesh of triangles (see fig.1), for representing the Capsule of Glisson, and a tetrahedric mesh for the Parenchyma. The triangular facets of the skin correspond to the external faces of the tetrahedra of the internal 3D mesh.

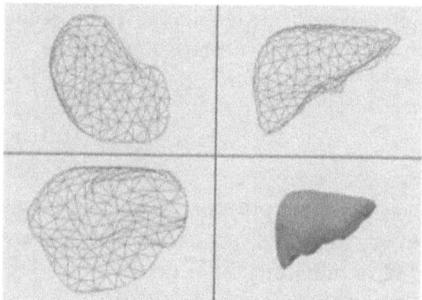

Figure 1. A 2D mesh, representing the Capsule of Glisson of the liver (the associated 3D mesh representing the Parenchyma is not shown in this picture).

2.2. Physical Component of the Model

The physical component of the model is used to compute the deformations of the liver resulting from the application of a set of external forces. These forces are either applied by the operator using the virtual tool (controlled using a haptic interface), or generated by the physical interactions with other virtual objects of the scene.

The physical component is constructed from the 2D and 3D meshes (geometrical component of the model), by associating point masses to the nodes of the previous meshes and by adding spring-damper connectors between appropriate local subsets of point masses (see [2]). In order to model the non-linear mechanical behavior of the liver, without increasing the algorithmic complexity of the approach, we have chosen to associate the following behavior to each spring :

$$\vec{F}_{spring} = \left(\lambda_1 d^3 + \lambda_2 d\right) + \mu \dot{p} \qquad (1)$$

where d is the relative deformation of the spring, that is to say $d = \frac{l-l_0}{l_0}$, with l is the length of the spring and l_0 its rest length, λ_1 and λ_2 are stiffness parameters, μ is a viscous parameter and $\dot{p}$ the relative velocity of the two particles connected by the spring. The choice of the first term of this law was directed by the fact that the few strain-stress data on the liver's mechanical behavior are showing linear behavior for small deformations (lower than 10%) and for bigger strain, the stress increases sharply. The sum of a linear and a cubic functions give a good approximation of this kind of behavior. The shape of the first term of this law is presented in the fig. 2. This non-linear relation make the springs to be relatively incompressible and unstrechable. The second term is a viscous term.

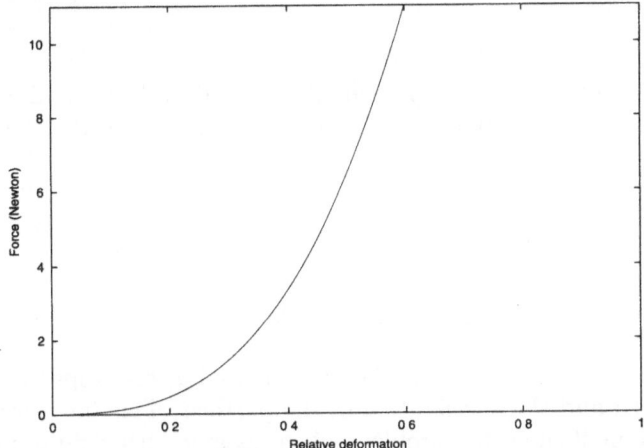

Figure 2. Plot of the springs force, as a function of their relative deformation, with $\lambda_1 = 50$ and $\lambda_2 = 0.35$.

In order to model the heterogeneousness of the liver, the connectors of the 2D mesh and of the 3D mesh are parameterized in such a way that they exhibit different mechanical characteristics :

- the Capsule of Glisson is modelised using high stiffness parameters (λ_1 and λ_2 of the Eq. 1) and low viscous parameter, allowing us to obtain a rigid elastic behavior which tend to bring back the Capsule of Glisson to its initial shape when no force is applied on it.

- for the Parenchyma, the stiffness parameters are tuned to be small in front of the viscous one, giving a plastic behavior (the inner material can easily be deformed, but does not go back quickly to its initial shape).

The combination of the two previous models (the elastic skin and the viscous volumetric internal material) gives us qualitatively and experimentally the required global behavior for the virtual liver (the behavior of a "sponge full of liquid covered by an elastic skin").

2.3. Integration of the Dynamic Equations

The integration method used in this simulator is the well-known explicit Newton-Euler. Using this method, point i of an object has the following update

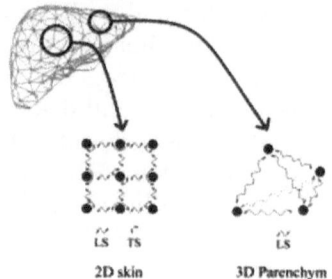

Figure 3. The hybrid mesh approach. A 2D and a 3D spring-damper meshes are used to model respectively the Capsule of Glisson and the Parenchyma.

formulae.

$$v_i^{t+\Delta t} = v_i^t + \Delta t a_i^t$$
$$x_i^{t+\Delta t} = x_i^t + \Delta t v_i^t \tag{2}$$

where Δt is the time step used and x_i, v_i and a_i are respectively the position, the speed and the acceleration of the particle. This integration method, known for it small time step problem when dealing with rigid objects, is good enough for our application, because we are dealing with relatively soft object. If object are stiffer, we can use implicit integration as demonstrated in [5].

2.4. Collision Detection

Further, to detect the interaction between objects in a scene, a collision detection algorithm must be used. It is well known that collision detection is a computationally expensive task and a potential bottleneck in every application that aims to achieve real-time performance. For this reason it is essential to optimize these functions to their maximal extent. [8] proposes an algorithm that obtains, in linear time, the points of contact on two deformable concave objects, as well as the direction of the contact and the volume of inter-penetration. This approach has the advantage that it may be used for two or more objects of any shape, but depending on their number and complexity may not guarantee real time performance.

However, for the special case of collision detection between a rigid object which has the shape of a parallelepiped and a deformable object of any shape the approach [9] may be used. It makes use of the OpenGL hardware to detect the polygons within a bounding box. If this bounding is superimposed onto the rigid object, the polygons in interaction will be detected. This approach has the advantage of being very fast due to hardware acceleration and is used in this application.

Processing physical interactions   Once a collision has been detected an appropriate response must be computed. There exist several collision response models but for deformable objects the penalty method seems to be the most appropriate (see [6] for more details). Note that the formulation presented is widely used in mechanics and has been verified physically. The force applied at a given point of an object where a collision has taken place is given by

$$\vec{F}_c = \begin{cases} (-\lambda v - \mu \dot{v} v)\vec{k} & \text{if } v < 0 \\ \vec{0} & \text{otherwise} \end{cases} \tag{3}$$

where λ is the rigidity factor of the collision, μ is a damping factor (which represents the dissipation of energy), v the volume of inter-penetration, and $\vec{k}$ the contact direction.

2.5. Haptic Interaction

As described in Section 2.4 the interaction force is calculated with respect to the interpenetration distance of two colliding bodies. This force is then used by the physical model to compute its update in state. Thus the rate at which force values can be supplied is limited by the execution time of collision detection and, thereafter, the update of the physical model.

Our aim is to provide force feedback through means of a haptic interface of type PHANToM[1]. These interfaces require a very high update rate for the force and typically this frequency is around 1KHz. However, the physical model is not able to provide force values at such rates.

Previously, to increase the frequency of the physical simulation [1] proposed a multi-resolution approach where only areas of haptic interest are simulated in detail. However, for complex objects this may not be sufficient and the resulting simulation still too slow. While [10] already suggested decoupling the haptic servo loop from the main application in the context of rigid bodies, we have extended this reasoning to deformable objects.

Our approach consists in making a first order approximation of the collision forces which can be calculated at a much higher frequency. The approximation is made through the use of a *local model* of the contact; in the case of a rigid tool interacting with a deformable object we make the (false) assumption that the surface of the deformable object remains in the same position between two updates of the physical model. The surface of the object that is in interaction with the tool is approximated by a simple geometric primitive such as a sphere or a plane. Thus the interpenetration distance between the tool and the object can be found at a much higher frequency (i.e. simpler and thus faster distance computation) and, therefore, the force values supplied at an increased rate. This, even though only gross approximations have been used, leads to much smoother haptic interaction (Figure 8) [4]. Thus we may divide this process into two categories:

- Model update Using the information about distance and the derivative of the distance, a simple *local model* of the objects surface is constructed. Thus this approximation is updated at the frequency of the physical model, and remains static in between those updates.

---

[1]see http://www.sensable.com

- **Haptic loop** Once the approximation of the surface is in place the distance between the haptic position and the surface is minimized analytically using Lagrangian multipliers. This distance multiplied by the stiffness constant yields the force values.

2.6. Changing the topology

An algorithm to tear the model has been implemented, it is presented in [3]. The global idea is to separate the elements of simulation (i.e. the tetrahedra) when they are stretched above a given threshold. This approach gives better results than when elements are removed, because the discretization is often coarse in real-time models. Thus when big elements are removed, one can see matter disappearing. Subdivision methods give more accurate results, but are not compatible with interactive constraint, because the number of elements to be simulated increases at each topology change, and the real-time constraint cannot always be assured.

Nevertheless, the limitation of our approach is that the topology changes are constrained by the initial topology of the model.

3. Implementation and Experimental Results

3.1. Architecture

The machine used for these tests was a biPentiumII 300, algorithms have been implemented in C++.

We have used as input a 2D mesh of the liver which has been pre-processed in order to reduce the number of facets, to smooth it, and to obtain the internal tetrahedric mesh (which has been calculated using the GHS3D[7] software of INRIA). The final 2D mesh we used is compound by 370 facets, and the 3D one by 1151 tetrahedra.

Then, we have used this 2D and 3D meshes to generate the spring-dampers network (see §2.2). For the purpose of the dynamic simulation, the model of the liver has been placed in an empty space, without gravity, but with a slight environmental viscosity. The four particles of a tetrahedron which is in the middle of the model are fixed. A virtual tool, simulated by a rigid object controlled in position by the operator, makes it possible to apply forces to the model of the liver, and to follow compliantly the external boundaries of the virtual liver.

3.2. Validation of the deformation model

To validate that the non-linear mechanical behavior of each connector leads to a global non-linear behavior, we have used a cylindrical model for virtual mechanical traction tests. The top of the cylindrical model (fig. 4) is fixed, and different forces are applied to all the bottom point-masses. The figure 5 presents the results of the tests. We can see that the behavior of the global cylindrical model presents a similar shape to those of each of its connectors. So, the non-linear behavior of the connectors is faithfully reproduced on the global model.

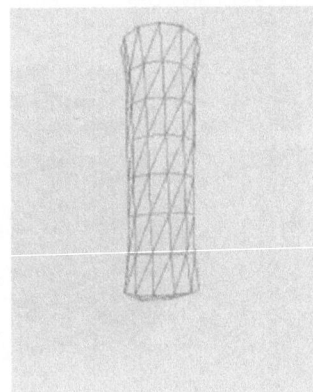

Figure 4. Cylindrical model at rest (left) and submitted to a six newton force (right).

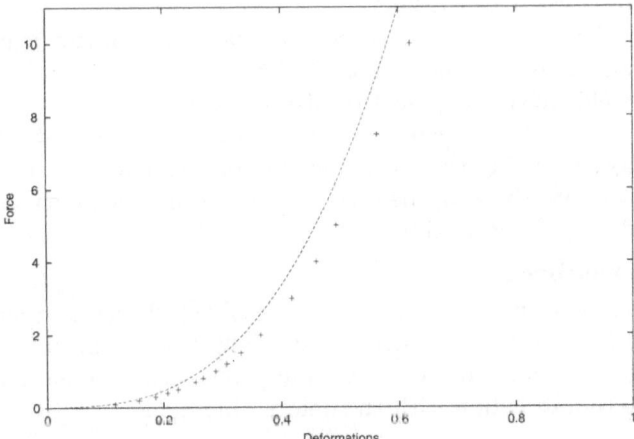

Figure 5. The dots curve is the mechanical law of the connectors (Eq. 1) and the cross were obtained by the virtual mechanical traction tests on the cylindrical model. Both curves presents the same non-linear shape.

3.3. Liver model

The model of the liver can be updated at a 150Hz frequency, and the local model gives a good haptic feedback. In the experiments, the simulations of the liver responses to various actions of the operator has shown qualitatively realistic behaviors : the liver was locally deformed under the effects of the forces applied using the rigid virtual tool and it went back to its initial shape as soon as no forces is applied to it. It was also possible to smoothly slide along its external surface, using the virtual tool.

Furthermore, if the operator, after having strongly pressed the liver, with-

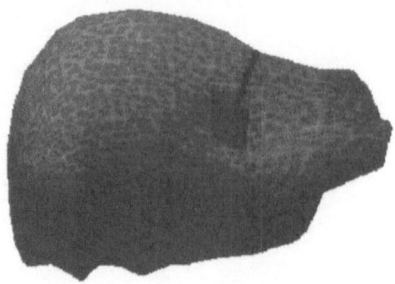

Figure 6. Interactive manipulation of the liver.

draws the tool, the liver didn't return quickly to its initial shape. This is due to the strong viscosity of the connectors of the Parenchyma, and simulates the malleable characteristic of the liver.

This model presents *qualitatively* a mechanical behavior which is closed to that of a real liver. It remains to find the exact numerical values of the parameters of elasticity and viscosity of the Parenchyma and the Capsule of Glisson. However, no force/displacement physical data is available yet, because of the difficulty in making the measurements *in vivo* (the liver being made up mainly of blood, its dynamic behavior is very different when not irrigated, because blood coagulates quickly).

3.4. Haptic feedback

Figure 7 shows how the PHANToM force-feedback device is used to interact with the model of the liver. Position and orientation of the stylus in the real world determine the location of the virtual probe. The simulation sends back the forces due to interaction with the deformable model.

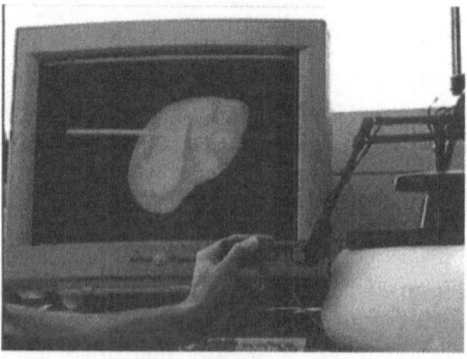

Figure 7. The liver model, being manipulated using the force feedback device.

Since the simulation frequency (Section 3.3) is insufficient to guarantee smooth haptic interaction the approach detailed in Section 2.5 is used. Figure 8

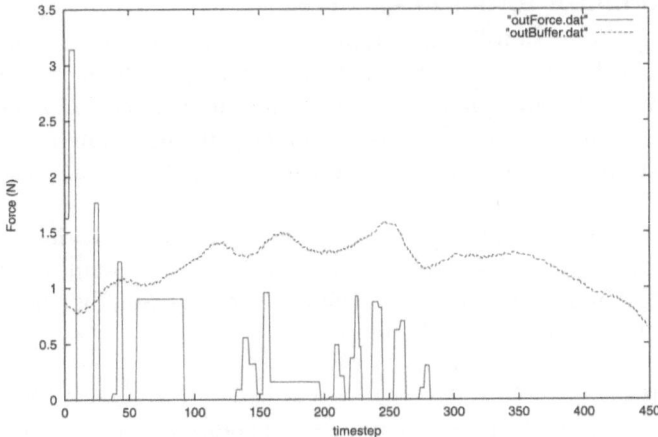

Figure 8. Evolution of the force over a time interval using the haptic local model (dashed), and without (solid).

shows the difference in the force feedback, for the interaction with a deformable model, with and without the local approximation. The harsh peaks in force without the local model propel the user (quite violently) away from the object thereby loosing contact, i.e. forces return to nil).

3.5. Changing the topology

Our current cutting algorithm works only on membrane models. It is possible to cut a dynamic model, using the PHANToM device to control the position and orientation of a virtual tool. The model is cut exactly on the tool trajectory (figure 9).

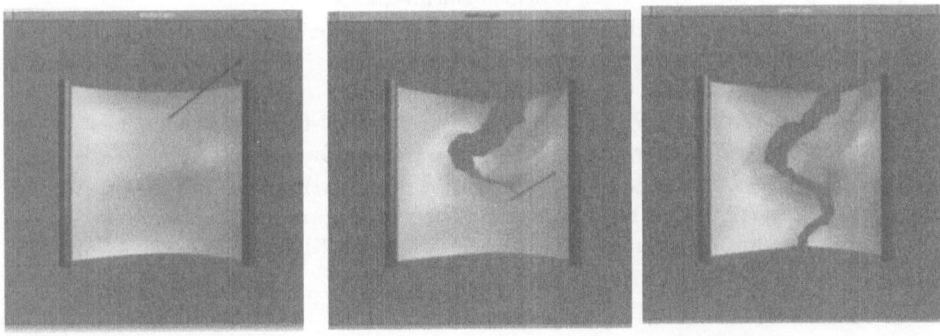

Figure 9. Interactive cut of a membrane model.

4. Conclusion and Perspectives

An heterogeneous, non-linear volumetric model of the liver based on a mass-spring model has been presented. In conformity with the reality, the model presents two different dynamic behaviors for the Parenchyma and the Capsule of Glisson. The optimized collisions detection algorithm makes possible to interact on the model in haptic real-time, using a virtual tool controlled in position.

However, it is important to remember that the elasticity and viscosity parameters of our model were chosen intuitively, to obtain a good *qualitative* behavior at a good simulation frequency, but real strain/displacement information is imperative to adjust the model.

References

[1] O.R. Astley and V. Hayward. Multirate haptic simulation achieved by coupling finite element meshes through norton equivalents. In *Proc. of the IEEE Int. Conf. on Robotics and Automation*, 1998.

[2] F. Boux de Casson and C. Laugier. Modeling the dynamics of a human liver for a minimally invasive simulator. In *Proc. of the Int. Conf. on Medical Image Computer-Assisted Intervention*, pages 1156–1165, Cambridge, U.K., September 1999.

[3] F. Boux de Casson and C. Laugier. Simulating 2d tearing phenomena for medical surgery simulators. In *Computer Animation*, Philadelphia, U.S.A., May 2000.

[4] D. d'Aulignac, R. Balaniuk, and C. Laugier. A haptic interface for a virtual exam of the human thigh. In *Proc. of the IEEE Int. Conf. on Robotics and Automation*, pages 2452–2457, San Francisco, CA (US), April 2000.

[5] D. d'Aulignac, C. Laugier, and M. C. Cavusoglu. Towards a realistic echographic simulator with force feedback. In *Proc. of the IEEE-RSJ Int. Conf. on Intelligent Robots and Systems*, pages 727–732, Kyongju (KR), October 1999.

[6] A. Deguet, A. Joukhadar, and C. Laugier. Models and algorithms for the collision of rigid and deformable bodies. In P. K. Agarwal, L. E. Kavraki, and M. T. Mason, editors, *Robotics: the algorithmic perspective*, pages 327–338. A K Peters, 1998. Proc. of the Workshop on the Algorithmic Foundations of Robotics. Houston, TX (US). March 1998.

[7] P.L. George and H. Borouchaki. *Delaunay Triangulation and Meshing - Applications to Finite Elements*. Éditions Hermès, 1998.

[8] A. Joukhadar, A. Wabbi, and C. Laugier. Fast contact localisation between deformable polyhedra in motion. In *Proc. of the IEEE Computer Animation Conf.*, pages 126–135, Geneva (CH), June june.

[9] J.-C. Lombardo, M.-P. Cani, and F.Neyret. Real-time collision detection for virtual surgery. In *Computer Animation*, Geneva Switzerland, May 26-28 1999.

[10] W.R. Mark, S.C. Randolph, M. Finch, J.M. Van Verth, and R.M. Taylor. Adding force feedback to graphical systems: Issues and solutions. In *Computer Graphics* Proceedings, Annual Conference Series, Proc. SIGGRAPH '96. ACM SIGGRAPH, 1996.

The Biomechanical Fidelity of Slope Simulation on the Sarcos Treadport Using Whole-Body Force Feedback

Rose Mills
University of Iowa, Computer Science Dept.
Iowa City, IA 52242-1419
rmills@blue.weeg.uiowa.edu

John M. Hollerbach and William B. Thompson
University of Utah, School of Computing
Salt Lake City, UT 84112-9205
jmh@cs.utah.edu, thompson@cs.utah.edu

Abstract: This paper addresses whether whole-body force feedback on treadmill-style locomotion interfaces can simulate the gravity forces experienced when walking on smooth inclines. By applying horizontal force feedback possible with the active mechanical tether of the Sarcos Treadport, it is shown that the biomechanics of walking are similar under conditions of real slope walking versus tether force walking. These biomechanical results complement previous psychophysical studies which yielded the same result, to conclude definitively that whole-body force feedback can realistically substitute for treadmill tilt.

1. Introduction

This paper addresses the issue of how well walking on sloped surfaces can be simulated on a locomotion interface using whole-body force feedback instead of actually tilting the walking surface. The Sarcos Treadport uses an active tether mechanism which both senses user position and applies forces to the user (Figure 1). This tether attaches to the user's back via a whole-body harness, and can apply a force along its linear axis. The tether can be used to simulate the extra gravity forces in slope walking by pulling or pushing on the user in the direction of walking. Because of its higher bandwidth, tether force can represent fast slope transients and is potentially a replacement for having a tilt mechanism at all. This has the added advantage of simplifying video displays which use mechanisms such as back-projected screens. Whole-body force feedback has other important uses as well, such as simulating hitting a wall or inertial forces during running [1].

When walking on a real slope, the gravity force f parallel to the slope that retards or assists walking is $f = mg\sin\theta$, where m is the user's mass, $g = 9.8m/s^2$ is gravity, and θ is the slope (Figure 2(A)). This gravity force can instead be applied by the mechanical tether to simulate slope walking (Figure

D. Rus and S. Singh (Eds.): Experimental Robotics VII, LNCIS 271, pp. 437–445, 2001.

Figure 1. The Sarcos Treadport with tether attachment to a user.

2(B)).

Previously we reported psychophysical results on the subjective equivalence of tether force for slope walking [7]. We asked subjects to walk on a tilted treadmill, then to walk on a flat treadmill but to adjust the tether force until it felt most like walking on the reference slope. We obtained a linear relationship between preferred tether force and slope, indicating psychological equivalence.

$$f = 0.65mg \sin \theta \qquad (1)$$

The fractional force preference of 65% was hypothesized to arise either from localized force application to the body through the harness or through a simplified mechanical model of the human as a lumped mass m. A similar fractional force preference was also found for inertial force display [1].

A more quantitative test would be to show biomechanical equivalence, i.e., the gait patterns are the same for the two situations. It is not unreasonable to expect a biomechanical correlate because a user has to lean against the tether force, in a manner which could conceivably be similar to leaning while walking on a slope. This paper presents such a biomechanical analysis.

Past research on the biomechanics of slope walking have employed various kinematic measures to quantify the change of gait with slope, such as leg joint angle ranges [4] and the knee-hip cyclogram [2]. We have examined a range of measures to deduce what is the best biomechanical correlate for slope, then used such a measure to show that the biomechanics of slope walking versus tether force walking are indeed similar.

2. Methods

Measurement of gait was done with the Northern Digital Optotrak System, which involves placement of active LED markers on the foot, calf, thigh, and hip (Figure 3). Special rigid bars for LED mounting were created to facilitate

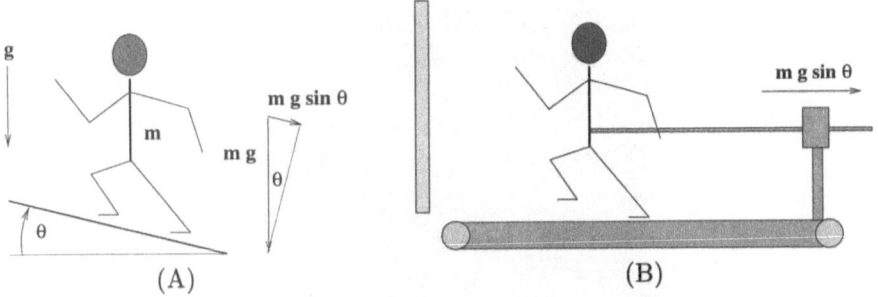

Figure 2. (A) Gravity force $mg \sin \theta$ opposes uphill walking. (B) Simulation of this gravity force with an active mechanical tether.

joint angle calculation by considering these bars as vectors representing absolute orientation of leg segments. Padding and straps were employed to ensure tight but comfortable coupling to the limbs.

Two different generations of Treadport were employed in this study. The second-generation Treadport has a redesigned belt drive and mechanical tether which are improvements over the first-generation Treadport [3], but does not yet have a functioning tilt mechanism. Therefore we employed the first-generation Treadport to generate a tilted walking surface and the second-generation Treadport to apply tether forces.

The gaits of six subjects, three male and three female, were measured both while walking on a tilted belt and while walking on a flat belt but with tether force application. The tether force is applied to a user via a whole-body harness to which the active mechanical tether attaches. The tether utilizes a linear drive consisting of a timing belt and geared electric motor [3], and is capable of exerting 315 N.

For the treadmill tilt experiments, the subject walked on the first-generation Treadport and the slope of the Treadport was varied randomly at two degree intervals between 6 degrees downhill and 14 degrees uphill. This range was dictated by the asymmetry of the tilting mechanism of the first-generation Treadport. For the tether force experiments, the subject walked on the second-generation Treadport and forces on the tether were varied randomly from -100 N to 45 N, depending on the subject's mass. A negative force corresponds to a force pulling the subject and therefore simulates a positive slope, and a positive force simulates a negative slope accordingly. Both Treadports were kept at a constant walking speed throughout the experiment.

3. Results

The data collected is similar across all of the subjects. Hence we present representative results of one specific subject (a female) to show the trends and characteristics that are common to all subjects in the experiment. Knee and hip angles were derived from the positions of the sensors and then plotted against one another. These plots are the knee-hip cyclograms found in Figures 4 and 5. Figure 4 shows how cyclograms change according to variation

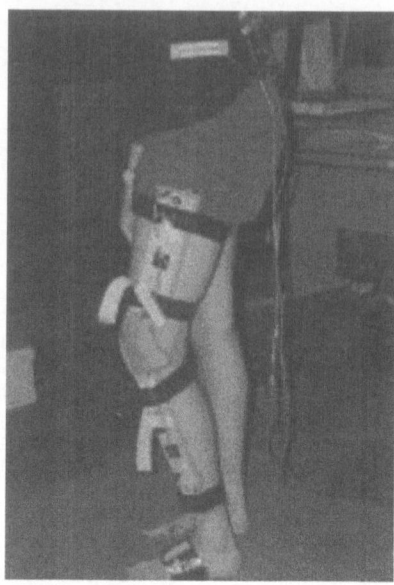

Figure 3. Marker attachment for leg joint angle measurements.

in slope. As the slope increases, the cusp of the cyclogram rotates clockwise, the knee and hip angle ranges widen, and the overall shape of the cyclogram becomes more oblong. The cusp happens at footfall, where the knee flexes almost elastically before straightening out to push off.

Figure 5 shows how the cyclograms change according to various tether forces applied to the subject. The cyclogram trends are similar to those of Figure 4; the cusp rotates clockwise, knee and hip angle ranges widen, and the overall shape of the cyclogram becomes more oblong.

These visual changes and trends can be captured quantitatively using moment-based analysis of the cyclograms and by joint angle ranges. Using this analysis the following statistics were calculated: hip range, knee range, ratio of hip range/knee range, ratio of knee range/hip range, area of cyclogram, circularity, eccentricity, orientation, and cusp orientation. These values are shown in Figures 6 and 7. Once again the general trends of the cyclogram are similar as the slope and tether force change from smaller slopes to larger slopes.

To find a relationship between tether force and the simulated slope angle, least squares equations were found for each of the properties of the cyclogram for all subjects. The most linear properties across both slope and force were the hip range, the knee/hip range ratio, the cyclogram orientation, and the cyclogram cusp orientation. Analysis of variance accounted for (VAF) for all subjects showed that hip range is consistently the most linear feature with slope or force, and hence is used in the subsequent analysis. Straight line fits were made to hip range versus force $HR = af + b$ and hip range versus slope $HR = c\theta + d$ for each subject. The approximation $\theta \approx \sin\theta$ is used, which only

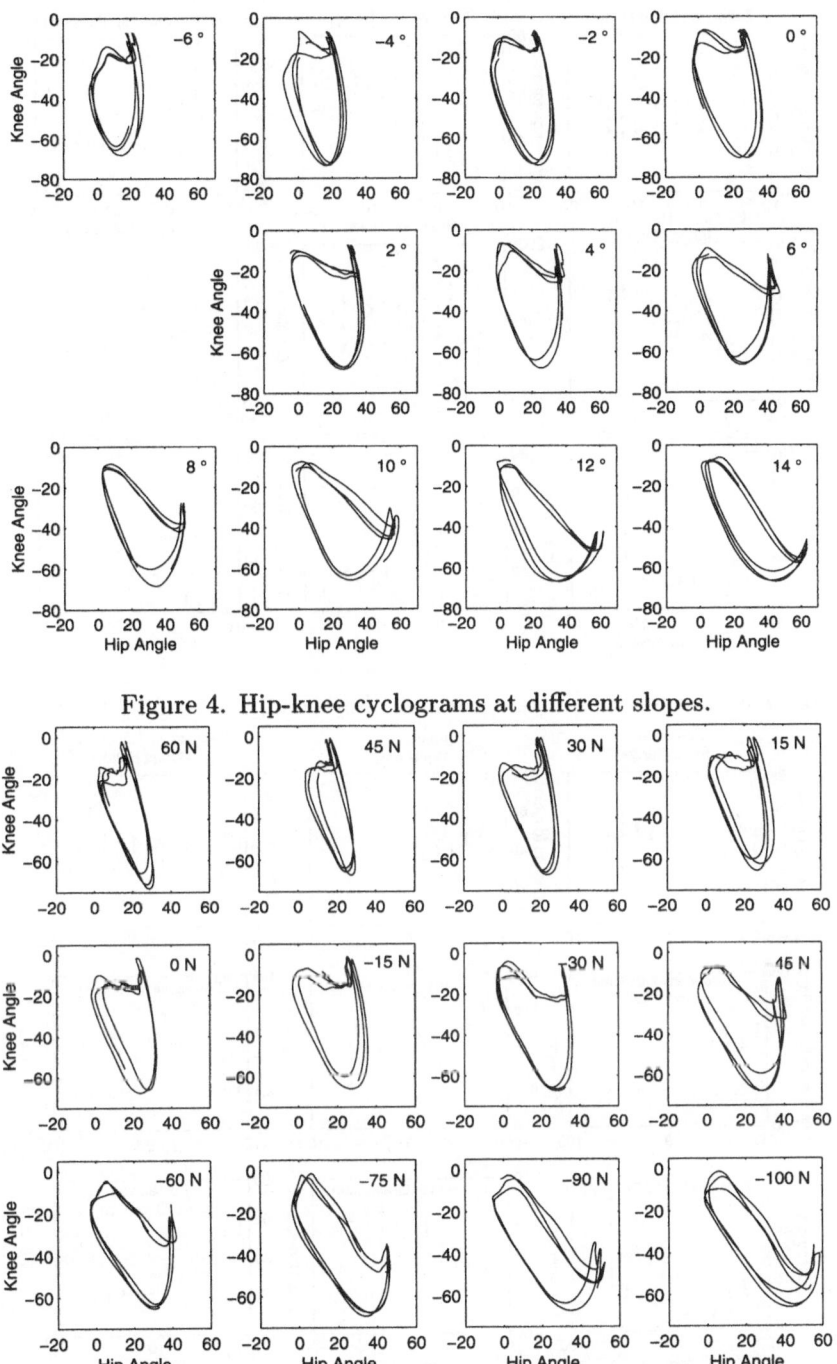

Figure 4. Hip-knee cyclograms at different slopes.

Figure 5. Hip-knee cyclograms at different tether forces.

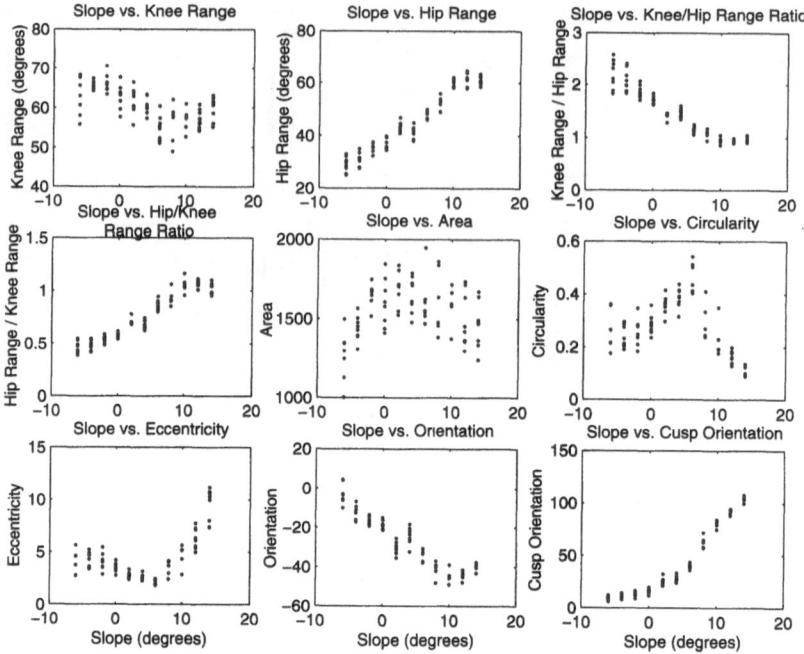

Figure 6. Properties of cyclograms as they change according to slope.

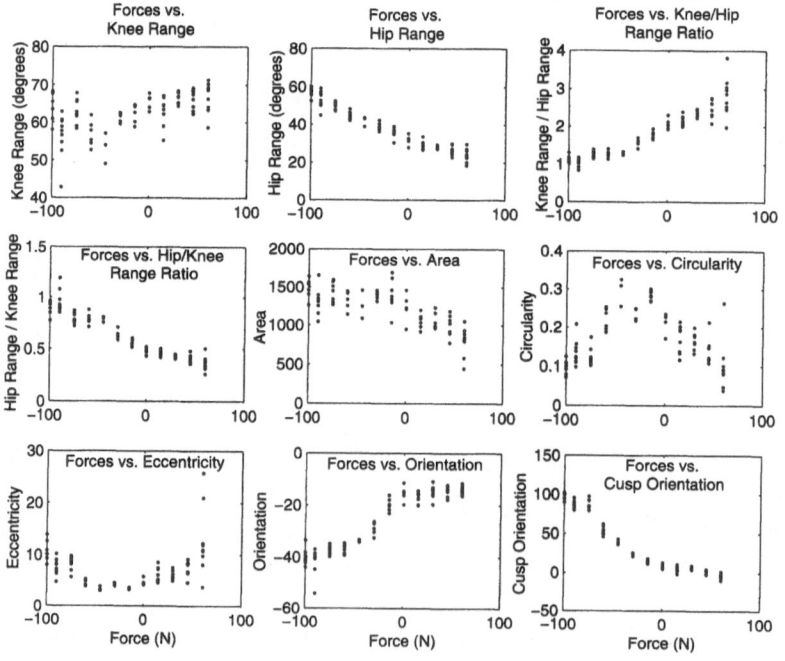

Figure 7. Properties of cyclograms as they change according to tether force.

Subject	c/a	$(d-b)/a$	$c/a/mg$
1	-437	2.0	0.647
2	-289	-15.5	0.520
3	-494	-19.8	0.756
4	-444	-17.9	0.526
5	-576	32.0	0.730
6	-480	3.6	0.663

Table 1. The slope c/a and intercept $(d-b)/a$ of the experimentally derived linear relation between tether force and slope, and the fractional force result $c/a/mg$.

has a 1% error at the maximum tilt of 14 degrees. Then a relation between the tether force and slope could be predicted as:

$$f = \frac{c}{a}\theta + \frac{d-b}{a} \qquad (2)$$

Table 1 shows the results for the six subjects. We are expecting a relation $f = mg\sin\theta$, so the intercept $(d-b)/a$ not being zero is an indication of the approximation of the linear fits.

As mentioned earlier, it was found from psychophysical experiments that there was a fractional force preference of 65% of the full predicted gravity force, i.e., $f = 0.65mg\sin\theta$ [7]. Dividing the slope c/a by a subject's weight mg indicates what would be any fractional force determined from biomechanics. The last column in Table 1 shows these fractional forces, which range from 52% to 73%. These results are in the vicinity of the average result 65% found from the psychophysical studies of [7].

4. Discussion

These results demonstrate that the horizontal tether force changes a person's gait in a manner that is similar to the gait changes of the person walking on different slopes. This means that an applied force from the tether is biomechanically a reasonable simulation of inclined surfaces. These results complement the previously reported psychophysical results which showed the same result [7]. In combination, the biomechanical and psychophysical results demonstrate conclusively that tether force can reasonably simulate walking on slopes.

There is an implication for treadmill design, because one can now choose between treadmill tilt and tether force to simulate slope. There are already reasons to include an active mechanical tether with treadmills when creating locomotion interfaces, such as inertial force display [1], the display of hitting objects, safety restrictions to range of forward motion on the treadmill surface, and accurate tracking of user position. One can then add to that list the accurate display of slope.

Although treadmill tilt of course displays slope realistically, there are some reasons against tilt implementations. The treadmill surfaces are large and heavy, especially the large 6-by-10 foot surface of the Sarcos Treadport II, and so the tilt mechanism adds cost and complexity to the design and will be slower than the fast-acting mechanical tether. When using a stationary CAVE visual display, a tilted platform would obscure portions of the screens. An alternative is to mount the CAVE on the treadmill (Noma, personal communication), although the size of the display will be necessarily limited. If projection onto the belt surface is contemplated, then the image will be distorted and will have to be compensated for by computation.

Several gait features were found to have approximately linear relationships with slope or tether force: hip range, the knee/hip range ratio, the knee/hip cyclogram orientation, and the knee/hip cyclogram cusp orientation. Of these the hip range was the more linear. Hip range was also noted as an important slope indicator by [4], although the plots of hip range versus slope were not as linear as what we found. Goswami had previously characterized higher-order moments of the knee/hip cyclograms as good descriptors of slope walking [2]. Our work has shown that the orientation of the knee/hip cyclogram as a whole and the orientation of just the cusp part of the knee/hip cyclogram provided good linear characterizations of slope walking.

The good linear fits of tether force to hip range and treadmill tilt to hip range allowed a prediction of tether force to treadmill tilt. By dividing the slope of the linear relation of tether force to treadmill tilt angle by each subject's weight, it was found that a fractional application of force between 52% and 73% of the expected amplitude $f = mg \sin \theta$ was appropriate to represent a particular slope θ. This result is consistent with the 65% fractional preference determined from psychophysical experiments [7]. Because the fractional forces were derived from biomechanics, it must indeed be the case that the proper tether force is not 100% of the predicted gravity force $f = mg \sin \theta$. As mentioned earlier, the cause of the fractional force must have something to do with the point-force application to the body by the tether or by the method of force distribution to the body by the harness. An exact explanation for the fractional force based on a mechanical analysis awaits future analysis.

An application of these results besides virtual reality is in the use of treadmills for legged robot research. For example, treadmills have been built for the running robots of [6] and [5]. Our results suggest that slopes could be simulated for the running robots by adding an active mechanical tether. In addition, the mechanical tether could supply realistic inertial forces to those robots [1].

Acknowledgments

This research was supported by NSF Grant IIS-9908675 and by NSF Grant CDA-96-2361. We thank Yangming Xu for assistance with Treadport I and Ryan Hayward for experimental assistance.

References

[1] Christensen RR, Hollerbach JM, Xu Y, Meek SG 2000 Inertial-force feedback for the Treadport locomotion interface. *Presence: Teleoperators and Virtual Environments.* 9:1-14.

[2] Goswami A 1998 A new gait paramterization technique by means of cyclogram moments: application to human slope walking. *Gait & Posture.* 8(1):15-36.

[3] Hollerbach JM, Xu Y, Christensen R, Jacobsen, SC 2000 Design specifications for the second generation Sarcos Treadport locomotion interface. *Haptics Symposium, Proc. ASME Dynamic Systems and Control Division, DSC-Vol. 69-2*, Orlando, Nov. 5-10, 2000, pp 1293-1298.

[4] Masmoudi L-A, Martin L, Cordier E, Blanchi, J-P 1999 Kinematics analysis of human slope walking. *1st Intl. Conf. on Science & Technology in Climbing and Mountaineering*. Leeds.

[5] Moghaddam M, Buehler M 1993 Control of virtual motion systems. *Proc. IEEE/RSJ Intl. Conf. on Intelligent Robots and Systems,* Yokohama, July 26-30, pp 63-67.

[6] Raibert MH 1986 *Legged Robots That Balance.* MIT Press, Cambridge, MA.

[7] Tristano D, Hollerbach JM, Christensen R 2000 Slope display on a locomotion interface. In: Corke P, Trevelyan J (eds) *Experimental Robotics VI,* Springer-Verlag London, pp 193-201.

A New Approach to the Control of a Hydraulic Stewart Platform

M. R. Sirouspour and S. E. Salcudean
University of British Columbia
Vancouver, BC, Canada
shahins@ece.ubc.ca,   tims@ece.ubc.ca

Abstract: Novel stable adaptive nonlinear controllers are proposed for position tracking control of hydraulic manipulators. The controllers were developed by using the backstepping approach and are based on realistic models that include rigid body dynamics, friction, and hydraulic actuator dynamics with valve orifice and spool displacement nonlinearities. Using Lyapunov analysis, it is shown that tracking errors are bounded and converge to zero in the absence of Coulomb friction and are bounded when Coulomb friction is present. The proposed techniques were used to control a hydraulic Stewart platform. Their effectiveness were confirmed by simulations and experiments.

1. Introduction

Hydraulic robots and machinery are widely used in the construction and mining industries, as well as in motion simulators. They have rapid responses and high power-to-weight ratios which suit these applications. Currently, linear controllers, e.g. P.D., are commonly used to control hydraulic robots, since they are simple to implement. Their level of performance is limited because of the highly nonlinear nature of the hydraulic actuator dynamics.

The high performance control of robot manipulators has been the subject of much research. Examples include computed torque techniques [1], passivity based techniques [2], adaptive control techniques [3], and robust control methods [4]. Provably stable controllers that account for rigid body and actuator dynamics have been developed for electrically-driven robots (e.g., [5]), but the authors are aware of only two reports of provably stable hydraulic robot controllers that account for both actuator and rigid body dynamics [6, 7]. While the problem of single cylinder hydraulic control has been studied analytically and experimentally [8, 9], other work in hydraulic robot control (e.g., [10] using singular perturbations, [11] using decentralized adaptive control, and [12] using pressure feedback control) does not include complete stability proofs and contains limited experimental and simulation results.

This paper addresses the high performance control of a hydraulic Stewart platform. Following the authors' earlier work in [9, 7], novel nonlinear position tracking controllers are proposed for hydraulic robots using *backstepping*. The controllers are augmented with adaptation laws to compensate for parametric uncertainties in the system dynamics. In order to avoid acceleration feed-

D. Rus and S. Singh (Eds.): Experimental Robotics VII, LNCIS 271, pp. 447–460, 2001.

back, the first controller is augmented with a passivity-based adaptive observer while the second one features a robust sliding type observer. The adaptive controller/adaptive observer is proven to be semi-globally asymptotically stable. The asymptotic stability of the adaptive controller/sliding observer is also shown via Lyapunov analysis. In the presence of friction in the actuators, it can be shown that the tracking errors become bounded.

These approaches are different from the method proposed in [6] mainly in the form of the observers and the way they are coupled with the controllers. This is especially evident in the case of the second controller that features a robust observer. The proposed controllers were implemented to control a hydraulic Stewart platform. The simulation and experimental results demonstrate excellent position tracking performance for these new methods.

Following this section, a short description of the system model is given in Section 2. The control approaches are described in Section 3. Section 4 presents the simulation results. Section 5 discuses implementation issues and experimental results. Conclusions are drawn in Section 6.

2. Rigid Body/Actuators Dynamics

The dynamic model of a robot manipulator driven by single-rod hydraulic actuators is composed of two parts, namely rigid body and hydraulics dynamics. The dynamics of n-link rigid body robots are governed by a second order nonlinear differential equation

$$D(q)\ddot{q} + C(q,\dot{q})\dot{q} + G(q) = \tau, \tag{1}$$

where $q \in R^n$ is the generalized joint position vector and $\tau \in R^n$ is the generalized joint torque vector. $D(q) \in R^{n \times n}$ is the manipulator mass matrix, $C(q,\dot{q}) \in R^{n \times n}$ a matrix containing coriolis and centrifugal terms and $G(q) \in R^n$ is a vector representing the gravitational effects. Equation (1) can also be rewritten in a *linear-in-parameters* form which is convenient for the application of adaptive control techniques:

$$D(q)\ddot{q} + C(q,\dot{q})\dot{q} + G(q) = Y(q,\dot{q},\ddot{q})\theta, \tag{2}$$

where $Y(q,\dot{q},\ddot{q})$ is a regressor matrix and $\theta \in R^m$ is the vector of unknown rigid body parameters. The rigid body dynamics of the UBC motion simulator, the experimental setup, are given in *Appendix B*. The leg dynamics have been ignored in deriving this model. It is possible to include these dynamics in the system model, however, the resultant model would become very complex and only little improvement may be achieved. Unlike electrically driven manipulators, hydraulic robots exhibit significant nonlinear actuator dynamics. These dynamics can be expressed in the following form (assuming a three-way valve configuration)

$$\dot{\tau} = f(q,\dot{q}) + g(q,\tau,u), \tag{3}$$

where $u \in R^n$ is the control command vector and $f, g \in R^n$ are nonlinear functions of q, $\dot{q}$ and τ. The expressions for f and g are given in *Appendix A*.

This equation can also be described in the linear-in-parameters format

$$\dot{\tau} = f_0(q, \dot{q})\gamma_1 + g_0(q, \tau, u)\gamma_2, \tag{4}$$

where $\gamma_1 = \begin{bmatrix} \gamma_1^1 & \cdots & \gamma_1^n \end{bmatrix}^T$, $\gamma_2 = \begin{bmatrix} \gamma_2^1 & \cdots & \gamma_2^n \end{bmatrix}^T$ are two sets of hydraulic parameters as defined in *Appendix A* and f_0, g_0 are defined by:

$$f_0(q, \dot{q}) = diag\{f_0^i(q^i, \dot{q}^i)\}, \quad g_0(q, \tau, u) = diag\{g_0^i(q^i, \tau^i, u^i)\} \tag{5}$$

This model addresses the nonlinearities in spool displacement and valve orifice.

Note that (1) describes the rigid body dynamics in joint-space coordinates. An alternative approach is to represent these dynamics in task-space coordinates. All of the arguments made in this paper are applicable to task-space coordinates with only minor changes in the derivation.

3. Control Approach

The system to be controlled is a third-order nonlinear system subject to parametric uncertainties both in rigid body and hydraulic dynamics. In this section two methods are proposed for the control of this system. The following Lemma [13] will be used in the stability proof.

Lemma 1: Consider the scalar function $\alpha = (\theta - \hat{\theta})^T(\rho - \dot{\hat{\theta}})$, with $\theta, \hat{\theta}, \rho \in R^n$ and $a^i \le \theta^i \le b^i$. Then, if $\dot{\hat{\theta}} = \kappa(a, b, \rho)\rho$ where $\kappa(a, b, \rho)$ is a diagonal matrix with entries

$$\kappa^i(a, b, \rho) = \begin{cases} 0 & \text{if } \hat{\theta}^i \le a^i, \rho^i \le 0 \\ 0 & \text{if } \hat{\theta}^i \ge b^i, \rho^i \ge 0 \\ 1 & \text{otherwise} \end{cases} \tag{6}$$

it follows that $\alpha \le 0$.

3.1. Adaptive Controller with Adaptive Observer

An adaptive controller is proposed for the control of hydraulic robots modeled by (1) and (3). An adaptive observer is also presented to avoid using acceleration feedback in the control law. Throughout this section, the following notation will be used:

$$\dot{q}_r = \dot{q}_d - \Lambda_1(\hat{q} - q_d) = \dot{q}_d - \Lambda_1(e - \tilde{q}) \quad \dot{q}_o = \dot{\hat{q}} - \Lambda_2(q - \hat{q}) = \dot{\hat{q}} - \Lambda_2\tilde{q}$$

$$s_1 = \dot{q} - \dot{q}_r = \dot{e} + \Lambda_1(e - \tilde{q}) \qquad\qquad s_2 = \dot{q} - \dot{q}_o = \dot{\tilde{q}} + \Lambda_2\tilde{q}, \tag{7}$$

where $\hat{q} \in R^n$ is the estimated value of q, $e = q - q_d$, and $\tilde{q} = q - \hat{q}$ are position tracking errors and observation errors, respectively. Furthermore, $\Lambda_1, \Lambda_2 > 0$ are diagonal. Note that in the definitions of $\dot{q}_r$ and $\dot{q}_o$, $\dot{q}$ has been replaced by $\dot{q}_d$ and $\dot{\hat{q}}$ which eliminates the need for acceleration feedback as it will be seen later.

Theorem 1: Consider the system described by (1), (3), the observer dynamics

$$\dot{\hat{q}} = z + \Lambda_2 \tilde{q}$$

$$\dot{z} = \hat{D}_2(q)^{-1}[\tau - \hat{C}_2(q,\dot{q})\dot{q}_o - \hat{G}_2(q) + L_p\tilde{q} + K_ds_1 + K'_ds_2] \tag{8}$$

and the controller obtained by solving the following simple algebraic equation

$$g_0(q,\tau,u)\hat{\gamma}_2 = \dot{\tau}_d - f_0(q,\dot{q})\hat{\gamma}_1 - \Gamma_\tau^{-1}s_1 - K_\tau\tilde{\tau} \tag{9}$$

where

$$\begin{aligned}\tau_d &= \hat{D}_1(q)\ddot{q}_r + \hat{C}_1(q,\dot{q}_r)\dot{q}_r + \hat{G}_1(q) - K_d(s_1 - s_2) - K_pe \\ &= Y(q,\dot{q}_r,\ddot{q}_r)\hat{\theta}_1 - K_d(s_1 - s_2) - K_pe\end{aligned} \tag{10}$$

with unknown rigid body parameter adaptation laws

$$\dot{\hat{\theta}}_1 = -\kappa_{\theta_1}\Gamma_1^{-1}Y^T(q,\dot{q}_r,\ddot{q}_r)s_1, \quad \dot{\hat{\theta}}_2 = -\kappa_{\theta_2}\Gamma_2^{-1}Y^T(q,\dot{q},\dot{q}_o,\ddot{q}_o)s_2 \tag{11}$$

where

$$Y(q,\dot{q},\dot{q}_o,\ddot{q}_o)\hat{\theta}_2 = \hat{D}_2(q)\ddot{q}_o + \hat{C}_2(q,\dot{q})\dot{q}_o + \hat{G}_2(q) \tag{12}$$

and with hydraulic parameter adaptation laws

$$\dot{\hat{\gamma}}_1 = \kappa_{\gamma_1}\Gamma_{\gamma_1}^{-1}\Gamma_\tau f_0(q,\dot{q})\tilde{\tau}$$

$$\dot{\hat{\gamma}}_2 = \kappa_{\gamma_2}\Gamma_{\gamma_2}^{-1}\Gamma_\tau\Pi(\frac{\tilde{\tau}}{\hat{\gamma}_2})\left(\dot{\tau}_d - f_0(q,\dot{q})\hat{\gamma}_1 - \Gamma_\tau^{-1}s_1 - K_\tau\tilde{\tau}\right) \tag{13}$$

where $\Pi(\frac{\tilde{\tau}}{\hat{\gamma}_2}) = diag\{\frac{\tilde{\tau}^i}{\hat{\gamma}_2^i}\}$. Then, if the conditions given below in (14,15) are satisfied, $\underline{0}$ is an asymptotically stable equilibrium point of the state $\tilde{x} = \begin{bmatrix} e^T & \tilde{q}^T & s_1^T & s_2^T & \tilde{\tau}^T \end{bmatrix}^T$.

$$(i) \quad \underline{\sigma}(K_p)\underline{\sigma}(L_p)\underline{\sigma}(\Lambda_1)\underline{\sigma}(\Lambda_2) > \frac{1}{4}\bar{\sigma}^2(K_p)\bar{\sigma}^2(\Lambda_1)$$

$$(ii) \quad \|\tilde{x}(0)\| \leq \sqrt{\frac{\alpha_m}{3\alpha_M}\left(\frac{\underline{\sigma}(K_d) - C_M\dot{q}_{dm}}{C_M\bar{\sigma}(\Lambda_1)}\right)^2 - \frac{V_{pM} - V_{pm}}{\alpha_M}} \tag{14}$$

where

$$\alpha_m = \frac{1}{2}\min\{D_m, \underline{\sigma}(K_p), \underline{\sigma}(L_p), \underline{\sigma}(\Gamma_\tau)\}$$

$$\alpha_M = \frac{1}{2}\max\{D_M, \bar{\sigma}(K_p), \bar{\sigma}(L_p), \bar{\sigma}(\Gamma_\tau)\} \tag{15}$$

$$V_{pm} \leq \frac{1}{2}\left(\tilde{\theta}_1^T\Gamma_1\tilde{\theta}_1 + \tilde{\theta}_2^T\Gamma_2\tilde{\theta}_2 + \tilde{\gamma}_1^T\Gamma_{\gamma_1}\tilde{\gamma}_1 + \tilde{\gamma}_2^T\Gamma_{\gamma_2}\tilde{\gamma}_2\right) \leq V_{pM}$$

Here, $\bar{\sigma}(.)$ and $\underline{\sigma}(.)$ denote the maximum and minimum singular values of their matrix argument, respectively, and $\dot{q}_{dm}$ is an upper bound on the norm of the desired velocity. The projection gains κ_{θ_1}, κ_{θ_2}, κ_{γ_1} and κ_{γ_2} are defined as in (6).

Remark: Condition (ii) in (14) specifies the boundary of the attraction region. Since this boundary can be enlarged arbitrarily, the system is semi-globally asymptotically stable. All the gain matrices are assumed to be diagonal and positive definite. Also note that the controller and the observer are using different sets of estimated parameters. They could also be modified to use the same parameter estimates. The use of projection gains as defined in *Lemma* 1, guarantees that the estimated parameters remain within predefined levels and therefore the control law is always defined. The proposed control does not require acceleration measurements since there are no velocity terms, $\dot{q}$, involved in τ_d in (10).

Proof : By substituting (10) into (1) the following error dynamics are obtained

$$D(q)\dot{s}_1 + C(q,\dot{q})s_1 + K_d s_1 + K_p e =$$
$$K_d s_2 - C(q, s_1)(\dot{q} - s_1) - Y(q, \dot{q}_r, \ddot{q}_r)\tilde{\theta}_1 + \tilde{\tau} \quad (16)$$

The observer closed-loop dynamics could also be written as

$$D(q)\dot{s}_2 + C(q,\dot{q})s_2 + K'_d s_2 + L_p \tilde{q} = -K_d s_1 - Y(q,\dot{q},\dot{q}_o,\ddot{q}_o)\tilde{\theta}_2 \quad (17)$$

where (1) and (8) have been used in deriving (17). Now, let the Lyapunov-like function V_1 be defined as:

$$V_1 = \frac{1}{2}s_1{}^T D(q)s_1 + \frac{1}{2}e^T K_p e + \frac{1}{2}s_2^T D(q)s_2 + \frac{1}{2}\tilde{q}^T L_p \tilde{q} + \frac{1}{2}\tilde{\theta}_1^T \Gamma_1 \tilde{\theta}_1 + \frac{1}{2}\tilde{\theta}_2^T \Gamma_2 \tilde{\theta}_2 \quad (18)$$

It can be shown that the derivative of V_1 along the trajectory of the closed-loop system is given by

$$\dot{V}_1 \leq -s_1{}^T K_d s_1 - s_2{}^T K'_d s_2 - e^T K_p \Lambda_1 e - \tilde{q}^T L_p \Lambda_2 \tilde{q}$$
$$+ e^T K_p \Lambda_1 \tilde{q} + s_1^T \tilde{\tau} - s_1^T C(q, s_1)(\dot{q}_d - \Lambda_1 e + \Lambda_1 \tilde{q})$$
$$\leq -\left(\underline{\sigma}(K_d) - C_M(\dot{q}_{dm} + \bar{\sigma}(\Lambda_1)\|e\| + \bar{\sigma}(\Lambda_1)\|\tilde{q}\|)\right)\|s_1\|^2 \quad (19)$$
$$- \underline{\sigma}(K'_d)\|s_2\|^2 - \underline{\sigma}(K_p)\underline{\sigma}(\Lambda_1)\|e\|^2 - \underline{\sigma}(L_p)\underline{\sigma}(\Lambda_2)\|\tilde{q}\|^2$$
$$+ \bar{\sigma}(K_p)\bar{\sigma}(\Lambda_1)\|e\|\|\tilde{q}\| + s_1{}^T \tilde{\tau} = H(\|e\|, \|\tilde{q}\|, \|s_1\|, \|s_2\|) + s_1^T \tilde{\tau}$$

In deriving (19), (16), (17), Lemma 1 and the adaptation laws given in (11) have been used. Note that

$$- \underline{\sigma}(K_p)\underline{\sigma}(\Lambda_1)\|e\|^2 - \underline{\sigma}(L_p)\underline{\sigma}(\Lambda_2)\|\tilde{q}\|^2 + \bar{\sigma}(K_p)\bar{\sigma}(\Lambda_1)\|e\|\|\tilde{q}\|$$
$$= -\begin{bmatrix} \|e\| & \|\tilde{q}\| \end{bmatrix} \begin{bmatrix} \underline{\sigma}(K_p)\underline{\sigma}(\Lambda_1) & -\frac{1}{2}\bar{\sigma}(K_p)\bar{\sigma}(\Lambda_1) \\ -\frac{1}{2}\bar{\sigma}(K_p)\bar{\sigma}(\Lambda_1) & \underline{\sigma}(L_p)\underline{\sigma}(\Lambda_2) \end{bmatrix} \begin{bmatrix} \|e\| \\ \|\tilde{q}\| \end{bmatrix} \quad (20)$$

The condition (i) given in (14) guarantees the positive definiteness of the above matrix. Furthermore, if the following condition is satisfied

$$\|e\| + \|\tilde{q}\| < \frac{\underline{\sigma}(K_d) - C_M \dot{q}_{dm}}{C_M \bar{\sigma}(\Lambda_1)} , \tag{21}$$

then one can write

$$H(\|e\|, \|\tilde{q}\|, \|s_1\|, \|s_2\|) \le -\alpha(\|e\|^2 + \|\tilde{q}\|^2 + \|s_1\|^2 + \|s_2\|^2) \tag{22}$$

with $\alpha > 0$. It is not difficult to show that if (ii) in (14) holds then (21) is also satisfied.

Following the backstepping approach, V_2, which is a Lyapunov function for the system dynamics, is defined as

$$V_2 = V_1 + \frac{1}{2}\tilde{\tau}^T \Gamma_\tau \tilde{\tau} + \frac{1}{2}\tilde{\gamma}_1^T \Gamma_{\gamma_1} \tilde{\gamma}_1 + \frac{1}{2}\tilde{\gamma}_2^T \Gamma_{\gamma_2} \tilde{\gamma}_2 \tag{23}$$

where $\tilde{\gamma}_1 = \begin{bmatrix} \tilde{\gamma}_1^1 & \cdots & \tilde{\gamma}_1^n \end{bmatrix}^T$ and $\tilde{\gamma}_2 = \begin{bmatrix} \tilde{\gamma}_2^1 & \cdots & \tilde{\gamma}_2^n \end{bmatrix}^T$ are the vectors of hydraulic parameter estimation errors. By taking the derivative of (23) and employing the control law given in (9) and after some manipulation one can show that

$$\dot{V}_2 \le H(\|e\|, \|\tilde{q}\|, \|s_1\|, \|s_2\|) - \tilde{\tau}^T \Gamma_\tau K_\tau \tilde{\tau} + \tilde{\gamma}_1^T \left[f_0(q, \dot{q}) \Gamma_\tau \tilde{\tau} - \Gamma_{\gamma_1} \dot{\tilde{\gamma}}_1 \right]$$
$$+ \tilde{\gamma}_2^T \left[\Gamma_\tau \Pi(\frac{\tilde{\tau}}{\tilde{\gamma}_2})(\dot{\tau}_d - f_0(q, \dot{q})\hat{\gamma}_1 - \Gamma_\tau^{-1} s_1 - K_\tau \tilde{\tau}) - \Gamma_{\gamma_2} \dot{\tilde{\gamma}}_2 \right] \tag{24}$$

Using the adaptation laws given in (13), the derivative of V_2 becomes

$$\dot{V}_2 \le -\gamma(\|e\|^2 + \|\tilde{q}\|^2 + \|s_1\|^2 + \|s_2\|^2 + \|\tilde{\tau}\|^2) \tag{25}$$

with $\gamma > 0$. Therefore the tracking errors converge to zero asymptotically.
Remark: The method can be easily modified to use the same set of parameters in the controller and the observer. In this case

$$\dot{\hat{\theta}} = -\kappa_\theta \Gamma^{-1} \left[Y^T(q, \dot{q}_r, \ddot{q}_r)s_1 + Y^T(q, \dot{q}, \dot{q}_o, \ddot{q}_o)s_2 \right] \tag{26}$$

3.2. Adaptive Controller with Sliding Observer

An adaptive controller featuring a sliding-type observer is introduced in this section. By comparison to the previous method, this new approach requires fewer computations. First the following notation is defined:

$$\dot{q}_r = \dot{q}_d - \Lambda e \qquad s = \dot{q} - \dot{q}_r = \dot{e} + \Lambda e, \tag{27}$$

where $e = q - q_d$ and q_d, $\dot{q}_d$, $\ddot{q}_d$ are the desired position, velocity and acceleration trajectories, and $\Lambda > 0$ is a diagonal matrix.

Theorem 2: Consider the system described by (1),(3) and the following observer:

$$z = \dot{\hat{q}}, \qquad \dot{z} = \Gamma_o \dot{\tilde{q}} + \Lambda_o sgn(\dot{\tilde{q}}) - W^T(q, \dot{q}_r, \hat{\theta})s + \ddot{\tilde{q}} \tag{28}$$

with

$$W(q, \dot{q}_r, \hat{\theta}) = -\hat{D}(q)\Lambda + \hat{C}(q, \dot{q}_r) - K_d, \quad \ddot{\bar{q}} = \bar{D}^{-1}\left[\tau - \bar{C}\dot{q} - \bar{G}\right] \qquad (29)$$

where $\bar{D}$, $\bar{C}$ and $\bar{G}$ are constant matrices and $\dot{\bar{q}} = \dot{q} - \dot{\hat{q}}$ is the velocity observation error. Let the control law be given by the solution u of the following algebraic equation

$$g_0(q, \tau, u)\hat{\gamma}_2 = \dot{\tau}_d - f_0(q, \dot{q})\hat{\gamma}_1 - \Gamma_\tau^{-1}s - K_\tau\tilde{\tau} \qquad (30)$$

with

$$\tau_d = \hat{D}(q)(\ddot{q}_r + \Lambda\dot{\hat{q}}) + \hat{C}(q, \dot{\hat{q}})\dot{q}_r + \hat{G}(q) - K_d(s - \dot{\hat{q}}) - K_p e \qquad (31)$$

and let the parameters be adapted according to the following laws

$$\dot{\hat{\theta}} = -\kappa_\theta \Gamma_\theta^{-1} Y^T(q, \dot{q}, \dot{q}_r, \ddot{q}_r)s \qquad (32)$$

and

$$\dot{\hat{\gamma}}_1 = \kappa_{\gamma_1}\Gamma_{\gamma_1}^{-1}\Gamma_\tau f_0(q, \dot{q})\tilde{\tau}$$
$$\dot{\hat{\gamma}}_2 = \kappa_{\gamma_2}\Gamma_{\gamma_2}^{-1}\Gamma_\tau \Pi(\frac{\tilde{\tau}}{\hat{\gamma}_2}) \cdot (\dot{\tau}_d - f_0(q, \dot{q})\hat{\gamma}_1 - \Gamma_\tau^{-1}s - K_\tau\tilde{\tau}) \qquad (33)$$

for the rigid body and hydraulic parameters, respectively, where κ_θ, κ_{γ_1} and κ_{γ_2} are projection gains as defined in (6). Then, $\underline{0}$ is an asymptotically stable equilibrium point of the state $\tilde{x} = \begin{bmatrix} e^T & s^T & \dot{\bar{q}}^T & \tilde{\tau}^T \end{bmatrix}^T$.

Remark : In Equation (31)

$$\ddot{q}_r + \Lambda\dot{\bar{q}} = \ddot{q}_d - \Lambda(\dot{q} - \dot{q}_d) + \Lambda(\dot{q} - \dot{\hat{q}}) = \ddot{q}_d - \Lambda(\dot{\hat{q}} - \dot{q}_d)$$
$$s - \dot{\bar{q}} = \dot{q} - \dot{q}_r - \dot{q} + \dot{\hat{q}} = \dot{\hat{q}} - \dot{q}_r \qquad (34)$$

Therefore, τ_d does not contain any velocity terms. This is a very important point since $\dot{\tau}_d$ appears in (30) and (33). In other words, the proposed control law does not require acceleration measurements.

Proof: By substituting (31) into (1) the following dynamics are obtained

$$D(q)\dot{s} + C(q, \dot{q})s + K_d s + K_p e = -Y(q, \dot{q}, \dot{q}_r, \ddot{q}_r)\tilde{\theta} - W(q, \dot{q}_r, \hat{\theta})\dot{\bar{q}} + \tilde{\tau} \qquad (35)$$

Define the Lyapunov-like function V_1 to be

$$V_1 = \frac{1}{2}e^T K_p e + \frac{1}{2}s^T D(q)s + \frac{1}{2}\dot{\bar{q}}^T\dot{\bar{q}} + \frac{1}{2}\tilde{\theta}^T\Gamma_\theta\tilde{\theta} \qquad (36)$$

It can be shown that the derivative of V_1 becomes

$$\dot{V}_1 = -s^T K_d s - e^T K_p \Lambda e - \dot{\bar{q}}^T\Gamma_o\dot{\bar{q}} - \dot{\bar{q}}^T\left[\ddot{\bar{q}} - \ddot{q} + \Lambda_o sgn(\dot{\bar{q}})\right]$$
$$+ \tilde{\theta}^T\left[\Gamma_\theta\dot{\tilde{\theta}} - Y^T(q, \dot{q}, \dot{q}_r, \ddot{q}_r)s\right] + s^T\tilde{\tau} , \qquad (37)$$

where (35) has been used. The adaptation laws in (32) render $\dot{V}_1$ into

$$\dot{V}_1 = -s^T K_d s - e^T K_p \Lambda e - \dot{\tilde{q}}^T \Gamma_o \tilde{q} + \Sigma + s^T \tilde{\tau} \tag{38}$$

where $\Sigma = -\dot{\tilde{q}}^T [\ddot{\hat{q}} - \ddot{q} + \Lambda_o sgn(\dot{\tilde{q}})]$. It is not difficult to show that

$$\|\ddot{\hat{q}} - \ddot{q}\| \le \sigma_0 + \sigma_1 \|\dot{q}\|^2 + \sigma_2 \|\tau\| + \sigma_3 \|\dot{q}\|. \tag{39}$$

Therefore, the following choice for Λ_o renders $\Sigma < 0$:

$$\Lambda_o = diag\{\Lambda_o^i\}, \qquad \Lambda_o^i = \lambda_0^i + \lambda_1^i \|\dot{q}\|^2 + \lambda_2^i \|\tau\| + \lambda_3^i \|\dot{q}\| \tag{40}$$

and $\lambda_k^i > \sigma_k$ for $i = 1, \cdots, n$, $k = 0, \cdots, 3$. The rest of the proof is the same as in *Theorem 1* and will not be presented here. In summary, the controllers proposed in this paper require position, velocity and actuator forces (pressures) to be measured.

3.3. Effect of Friction

In the controllers proposed in this paper, friction in the hydraulic actuators has been neglected. It is easy to handle viscous friction since it acts as additional damping in the system. It is also straightforward to show that in the presence of Coulomb friction the tracking errors do not converge to zero but remain bounded. The error bounds can be reduced by increasing the gains. The proof will be omitted here.

4. Simulation Results

Simulations have been conducted to investigate the effectiveness of the proposed methods and also to obtain guidelines for performing the experiments. For this purpose the dynamic model of a six-degree-of-freedom hydraulic Stewart platform (see *Appendix B*) and the controller were simulated using the *Matlab Simulink^{TM}* toolbox. A task-space control strategy was adopted because of the simpler form of the dynamics in these coordinates. Since in practice the actuator lengths are measured, the forward kinematics must be computed on-line. This was done using Newton's method. The proposed controllers were modified slightly in order to be used in task-space coordinates.

The system parameters were chosen close to those of the experimental setup and are given in Table 1. The controllers performed similarly in simulation and only the results obtained with the adaptive controller/sliding observer are presented here. The tracking errors for a reference trajectory composed of $x_d = 0.02 \sin(2\pi t) + 0.01 \sin(4\pi t) + 0.01 sin(6\pi t)$, $y_d = 0$, $z_d = 0.02 \sin(2\pi t) + 0.01 \sin(4\pi t)$, $\psi_d = 0.0873 \sin(2\pi t) + 0.0349 \sin(4\pi t)$, $\theta_d = 0.0524 \sin(2\pi t) + 0.0175 \sin(4\pi t)$, $\phi_d = 0.0524 \sin(2\pi t) + 0.0175 \sin(4\pi t)$ are shown in Figure 1(a). Positions and angles are expressed in meters and radians, respectively. The tracking errors clearly converge to zero. The profiles of the parameter estimates are given in Figure 1(b). The parameter adaption laws were activated after $t = 0.5 sec$. The parameters converge to their actual values as seen in this figure even though the parameter convergence is not guaranteed in theory.

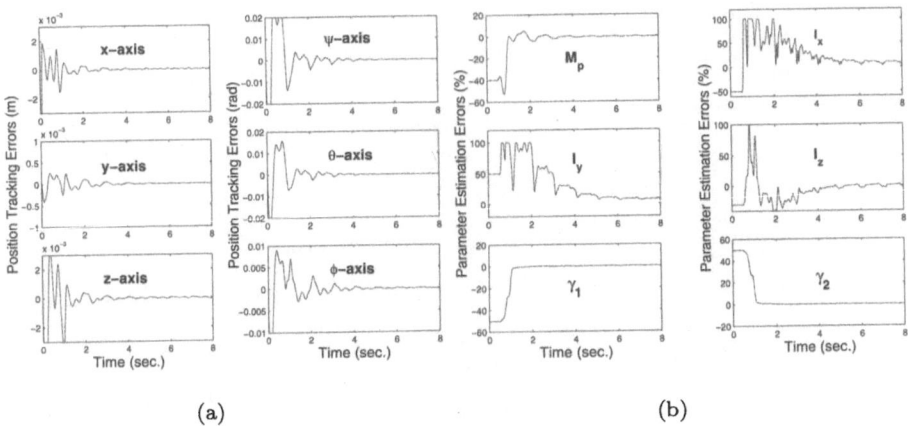

(a)   (b)

Figure 1.   Simulation results.   (a) Position tracking errors.   (b) Parameter estimation errors.

5. Experimental Results

The control methods proposed in this paper were also evaluated experimentally. The University of British Columbia motion simulator (Figure 4) was chosen for this purpose [14]. This simulator uses six 1.5 inch bore, 54 inch stroke hydraulic jacks. The jacks are anchored by roller-bearing U-joints in 120 degree symmetric configurations on the base and platform. Each cylinder is capable of exerting forces in excess of 4000 N at 1 m/s, and over 8000 N at zero rod speed. The hydraulic actuation system is equipped with *Rexroth 4WRDE* three-stage proportional valves connected in a three-way configuration. Low friction Teflon seals are used in the hydraulic cylinders. The installed sensors measure the actuator lengths, the valve spool positions, and the pressures both in the control and supply sides of the cylinders. High bandwidth valves with a bandwidth of around 80Hz have been used in the setup so the dynamics of the valves may be ignored. In order to synthesize the control command an algebraic equation must be solved (Equation (9) or (30)). In practice, the valve spool positions are sensed and employed in the implementation of the controllers. The actuator velocities which are needed in the control law are estimated from the measured actuator lengths using fixed gain Kalman filters. Off-line experiments were performed to identify the initial values of the parameter estimates.

The computational setup was a PC running $VxWorks^{TM}$ *5.4* and a Sparc 1e board running $VxWorks^{TM}$ *5.2* (see Figure 4). The Sparc 1e performs the I/O and safety functions and the controller runs on the PC. The controller was implemented using the *Matlab Real Time Workshop^{TM}* toolbox targeting $Tornado^{TM}$ *2.0*. Data between the PC and the VME board are communicated trough a custom parallel I/O communication protocol. Using this setup a control frequency of 512 Hz was successfully achieved. The same controller block used in the simulation studies was utilized to control the platform. This is a great advantage of the computational approach adopted in this paper.

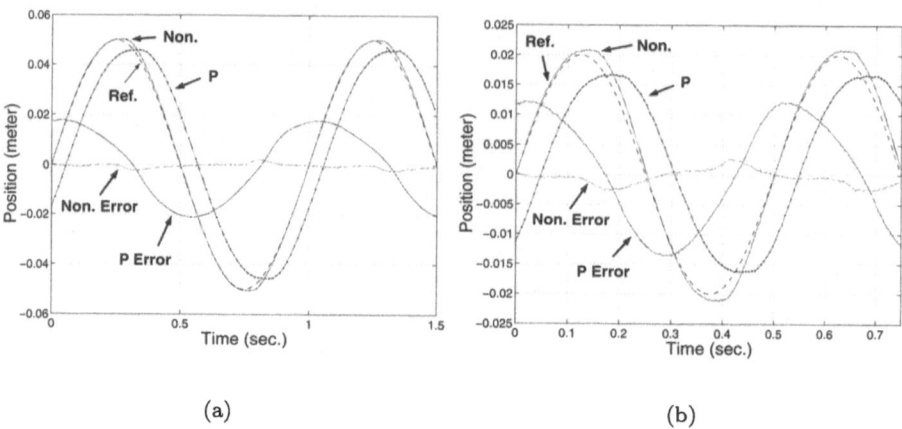

Figure 2. Position tracking along z coordinate (experiment). (a) 1HZ reference trajectory. (b) 2HZ reference trajectory.

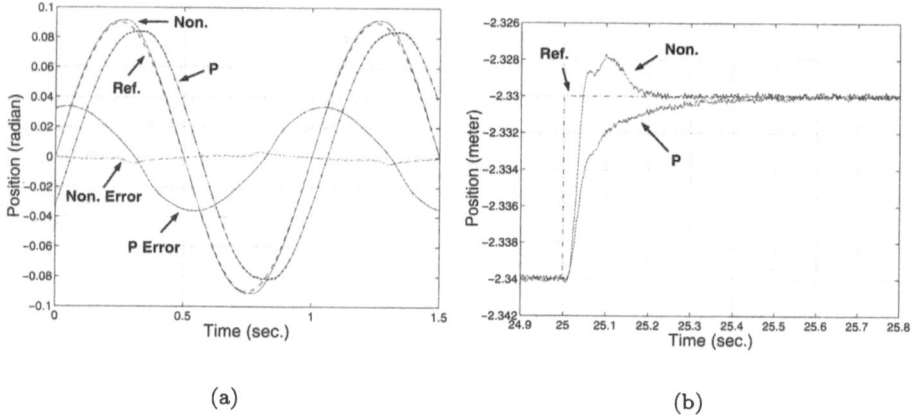

Figure 3. Experimental results. (a) 1HZ position tracking along ψ coordinate. (b) Step response along z coordinate.

Only the results of the experiments with adaptive controller/sliding observer are presented here while similar performance was observed for the other controller. Figure 2(a) shows the tracking behavior of the nonlinear controller compared with that of a well-tuned P controller in tracking the reference trajectory $z_d = -2.34 + 0.05 \sin(2\pi t)$ meter (the bias is not shown). The maximum tracking errors are 4% and 43% for the nonlinear and P controller, respectively. The response of the system to a 2Hz reference trajectory was also examined and is presented in Figure 2(b). In this case $z_d = -2.34 + 0.02 \sin(4\pi t)$ whereas the maximum tracking errors are 14% and 69%. Similar results were obtained in the other coordinates. For example, Figure 3(a) shows the tracking results along the ψ axis where $\psi_d = 0.09 \sin(2\pi t)$ with 4% and 41% maximum tracking error for the nonlinear and P controller, respectively. In all of these cases the

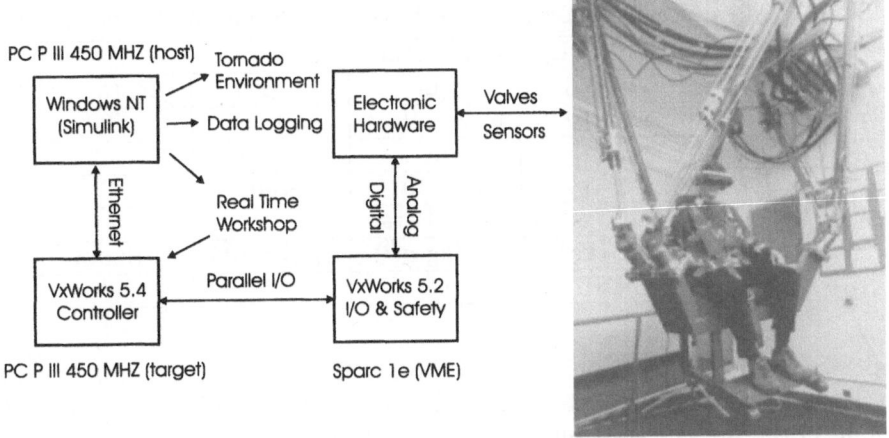

Figure 4. The experimental setup.

proposed adaptive nonlinear controller clearly outperforms the well-tuned P controller and exhibits excellent tracking performance.

During the experiments, the estimated parameters did not converge to fixed values, contrary to what was observed in simulations. Friction is an important factor which could introduce tracking errors and prevent the parameters from converging. The proposed controllers may be interpreted as cascade combinations of passivity-based position controllers and actuator force controllers. The very stiff dynamics of hydraulic actuators make the force (pressure) control loop sensitive to velocity estimation errors (or velocity measurement noise) and pressure measurement noise. This limits the level of the pressure feedback gains which may deteriorate the force tracking and subsequently parameter estimation, especially for the hydraulic parameters. Other factors such as unmodeled dynamics and insufficient excitation could also prevent parameter convergence. Moreover, it should be stressed that the parameter convergence is not even guaranteed in theory, so the experimental results do not contradict the theoretical arguments. The adaptation was found to be quite helpful in obtaining excellent tracking performance, which is the main goal of this research. In fact, it appears that parameters are adapted in way that reduces the tracking errors. The projection gains used in the adaptation laws proved effective in preventing the large parameter swings that can occur especially during start-up transients. The step response of the controller along the z axis is also compared with that of the P controller in Figure 3(b). As it can be seen, the nonlinear controller exhibits a much faster response with some overshoot.

6. Conclusions

This paper addressed the high performance position tracking control of hydraulic manipulators. Novel adaptive nonlinear controllers were proposed using the backstepping technique. Rigid body and hydraulic actuator models were incorporated in the design. These controllers feature novel adaptive and sliding

Table 1. The system parameters used in the simulations and experiments.

Hydraulic Parameters				
Parameter	A (m^2)	a (m^2)	L (m)	P_s (psi)
Value	1.14×10^{-3}	6.33×10^{-4}	1.37 m	1500
Parameter	d (m)	c	β (Mpa)	-
Value	55.4×10^{-6}	1.5×10^{-4}	700	-
Rigid Body Parameters				
Parameter	M_p (kg)	I_x (kg.m^2)	I_y (kg.m^2)	I_z (kg.m^2)
Value	250	45	45	43

type observers to avoid acceleration feedback. The tracking errors were shown to converge to zero asymptotically (they are bounded by a gain-controlled bound in the presence of Coulomb friction) using Lyapunov analysis. Simulation and experimental data obtained using the UBC hydraulic Stewart platform are excellent and demonstrate the effectiveness of the proposed approach.

References

[1] Luh J, Walker M, Paul R 1980 Resolved acceleration control of mechanical manipulators. *IEEE Tran. Tran. Automat. Cont.* 25:468-474.
[2] Berghuis H, Nijmeijer H 1993 A passivity approach to controller-observer design for robots. *IEEE Tran. Robot. Automat.* 9:740-754.
[3] Ortega R, Spong M W 1989 Adaptive motion control of rigid robots: A tutorial. *Automatica* 25:877-888.
[4] Spong M W 1992 On the robust control of robot manipulators. *IEEE Tran. Automat. Cont.* 37:1782-1786.
[5] Su C -Y, Stepanenko Y 1998 Redesign of hybrid adaptive/robust motion control of rigid-link electrically-driven robot manipulator. *IEEE Tran. Robot. Automat.* 14:651-655.
[6] Bu F and Yao B 2000 Observer-based coordinated adaptive robust control of robot manipulators driven by single-rod hydraulic actuators. In: *Proc. IEEE Int. Conf. Robot. Automat.*, pp. 3034-3039.
[7] Sirouspour M R, Salcudean S E 2000 Nonlinear control of hydraulic robots. *conditionally accepted by IEEE Tran. Robot. Automat.*.
[8] Sohl G A, Bobrow J E 1999 Experiments and simulations on the nonlinear control of a hydraulic servosystem. *IEEE Tran. Cont. Syst. Tech.* 7:238-247.
[9] Sirouspour M R, Salcudean S E 2000 On the nonlinear control of hydraulic Servosystems. In: *Proc. IEEE Int. Conf. Robot. Automat.*, pp. 1276-1282.
[10] d'Anrrea-Novel B, Garnero M A, Abichou A 1994 Nonlinear control of a hydraulic robot using singular perturbations. In: *Proc. IEEE Int. Conf. Sys. Man and Cyber.*, pp. 1932-1937.
[11] Edge K A, Gomes de Almeida F 1995 Decentralized adaptive control of a directly driven hydraulic manipulator part 1: theory. *Instn. Mech. Engrs.* 209:191-196.
[12] Li D, Salcudean S E 1997 Modeling, simulation and control of a hydraulic Stewart platform. In: *Proc. IEEE Int. Conf. Robot. Automat.*, pp. 3360-3366.
[13] Zhu W -H, De Schutter J 1999 Adaptive control of mixed rigid/flexible joint robot manipulators based on virtual decomposition. *IEEE Tran. Robot. Automat.*

15:310-317.

[14] Salcudean S E, Drexel P A, Ben-Dov D, et. al 1994 A six degree-of-freedom, hydraulic, one person motion simulator. In: *Proc. IEEE Int. Conf. Robot. Automat.*, pp. 859-864.

[15] Merrit H E 1967 *Hydraulic control systems*. Prentic-Hall Inc., New Jersey.

Appendix A

The dynamics of a typical hydraulic actuator are presented in more detail in this Appendix. A three-way valve configuration is assumed to be used in the actuators as shown in Figure 5. For such a configuration, the control pressure dynamics are governed by [15]

$$\frac{V_t}{\beta}\dot{p}_c = q_l + c_l(p_s - p_c) - \dot{V}_t \tag{41}$$

where V_t is the trapped fluid volume in the control side, β is the effective bulk modulus, p_c is the control pressure acting on the control side, p_s is the supply pressure acting on the rod side, q_l is the load flow, and c_l is the coefficient of total leakage. The load flow, q_l, is a nonlinear function of the control pressure and the valve spool position and is given by

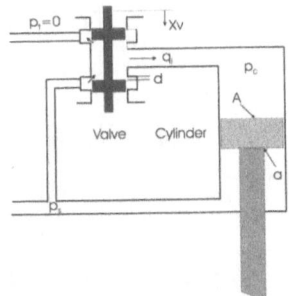

$$q_l = \begin{cases} c(u-d)\sqrt{p_c} & u < -d \\ c(u+d)\sqrt{p_s - p_c} + c(u-d)\sqrt{p_c} & -d \le u \le d \\ c(u+d)\sqrt{p_s - p_c} & u > d \end{cases} \tag{42}$$

Figure 5. A typical three-way valve configuration.

and $c = c_d w \sqrt{\frac{2}{\rho}}$ where c_d is the effective discharge coefficient, w is the port width of the valve, ρ is the density of the fluid, d is the valve underlap length and u is the valve spool position which is the control command. Note that the actuator output force is $\tau = p_c A - p_s a$. Therefore, using (41) and (42), the dynamics of i'th hydraulic actuator can be written in the following form (assuming $c_l \approx 0$)

$$\dot{\tau}^i = -\frac{A\beta^i \dot{q}^i}{q^i - L} + \frac{\beta^i}{q^i - L}q_l^i(\tau^i, u^i) = f^i(q^i, \dot{q}^i) + g^i(q^i, \tau^i, u^i) \tag{43}$$

For a Stewart platform, there are six actuators driving the system. The actuator subsystem dynamics can be represented in matrix form as in (3).

Note that (43) can be rewritten in the following form which is suitable for adaptive control.

$$\dot{\tau}^i = \gamma_1^i f_0^i(q^i, \dot{q}^i) + \gamma_2^i g_0^i(q^i, \tau^i, u^i) \tag{44}$$

where $\gamma^i = \begin{bmatrix} \beta^i & \beta^i c^i \end{bmatrix}^T$, $f_0^i = -\frac{A^i \dot{q}^i}{q^i - L^i}$, and $g_0^i = \frac{q_l^i}{c_i(q^i - L^i)}$ (does not depend on c_i see (42)). These equations can be written in matrix form as in (4).

Appendix B

The Stewart platform is a parallel manipulator widely used in conventional motion simulators. The simplified rigid body dynamics of a typical Stewart platform are presented here.

In task-space coordinates, the dynamics of the platform are governed by (neglecting the leg dynamics):

$$D(q)\ddot{q} + C(q,\dot{q})\dot{q} + G = (JL)^T \tau \qquad (45)$$

where $q = \begin{bmatrix} x & y & z & \psi & \theta & \phi \end{bmatrix}^T$ is position of the platform with respect to a fixed frame and ϕ, θ and ψ are the platform roll, pitch and yaw angles, respectively. Furthermore, J is the manipulator Jacobian matrix and L is a function of θ and ϕ.

$$L = \begin{bmatrix} I_{3\times3} & 0 \\ 0 & T \end{bmatrix}, \qquad T = \begin{bmatrix} \cos(\theta)\cos(\phi) & -\sin(\phi) & 0 \\ \cos(\theta)\sin(\phi) & \cos(\phi) & 0 \\ -\sin(\theta) & 0 & 1 \end{bmatrix} \qquad (46)$$

Finally, $D(q)$, $C(q,\dot{q})$ and G have the following forms:

$$D(q) = \begin{bmatrix} M_p I_{3\times3} & 0 \\ 0 & T^T(^b I_p)T \end{bmatrix}, \qquad C(q,\dot{q}) = \begin{bmatrix} 0 & 0 \\ 0 & c_{22} \end{bmatrix}$$

$$c_{22} = T^T Skew(^b\omega_p)^b I_p T + T^T(^b I_p)\dot{T}, \qquad G = \begin{bmatrix} 0 & 0 & M_p g & 0 & 0 & 0 \end{bmatrix}^T \qquad (47)$$

where ω is the angular velocity of the platform.

In the above equations, $^b I_p$ is the platform inertia matrix with respect to the base frame and is given by

$$^b I_p = R\,^p I_p R^T, \qquad ^p I_p = \begin{bmatrix} I_x & 0 & 0 \\ 0 & I_y & 0 \\ 0 & 0 & I_z \end{bmatrix} \qquad (48)$$

and R is a rotation matrix representing the coordinates of the platform-attached base vectors in the base frame. Note that (45) is not exactly as (1). However, since J is a function of platform position and is known, the controllers can be easily modified to be used in this case. Moreover, the rigid body dynamics may be written in a linear-in-parameters form

$$D(q)\ddot{q} + C(q,\dot{q})\dot{q} + G = Y_{6\times4}(q,\dot{q},\ddot{q})\theta , \qquad (49)$$

where $\theta = \begin{bmatrix} M_p & I_x & I_y & I_z \end{bmatrix}^T$ is a vector of kinematic and dynamic parameters. The detailed expressions of the elements of Y are long but fairly straightforward to derive and will not be presented here.

Design of life-size haptic environments

Yoky Matsuoka
Carnegie Mellon University
Pittsburgh, PA 15213
yoky@cs.cmu.edu

Bill Townsend
Barrett Technology, Inc.
Cambridge, MA 02141
wt@barrett.com

Abstract: Brake-actuated haptic devices, a subset of passive haptic devices, are not yet common, so their capabilities and limitations are only superficially understood. In this paper, we identify an optimal kinematics, and introduce a prototype of a life-size brake-actuated haptic device. With the new device, we conducted experiments with human subjects to evaluate the efficacy of passively created virtual environments. We identified key performance drawbacks that modern robotics theories fail to anticipate and overcome, and investigated design methods that can avoid such drawbacks.

1. Introduction

Motor-actuated devices are widely used in haptics research and advanced teleoperator masters - capable of creating a rich set of virtual environments in the former and recreating physical environments in the latter. Motors, however, introduce the risk of a device striking or overpowering users within reach. With rare exception, safety concerns limit motor-actuated haptic devices to small workspaces.

The replacement of energetic servo-motors with passive actuators such as brakes improves the inherent safety of the system, and expands the range of motion while still allowing us to simulate virtual objects (i.e. to constrain the user's motion inside an object). Safety is essential for acceptance in new applications such as medical procedures, quantitative rehabilitation, advanced exercise training, and entertainment. Even if the device experiences a power, hardware, or software failure, a brake-actuated device is inherently incapable of exceeding the kinetic energy that a user supplies in each motion. Using currently available engineering technology, the workspace of brake-actuated haptic devices can expand to cover whole-body movements without risking a user's safety. With the expanded workspace, a variety of large movements can be trained and quantitatively analyzed.

In this paper, we describe a novel design of a brake-actuated device, and present theoretical and experimental comparisons to active devices. Theoretically, we demonstrate that brake-actuated devices surprisingly require at

D. Rus and S. Singh (Eds.): Experimental Robotics VII, LNCIS 271, pp. 461–470, 2001.

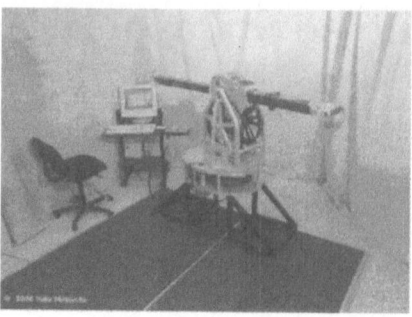

Figure 1. The brake-actuated spherical haptic device we constructed.

least one prismatic joint in order to produce the simplest environments such as viscous fields. A key difference from active devices is that brakes applied at different joints cannot be added vectorially. As a result, it is difficult to prevent the user from penetrating into an object, and at the same time allow motion in collision free directions ("sticky walls" problem discussed below).

In experiments with our prototype device, we evaluated the machine's efficacy with human subjects, exploring various force fields, including damping, stiffness, inertia, and shape (concave, convex) of virtual objects. Analysis confirms the anticipated problems. Relying on the characteristics of human motor perception, we propose a technique to overcome the device limitations.

2. Prototype Design of an Optimal Human-Scale Brake-Actuated Haptic Device

In 1995, we set out to design and build the best haptic prototype under the following constraints:

1. Life size - workspace size on the order of cubic meters, forces on the order of tens of kilograms, and fully spatial 6 degrees of freedom (DOF) kinematics.

2. Safe - passive actuation only.

3. Economically practical - with today's mechanisms and materials.

Analysis led us uniquely to the device illustrated in Figures 1, with brake-actuated spherical kinematics for regional motions terminating at an unactuated 3-axis wrist that supports a user hand grip.

In our analysis of the machine design, we begin by defining key terms. "Servo-actuation" is the ability to modulate joint torques under computer control in order to create the haptic force field that a user experiences. "Energetic servo-actuation" is capable of channeling power from an unlimited external source of energy into the joint torques. "Passive servo-actuation" is physically incapable of channeling power from any external source into the joint torques. "Brake servo-actuation" is a subset of passive servo-actuation that uses servo-brakes and does not rely on other passive mechanisms. "Sticky-walls" are unique to brake-actuated haptic devices, where a collision with a wall traps the user behind the wall. There are workarounds, as we discuss in this paper.

Ruling out energetic actuation for its life-threatening safety concerns leaves only passive actuation. One technique was explored by Troccaz et al. [1] who used overrun clutches to brake joint movements. These overrun clutches were not servo-actuated and the applied torque purely depended on the joint angles and speed. Another technique for controlling torques passively was through steerable joints. Substantial progress has been made in building a wide array of steerable robotic joints, collectively known as Programmable Constraint Machines (PCMs) and embodied in cobots [2],[3]. These PCMs have the fundamental advantage of eliminating "sticky walls" by actively aligning the constraint direction with the desired motion. Their disadvantage is the sizable side-slip that traction devices exhibit with existing designs and material technology. That disadvantage however may be overcome when new materials become available in the future.

2.1. Kinematic Design Considerations

Our approach to expanding the safe use of haptics to life-size was to replace the motors driving each joint axis of a device with similarly sized servo-brakes. We wish to set the kinematics of the robot in such a way that the end point only opposes an externally imposed motion, so to satisfy,

$$v^T f = v^T(-cv) = -cv^2 < 0 \qquad (1)$$

where v is the end-point velocity vector, f is the force vector opposing the external motion, and c is a constant. Given that the Jacobian matrix, J, is invertible and relates endpoint velocities to joint forces τ and velocities $\dot{\theta}$, we can express (1) as

$$v^T f = \dot{\theta}^T J^T (J^T)^{-1}\tau = \dot{\theta}^T \tau < 0 \qquad (2)$$

For energetically actuated robots, (2) is sufficient because terms from each joint i can simply add up vectorially as long as the entire sum is less than zero for all joints n:

$$\dot{\theta}^T \tau = \sum_i^n \dot{\theta}_i \tau_i < 0 \qquad (3)$$

However, if every joints in a robot are actuated by brakes, none of the joints can produce any active force, thus,

$$\dot{\theta}_i \tau_i < 0 \qquad (4)$$

for each joint i, making (2) invalid for brake actuated devices.

The goal is to control forces in the Cartesian space of the user. So a logical first concept is a Cartesian machine with three orthogonal prismatic joints, such as employed on vertical milling machines. For each Cartesian joint i,

$$\dot{\theta}_i \tau_i = \dot{\theta}_i (J^T f)_i = \dot{\theta}_i (J^T(-cv))_i = \dot{\theta}_i j_i^T(-cJ \dot{\theta}) = -c \dot{\theta}_i j_i^T \sum_i^n j_i \dot{\theta}_i \qquad (5)$$

for all n joints in the robot, where j_i is the ith row of J. When all joints are orthogonal as in the Cartesian coordinates,

$$\dot{\theta}_i \, \tau_i = -c \, \dot{\theta}_i \, j_i^T \sum_i^n j_i \, \dot{\theta}_i = -c \, \dot{\theta}_i^2 \, j_i^T j_i < 0 \qquad (6)$$

Therefore, a Cartesian machine with three orthogonal prismatic joints can oppose an externally imposed motion with brakes. In fact, (6) is true for any of a large number of mathematically orthogonal coordinate frames. But realistically, robotic orthogonal kinematic arrangements are limited to Cartesian, cylindrical, and spherical with spatial mechanisms listed in Table 1.

Kinematic Arrangement	No. of Revolute Joints	No. of Prismatic Joints
Cartesian	0	3
Cylindrical	1	2
Spherical	2	1

Table 1. Spatial mechanisms for orthogonal kinematics

Unfortunately, orthogonal kinematics cannot be reasonably constructed without including at least one prismatic joint. In serial-link robotics, kinematic structures with prismatic joints tend to be more massive for a given work volume than purely revolute structures. A massive structure may be acceptable for a CNC machine in order to achieve the high stiffness needed to cut through metal precisely, but the accompanying inertia and friction, each of which degrade backdrivability, is not acceptable for haptic devices.

So why not build a fully revolute brake-actuated haptic device, and eliminate all prismatic joints? In fact, most serial-link robots do just that, abandoning orthogonal kinematics without penalty [4]. For non-orthogonal kinematics, the summation in (6) does not simplify but rather expands to three independent terms (for a 3 DOF system) as:

$$\dot{\theta}_i \, \tau_i = -c \, \dot{\theta}_i \, j_i^T \sum_i^n j_i \, \dot{\theta}_i = -c \, \dot{\theta}_i^2 \, j_i^T j_i - c \, \dot{\theta}_i \dot{\theta}_j \, j_i^T j_j - c \, \dot{\theta}_i \dot{\theta}_k \, j_i^T j_k < 0 \qquad (7)$$

Because there are conditions for joint velocities that make (7) to be greater than 0 (i.e. for any $\dot{\theta}_k < -(\frac{\dot{\theta}_i^2 j_i^T j_i + \dot{\theta}_i \dot{\theta}_j j_i^T j_j}{\dot{\theta}_i j_i^T j_k})$), non-orthogonal robots cannot always oppose an externally imposed motion with brakes.

As an example, Figure 2 shows a non-orthogonal kinematics. The proposed straight motion could be opposed using an energetic device, but not for a brake-actuated device; the motion cannot proceed from A to AB nor from BC to C, but it is permissible from AB through B to BC. These permissible areas shift and rotate even for straight-line arm motions, so that most trajectories through the workspace cannot be supported.

2.2. Prototype with Spherical Kinematics

Given that we cannot build a simple all-revolute, brake-actuated device, we settle for one of the three orthogonal kinematic structures of Table 1. And

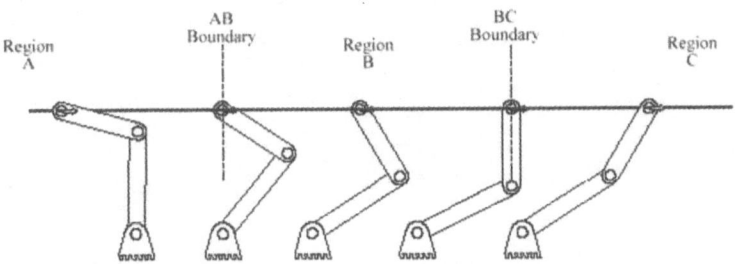

Figure 2. An illustration of how brake-actuated devices fail to create simple environments such as viscous fields. For a brake-actuated device, Regions A and C are impossible to attain.

given that revolute joints are preferable over prismatic joints, we chose the option with only one prismatic joint - spherical kinematics.

The prototype is comprised of backdrivable cable drives free of backlash and friction. The revolute joints are on intersecting yaw and pitch axes with 180 and 90 degrees of joint range respectively. A cable drive is also used to drive the prismatic axis, with 1.1 meters of active stroke. All three axes mutually intersect at the kinematic center of the device and all three drives are terminated at one of three servo-controlled magnetic particle brakes. A high-resolution optical incremental encoder is mounted on the free shaft of each servo-brake. An unpowered and unsensored three-axis handle supports the user's grip at the end of the prismatic link. The machine allows for large, sweeping motions from its user with loads up to 50 kg.

The control of the three particle brakes is done via current amplifiers and an I/O board with encoder counters mounted on the bus of a standard Pentium PC running Windows 95. The machine controller converts the encoder readings to Cartesian coordinates and can respond to a user's musculo-skeletal changes in force, position, velocity, acceleration, power, work, and range of motion in real time. LabView graphical interface software allows the user to specify the training variables in a simple manner. A foot pedal is installed within the user's workspace to make fine adjustments or to send commands during training without stopping the motion.

Finally, to prove that the workspace can provide continuous control free of the type of discontinuities predicted for the all-revolute concept, we attached a weight at the end tip of the robot and dropped it in a viscous field as shown in Figure 3. The weight fell in a straight line with very little variation from a perfect downward trajectory, while the prismatic joint changed the velocity direction.

3. Experimental Evaluation of Brake-Actuated Virtual Environments

With the spherical life-scale device we designed and constructed, we now explore whether an effective virtual environment can be constructed. Previously,

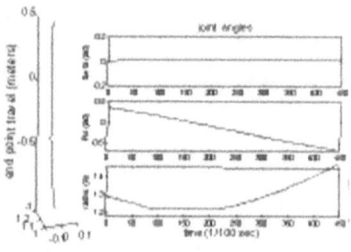

Figure 3. We attached a weight at the end of the robot and dropped it in the force field. The weight dropped straight in the Cartesian coordinates while one of the joints changed the velocity direction.

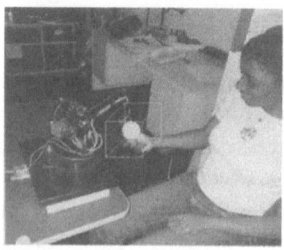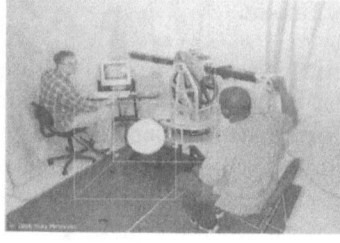

Figure 4. Left: The energetically actuated robot (PHANToM) with a virtual ball overlayed. Middle: Our brake actuated robot with a virtual ball. Right: The simulated environment.

purely dissipative devices were used only to damp human limb movements. To suppress intentional tremor, Rosen and Baiges [5] have developed a tabletop brake-actuated device. It has been demonstrated that mechanical damping on effected joints reduces the amplitude of tremor while allowing slower voluntary movements [6]. The same research group produced viscous joystick [7] and wearable [8] devices to suppress tremor.

We are interested in taking a step further to create a rich set of virtual environments with virtual objects and movement guidance paths. Here, we examine virtual objects with three different force characteristics, namely inertial, viscous, and stiff forces, by conducting simple experiments with human subjects. Inertial, viscous, and stiff forces are proportional to the acceleration, velocity, and position of the machine handle respectively: $F = Ia$, Bv, or $K(x - x_0)$, where I is a inertial constant, B is a damping constant, and K is a stiffness constant. We are interested in understanding how each force field affects the hand movement, and whether these force fields can prevent movements into a certain region (i.e. inside objects). To compare with equivalent objects created by an energetically actuated device, we used PHANToM to also produce inertial, viscous and stiff virtual objects, as shown in Figure 4.

Four human subjects participated in this study. The computer screen displayed a visual feedback of their hand location in the workspace. There were two targets within the workspace, and subjects were instructed to move from one target to another with a sound cue. For the energetically actuated device, the movement was 10cm long and had to be executed within 650 milliseconds. For the brake-actuated device, the movement was 100cm long and was executed within 1.20 seconds.   To establish a baseline movement for all subjects, we first recorded ten hand movements for each subject for each machine without any virtual objects in the workspace. The baseline movements were straight between two points, and their velocity profiles had a single bell-shaped curve for both devices.

Once the baseline movements were established, we placed a virtual sphere in the baseline hand path as shown in Figure 4 (2.5cm radius for the energetic device and 25cm for the brake-actuated device). The virtual object was introduced unexpectedly, and visual feedback was suppressed to assure that subjects did not alter their movement due to the visual information.

First, we studied viscous objects placed in the middle of the hand path. With low viscosity[1], subjects continuously executed straight movements through the virtual objects for both devices, as shown in Figure 5. The effect of the viscous objects was obvious in the velocity profiles compared to the baseline profiles; the velocities were reduced while the hand was in the object, and resumed to finish the original bell-shape curve after the hand exited the object. Because the end-point moving direction is always consistent with the velocity vector, both energetic and passive devices produced viscous force fields accurately.   However, this was not true for virtual objects with inertial force fields.   When the end-point movement was decelerating, the brake-actuated robot could not produce assistive force as for the energetic device. As a result, accurate inertial force fields could not be created with brake-actuated devices. In both devices, viscous and inertial objects acted as a tool to slow down movements but the hand trajectories were not deviated from the intended paths.

As a last set of force fields, we created an object with stiff force fields in the hand path. Stiff objects apply forces that are proportional to the distance from the boundary of the object. The deeper the penetration, the stronger the force applied. With the brake-actuated device, the subject's hand still projected a straight movement, as shown in Figure 6. When the stiffness was set low, the hand went through the object, and when the stiffness was high the hand stopped just inside the object. When the hand was stopped inside the object, the resistive force was strong for all moving directions, causing subjects to experience "sticky walls".

In contrast, when the energetically actuated device created stiff objects, the hand changed its course. The hand penetrated the object slightly then slipped off to the side of the object before continuing to the target. This type of movement deviation was observed for a variety of stiffness, and subjects perceived a wide range of object hardness corresponding to the stiffness of the object. This is the behavior we want to reproduce using brake-actuated device, but we must first get around the "sticky walls" problem.

As a first attempt to remove the sticky walls, the velocity vector orientation

---

[1] When the viscosity was set high, control instability was experienced for both devices when the control loop frequency was 1kHz.

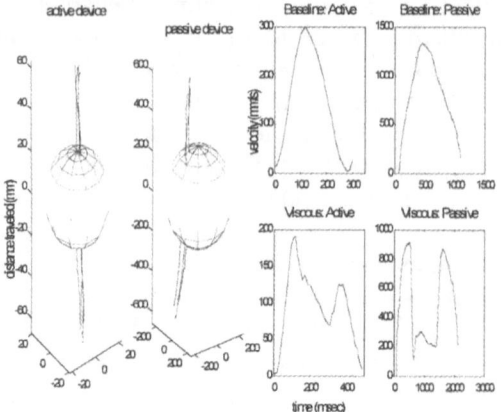

Figure 5. Left: Hand trajectories in baseline, viscous and inertial force fields. Hand goes through the virtual object for both active and passive environments. Top Right: Baseline velocity profiles. Right Bottom: Velocity profiles with a viscous object.

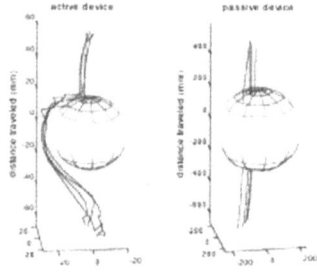

Figure 6. Hand paths while interacting with objects with stiff force fields. Using the energetic robot, hand paths were diverted at the surface of the object.

was monitored throughout the program. When the hand penetrates the object, ideally, the hand should be freed away from the object surface. Therefore, we set a condition to turn off the forces if the velocity vector did not point into the object. This technique was effective at a high velocity, but singularities occurred when the velocity vector did not specify the moving direction (i.e. velocity equaled to zero). At singularities, without predicting the next moving direction prior to the move, brakes could never respond correctly.

Furthermore, monitoring the velocity vector did not cause any movement diversion that we desire to produce. How can we make an environment that imposes a restriction into a selected region? As a way to constrain movements into a certain path, we could eliminate a joint from the device and make a two DOF robot. If one of the brakes on the device is locked, the simulated environment should be reduced to a plane as if eliminating a joint. If two of the brakes are locked, the environment should be further restricted to a line.

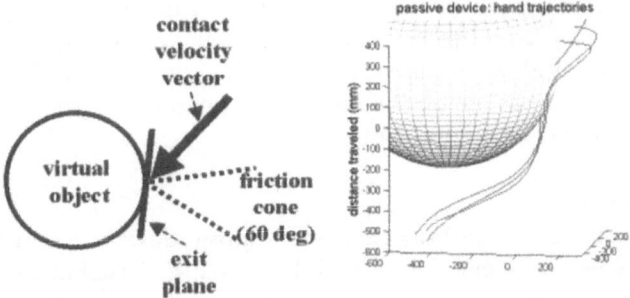

Figure 7. By aligning machine axes with exit plane of the virtual object, passively created stiff objects could divert hand paths as shown on the right.

Even though these planes and lines are restricted to the machine constraints, if there are ways to use this locking technique with respect to virtual objects, we may be able to divert movements using purely dissipative devices.

To combine the locking technique with virtual objects, an object contact point was first defined as illustrated in Figure 7. We defined a contact velocity vector as the hand velocity at contact. In addition, we defined a tangential plane with respect to the object at the contact point. This tangential plane is a collection of tangential vectors, defined as exit vectors, that the hand movement should follow immediately after contacting the object. Friction cone was defined to be 60 degrees wide. When the contact velocity vector was within the friction cone, then the movement was not deviated. If the contact velocity vector was outside of the friction cone, *and* any of the exit vectors was closely aligned with one of the machine axes, then an appropriate joint was locked to move the hand along the exit plane. After the hand deviation was completed, the joint was unlocked. Figure 7 shows an example where hand trajectories were diverted from an object using our brake-actuated device.

We tested various movement cases with a virtual sphere, and found that we cannot consistently align the machine axes with the exit plane. As a way to overcome this issue, we created a workaround that utilizes the discrepancy in human visual and proprioceptive perception. When humans are given perturbed visual information, their motor behavior adjusts to match the visual input. Due to this perceptual gap, some haptic inaccuracy may be tolerated if the visual feedback is consistent. In order to use this gap, the subject's movements were projected for 50 milliseconds in advance using the movement velocity profiles. This projection estimated the contact point with the object and the exit vector. If the projected exit vector did not align well with the machine axes, the closest point along the movement trajectory with an alignment was searched. When the point was found, the object was moved to that location and the visual feedback sped up or slowed down proportionally to match the new object location. With this search algorithm, we were able to divert a hand movement on a virtual sphere for all cases.

A limitation with this technique was the shape of the object. A sphere was a convenient example because the tangential plane was always outside of the object. However, problems arose when the entire tangential plane was in the object (i.e. concave objects). Therefore, concave objects had spots that could not be declared as a contact point. The same search algorithm was implemented to find the nearest non-concave points, but the technique was ineffective for large concave surfaces.

4. Conclusions

Using the optimal life-scale brake-actuated haptic device, we created a large and safe haptic environment. We were able to create virtual objects and guide movement path in this new environment, but the non-energetic nature of the brakes imposed limitations on the virtual environments. Perhaps, the most exciting part of this investigation was the development of algorithms utilizing human movement and perceptual information, but there were still limitations that we could not eliminate.

We intend to improve the design by adding force sensors in the handle. If we could detect the movement direction before the device joints starts moving, then we may be able to eliminate the sticky walls. Furthermore, we plan to add weights and springs to produce negative inertial and stiff forces for some applications.

5. Acknowledgments

We would like to thank Dr. Pierre Dupont for his insightful inputs.

References

[1] Troccaz J, Lavallee S, Hellion E 1993 Padyc: A passive arm with dynamic constraints. *Int'l. Conf. on Advanced Robotics*, pp.361–366.

[2] Colgate J E, Peshkin M A, Wannasuphoprasit W 1996 Nonholonomic haptic display. *Proc of the IEEE Int'l Conf. on Robotic and Automation*.

[3] Peshkin M A Colgate J E 1999 Cobot architecture. *IEEE Tans. Robotics and Automation*.

[4] Townsend W T 1988 The effect of transmission design on force — controlled manipulator performance. PhD thesis, Massachusetts Institute of Technology

[5] Baiges I, Rosen M 1989 Development of a whole-arm orthosis for tremor suppression. *Proc. of the 12th Annual Conf. on Rehab. Technology*. p.290-1.

[6] Aisen M L, Arnold A, Baiges I, Maxwell S, Rosen M 1993 The effect of mechanical damping loads on disabling action tremor. *Neurology. 43(7):1346-50.*

[7] Beringhause S, Rosen M, Huang S 1989 Evaluation of a damped joystick for people disabled by intention tremor. *Proc. of the 12th Annual Conf. on Rehab Technology. p.41-2.*

[8] Kotovsky J, Rosen M J 1998 A wearable tremor-suppression orthosis. *J. of Rehabilitation Research and Development. 35(4):373–387.*

Micro Nafion Actuators for Cellular Motion Control and Underwater Manipulation

Michael Y. F. Kwok, Wenli Zhou, Wen J. Li, and Yangsheng Xu
Centre for Micro and Nano Systems
The Chinese University of Hong Kong
Shatin, N. T., Hong Kong
{yfkwok,wlzhou,wen,ysxu}@acae.cuhk.edu.hk

Abstract: The manipulation of biological objects is a key technology necessary for many new applications in Bio-MEMS. In this paper, we will report on our preliminary experimental work in using an Ionic Conducting Polymer Film (ICPF) to develop a biological cell robotic gripper. The ability of ICPF actuators to give large deflection with small input voltage (~5V) will allow many new applications to be developed, spanning from biology to underwater MEMS and artificial muscles. A laser micromachining process is introduced to fabricate arrays of ICPF griping devices, which can be potentially integrated onto a PCB board to develop a micro manipulation system. Individual multi-finger grippers with dimensions of 200μm x 200μm x 3000μm for each finger were realized. We will report on the design, fabrication procedures, and operating performance of these micro-grippers. Further development in the reduction of size of these actuators will enable effective control of underwater micro objects and lead to new frontiers in cellular manipulation.

1. Introduction

Many micromachined actuators now exist which operate using electrostatic, thermal, magnetic, or pneumatic control principles. However, almost all of these micro actuators cannot be used in any biological applications due to one hindrance: they must operate in a dry-environment. Although pneumatic micro grippers were ingeniously used under water to capture biological cells recently [1], slow frequency response and the inability to control individual appendages of the grippers impede these micro grippers from gaining general acceptance from the biological community. Conjugated polymers such as polypyrrole are also under investigation as aqueous microactuators (as reported in [2] and [3]) because they can change volume to deliver significant stresses and strains when electro-activated. However, an electrolyte solution is needed as an ion source or sink to activate this material, and hence, using polypyrrole will limit the medium of operation

This project was partially supported by the Hong Kong RGC Earmarked Grant (No. CUHK 4206/00E).

D. Rus and S. Singh (Eds.): Experimental Robotics VII, LNCIS 271, pp. 471–480, 2001.

for these aqueous actuators. Nevertheless, polypyrrole offers certain advantages over other electro-activated materials, and should be further investigated to build underwater micro-manipulation devices.

ICPF is a sandwich of a film of perfluorosulfonic acid polymer that is between two thin layers of metal film such as gold, which serve as metallic electrodes.   Strips of ICPF can give large and fast bending displacement in the presence of a low applied voltage in wet condition. However, specially coated ICPF actuators can also be made to operate in dry condition.   Thus, ICPFs have a high potential to be incorporated into sensors or actuators where a large displacement is desired. ICPFs have been investigated widely in the past decade, but only as macro actuators [4][5].   There are some developmental work in progress to use ICPF for micro applications [6], but from literature survey and to the best of our knowledge, ICPF microactuators for micro-manipulation have not yet been reported.   Comprehensive micromechanical studies on the motion of ICPF actuators are also non-existent at the time of this publication.

In this paper, we report on a fabrication process that uses laser-micromachining to produce ICPF actuators with width dimension less than 500μm. Hence, a new breed of micro-scale actuators is introduced to the MEMS community: actuators that can be actuated in an aqueous environment with large deflection, while consuming relatively low actuation voltage. In addition, laser-micromachining technique offers a relatively fast and inexpensive fabrication method, and will potentially give cheap and pseudo-batch-fabricable ICPF micro actuators.   We have initiated an effort to create micro-cellular-manipulators by using laser-micromachining to process a commercial perfluorosulfonic acid polymer (Nafion®) [7].   Our goal is to eventually create an array of micro actuators capable of operating in biological fluids (see conceptually drawing in Figure 1).   Details of the fabrication procedures and initial experimental results from our micro underwater actuators are presented in the following sections.

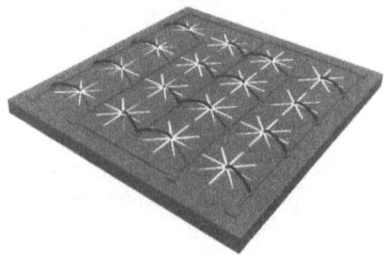

Figure 1. Conceptual illustration of an array of micro actuators that will operate in an aqueous solution.   The wirebonding wires and electrical connections will be coated before aqueous operations.

2.   Fabrication Process for the Nafion ICPF

The development of ionic polymer-metal composites actuators requires an interdisciplinary study in chemistry, materials science, controls, and robotics.   For fabrication, the poor surface adhesion of any metal coating sandwiching the polymer was an obstacle in making controllable and stable Nafion actuators. Metal deposited on the polymer surface will easily crack and peel off if there is no appropriate surface pre-treatment. Bar-Cohen et al. [8] reported workable solutions by using a chemical etchant (*Tetra-etch®*) to etch the surface or by introducing a seed layer between the metal and the polymer. For our work, we have developed an alternative and simpler method to over come the peeling and cracking problem of using gold coatings.

2.1   Metal Deposition

We chose the Nafion 117 produced by Dupont to create our ICFP actuators. Chromium, platinum and silver coating compounds were tested as a seed layer. However, due to the residual stress between the seed layer and the gold electrodes, cracks generally exist when these seed layers were used. Also, when actuators fabricated with these seed layers were tested, the metal electrodes generally peeled off after a sufficiently high voltage was applied, i.e., ~7V. This led us to shift to another process, which is described below.

The following process was used in our laboratory to produce reliable Nafion actuators. First, the Nafion should be roughed by fine sand paper (Class 1500). Then, the sample should be cleaned with HCl to remove impurities, followed by DI water rinse and Nitrogen drying. Then, a seed layer (about 0.4μm) of gold should be deposited on both sides of the polymer film using E-Beam evaporation. Then, about a 2μm thick of gold should be deposited on top of the seed layer by chemical electroplating (see Section 2.2). A satisfactory adhesion could be achieved between the Nafion and Au layers based on the above fabrication procedures in our laboratory. These gold-polymer composites can withstand a high voltage (20V) without the electrodes peeling off. The results of using E-beam evaporated Cr seed layer and Au seed layer are compared in Figure 2.

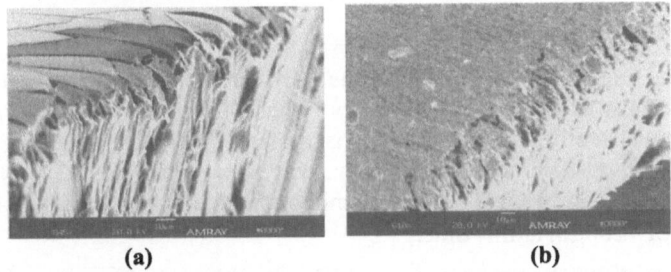

(a)   (b)

Figure 2. (a) SEM photo showing cracks due to residual stress between seed layer (film of Cr) and film of Au on the Nafion 117 polymer surface. (b) Au thin film with good adhesion on the surface of Nafion 117 polymer film. The Nafion polymer surface has already been processed with Class 1500 sand paper.

2.2   Electroplating Au on Nafion Polymer

After depositing a thin seed layer of Au by E-beam evaporation, the sample was processed to further increase the thickness of Au by chemical electroplating. A thicker Au layer is needed to increase electrical conductivity. Using *Gold Elconac 138* electroplating solution provided by Chartermate International Ltd., the deposition rate of Au could be calibrated with respect to time.   The Nafion with Au seed layers were pre-cleaned with *Copper Wet-1150* surface cleaning solution to remove any dust or oil on the surface to ensure good adhesion with the electroplated Au.   Identical metal alloy strips were used to serve as a calibrating medium.   Each strip was partially covered with an electrically insulating tape and electroplated in the *Elconac 138* with different time durations.   Afterwards, the step-height between the electroplated Au and the original alloy surface was measured using an *Alpha-Step® 500* surface profiler.   We have consistently produced reliable actuators after 10min (~2.25um) of electroplating using the following parameters:

 - Stirred 400ml of *Elconac 138* solution at room temperature.
 - 10cmx7cm wire grid plated with 2.5µm platinumized titanium as anode.
 - Apply 2.5V input at 20mA between and anode and cathode.

3.   Laser Micro-Fabrication Process

Both CO_2 and Nd:YAG lasers were explored as a micromachining tool to micro-fabricate the Nafion 117 polymer film. Lasers provide a fast and convenient way to produce small-scale Nafion actuators for mechanical testing. Since the Nafion is a polymer, our initial concern was that it would not absorb enough Nd:YAG laser energy to allow laser ablation, as CO_2 lasers are normally recommended to cut polymers. Nevertheless, we wanted to explore the possibility of using Nd:YAG lasers as a micromachining tool for polymers due to the fact that it has a shorter beam wavelength (1.06µm compare to 10.6µm of CO_2 lasers), which will allow more focused beam spot and higher energy density to enable a more precisely cutting process.

3.1 CO_2 Laser-micromachining

We have used the Electrox CO_2 laser System, which is designed for cutting and masking organic materials, to process the Nafion polymer. The system can definitely be used to cut Nafion consistently if 7.5W of power is applied. However, fibre-like residue usually accompanies the laser-processed polymer structures (see Figure 3). Nonetheless, this CO_2 laser system can be used to reliably micromachine Nafion structures with minimum feature size of ~200µm at this time.

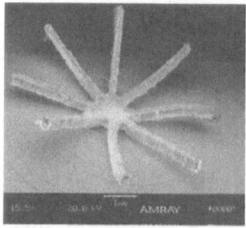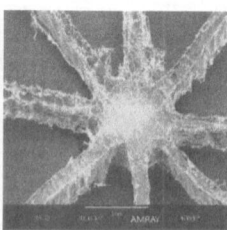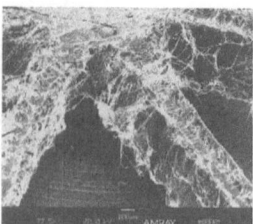

Figure 3. SEM pictures of Nafion structures laser-fabricated by a CO_2 laser system. The width of each arm is ~200µm.

3.2   Nd:YAG Laser-Micromachining

The melting point of Nafion is lower than that of metals, so its damage threshold is lower than that of metals. This means that lower laser energy intensity is required for cutting the polymer. However, Nafion is a transparent material for Nd:YAG laser beam, which means that Nafion has a very low absorptivity to Nd:YAG laser energy. Therefore, higher power is required from the laser system to cut this polymer than cutting metals such as copper.   On the other hand, Nafion has lower thermal conductivity than metals so that the thermal diffusion in Nafion during laser cutting is slow, causing possible burning of the polymer if the power is set too high. Consequently, an appropriate power level had to be found that will cut the Nafion but will not burn it during the laser micromachining process.

We have used the Electrox Nd:YAG Laser system successfully to micromachine the Nafion polymers. We have found that using an aperture size of 1.5mm and ~5W of laser power (70% input power), we can consistently micromachine the Nafion polymers. A sample of a Nafion polymer structure cut by this system is shown in Figure 4. Clearly, the fibre-like residues are not visible as in the case for CO_2 laser processing. Also, the edges can be more precisely laser-machined (compare to Figure 3).

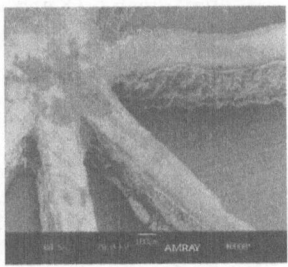

Figure 4. SEM picture of a Nd:YAG laser processed Nafion structure with each leg-width of 200μm.

4.   Mechanical Properties of Nafion

In order to design functional micro underwater actuators using Nafion 117, some fundamental studies on the mechanical behaviour were performed for these polymers. Since these actuators are electro-activated devices made of composite materials that may undergo large deflections, close-form solutions for modelling the behaviour of ICPF actuators are very complicated, and consequently, there is currently no generally accepted model to describe the motion of Nafion actuators as a function of voltage. Nevertheless, Shahinpoor et al. [5], and Kanno et al. [9] are striving to produce a general workable model for ICPF actuators presently.

Since no micro-scale mechanical properties of Nafion polymers have been reported (to the best of our knowledge) we have set up an in-situ measurement system to observe and quantify the deflection of the laser fabricated polymer structures. A CCD camera was linked to Snapper®, which was then connected to the computer graphics interface card of a

PC. In a water tank with transparent wall, we attached a transparency with predefined position grids, which allowed the motion of the polymer actuators to be quantified if images of the actuators could be captured with the superposition of these grids. The setup is shown in Figure 5. The motion of the actuators were digitally recorded with the grids superimposed in the background. The recorded files were then played back to find the tip deflection and velocity of the actuators.

To find the Young's modulus of our Nafion 117 actuators, we applied different forces to Nafion cantilevers and measured the tip deflection as a function of force. Forces were applied using magnet cubes (1mmx1mmx1mm) with mass of 7.4×10^{-3}g each. So, by counting the number of magnets attached we know the force applied at the tip each cantilever. Using the predefined grid scale from the captured image, we calculated the Young's modulus (E) for these actuators by noting that, for a cantilever clamped at one end, the deflection δ of the suspended end can be related to the applied force F at the suspended end and E by $E = Fl^3/3I\delta$, where l is the beam length and I is the moment of inertia about the neutral axis, which is equal to $wt^3/12$ for a rectangular cross-sectional beam (w is the width, and t is the thickness of the beam). The average E is 1.32×10^8N/m^2 for our Nafion actuators with 2μm Au layers on each side, which is close to the value of 2.2×10^8N/m^2 given by Kanno et al. [9] using laser deflection measurements for ICPF with dimensions of 10mmx2mmx184μm. In calculating E, the linearity between the loading and deflection of the beam obeys Hooke's law as shown in Figure 6 (3mmx14.5mmx180μm cantilever).

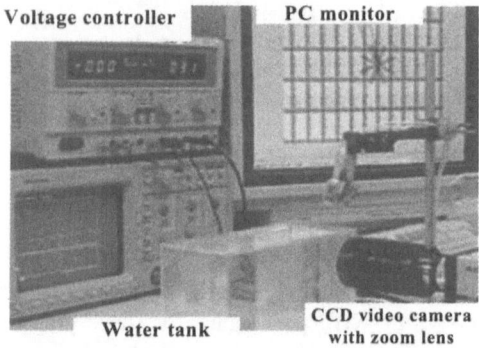

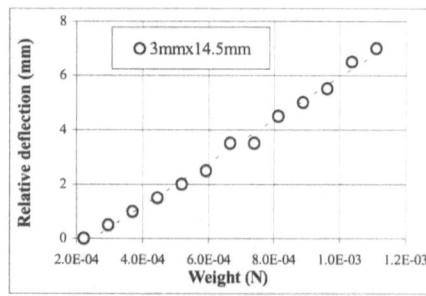

Figure 5. Picture of the in-situ monitoring setup for observing the motion of Nafion actuators.

Figure 6. Experimental results of the cantilever tip deflection due to loading (Nafion with ~2μm Au on both sides).

5.   Performance of Nafion Actuators

Experimental results from testing Nafion actuators with various geometrical variations are presented below. For actuation of the Nafion actuators, we applied a voltage across the electrodes on the polymer, and the actuator will bend towards their anode side.

5.1   Actuation of Nafion Actuators

The typical motion of Nafion actuators is shown in Figure 7 and can be described as a circular path if its tip deflection is traced from the unactuated vertical position. In this paper, "deflection" is defined as the path distance measured from the original unactuated tip position to a new tip position of interest.

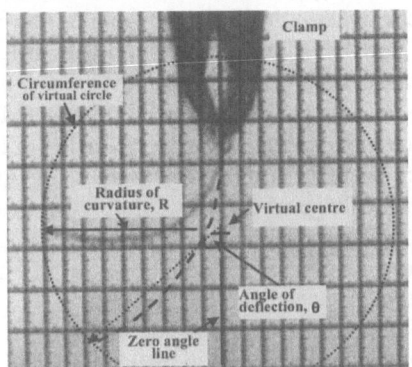 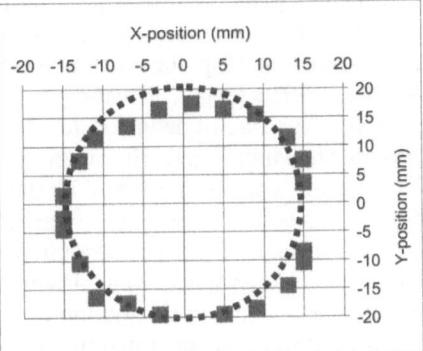

Figure 7. (Left) Definitions of parameters used to quantify the deflection of an ICPF strip actuator.   (Right) Tip position of ICPF strip during actuation.

5.2   Parametric Experimental Study of the Nafion Actuators

Several experimental studies were performed to understand the motions of the Nafion actuators. We have varied the geometries of the actuators and also the applied potential across the electrodes to examine the deflection rate of the actuators.

Keeping the length of the actuator to 24mm and applied a constant voltage across the electrodes, the time needed to complete a deflection cycle decreases with the reduction of width dimension. The experimental result is shown in Figure 8. A *complete cycle time* is defined as the time required for the actuator tip to move from its original position to a maximum deflection position to the left ($\theta \sim 180°$ as defined in Figure 7), and then move to its maximum deflection position to the right ($\theta \sim -180°$), and finally back to its original unactuated position ($\theta = 0°$). Experimentally, 4.5V DC was applied across the electrodes on the Nafion strip surface, which caused it to eventually bend to a maximum position on the anode side (left-hand side in this case). Then by reversing the polarity of the electrodes, thus, previously positive anode now becomes the negative cathode, a maximum right deflection is eventually obtained. Finally, voltage is shut off and the strip bends back to its original position to complete one bending cycle.

As shown in Figure 8, the response of the actuators clearly is not a linear function of the width of the actuators, although small widths do give faster response. However, it seems as if an optimal width can be found to maximize time response and bending deflection for given fixed parameters of voltage, length, and thickness, because reducing the width from 1mm to 0.5mm did not yield much improvement relative to width reduction from 1.5mm to 1mm. It should be noted that commercially available Nafion films are ~180μm thick, so scaling the width below 180μm would not be a good mechanical design for ICPF actuators made using commercial Nafion films.

Experiments with varying lengths were also carried out. As shown in Figure 9, rate of actuation for strips of ICPF with lengths of 8mm, 16mm and 24mm were tested (each strip was 1mm wide, and a 4.5V DC potential was applied across the electrodes). As indicated in the figure, for 8mm long strip, the maximum deflecting angle was 110° (13mm); for 16mm long strip, the maximum deflecting angle was 140° (28mm); and finally, for 24mm long strip, the maximum deflecting angle was 155° (48mm).

Apart from the length, actuating voltage can also affect the maximum deflection angle as well as the rate of actuation for the actuators. Voltage tests were carried out and the results are shown in Figure 10. The frequency response and maximum tip deflection of the Nafion actuators are both affected by the input voltage as shown in the Figure. The tip deflection as a function of driving voltage frequency is shown in Figure 11. Note the results are still inconclusive for deflection response versus driving frequency, as the effects of different geometric parameters and driving potentials need to be investigated.

In summary, although the motion of Nafion actuators are still not well understood, we can now consistently make micron-scale wide Nafion actuators of different mechanical designs using our laser-machining process. An example of an underwater grasp-manipulator made of 2-legs ICPF actuator is shown in Figure 12.

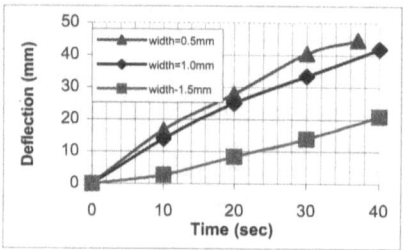

Figure 8. Rate of actuation for different widths of strip actuators (4.5V, l=24mm).

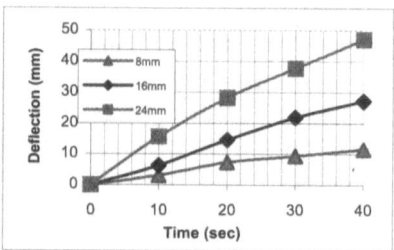

Figure 9. Rate of actuation for different lengths of strip actuators (4.5V, l=24mm).

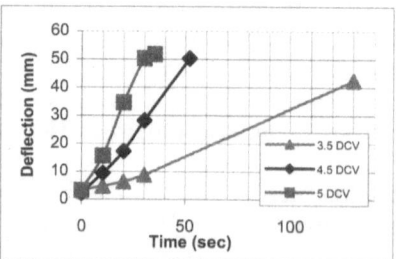

Figure 10. Deflection due to different applied voltage (4.5V, l=24mm, w=1mm).

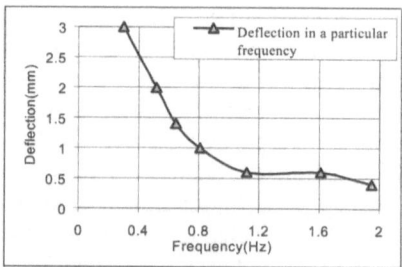

Figure 11. Actuator deflection versus driving voltage frequency (4.5V, l=24mm, w=1mm).

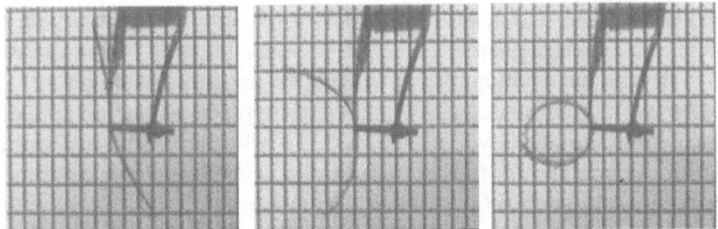

Figure 12. A 2-legs ICPF actuator which can be controlled to undergo a grasping motion under 4.5V in water.

5.3   Actuation of Micro-Scale Nafion Actuators

We have also successfully actuated actuators less than 500µm wide under water. To the best of our knowledge, these are the smallest reported underwater Nafion actuators to date.   The smallest actuators we have successfully tested are with dimensions of $w=300\mu m$, $l=3000\mu m$, $t=200\mu m$, using 15V DC voltage (see Figure 13). We have found that these actuators have a ratio of tip-deflection/length smaller than the meso-scale actuators.   This is due to the greater spring constant k presented by the shorter length dimensions, i.e., k scales with w/l^3.   Hence, w must be reduced significantly if a micro-scale Nafion actuator is to have large deflections.   We are currently developing a new in-situ monitoring system to observe and quantify these Nafion actuator motions.   Also, we are calibrating our laser system to improve the cutting resolution and hence reduce the minimum feature size of the Nafion actuators.

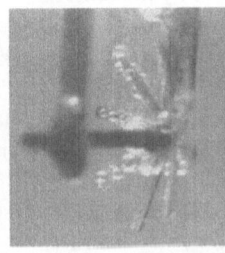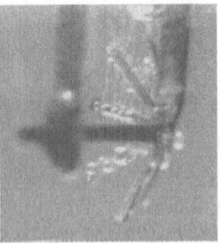

Figure 13. Time-sequence images of micro underwater Nafion actuators. The actuator shown has the dimensions of w=500mm, l=4000mm, t=200mm for each leg. The actuator was actuated with 15V input voltage with 50mA current to a tip deflection of ~1mm.

6.   Conclusion

We have successfully micro-fabricated Nafion ICPF actuators using CO_2 and Nd:YAG laser systems. A simple process was also developed to fabricate these actuators using Au electroplating. Features as small as 200µm were micro-fabricated successfully with a Nd:YAG laser system, and actuators with dimensions of $w=300\mu m$, $l=3000\mu m$, $t=200\mu m$ were successfully actuated under water with 15V DC voltage. We have also perform parametric experiments to understand the behavior of Nafion actuators with variations of applied voltage and actuator geometries. The knowledge gained from these experiments will allow us to design and develop micron scale ICPF actuators in the future.

Our future work includes designing and creating practical Nafion actuators using the laser-micromachining process developed from this work. In parallel, we will also develop other fabrication techniques to further reduce the widths and thickness of the Nafion polymer structures. The successful development of these actuators will enable effective and fast control of underwater micro objects and lead to new applications in cellular manipulation.

Acknowledgement

We would like to thank Dr. W. Y. Cheung for his help in the metal deposition and The Electronic Engineering Department of The Chinese University of Hong Kong for allowing us to use its cleanroom equipment. Appreciations are also due to A&P Instrument Co., Ltd. for assisting us in cutting the Nafion with CO_2 laser and in setting up our Nd:YAG laser system. This work was funded by the Research Direct Grant (Grant no. 2050173) of The Chinese University of Hong Kong and the Hong Kong Research Grants Committee (RGC Grant no. CUHK 4206/00E).

References

[1]   Ok J, Chu M, Kim C J 1999 Pneumatically driven microcage for micro-objects in biological liquid. Proceedings of IEEE MEMS 459-463.

[2]   Jager E W H, Inganas O, Lundstrom I 2000 Microrobots for micron-size objects in aqueous media: potential tools for single-cell manipulation. *Science* Vol 288 2235-2238.

[3]   Smela E 1999 Microfabrication of PPy microactuators and other conjugated polymer devices. *J. Micromech. Microeng.* Vol 9 1-18.

[4]   Bar-Cohen Y 2000 Smart Structures and Materials 2000: Electroactive Polymer Actuators and Devices. Proceedings of SPIE, Vol. 3987.

[5]   Shahinpoor M, Bar-Cohen Y, Harrison J O, Smith J 1998 Ionic Polymer-metal Composites (IPMCs) as biomimetic sensors, actuators and artificial muscles - a review. *Smart Mater. Struct.* R15-R30.

[6]   Guo S, Fukuda T, Nakamura T, Arai F, Oguro K, Negoro M 1996 Micro active guide wire catheter system-characteristic evaluation, electrical model and operability evaluation of micro active catheter. Proceedings of IEEE International Conference on Robotics and Automation Vol.3:2226 –2231.

[7]   Kwok M Y F, Qin J S J, Li W J 2000 Micro nafion actuators for cellular motion control and manipulation. Proceedings of 3rd Asian Control Conference 622-627

[8]   Bar-Cohen Y, Leary S, Shahinpoor M, Harrison J O, Smith J 1999 Electro-Active Polymer (EAP) actuators for planetary applications. Proceedings of SPIE 3669-05

[9]   Kanno R, Tadokoro S, Takamori T, Oguro K 1996 3-Dimensional dynamic model of Ionic Conducting Polymer Gel Film (ICPF) actuator. Proceedings of IEEE International Conference on Systems, Man and Cybernetics 2179 –2184.

Control of an under actuated unstable nonlinear object

Nils A. Andersen, Lars Skovgaard and Ole Ravn

Department of Automation, Building 326, Technical University of Denmark,
DK-2800 Lyngby, Denmark,
{naa,or}@iau.dtu.dk

Abstract: This paper presents a comprehensive comparative study of several non-linear controllers for stabilisation of the under actuated unstable nonlinear object known as the Acrobot in the literature. The object is a two DOF robot arm only actuated at the elbow. The study compares several control algorithms from the literature and a new algorithm developed during the study. The comparison is based on both simulation and real experiments for all controllers.

1. Introduction.

During the last decade the Acrobot has been used as example object in several studies of non linear control. There are several reasons for this. First of all having only two degrees of freedom the complexity of the equations of motion is low enough to allow analytical results of the proposed non linear methods and in spite of the low order of the object the non linearities are very decisive for the behavior which means that classical LQR-controllers designed using a linearization of the object have a very limited region of stability around the linearization point. Secondly stabilizing the Acrobot can be seen as a prestudy to bipedal walking machines as it simulates balancing of a person having no feet (standing on stilts) using only hip movements. As mentioned above the Acrobot has been the subject of several studies but most of these present only one algorithm and often only theoretical or simulation results. The main purpose of this study is to perform systematic comparison of several algorithms for stabilisation of the acrobot using both simulation and real experiments. The importance of testing different algorithms on the same physical acrobot is emphasized by the fact that the actual configuration of the acrobot (lengths and masses of the two arms) has great influence of the performance of different controllers thus making it difficult to compare results obtained on different physical systems directly. In some configurations the performance of a linear controller is equal to the performance of some non linear controllers.

The outline of the paper is as follows. In section 2 the basic equations for the acrobot are given and the difference between torque control and acceleration control is explained. In section 3 a short description of the five investigated control algorithms is presented along with simulation results for the ideal Acrobot i.e. simulation without physical constraints of the actuator and measuring systems and feedback from all states. Section 5 contains simulations of the full system including the dynamics and constraints of the actuator and quantization of the angle measurements and sampling of the digital controller. It also contains the measurements on the real system. Section 6 gives a description of the experimental setup and describes the implementation of the acceleration control.

D. Rus and S. Singh (Eds.): Experimental Robotics VII, LNCIS 271, pp. 481–490, 2001.

2. The Acrobot.

The Acrobot is a two DOF planar robot. The first arm is connected to the fixed environment with a rotational joint, joint 1. The second arm is connected to the first arm through a second rotational joint, joint 2. Only joint 2 is actuated which means that the torque of joint 1 is zero. Figure 1 shows the acrobot with the definitions of the joint angles and some basic physical parameters.

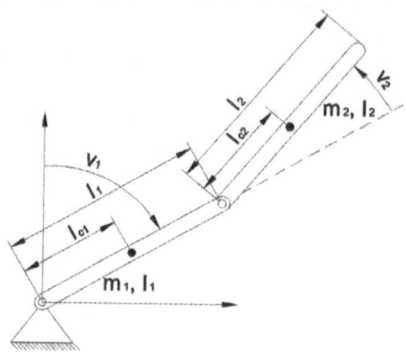

Figure 1. Acrobot model

As the torque of joint 1 is zero the equations of motion considering the torque of joint 2 as input are

$$-d_{11}\ddot{v}_1 + d_{12}\ddot{v}_2 + h_1 + \phi_1 = 0$$
$$-d_{21}\ddot{v}_1 + d_{22}\ddot{v}_2 + h_2 + \phi_2 = \tau$$
(1)

Where

$$a_1 = m_1 l_{c1}^2 + m_2(l_1^2 + l_{c2}^2) + I_1 + I_2$$
$$a_2 = m_2 l_1 l_{c2}$$
$$a_3 = m_2 l_{c2}^2 + I_2$$
$$a_4 = g(m_1 l_{c2} + m_2 l_1)$$
$$a_5 = g m_2 l_{c2}$$

$$d_{11} = a_1 + 2a_2 \cos(v_2)$$
$$d_{22} = a_3$$
$$d_{12} = d_{21} = a_3 + a_2 \cos(v_2)$$
$$h_1 = -a_2 \sin(v_2)(\dot{v}_2^2 - 2\dot{v}_1\dot{v}_2) \quad (2)$$
$$h_2 = a_2 \sin(v_2)\dot{v}_1^2$$
$$\phi_1 = a_4 \sin(v_1) + a_5 \sin(v_1 - v_2)$$
$$\phi_2 = a_5 \sin(v_1 - v_2)$$

If the angular acceleration $\ddot{v}_2$ is considered as input the equations reduce to

$$-d_{11}\ddot{v}_1 + d_{12}u + h_1 + \phi_1 = 0$$
$$\ddot{v}_2 = u$$
(3)

which is the form used in this study. This can be thought of as a mere mathematical reformulation of the problem as the torque can be calculated knowing u and the present state of the Acrobot but it can also be seen as a change of control paradigm from torque control to acceleration control with great impact on the achievable control performance. The acceleration control of joint two can be implemented with a high precision position servo calculating position and speed references digitally from the acceleration given by the controller. Using an analog high bandwidth speed controller in the position servo makes it possible to suppress ill described non linearities such as coulomb friction along with some other non linearities thus obtaining a control object which is near to the mathematical description assuring a greater likelihood for success with the non linear controllers found based on the ideal equations.

Several authors have found that the actual physical configuration of the the acrobot has great impact on the influence of the non linearities on the control. Timcenko [1] found that if $a_1 = 2 a_3$ her non linear controller was equal to a linear controller. In this study the physical dimensions have been chosen so that this equality is far from fulfilled thus emphasizing the non linearities of the Acrobot.

3. The investigated stabilisation algorithms.

In this section a short description of the investigated algorithms are given. For a in depth presentation the reader is referred to the references. Five different algorithms have been investigated. For each controller a region of attraction (ROA) and initial state time response are shown. These are based on simulation results with the ideal acrobot. The ROA is the 4 dimensional space of initial states from which the controller is able to stabilize the system. As the ROA is found by simulation only the two dimensional slice with initial angular speeds equal to zero is found.

3.1.   LQR-controller.

The system has been linearized at the vertical equilibrium point and a classical LQR-controller has been designed. The performance of this controller is used to emphasize the poor results of a linear controller for the non linear system and as a reference for the improvements obtained by the non linear controllers. Figure 2 shows the time response for the largest value of v_1 in the ROA and it is seen that the linear controller only is able to stabilize the system from 1.2 deg. This is also seen in figure 3 which shows a very small ROA.

3.2.   Pseudolinearization.

Pseudolinearization is described in [2]. The method is used for systems with more equilibrium points. This makes it suited for the Acrobot. Bortoff [3,4] states that it is not possible to pseudolinearize the Acrobot, so he presents a further development of the method where numerical approximation with splines are used. The method is demonstrated on the Acrobot. In this study an analytical solution to the pseudolinearization of the Acrobot has been found. This has been possible due to the simpler equations of motion obtained considering the angular acceleration of v_2 as input instead of the torque of joint two. The algorithm has been tested by simulation and

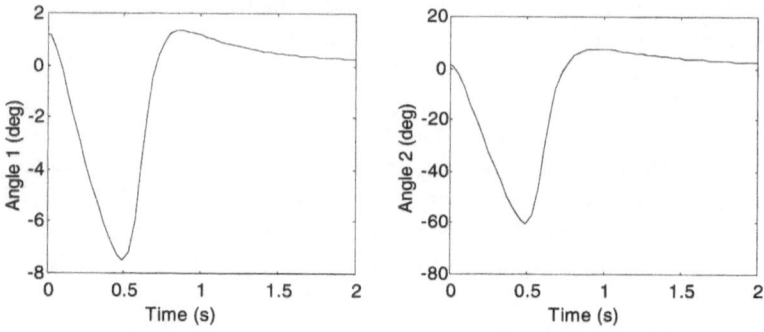

Figure 2. Initial state time response of the LQR controller

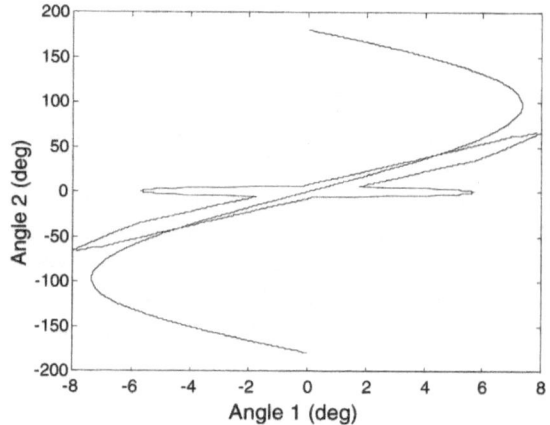

Figure 3. ROA for the LQR controller

experiment. Figure 4 shows an initial state time response for the largest possible value of $v_1 = 3.4$ deg. It is seen that this angle is three times greater than for the linear controller.

3.3.   Partial feedback linearization and linearization of the remaining non linear system.

Timcenko [1] and Olfati-Saber and Megretski [5] describe a solution where a non linear transformation is found that transforms the system to another 'less non linear system'. This system is then linearized and a LQR controller is designed for the linearized system. The method has been tested by simulation in these investigations. In this study the method has been used on the acceleration controlled Acrobot and tested both by simulation an experiment. Figure 5 shows an initial state time response for the largest possible value of $v_1 = 4.6$ deg. It is seen that this angle is four times

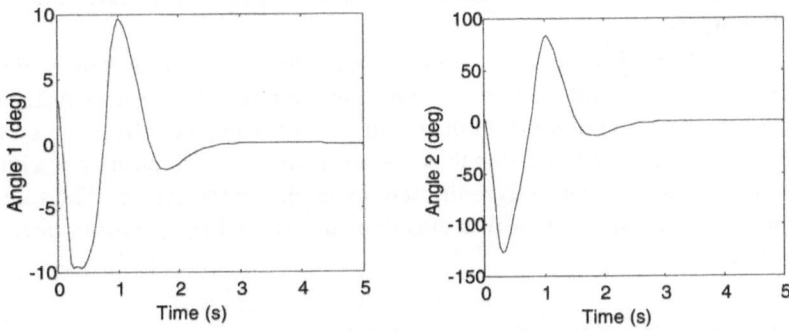

Figure 4. Initial state time response of the pseudolinearized controller

greater than for the linear controller. The ROA in figure 6 shows this improvement is general for all initial values.

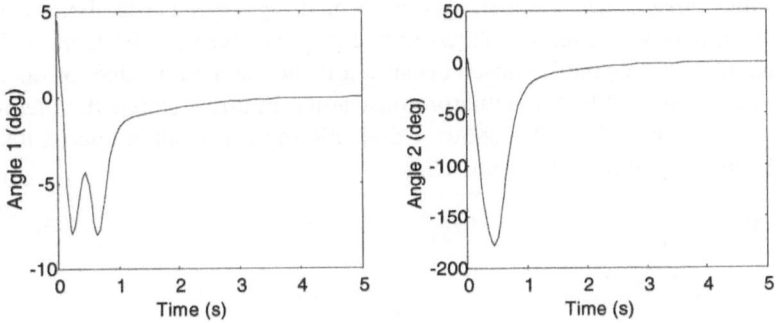

Figure 5. Initial state time response of the Timcenko controller

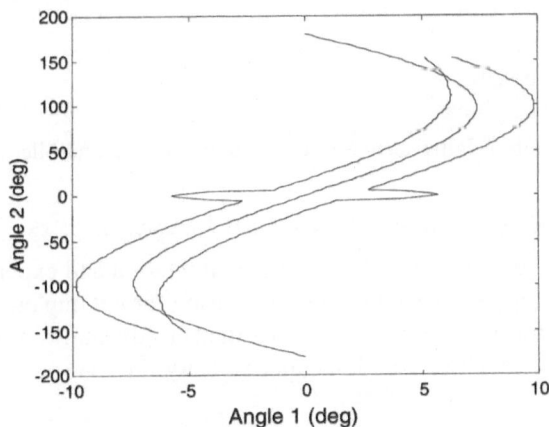

Figure 6. ROA for the Timcenko controller

3.4. Partial feedback lineariztion with relative degree 2, unstable zero dynamics.

De Luca and Oriolo [6] proposes a three step controller for the Acrobot. First step controls the swing up until the lower arm is pointing upwards, the second step brings the system near to the equilibrium manifold and third step stabilizes the system in vertical position. In the second step they use a partial feedback linearization with relative degree 2. They do not analyze the zero dynamics of the system. The simulation in the present study shows that it is unstable so this controller has not been tested experimentally.

3.5. Partial feedback linearization with relative degree 3, stable zero dynamics.

As it has been proved that the Acrobot is not feedback linearizable the maximum relative degree for an output function is three. Following the methods of [7] an output function with relative degree 3 and stable zero dynamics has been found. The reason for success probably is the reduced complexity of the acceleration controlled system. Based on this transformation a controller has been designed and tested both with simulation and experiment. Figure 7 shows a time response for $v_1 = 10$ degrees. This is not the maximum value but is chosen because it is the maximum value for the experimental system limited by the actuator constraints. Figure 8 shows the theoretical ROA of the new controller. This shows a dramatic improvement of approximately a factor 40 compared to the linear controller.

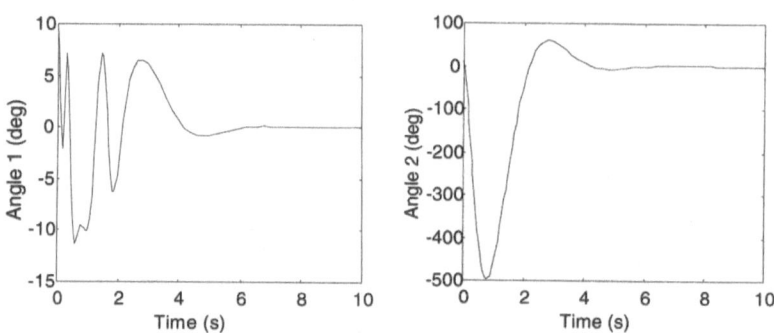

Figure 7. Initial state time response of the new controller

4. Comparison based on simulation and experiments.

All the controllers have been tested using both simulation and experiment except 4 which has only been tested by simulation. The reason for not implementing 4 experimentally was that the simulation showed that the zero dynamics of the system was unstable thus ending with a very high angular velocity $\dot{v}_2$.

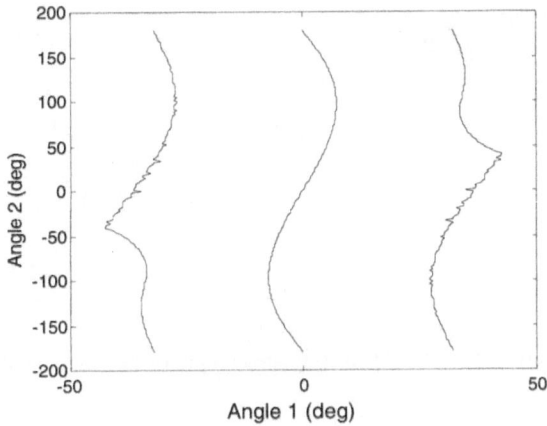

Figure 8. ROA of the new controller

In the experiments the special start configuration with $v_1=v_2$ has been used and the largest value from which the controller can stabilize the system has been found. The results are shown in table 1.

TABLE 1.   Catching zone

Controller	Theoretical	Simulated	Measured
Linear	1.2^o	-1.0^o-1.5^o	-0.8^o-1.3^o
New	41°	10°	12°
New (-180)	4.5^o	5.2^o	7.0^o
Timcenco	4.6^o	5.6^o	5.6^o
Bortoff	3.4^o	3.4^o	4.0^o

The third row, new (-180) is the limitation of the new controller if v_2 must stay within +- 180 degrees. The values demonstrate a good correspondence between theory (simulation of ideal system), simulation (with physical constraints) and measurements. The big difference between theoretical and simulated values for the new controller is due to motor constraints in the physical system. It is clear that all the non linear controllers have much better performance than the linear controller and that the new controller is clearly the best from a ROA view. Figure 9 to 12 show simulated (thin line) and measured (bold line) initial state time responses for the four stable controllers. Simulation and experiments have shown that the exact form of the time response is very sensitive to small moments on joint 1, so the differences between measured and simulated response is probably mostly due to the moment caused by the wires on the acrobot.

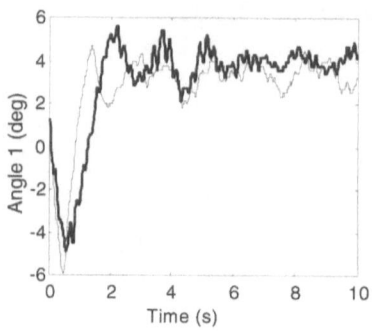
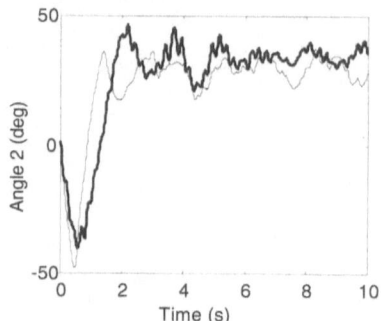

Figure 9. Measured and simulated output of the linear controller

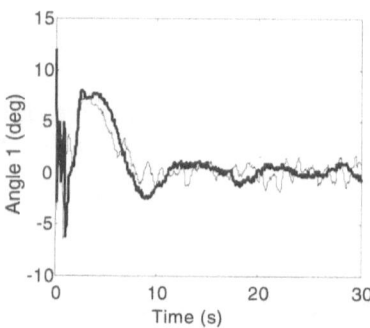
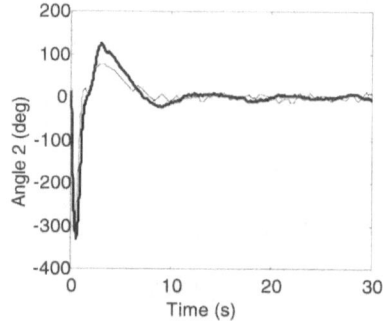

Figure 10. Measured and simulated output of the new controller

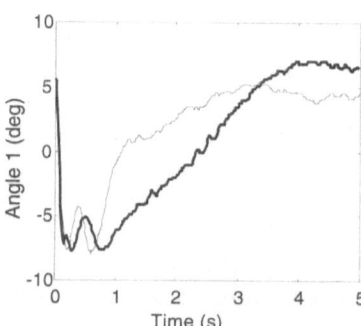
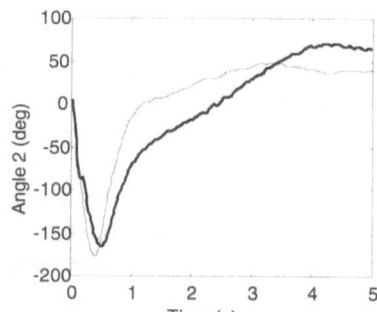

Figure 11. Measured and simulated output of the Timcenco controller

5. Experimental setup.

The implemented acrobot is shown on Figure 13. The lower arm consists of two parts between which the upper arm swings. The advantage of this is that the upper arm will not cause any torsion moment on the lower arm during operation thus allowing a

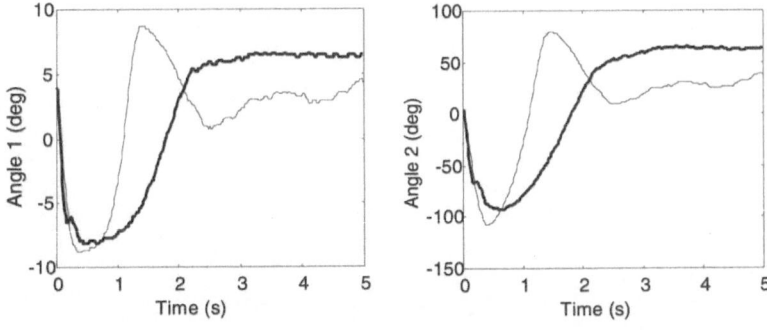

Figure 12. Measured and simulated output of the Bortoff controller

lighter construction. The disadvantage is that the upper arm must be shorter than the lower thus limiting the possible configurations in the investigation. The physical parameters of the Acrobot is given in table .

TABLE 2.   Parameters for the Acrobot

	Mass	Inertia	Length
Arm 1	3.49 kg	$91.3\ 10^{-3}\ \mathrm{kgm}^2$	438 mm, 102 mm
Arm 2	1.03 kg	$17.5\ 10^{-3}\ \mathrm{kgm}^2$	410 mm, 100 mm

The joint between the upper and the lower arm is driven by a 12 V DC-motor (Minimotor type 3757 CR) with tachogenerator through a two stage tooth belt drive giving a total gearing ratio of 20. The tooth belt drive has been chosen because of its low backlash. The motor and the first drive stage is mounted on the lower arm near joint 1 to reduce the forces on this joint. The angles of joint 1 and joint 2 are measured by optical encoders with 500 lines thus giving a resolution of 0.18 degrees. The reference for the power amplifier is given using a 12-bit DA-converter. The controllers are implemented in C on a 100 MHz 80486DX4 PC using a real time operating system developed at Institute of Automation based on OSkit. A sample frequency of 64 Hz is used acceleration control is implemented by finding speed and position references digitally from the acceleration given by the control algorithm. These references are the given to a high precision position servo based on the encoder feedback and a high bandwidth analog speed controller using the motor tacho feedback.

6. Conclusion.

Five algorithms for stabilization of the Acrobot have been investigated. Three of these have been presented previously by other authors, but two of them have been verified experimentally for the first time in this study. A new algorithm based on partial feedback with relative degree of 3 and stable zero dynamics has been developed an tested both by simulation and experiment and shows far superior performance with respect to stability region (ROA).

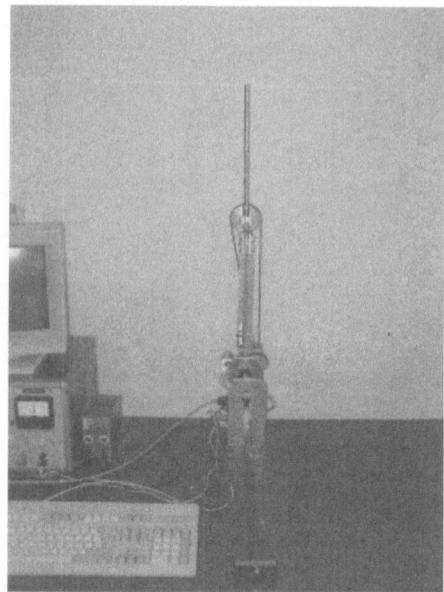

Figure 13. The Acrobot

7. References.

[1] Timcenko,Olga (1999) Hybridization of Classical and Fuzzy Control for an Underactuated Bipedal Walking. Odense, Mærsk Mc-Kinney Møller Instituttet for produktionsteknologi, Syddansk Universitet,1999

[2] Reboulet, C.,Champetier, C. (1984) A new method for linearizing non-linear systems: pseudolinearization. International Journal of Control, vol. 40, no. 4,1984, pp. 631-638

[3] Bortoff, Scot A. (1994). Advanced Nonlinear Control Using Digital Signal Processing. IEEE Transactions on Industrial Electronics, vol. 41, no. 1,1994, pp. 32-39

[4] Davison, Daniell E.,Bortoff, Scott A. (1994) Enlarge Your Region of Attraction Using High-Gain Feedback. Proceedings of the 33rd IEEE Conference on Decision and Control, Lake Buena Vista, FL., 1994, pp. 634-639

[5] Olfati-Saber, Reza, Megretski, Alexandre (1998). Controller Design for a Class of Underactuated Nonlinear Systems. Proceedings of the 37th IEEE Conference on Decision and Control, Tampa, Florida. 1998, s. 4182-4187

[6] De Luca, Alessandro, Oriolo, Giuseppe (1998). Stabilization of the Acrobot via Iterative State Steering. Procedings of IEEE international Conference on Robotics & Automation, Leuven, Belgium, 1998, pp.3581-3587

[7] Isidori, Alberto (1995) *Nonlinear Control Systems, 3 ed.* London, Springer-Verlag,

Singularity Handling on Puma in Operational Space Formulation

Denny Oetomo, Marcelo Ang Jr.
National University of Singapore
Singapore
d_oetomo@yahoo.com
mpeangh@nus.edu.sg

Ser Yong Lim
Gintic Institute of Manufacturing Technology, Singapore
sylim@gintic.gov.sg

Abstract: A solution to the singularity problem is presented from the approach of the operational space formulation which involves both motion and force control. A brief summary of the Operational Space Formulation and how singularity presents a problem is explained. The inverse of the Jacobian at singular configuration is handled by removing the degenerate components of the motion, therefore manipulator is made redundant to the task. Khatib's [3] dynamically consistent inverse was then used to invert the Jacobian and to create a null space motion to escape from the singular configuration in the case that the desired path lies in the degenerate direction. The algorithm was implemented on the PUMA 560 manipulator, and the results are presented.

1. Introduction

Conventional control algorithms of a robot manipulator are often found to use the solution of inverse kinematics or inverse Jacobian to transform various tasks into desired joint motions. These methods were found to be insufficient in the vicinity of a singular configuration, as they produce high joint velocity and large error in end-effector's trajectory.

Various ways have been devised to handle the problem of singularity, starting from the simple approach of switching into joint space control. Others try to avoid going near the singular configuration by maximizing the manipulability of the end-effector at all time [7] [11]. However, this is not always possible when the manipulator is not redundant with respect to the task.

Another idea in handling singularity was to eliminate the degenerate component(s) of motion and therefore, there will be no large joint velocity generated when the manipulator goes into the vicinity of singularity [3] [4] [5]. Chiaverini and Egeland [5] used partial Jacobian matrix, with rows along the singular direction eliminated, was inverted using pseudo inverse to obtain joint velocity from velocity in task space. Aboaf and Paul [1] bounded the rates of excessive joints and compensated for the error by a modified inverse Jacobian to

D. Rus and S. Singh (Eds.): Experimental Robotics VII, LNCIS 271, pp. 491–500, 2001.
© Springer-Verlag Berlin Heidelberg 2001

calculate the remaining joint rates. Cheng et al [4] analyzed the singularities in PUMA and decomposed the workspace into achievable and non-achievable directions. The non-achievable direction was released, Compact QP method was applied to minimize the tracking error of the end-effector, and SVD was used to obtain the Jacobian inverse and a weight system to regulate the trade off between exactness and feasibility of motion.

The experiment described in this report was conducted on the platform of Operational Space Formulation [9]. In this scheme, motion is generated not by kinematics-based method, but by the inertial parameters of the manipulator. The necessity of using the inverse of Jacobian comes from the calculation of the dynamic model of the manipulator. In comparison to the conventional method of kinematics-based control, singularity handling in Operational Space Formulation covers the problem both in motion and force control. The proposed idea is to handle singularity by transforming the Jacobian and the generalized forces into the frame whose one or more of the axes represent the degenerate direction(s) of singularity. The degenerate component is then removed resulting in a lower dimension, full rank Jacobian that is redundant with respect to the task. The dynamically consistent inverse [3], which is a generalised inverse weighted by inertial matrix, is used to invert the Jacobian. Null space motion [9] was also used to reconfigure the manipulator to escape from singularity into a path that happens to lie in the degenerate direction, which we have removed. The experimental result on PUMA 560 is presented.

2. Operational Space Formulation

Operational Space Formulation [9] was devised to enable a unified approach on the force and motion control of a manipulator. In this scheme, we would be able to resolve the singularity issue in both motion and force control. In the scope of this experiment, we are more concerned about the motion control part of the algorithm. Motion of the end-effector is generated by:

$$\Gamma = J^T F + N\Gamma_0 \tag{1}$$

where Γ is the joint torque, J is the Jacobian matrix, and F is the desired force obtained by:

$$F = \hat{\Lambda}\ddot{x} + \hat{\mu}(x,\dot{x}) + \hat{p}(x) \tag{2}$$

where Λ is the pseudo kinetic energy matrix, $\hat{\mu}(x,\dot{x})$ is the Coriolis and Centrifugal matrix, and $\hat{p}(x)$ is the gravity compensation matrix. N is the null space of the manipulator defined as $N = [I - \bar{J}J]$, and Γ_0 is the null space torque.

$\ddot{x}$ is the linear dynamic behaviour defined by:

$$F^* = I\ddot{x}_d - k_v(\dot{x} - \dot{x}_d) - k_p(x - x_d) \tag{3}$$

where $\ddot{x}_d, \dot{x}_d, x_d$ are the desired acceleration, velocity, and position of the end-effector at the given time, and $\ddot{x}, \dot{x}, x$ are the actual acceleration, velocity, and position of the end-effector. It should be noted, that throughout this experiment, the task given would be to control both position and orientation in 3D space (6 DOF task in motion control).

3. Dynamic Formulation

The need for inverse of the Jacobian in the dynamic formulation comes from the derivation of the Λ, μ, and p matrices. These are the pseudo kinetic energy, Coriolis, and Centrifugal matrices in task space, and were all derived from the joint space equivalent. They are derived in [9] and [10] as the following:

$$\Lambda(x) = J^{-T}(q)A(q)J^{-1}(q)$$
$$\mu(x\dot{x}) = [J^{-T}(q)B(q) - \Lambda(q)H(q)][\dot{q}\dot{q}] \qquad (4)$$
$$p(x) = J^{-T}(q)g(q)$$

where A(q) is the kinetic energy matrix, B(q) and H(q) are matrices of the Coriolis and Centrifugal forces, and g(q) the gravitational compensation matrix.

For further explanation of the Operational Space Formulation, please refer to [8], [9] and [10].

4. Singularity Identification in PUMA

Singularity is defined as the configuration where the values of joint position causes the Jacobian to become singular (the determinant of the Jacobian is zero) or when the manipulator loses degree(s) of freedom [11]. The degenerate direction is defined as the degree of freedom that the manipulator loses in the singularity, or a motion that the manipulator can not execute while in singularity. It is also represented by the component of the Jacobian matrix that causes it to be singular.

Khatib [10] categorizes singularities into two types. Type 1 is when null space torque creates motion in the degenerate direction. Type 2 is when null space torques affects only the internal joint motions, or when the null space motion moves (shifts) the degenerate direction.

In PUMA, or many PUMA-like manipulators (or referred to as anthropomorphic with spherical wrist manipulator in [12]), there are three singularities: wrist lock (type 2), elbow lock (type 1), and head lock (type 2). These singularities can happen individually, or as a combination of two of even three at the same time.

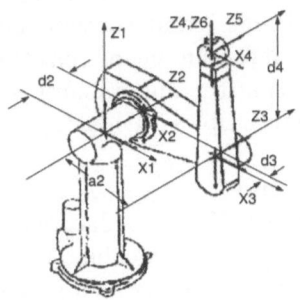

Figure 1. The PUMA 560 at zero position, by Craig's modified DH parameter [6]

Table 1. THE MODIFIED DH PARAMETERS

i	α_{i-1}	a_{i-1}	d_i	ϑ_i
1	0	0	0	ϑ_1
2	-90	0	d2	ϑ_2
3	0	0	d3	ϑ_3
4	90	0	d4	ϑ_4
5	-90	0	0	ϑ_5
6	90	0	0	ϑ_6

Mathematically, singularity occurs when the determinant of the Jacobian matrix approaches zero, i.e.: $Det(J) = 0$ or $Det(JJ^T = 0)$ for manipulators with non-square Jacobians [11].

For PUMA 560 with 6 DOF, the Jacobian is a square 6 x 6 matrix, which can be partitioned into:

$$J = \begin{bmatrix} J_{11} & J_{12} \\ J_{21} & J_{22} \end{bmatrix}$$ (5)

By defining the control point to be at the wrist, we will obtain a Jacobian matrix with $J_{12} = 0_{3x3}$. With the frame assignment shown in Figure 1, and modified DH parameters according to Craig's [6] (See Table 1), the determinant of PUMA is shown as:

$$Det(J) = Det(J_d) = Det(J_{11})Det(J_{22});$$
$$Det(J_{11}) = -a_2(d_4C_3 - a_3S_3)(d_4S_23 + a_2C_2 + a_3C_23);$$ (6)
$$Det(J_{22}) = -S_5;$$

(Cheng et al[4])

where $a_2 = 0.4318(m), a_3 = -0.0203(m), d_2 = 0.2435(m)$, $d_3 = -0.0934(m), d_4 = 0.4331(m)$.

When the determinant equals zero, Equation 6 represents the elbow, head, and wrist singularities respectively.

5. Identifying the Singular Direction

When singularity occurs, there is a row(s) in the Jacobian - when it is transformed onto the correct frame - that contains only zeros. By definition that $\dot{x} = J\dot{\vartheta}$, a zero row in the Jacobian means that there is a direction (or a degree of freedom) in task space that the manipulator is unable to move into. This is the case in singular configuration, as the manipulator loses a degree of freedom at these points.

When Head Lock occurs, $a_2.C_2 + a_3.C_23 + d_4.S_23 = 0$ (Equation 6). Equation(7) shows the top half of the Jacobian ^4J. From this equation, it is shown that the second row of the Jacobian is zero, which corresponds to the translation along Y-axis of Frame{1} which is the degenerate direction.

$$^1J = \begin{bmatrix} -d_2 - d_3 & d_4C_{23} - a_2S_2 - a_3S_{23} & d_4C_{23} - a_3S_{23} & 0 & 0 & 0 \\ a_2C_2 + a_3C_{23} + d_4S_{23} & 0 & 0 & 0 & 0 & 0 \\ 0 & -a_2C_2 - a_3C_{23} - d_4S_{23} & -a_3S_{23} - d_4S_{23} & 0 & 0 & 0 \end{bmatrix}$$ (7)

For the Wrist Lock, there is no entire row of zeros in the J matrix (Equation

(8)).

$$
{}^4J = \begin{bmatrix}
(a_2C_2 + d_4S_{23})S_4 + C_{23}(-(d_2+d_3)C_4 + a_3S_4) & C_4(d_4 - a_2S_{23}) & d_4C_4 & 0 & 0 & 0 \\
(a_2C_2C_4 + d_4C_4S_{23} + C_{23}(a_3C_4 + (d_2+d_3)S_4) & -(d_4 + a_2S_3)S_4 & -d_4S_4 & 0 & 0 & 0 \\
-(d_2+d_3)S_{23} & -a_3 - a_2C_3 & -a_3 & 0 & 0 & 0 \\
-C_4S_{23} & S_4 & S_4 & 0 & 0 & S_5 \\
S_{23}S_4 & C_4 & C_4 & 0 & 1 & 0 \\
C_{23} & 0 & 0 & 1 & 0 & C_5
\end{bmatrix}
$$

(8)

When $(\vartheta_5 = 0)$, a row of zero only appears at the last three elements of the fourth row of ${}^4J$:

$$
{}^4J_{22} = \begin{bmatrix} 0 & 0 & 0 \\ 0 & 1 & 0 \\ 1 & 0 & 1 \end{bmatrix}
$$

(9)

This means that it is still possible to rotate around the X-axis of Frame{4} in wrist singularity, but it is produced by the first three joints, which would also change the position of the end effector (i.e. not possible in 6 DOF).

Therefore, the first row of ${}^4J_{22}$, (or the fourth row of ${}^4J$) is the degenerate direction, representing the rotation around X-axis of Frame{4} (see Figure 1 and 4). Collapsing the Jacobian is then done by eliminating the fourth row of ${}^4J$.

Elbow singularity is shown by projecting the Jacobian and the task space forces onto Frame{B} which is not one of the frames in our DH assignment (see Figure 2 for the frame assignment, and Equation (10) for the resulting Jacobian). At this configuration, the singular direction is found to fall along the line connecting the wrist point to the origin of base frame.

$$
{}^BJ[1][1] = \frac{(d_2+d_3)S_2(d_4C_3 - a_3S_3)}{(a_2+a_3C_3+d_4S_3)D}
$$

$$
{}^BJ[1][2] = \frac{(d_4C_3 - a_3S_3)(a_2+a_3C_3+d_4S_3)^2 D}{a_2^2 + d_2^2 + 2d_2d_3 + d_3^2 + 2a_2a_3C_3 + a_3^2C_3^2 + 2a_2d_4S_3 + d_4^2S_3^2 + a_3d_4Sin[2q3]}
$$

$$
{}^BJ[1][3] = \frac{d_4C_3 - a_3S_3}{D}
$$

$$
{}^BJ[2][1] = (d_2+d_3)S_2; \quad {}^BJ[2][2] = a_2 + a_3C_3 + d_4S_3
$$

$$
{}^BJ[2][3] = a_3C_3 + d_4S_3
$$

$$
{}^BJ[3][1] = \frac{C_2(a_2^2 + d_2^2 + 2d_2d_3 + d_3^2 + a_2a_3C_3 + a_2d_4S_3) + (a_2+a_3C_3+d_4S_3)(a_3C_{23}+d_4S_{23})}{(a_2+a_3C_3+d_4S_3)D}
$$

$$
{}^BJ[3][2] = \frac{(d_2+d_3)(d_4C_3 - a_3S_3)}{(a_2+a_3C_3+d_4S_3)D}; \quad {}^BJ[3][3] = \frac{(d_2+d_3)(d_4C_3 - a_3S_3)}{(a_2+a_3C_3+d_4S_3)D}
$$

where:

$$
D = \sqrt{1 + \frac{(d_2+d_3)^2}{(a_2+a_3C_3+d_4S_3)^2}}
$$

(10)

Frame{B} is obtained by rotating Frame{2} by angle β, which is defined as:

$$
\beta = Tan^{-1}\left[\frac{d_2+d_3}{a_2+a_3C_3+d_4S_3}\right]
$$

(11)

From (6), it is shown that $-a_2(d_4C_3 - a_3S_3) = 0$ at elbow singularity. Therefore, the first row of ${}^BJ_{11}$ is a zero row $({}^BJ[1][1] = {}^BJ[1][2] = {}^BJ[1][3] = 0)$. This shows that the degenerate direction lies along the X-axis of Frame{B} (see Figure 2). Equation 10 only shows the elements ${}^BJ$ from the top left quadrant, because the top right quadrant is a zero matrix.

6. Removing the Degenerate Component

The idea of removing the degenerate component is done by removing the row(s) of the Jacobian matrix and elements of the task space force F (see Equation (2))

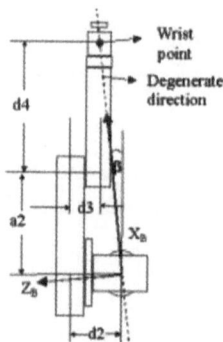

Figure 2. PUMA, from top view, shows the degenerate direction at elbow singularity, expressed in Frame{B}, which is derived from rotating Frame{2} by angle β

that represent the degenerate direction(s) of motion. To do so, the Jacobian matrix and force vector need to be expressed in the frame in which one of the axes represents the direction of singularity (degenerate direction).

Force vector in task space is obtained from the control law to represent the virtual force that 'pulls' the end effector to the desired position and orientation (see Equations (2), (3)). The force vector is then resolved into the correct frame, that represents the singular direction as one of the axes of the frame. The Jacobian is also resolved into the same frame, and the degenerate components in the Jacobian and the force vector are removed. The transpose of the new collapsed Jacobian and the force vector are multiplied to obtain the joint torque vector (Equation (1)).

For the case of PUMA 560, the degenerate direction of motion in wrist singularity, for example, is represented by the rotation around the X-axis of Frame{4}. The relationship between the generalized forces, expressed in the frame that best represent the singular direction, is therefore (without the Null Space component:

$$\Gamma = {}^4J^{T4}F \tag{12}$$

where

$$ {}^4J^T = {}^0R_4^{T4}J \quad and \quad {}^4F = {}^0R_4^{T4}J \tag{13}$$

^4F is the force in the Cartesian axis represented in Frame{4}, obtained from the control law with the fourth element (rotation around X-axis) removed, Γ is the torque sent out to each joint, and ^4J is the Jacobian expressed in Frame{4}, with the fourth row removed.

7. Moving Through Singularities

Motion from a singular configuration can be divided into of feasible and non-feasible directions. Non-feasible direction is one that requires motion in the singular (degenerate) direction.

In our experiment, as explained above, the motion in the degenerate direction has been disabled through the removal of the elements of the Jacobian matrix and task space forces representing the direction of singularity.

In the case of a feasible path, the end-effector was found to move through the singular configuration with error in position and orientation not significantly larger than that of motion through non-singular configuration. (Compare Figures A1 and A2 which represent tracking error in non-singular motion, with Figures A3 and A4 which represent tracking error of motion through a wrist singularity along a feasible path).

A path is non-feasible when the desired trajectory lies along the degenerate direction of the manipulator. In this experiment, the null space motion was then utilized in handling such motion. As the Jacobian was collapsed to be of lower rank, the manipulator is now regarded as 'redundant' with respect to the task. Null space torque can then be generated to reconfigure the manipulator to move in the non-feasible direction. Different potential functions can be used to determine the Null space torque, as long as the objective is to shift the degenerate direction out of the way of the desired trajectory.

Khatib [9] showed that the relationship between the generalized forces incorporating the null space torque, as in Equation 1. The $\bar{J}$ is the inverse of the Jacobian matrix, obtained by:

$$\bar{J} = A^{-1}(q)J^T(q)\Lambda(q) \tag{14}$$

where the components are defined in (4).

This is described as the dynamically consistent inverse [3], which is a generalised inverse weighted with mass (inertial) matrix. This inverse has been shown to produce no operational space acceleration when used to project joint torques into the null space[3].

The following sections show the result of the implementation the algorithm on PUMA 560. It is shown that in most cases, the error generated is not larger than that found in non-singular motion, except for some trade off shown in the cases of desired trajectory lying along non-feasible path, where exactness of orientation tracking was sacrificed to make the motion feasible.

7.1. Type 1 Singularity

Type 1 Singularity, or in the case of PUMA, the elbow lock, is one where null space torque would generate motion in the singular direction.

This means, for the case of PUMA, null space motion of joint 3 would generate motion in the singular direction (see Figure 2 for singular direction). Comparing the tracking error (position and orientation) of the manipulator moving out of elbow singularity into the degenerate direction with that of non-singular motion, no significant increase in position and orientation error is observed. (Compare Figure A7 and A8 to Figure A1 and A2).

7.2. Type 2 Singularity

A common type 2 singularity in PUMA is when the wrist joint is straightened ($\vartheta_5 = 0$). The non-feasible path is when it contains the component of degenerate direction, i.e. it requires the end-effector to turn around the X-axis of Frame{4}, or if the desired trajectory lies on the YZ plane of Frame{4}. The other one is the head lock, where the wrist point lies along the Z axis of

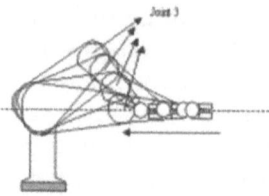

Figure 3. shows Puma-like manipulator moving out of elbow (and wrist) singularity, following the path which lies in the degenerate direction

the Base Frame. The degenerate direction is the Y-axis of Frame{1}. It is described that, as singularity occurs, at least two frames of the manipulator would line up. One of these two frames can therefore be used to create a null space motion to escape the singular position. In the case of PUMA's wrist, null space torque was used to move joint 4 (see Figure 4). This would then shift the plane that contains the non-feasible-path (the YZ plane) out of the way of the desired path (Figure 5). Similarly, in the case of PUMA head lock,

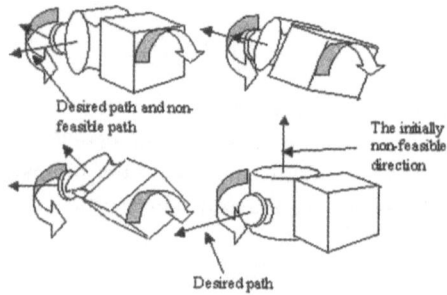

Figure 4. Null space torque is used to turn joint 4, so that the YZ plane of Frame{4} is shifted out of the way of the desired trajectory.

a null space motion of joint 1 is required to shift the the degenerate direction out of the desired path. The result of the experiment is shown in Figures A1 to A8. Figures A1 and A2 show the position and orientation error in end-effector tracking a non-singular path, while Figure A3 and A4, a singular, but feasible path. Figure A5 and A6 is of the end-effector escaping the wrist lock (type 2 singularity) into non-feasible path, and Figure A7 and A8 of the end-effector moving from an elbow singularity into a non-feasible path. The only significantly larger error encountered is in the orientation error of the manipulator as it escapes from a wrist lock into a path that lies in degenerate direction. It is because null space motion was required to turn joint 4 before the end-effector was able to track the desired trajectory, hence the large initial orientation error (see Figure 4).

8. Implementation Issues

Jerkiness in due to a switching of control was experienced as expected. This is especially obvious when the manipulator moves in the degenerate direction by null space torque to escape elbow singularity as in this motion it is not motion-controlled. It therefore does not follow the desired trajectory while in the singular region.

The chosen solution in the experiment was to take the end effector at the point (time and position) of getting out of the singular region as the start of a new trajectory, with the same goal as before, compensating for the time it took for the manipulator to reach this point from the starting point. This was found to produce a smooth transition between the control algorithms.

9. Conclusion

In this experiment, singularity handling in torque-controlled manipulator, based on Operation Space Formulation was explored. It was shown that by removing the degenerate component(s) of motion, control of the manipulator through singular configuration was possible, with tracking error no larger than that of a motion through non-singular path. Motion in the degenerate direction was made possible by motion of redundant joint to move the degenerate direction away from the desired path for type 2 singularities. For type 1 singularity, by definition, Null space torque would produce motion in the degenerate direction. A certain trade off between exactness and achievability was necessary in moving out of a singular configuration into a non-feasible path.

Acknowledgement

The authors would like to acknowledge and to thank Professor Oussama Khatib for his guidance in his capacity as the advisor to the project.

References

[1] E.W. Aboaf and R.P. Paul 1987 Living with the Singularity of Robot Wrists. *IEEE Intl. Conf. for Robotics and Automation* pp 1713-1717.

[2] B. Armstrong, O. Khatib, J. Burdick 1986 The Explicit Dynamic Model and Inertial Parameters of the PUMA 560 Arm. *IEEE Intl. Conf. Robotics and Automation* pp 510-518.

[3] K. Chang and O. Khatib 1995 Manipulator Control at Kinematic Singularities: A dynamically consistent Strategy. *Proc. IEEE/RSJ Int. Conference on Intelligent Robots and Systems* Pittsburgh, vol. 3, pp. 84-88.

[4] F.T. Cheng et al 1997 Study and Resolution of Singularities for a 6-DOF PUMA Manipulator. *IEEE Trans. Systems, Man and Cybernetics* Part B, vol.27 2, pp:332-343.

[5] S. Chiaverini and O. Egeland 1990 A Solution to the Singularity Problem for Six-joint Manipulators. *Proc. IEEE for Robotics and Automation* vol 1 pp 644-649.

[6] John J. Craig 1989 *Introduction to Robotics, Mechanics and Control.* 2nd ed, Addison-Wesley.

[7] P. Hsu, J. Hauser, S. Sastry 1988 Dynamic Control of Redundant Manipulators. *IEEE Int;. Conf. Robotics and Automation* vol. 1, pp 183-187.

[8] Jamisola, R., Ang, M, Jr., Lim, T.M., Khatib, O., Lim, S.Y.1999 Dynamics Identification and Control of an Industrial Robot. *The Ninth Intl. Conf. On Advanced Robotics* pp 323-328.

[9] O.Khatib 1987 A Unified Approach for Motion and Force Control of Robot Manipulators: The Operational Space Formulation. *IEEE J. Robotics and Automation* vol. RA-3, no. 1, pp 43-53.

[10] O.Khatib 1996 Advanced Robotics Lecture Notes, Stanford University.

[11] Y. Nakamura 1991 *Advanced Robotics - Redundancy and Optimization* Addison-Wesley.

[12] L. Sciavicco and B. Siciliano 1990 *Modeling and Control of Robot Manipulators* McGraw-Hill.

Appendix: Experimental Results

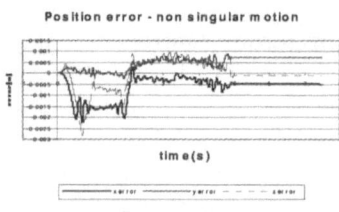

Figure A1.

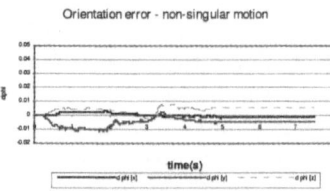

Figure A2.

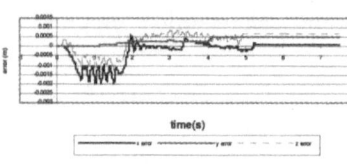

Figure A3.

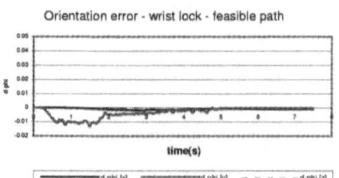

Figure A4.

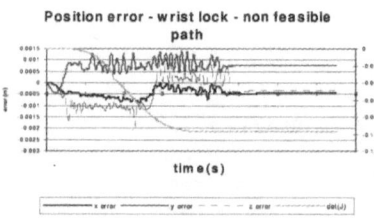

Figure A5.

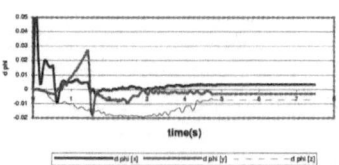

Figure A6.

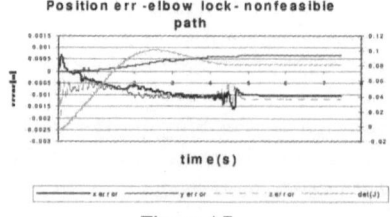

Figure A7.

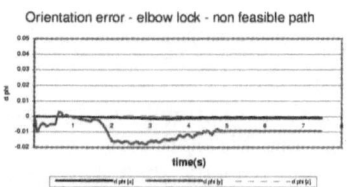

Figure A8.

Autonomous Rover Navigation on Unknown Terrains Functions and Integration

Simon Lacroix, Anthony Mallet, David Bonnafous
Gérard Bauzil, Sara Fleury, Matthieu Herrb, and Raja Chatila
LAAS/CNRS
7, av. du Colonel Roche
F-31077 Toulouse Cedex 4 - France
Firstname.Lastname@laas.fr

Abstract: Autonomous long range navigation in partially known planetary-like terrain is an open challenge for robotics. Navigating several hundreds of meters without any human intervention requires the robot to be able to build various representations of its environment, to plan and execute trajectories according to the kind of terrain traversed, to localize itself as it moves, and to schedule, start, control and interrupt these various activities. In this paper, we briefly describe some functionalities that are currently running on board the Marsokhod model robot Lama at LAAS/CNRS. We then focus on the necessity to integrate various instances of the perception and decision functionalities, and on the difficulties raised by this integration.

1. Introduction

To foster ambitious exploration missions, future planetary rovers will have to fulfill tasks described at a high abstraction level, such as **"reach the top of that hill"** or **"explore this area"**. This calls for the ability to navigate for several hundreds of meters, dealing with various and complex situations, without any operator intervention. Such an ability is still quite an open challenge: it requires the *integration* and *control* of a wide variety of autonomous processes, ranging from the lowest level servoings to the highest level decisions, considering time and resource constraints.

We are convinced that no simple autonomy concept can lead to the development of robots able to tackle such complex tasks: we believe in the efficiency of *deliberative* approaches [4], that are able to plan and control a variety of processes. Following such a paradigm and according to a general economy of means principle, we want the robot to autonomously *adapt* its decisions and behavior to the environment and to the task it has to achieve [5]. This requires the development of:

- Various methods to implement each of the perception, decision and action functionalities, adapted to given contexts;
- An architecture that allows for the integration of these methods, in which deliberative and reactive processes can coexist;

D. Rus and S. Singh (Eds.): Experimental Robotics VII, LNCIS 271, pp. 501–510, 2001.

• Specific decision-making processes, that dynamically select the appropriate decision, perception and action processes among the ones the robot is endowed with.

In this paper, we present the current state of development of the robot Lama, an experimental platform within which our developments related to autonomous long range navigation are integrated and tested. We especially focus on the necessity to integrate various implementations of each of the main functionalities required by autonomous navigation (*i.e.* environment modeling, localization, path and trajectory generation). After a brief description of Lama and its equipment, the rest of the paper is split in two parts: the first part briefly presents the main functionalities required by long range navigation we currently consider (terrain modeling, path and trajectory planning, rover localization), while the second part insists on the problems raised by the *integration* of these functionalities.

2. The Robot Lama

Lama is a 6-wheels Marsokhod model chassis [10] that has been totally equipped at LAAS[1]. The chassis is composed of three pairs of independently driven wheels, mounted on axles that can roll relatively to one another, thus giving the robot high obstacle traversability capacities. Lama is $1.20m$ wide, its length varies from $1.60m$ to $2.20m$, depending on the axles configuration ($1.90m$ in its "nominal" configuration), and weighs approximately $160kg$. Each motor is driven by a servo-control board, and its maximal speed is $0.17m.s^{-1}$. Lama is equipped with the following sensors:

Figure 1. The robot Lama on the experimentation site

• Each wheel is equipped with a high resolution optical encoder, allowing fine speed control and odometry;

• Five potentiometers provide the chassis configuration;

• A 2 axes inclinometer provides robot attitude, a magnetic fluxgate compass and an optical fiber gyrometer provide robot orientation and rotational speed;

• A first stereo bench on top of a pan and tilt unit, is mounted on a $1.80m$ mast rigidly tied to the middle axle. This bench has a azimuthal field of view of approximately $60°$, and is mainly dedicated to goal and landmarks tracking;

• A second stereo bench, also supported by a PTU, is mounted upon the front axle, at elevation of $0.80m$. It has a azimuthal field of view of approximately $90°$, and is mainly dedicated to terrain modeling in front of the robot;

• A differential carrier-phase GPS receiver[2] is used to qualify the localiza-

---

[1]Lama is currently lent to LAAS by Alcatel Space Industries.
[2]currently lent to LAAS by CNES.

tion algorithms.

All the computing equipment is in a VME rack mounted on the rear axle of the robot. The rack contains four CPUs (two PowerPcs and two 68040) operated by the real-time OS VxWorks.

The terrain on which we test the navigation algorithms is approximately $100 \times 50m^2$. It contains a variety of terrain types, ranging from flat obstacle-free areas to rough areas, including gentle and steep slopes, rocks, gravel, trees and bushes.

Navigation functionalities

3. Environment Modeling

Perceiving and modeling the environment is of course a key capacity for the development of autonomous navigation. Environment models are actually required for several different functionalities: to plan paths, trajectories and perception tasks (section 4), to localize the robot (section 5), and also to control the execution of trajectories. We are convinced that there is no "universal" terrain model and that we must build different *multi-layered and heterogeneous* representations adapted to their use.

3.1. Qualitative Model

We developed a method that produces a description of the terrain in terms of *navigability classes*, on the basis of stereovision data [7]. Most of the existing contributions to produce similar terrain models come to a data segmentation procedure (*e.g.* [15, 9]), that produce a *binary* description of the environment, in terms of traversable and non-traversable areas. Our method is a classification procedure that produces a probabilistically labeled polygonal map, close to an occupancy grid representation. It is an *identification* process, and does not require any threshold determination (a tedious problem with segmentation algorithms).

Our method relies on a specific discretisation of the perceived area, that defines a *cell image* (figure 2). It "respects" the sensors characteristics: cell resolution decreases with the distance according to the decrease of the data resolution.

Every cell is labelled with a supervised Bayesian classifier: a probability for each cell to correspond to a pre-defined traversability class is estimated. Figure 3 shows a classification result, with two terrain classes considered (flat and obstacle). There are several extensions to the method: the discretisation can be dynamically controlled to allow a finer description, and the classification results can be combined with a terrain physical nature classifier using texture or color attributes. One of its main advantages is that thanks to the probabilistic description, local maps perceived from different viewpoints can be very easily merged into a global description. The produced terrain model can be either used to generate elementary motions on rather obstacle-clear terrains (section 4.1), or to reason at the path level (section 4.3).

 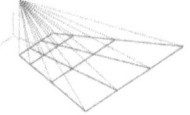    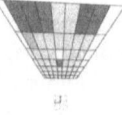 

Figure 2. Discretisation of a 3D stereo image. Left: regular Cartesian discretisation in the sensor frame; right: its projection on the ground (the actual discretisation is much finer).

Figure 3. An example of classification result. From left to right: image from stereo, partial probabilities of the cells to be an obstacle (represented as gray levels), and reprojection of the cells in the sensor frame, after the application of a symmetric decision function.

Figure 4. A digital elevation map built by Lama during a 20 meter run using 50 stereovision pairs.

Figure 5. Landmarks (black + signs) detected on the locally built digital elevation maps (left), and reprojected in the camera frame (right). Landmarks are here from 3 to 10 meters away.

3.2. Digital Elevation Map

Although there has been several contributions to the problem of building digital elevation maps ([11, 2]), we think that it has still not been addressed in sufficiently satisfactory way: the main difficulty comes from the uncertainties on the 3D input data, that can be fairly well estimated, but hardly propagated throughout the computations and represented in the grid structure.

However, a quite realistic model can be easily built by computing the mean elevation of the data points on the grid cells, using only the points that are provided with precise coordinates. With our stereovision algorithm for instance, 3D points whose depth is below $15m$ can be used to build a realistic $0.05 \times 0.05m^2$ cell digital elevation map. Provided the robot is localized with a precision of the order of the cell size, data acquired from several view-points can be merged into a global map (figure 4). This model is then used to detect landmarks (section 3.3) and to generate elementary trajectories (section 4.2).

3.3. Finding Landmarks

An efficient way to localize a rover is to rely on particular elements present in the environment, referred to as *landmarks* (section 5.3). Several authors presented 3D data segmentation procedures to extract salient objects, and our first attempts were based on a similar principle [3]. However, such techniques are efficient only in simple cases, *i.e.* in scenes were sparse rocks lie on a very flat terrain, but rather fragile on rough or highly cluttered terrains for instance.

To robustly detect such local peaks, we are currently investigating a technique that relies on the computation of similarity scores between a digital elevation map area and a pre-defined 3D peak-like pattern (a paraboloid for instance), at various scales. First results are encouraging (figure 5), and the detected landmark could be used to feed a position estimation technique (section 5.3).

4. Trajectory generation

A generic trajectory planner able to deal with any situation should take into account all the constraints, such as rover stability, rover body collisions with the ground, kinematic and even dynamic constraints. The difficulty of the problem calls for high time-consuming algorithms, which would actually be quite inefficient in situations where much simpler techniques are applicable. We therefore think it is worth to endow the rover with various trajectory generation algorithms, dedicated to the kind of terrain to traverse. Section 7 describes how they are actively started and controlled.

4.1. On easy terrains

On easy terrains, *i.e.* rather flat and lightly cluttered, dead-ends are very unlikely to occur. Therefore, the robot can efficiently move just on a basis of a goal to reach[3], and of a terrain model that exhibits non-traversable areas, using techniques that evaluates elementary motions [7].

To generate motions in such terrains, we use an algorithm that evaluates circle arcs on the basis of the global qualitative probabilistic model. Every cycle, the algorithm is

Figure 6. A set of circle arcs to evaluate on the global probabilistic model (left), and reprojection of the arcs in the camera view (right)

run on an updated terrain model. It consists in evaluating the interest (in terms of reaching the goal) and the risk (in terms of terrain traversability) of a set of circle arcs (figure 6). The arc that maximizes the interest/risk ratio is chosen.

4.2. On rough terrains

On uneven terrain, the notion of obstacle clearly depends on the capacity of the locomotion system to overcome terrain irregularities, and on specific constraints acting on the placement of the robot over the terrain. These constraints are the stability and collision constraints, plus, if the chassis is articulated, the configuration constraints (figure 7). To evaluate such constraints, the probabilistic qualitative model is not anymore sufficient and we use a digital elevation map.

We developed a planner (a stripped down version of [8]) that evaluate elementary trajectories, in a way very similar to section 4.1 : a set of circle arcs is produced, and for each arc, a discrete set of configurations are evaluated. Each

---

[3]not necessarily the distant global goal, it can be a formerly selected sub-goal - see section 4.3

arc is then given a *cost* that integrates the elementary costs ("dangerousness") of the successive configurations it contains, the arc to execute being the one that maximizes the interest/cost ratio.

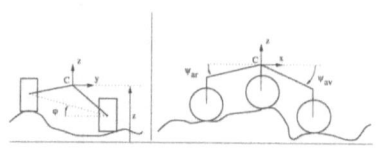

Figure 7. The chassis internal configuration angles checked on the digital elevation map.

Figure 8. A trajectory resulting from the application of the rough terrain local planner (approximately 30 cycles).

4.3. Planning Paths

The two techniques described above are not able to efficiently deal with highly cluttered areas and dead-ends. For that purpose, we use a *path* planner, that reasons on the global qualitative model to find sub-goals and perception tasks [12].

The global qualitative model, which is built upon a bitmap structure, includes a set of regions and a graph that connects them. A search algorithm provides an *optimal* path to reach the global goal The path is then analyzed to produce a sub-goal to reach: it is the last node of the path that lies in a traversable

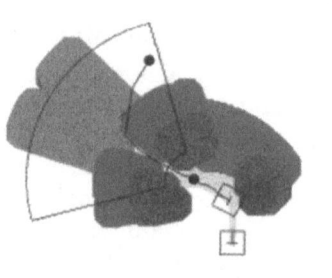

Figure 9. A result of the navigation planner on the qualitative model.

known area. The "optimality" criterion takes here a crucial importance: it is a linear combination of time and consumed energy, weighted by the terrain class to cross *and the confidence of the terrain labeling.* Introducing the labeling confidence in the crossing cost of an arc amounts to *implicitly* consider the modeling capabilities of the robot.

5. Localization

A position estimate is not only necessary to build coherent environment models, it is also required to ensure that the given mission is successfully being achieved, or to control motions along a defined trajectory. Robot self-localization is actually one of the most important issue to tackle in autonomous navigation.

One can distinguish various algorithm categories that compute the robot's position: *(i) motion estimation* techniques, that integrate data at a very high pace as the robot moves (odometry, inertial navigation, visual motion estimation - sections 5.1 and 5.2), *(ii) position refinement* techniques, that rely on the matching of landmarks perceived from different positions, and *(iii) absolute localization* with respect to an initial global model of the environment. All these algorithms are complementary, and provide position estimates with different characteristics. We are convinced that an autonomous rover must be endowed with at least one instance of each category.

5.1. Odometry on Natural Terrain

We use 3D odometry with Lama by incorporating the attitude informations provided by the 2 axes inclinometer[4] to the translations measured by the encoders of the central wheels. Due to skid-steering, the angular orientation measured by the odometers is not reliable: the information provided by the integration of the gyrometer data is much better, and do not drift significantly before a few tens of minutes.

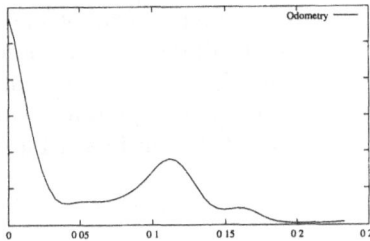

Figure 10.   Histogram of odometry errors measured every 0.5m steps during a 50 meter run with Lama on various kinds of terrain.

To have a quantitative idea of the precision of odometry, we gathered some statistics, using a carrier-phase DGPS as a reference. Figure 10 show that odometry can hardly be modeled an estimator with Gaussian uncertainties: some *gross errors* (due to lateral slippages) actually occur quite often. We are currently investigating the possibility to analyze on line a set of proprioceptive data in order to be able to dynamically qualify the odometry, and especially to detect such errors. These data are the 6 wheel encoders, the measured currents, the two attitude parameters and the five chassis configuration parameters.

5.2. Visual motion estimation

We developed an exteroceptive position estimation technique that is able to estimate the 6 parameters of the robot displacements in any kind of environments, provided it is textured enough so that pixel-based stereovision works well: the presence of no particular landmark is required [13]. The technique computes the motion parameters between two stereo frames on the basis of a set of 3D point to 3D point matches, established by tracking the corresponding pixels in the image sequence acquired while the robot moves.

We pay a lot of attention to the selection of the pixels to track: in order to avoid wrong correspondences, one must make sure that they can be faithfully tracked, and in order to have a precise estimation of the motion, one must choose pixels whose corresponding 3D points are known with a good accuracy. Pixel selection is done in three steps: an *a priori* selection is done on the basis of the stereo images; an empirical model of the pixel tracking algorithm is used to discard the dubious pixels during the tracking phase; and finally an outlier rejection is performed when computing an estimate of displacement between two stereo frames (*a posteriori selection*).

We evaluated the algorithm on many runs (totalizing several hundreds of meters) and it gives translation estimates of about 4% of the distance. Work related to this algorithm is still under way, with the goal of reaching a precision on translation estimates of about 1%.

5.3. Landmark Based Localization

We are currently investigating Set Theoretic approaches for landmark-based localization [14]. No statistical assumptions are made on sensor errors: the

---

[4]after the application of a slight smoothing filter on its data.

only hypothesis is that errors are bounded in norm. Estimates of the robot and landmarks positions are derived in terms of *feasible uncertainty sets*, defined as regions in which the robot and the landmarks are guaranteed to lie, according to all the available informations.

Some simulation results using realistic bounds, using the landmarks detection algorithms presented in section 3.3 are promising. The integration of these algorithms on board Lama is currently under way.

Integration

Our research group has been working for several years on the definition and development of a generic software and decisional architecture for autonomous machines [1]. This architecture has been successfully instantiated in multi-robot cooperation experiments, indoor mobile robotics experiments, and autonomous satellite simulations [6]. Within this architecture, we addressed the problem of running concurrent localization algorithms (section 6) and we also implemented navigation strategies (section 7).

6. Integration of Concurrent Localization Algorithms

To tackle the coexistence of several localization algorithms in a generic and reconfigurable way, we developed a particular module named PoM (position manager), that receives all the position estimates produced by localization as inputs, and produces a single consistent position estimate as an output. PoM addresses the *Sensor geometrical distribution* issue (section 6.1) as well as the *localization modules asynchronism* and the *fusion of the various position estimates* (section 6.2).

6.1. Internal situation

Distributing geometrical information is not satisfying: some information is hard-coded and duplicated within the modules, and it complicates the porting of a module to another robot or another sensor. For that purpose, we use InSitu, a centralized geometrical description of a robot. InSitu reads a configuration file upon startup, and provides the necessary frame coordinates to any module when the robot navigates. The configuration file is the textual description of a geometrical graph: the nodes are frames coordinates that need to be exported.

Every data acquisition module reads the frame configuration which it relates to, and associates to its data a "tag" structure. Thanks to this tagging, clients using such data do not have to care from where the data comes, since all the necessary geometrical and time information is contained in the data itself. The tag is propagated along with the data between modules, thus making inter-module data communication very flexible.

6.2. Position Management

Positions computed by the various position estimators are always produced with some delay, that depends on the computation time required to produce a particular position. PoM maintains the time consistency between the various position estimates by managing one time chart for each position estimator handled by PoM, plus one particular chart for the *fused* position (fusion of every position estimator). Actually, no data fusion algorithm is currently implemented within PoM: given the individual position estimator characteristics, we indeed consider that a *consistency check* (fault detection) has to be performed formerly to any fusion. Up to now, an estimate selection is performed on the basis of a confidence (real value between 0.0 and 1.0, 1.0 being the best) that is hard-coded for each position estimator.

7. Navigation strategies

The integration of the various algorithms presented in the first part of the paper requires specific decisional abilities, that are currently instantiated as Tcl scripts. The following simple strategy is currently applied: the three environment models (qualitative map, digital map and landmark map) are *continuously* updated every time new data are gathered, and the two integrated localization algorithms (odometry and visual motion estimate) are also continuously running[5]. The selection of the trajectory generation algorithm is the following: given a global goal to reach, the easy terrain algorithm is applied until no feasible arcs can be found. In such cases, the rough terrain algorithm is applied. It is run until either the easy terrain algorithm succeeds again, or until no feasible arcs are found in the elevation map. In the latter case (that can be assimilated to a dead end), the path planning algorithm is run, to select a sub-goal to reach. The whole strategy is then applied to reach the sub-goal, and so on.

8. Conclusion

We insisted on the fact that to efficiently achieve autonomous long range navigation, various algorithms have to be developed for each of the basic navigation functions (environment modeling, localization and motion generation). Such a paradigm eventually leads to the development of a complex integrated system, thus requiring the development of integration tools, at both the functional and decisional levels. We are convinced that such tools are the key to implement efficient autonomy on large time and space ranges.

There are however several open issues. Among these, we believe that the most important one is still localization. In particular, the system must be robust to extremely large uncertainties on the position estimates, that will eventually occur: this requires the development of landmark *recognition* abilities to tackle the data association problem, and also the development of terrain model structures that can tolerate large distortions. Note that both problems should benefit from the availability of an initial terrain map, such as provided by an orbiter, whose spatial consistency is ensured. Indeed, the development of algorithms that match locally built terrain models with such an initial map would guarantee bounds on the error of the position estimates.

References

[1] R. Alami, R. Chatila, S. Fleury, M. Ghallab, and F. Ingrand. An architecture for autonomy. *Special Issue of the International Journal of Robotics Research*

---

[5]landmark-based localization integration is under way, but is should also be continuously run, while *actively* controlling image acquisition.

on *Integrated Architectures for Robot Control and Programming*, 17(4):315–337, April 1998. Rapport LAAS 97352, Septembre 1997, 46p.

[2] P. Ballard and F. Vacherand. The manhattan method : A fast cartesian elevation map reconstruction from range data. In *IEEE International Conference on Robotics and Automation, San Diego, Ca. (USA)*, pages 143–148, 1994.

[3] S. Betge-Brezetz, R. Chatila, and M.Devy. Object-based modelling and localization in natural environments. In *IEEE International Conference on Robotics and Automation, Nagoya (Japan)*, pages 2920–2927, May 1995.

[4] R. Chatila. Deliberation and reactivity in autonomous mobile robots. *Robotics and Autonomous Systems*, 16(2-4):197–211, 1995.

[5] R. Chatila and S. Lacroix. A case study in machine intelligence: Adaptive autonomous space rovers. In A. Zelinsky, editor, *Field and service Robotics*, number XI in Lecture Notes in Control and Information Science. Springer, July 1998.

[6] J. Gout, S. Fleury, and H. Schindler. A new design approach of software architecture for an autonomous observation satellite. In *5th International Symposium on Artificial Intelligence, Robotics and Automation in Space, Noordwijk (The Netherlands)*, June 1999.

[7] H. Haddad, M. Khatib, S. Lacroix, and R. Chatila. Reactive navigation in outdoor environments using potential fields. In *International Conference on Robotics and Automation, Leuven (Belgium)*, pages 1232–1237, May 1998.

[8] A. Hait, T. Simeon, and M. Taix. Robust motion planning for rough terrain navigation. In *IEEE/RSJ International Conference on Intelligent Robots and Systems, Kyongju (Korea)*, pages 11–16, Oct. 1999.

[9] L. Henriksen and E. Krotkov. Natural terrain hazard detection with a laser rangefinder. In *IEEE International Conference on Robotics and Automation, Albuquerque, New Mexico (USA)*, pages 968–973, April 1997.

[10] A. Kemurdjian, V. Gromov, V. Mishkinyuk, V. Kucherenko, and P. Sologub. Small marsokhod configuration. In *IEEE International Conference on Robotics and Automation, Nice (France)*, pages 165–168, May 1992.

[11] I.S. Kweon and T. Kanade. High-resolution terrain map from multiple sensor data. *IEEE Transactions on Pattern Analysis and Machine Intelligence*, 14(2):278–292, Feb. 1992.

[12] S. Lacroix and R. Chatila. Motion and perception strategies for outdoor mobile robot navigation in unknown environments. In *4th International Symposium on Experimental Robotics, Stanford, California (USA)*, July 1995.

[13] A. Mallet, S. Lacroix, and L. Gallo. Position estimation in outdoor environments using pixel tracking and stereovision. In *IEEE International Conference on Robotics and Automation, San Francisco, Ca (USA)*, pages 3519–3524, April 2000.

[14] M. Di Marco, A. Garulli, S. Lacroix, and A. Vicino. A set theoretic approach to the simultaneous localization and map building problem. In *39th IEEE Conference on Decision and Control, Sydney (Australia)*, Dec. 2000.

[15] L. Matthies, A. Kelly, and T. Litwin. Obstacle detection for unmanned ground vehicles: A progress report. In *International Symposium of Robotics Research, Munich (Germany)*, Oct. 1995.

Map Building and Localization
for Underwater Navigation

Somajyoti Majumder, Julio Rosenblatt, Steve Scheding, Hugh Durrant-Whyte
Australian Centre for Field Robotics
University of Sydney, Australia
{som, julio, scheding, hugh}@acfr.usyd.edu.au

Abstract: A framework for underwater navigation by combining raw information from different sensors into a single scene description is presented. It is shown that features extracted from such a description are more robust than those extracted from an individual sensor. These descriptions are then combined from many consecutive scenes to form the basis of a new method of map building and representation as a multilevel feature space using probability theory. Comparison of maps previously created and stored in this format against newly acquired scene descriptions are then used for vehicle localization. Experimental results from offshore trials, as well as simulation results, are presented.

1. Introduction

The development of a fully autonomous underwater vehicle is an extremely challenging issue. A key problem is being able to navigate in a generally unknown and unstructured terrain given the limited sensing opportunities available in the underwater domain [1,2,5,6,8]. In principle, by successively integrating information from a number of sensors, a map of landmarks can be constructed and be used to navigate a sub-sea vehicle [3,9]. In practice, no single sensor in the sub-sea environment can provide the level of reliability and the coverage of information necessary to perform this type of navigation. It is therefore necessary to use a number of sensors and combine their information to provide the necessary navigation capability.

Representation of natural outdoor terrain is another important issue, on which very little work is reported in the literature. In general most of the schemes used for terrain modeling and reconstruction involve either meshing or recursive synthesis, which are more suitable for off-line map generation. These methods also suffer from scaling and interpolation errors. The computational cost and complexity for extracting information from an arbitrary map location is quite high. As a result such terrain maps are often inaccurate and cannot be used for navigation.

In this paper, a method is described using blobs and blob-like patches as scene descriptors to segregate feature information from background noise and other errors. A multi-layered data fusion scheme is then used to combine information from the sensors. The general principle is that all sensor information is projected into a common state-space before the extraction of features. Once projection has occurred, feature extraction and subsequent processing is based on a combined (multi-sensor) description of the environment described here as the sensor map . This paper also presents a new method of map building and representation, where the map is

D. Rus and S. Singh (Eds.): Experimental Robotics VII, LNCIS 271, pp. 511–520, 2001.

considered as a mapping of the real world into a multilevel feature space. Then, using probability theory and random variables it can be shown that the map is essentially a conditional probability estimation or maximum likelihood estimation of detectable features. Therefore, the resulting feature map is a multimodal distribution. Gaussian distributions were selected due to their compactness in representation and other useful properties in the transformation mathematics; however any distribution can be applied.

Sensor maps created during a vehicle run can then be compared against maps created and stored from previous missions, and improved localization can be achieved by maximizing correlation measures.

2. Sensor Fusion

At the core of the present sensor fusion model lies the concept of a single composite scene description, its representation and projection to the sensor space. This concept is best defined as a sensor map as depicted in Figure 1. The sensor map is no more than a composite description of all sensor information. However, because all sensor information is projected into one representation, it is also possible to project (at least partially) one sensor's information directly to the information representation of the other sensor(s).

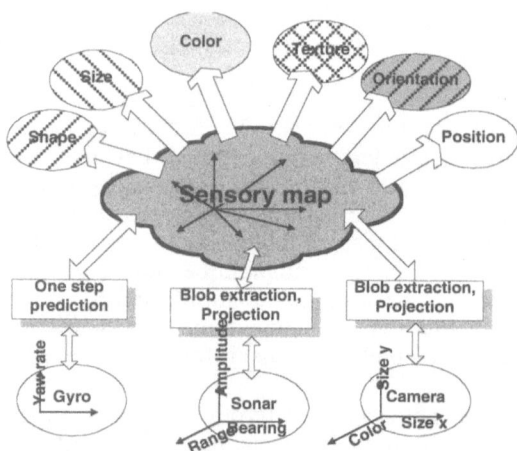

Figure 1.   Sensor map: a representation

The logic behind this representation is to fuse sensor information before feature extraction so as to provide robustness and to base subsequent decisions on the maximum amount of sensor information. This should be compared with the normal data fusion approach of extracting features from a single sensor's information and then fusing the features extracted from many sensors. The reason behind this approach is that in a sub-sea environment individual sensor information is so poor it is difficult to reliably extract features directly without some closely correlated information from a second sensor. Sometimes it is necessary to do some pre-processing or smoothing of the data before fusion. This is an important issue that may directly influence the quality of the sensor map. In case of a very large number of sensors, the dimension or states of the sensor map will still be finite due to the fact that many of these sensors will have common states.

3. Problems in Terrain Modeling

The task of natural terrain modeling involves a number of issues and the quality of the final terrain map is a function of how these issues are addressed. The feature and terrain representation and feature density are of course two issues of paramount concern. Other highly important issues are that the transform is arbitrary or unknown, that the (large) data sets have varying resolution, and that there exists occlusion or shadows. These latter issues are mainly due to the fact that the platform used for data collection is situated within the environment rather than far outside. Therefore, as the observer or platform moves closer to the object, the main features decompose into multiple sub-features. Although these sub-features may provide more insight into the surface characteristics and the contents of the object or terrain, they also create anomalies in data association and tracking. Another important aspect in this respect is resolution or scale. Generally, a high scale represents the global view, disregarding the minute local features whereas at low scale finer details or sub-features become dominant.

Another common source of error in natural terrain sensing is the inability of the sensor to see the whole terrain due to occlusion. The traditional Cartesian space based approaches are unsuitable to discern the occluded areas due to sparseness and interpolation error. This further creates problems in identification on upper or lower limit of the terrain. Besides, each sensor has its own inherent inaccuracies in measurements, each being most useful under different sets of operating conditions. A detailed analysis of various errors, such as shadowing, sparseness, and measurement error, is reported in [7]. These factors limit the completeness and hence the utility of the terrain map. Therefore, a single sensor based Cartesian terrain map often contains inadequate information regarding the terrain and falls short of the requirement for a representation which can be utilized for navigation.

4. Feature Map

In this work, a feature is defined as a sensor detectable entity represented by a two dimensional spatial region or its transformation, which satisfies some well defined mathematical constraints. Therefore, any sensor detectable entity can be considered a feature. A few important features that can be used for map building are size, shape, color, texture, etc. Some are more reliable than others under some specific operating conditions, but a combination of them will always result in more precision/robustness than any individual one.

Statistically, a feature map can be best explained as a conditional probability or probability density function for a given location of any observable feature over entire period of time. Features are represented by joint Gaussian distributions, using Bayesian updates to estimate the posterior distribution conditioned on all given previous information. The volume or area under the distribution is proportional to feature likelihood. Gaussian distributions were selected due to their compactness in representation and other useful properties in the transformation mathematics; however any distribution can be applied.

If the same features are observed in successive scans, then they are combined to form a multimodal distribution. This is depicted in Figure 2, which shows how features are combined to form a feature map. The top image shows that two adjacent specially localized features combine to form a multimodal distribution where as the bottom one shows that even after combining they remain isolated. In this sense this approach is a logical extension of probabilistic maps [11] and occupancy grids [4]. One of the major advantages of this approach is that it can handle any number of features resulting a multi-level map, yet its representation is compact due to the fact that only two scalars, mean and variance, are need to represent a joint Gaussian distribution. Combining features in this way over the vehicle s course results in a feature map such as that shown in Figure 3.

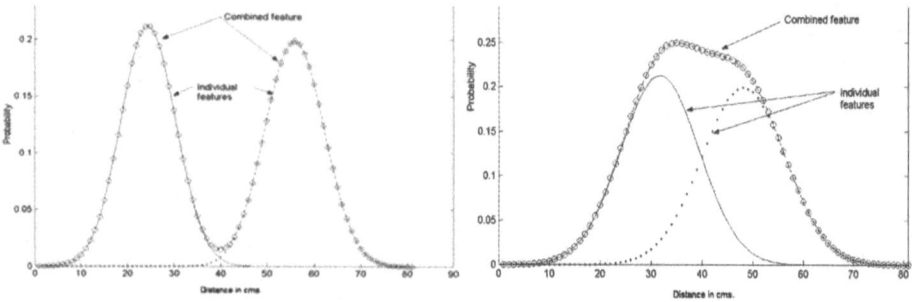

Figure 2.   Probabilistic representation scheme as sums of Gaussian distributions

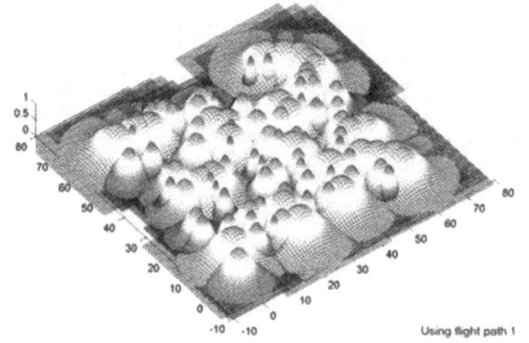

Figure 3.   Feature map

4.1. Data registration

Registration refers to the correct alignment of two image frames or data sets by rotating and translating one over the other so that corresponding features of the two images match. For the purpose of registration, two methods were applied: pair-wise matching, and global matching with respect to the first image. The former generates a reliable local estimate, whereas the latter is more suitable for a global estimate. In case of global matching, the matched entity disappears from the processing window after a few frames since the window is small and the environment does not contain many distinctive features. Therefore, pair-wise matching is used and instead of using one window, eight small (25x25 pixels) windows are taken from the first image as a template. The second image is then searched until the best match for the template is found. To reduce the orientation search space, rotational matching is limited by the angular rotation obtained from the integration of gyro data within a 15° orientation change. For slow speed linear operation of the Oberon vehicle, instantaneous orientation has been obtained by integrating gyro data.

5. Implementation

A three level computational framework has been developed to generate the feature maps discussed here. Figure 4 depicts the various data processing levels to achieve the terrain map from the raw data.

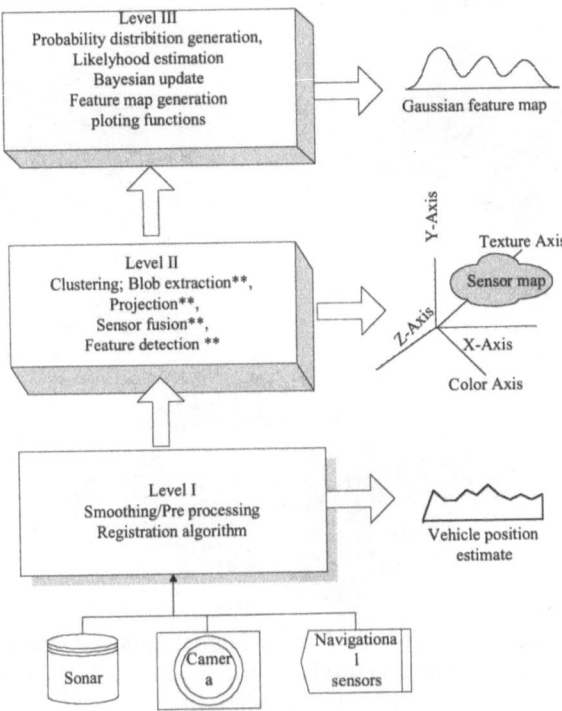

Figure 4.   Computational framework

Level I: This is basically an alignment and registration level that generates the necessary position and orientation estimates from the sensor measurements.

Level II: Sensor data from both sonar and vision is fused to create a composite image from successively aligned scans using blob detection, clustering and projection method.

Level III: This is the actual map building and representation level that generates 3D Gaussian distribution map and updates the posterior distribution using Bayes rule for each identified feature parameter.

This cycle can be repeated as many times based on the number of required feature maps. The final terrain is the combination of this individual feature maps. Further information on items marked with ** are available in [10].

6. Vehicle Localization

Having mapped an area using these methods, the stored representation of the terrain can be used in successive missions for the purpose of vehicle localization. A newly acquired map created from current sensor data, as shown in the upper left of Figure 5, can be matched against an *a priori* map as on the right. This matching isperformed by maximizing the correlation between the two maps. Dead reckoning position data is used to determine a search window for the matching process, thereby reducing the computational cost and decreasing the chances of encountering local maxima. The correlation measure provides a well-defined peak, as can be seen at the bottom of Figure 5, thus establishing the vehicle s location relative to the reference frame of the standard map.

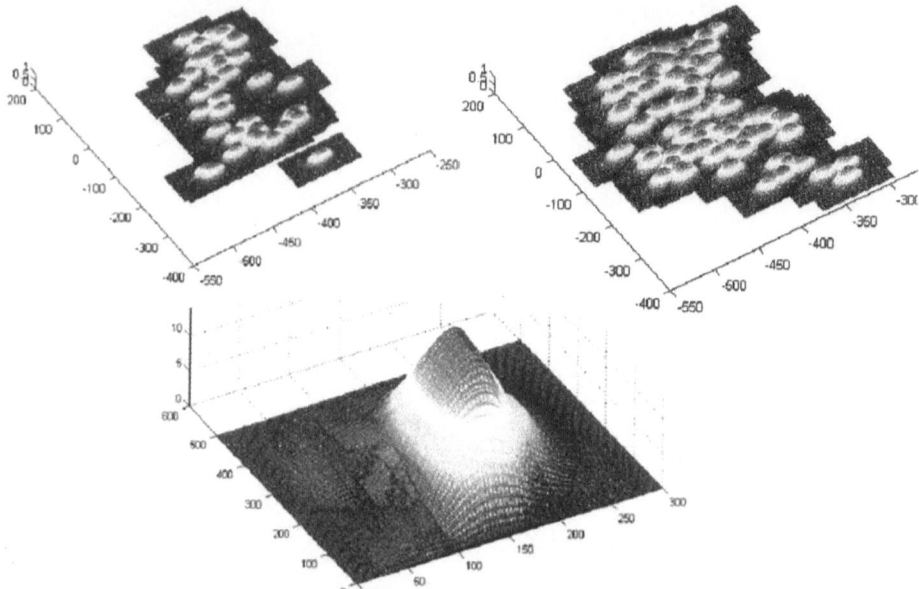

Figure 5.   Feature map from current vehicle mission and stored map from a previous vehicle mission are correlated and produce a distinct maximum for localization.

7. Experimental Setup

For experimental verification of the concept described in Sections 2 and 3, test data has been collected from several offshore trials using the Remotely Operated Vehicle Oberon , developed at the Australian Centre For Field Robotics of the University of Sydney. Oberon is shown in Figure 6. Further details of the experimental platform are available in [12]. Sensors used in the work described here are a 585 kHz or 1210 kHz (user selectable) pencil-beam sonar, a 675 kHz fan beam sonar, a specially constructed color camera, a fiber optic gyro and a pressure transducer for depth.

An offshore test site with natural terrain features was selected for the experiments. During the trial, Oberon was allowed to follow a course parallel to the shoreline facing the sea followed by a similar course in opposite direction. The length of each run was approximately 25 meters. The experimental data consists of gyro, depth, sonar, and camera data. Two forward-looking sector scan sonars used here are mounted at an angle of 45°, pointing towards the sea floor. The upper one has a narrow pencil-beam whereas the other one is a fan-beam sonar. Therefore the sonar objects belong to a narrow window within the respective camera image.

Figure 6.   Oberon AUV during offshore trials

To combine information form Sonar to Vision and vice versa, the method of perspective transformation has been applied. The location of forward looking sonar and uncalibrated camera is also shown in Figure 6. Camera calibration was not essential considering the intended accuracy of the system. It is evident that the sonar window is a part of the visual window separated by the distance between the camera and sonar center. In order to build the sensor map from offshore data, either all or a useful part of it needs to be projected into the common state space.

To minimize the effect of outlier data and to limit the required computation, a structure, termed the scene descriptor, was introduced to segregate important scene information from the background. Blobs or blob-like structures are suitable for this purpose. Blobs are defined in computer vision as soft irregular shaped objects, the main purpose of which is to segregate the useful foreground information with all its characteristics from the background, such that any feature extraction algorithm can extract appropriate features such as color, texture, or shape from them.

8. Results

Offshore data after blob analysis and subsequent image plane fusion using the projection method described is shown in Figure 7 for a single scan where both of the sensors are looking ahead and are tilted to an angle of 45° downwards to view the terrain. It also shows the various stages of processing. The uppermost one is the sub-sea image of natural terrain devoid of tractable features. The second image from top shows the perspective view of the actual terrain, seen by a co-located sonar. It further shows the camera viewing field within the sonar domain. The third image is the sensor map that combines both sonar and vision where sonar objects are shown on the top part of image. The fourth one from the top shows the detected objects with combined features in the scene that are corroborated by both sonar and vision.

Another four sensor maps, which are generated from a 25 meter long offshore test run, are shown in Figure 8. Sonar objects are combined with camera and plotted as dark points on the top part of the image. These combined features form the basis of our map building process, which is represented by a bivariate Gaussian distribution to generate the feature map, depicted in Figure 9. Where size of the rock and variance of sonar returns (a measure of texture) are used as a feature metric.

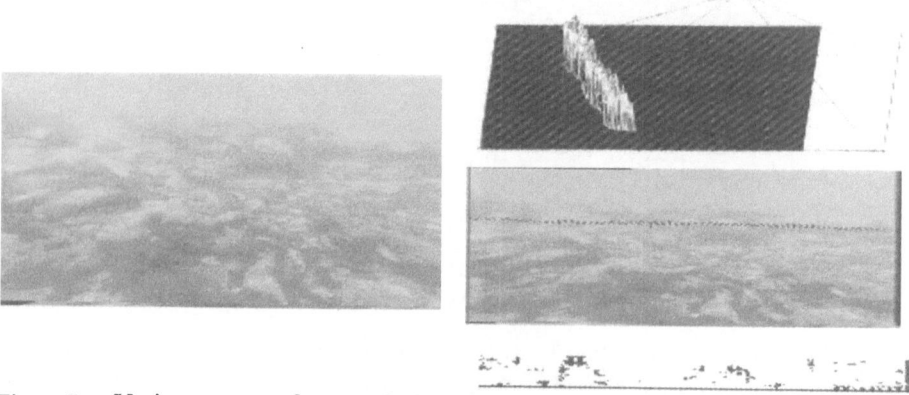

Figure 7.   Various stages of sensor fusion: camera image, sonar image, sensor map, and detected objects

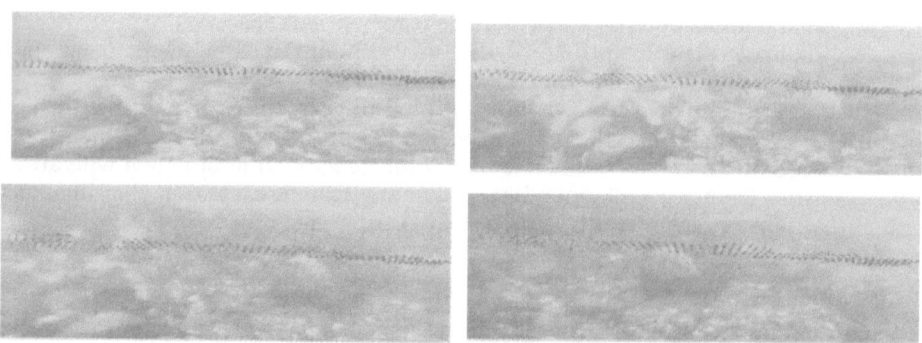

Figure 8.   Successive sensor maps at 2 second intervals

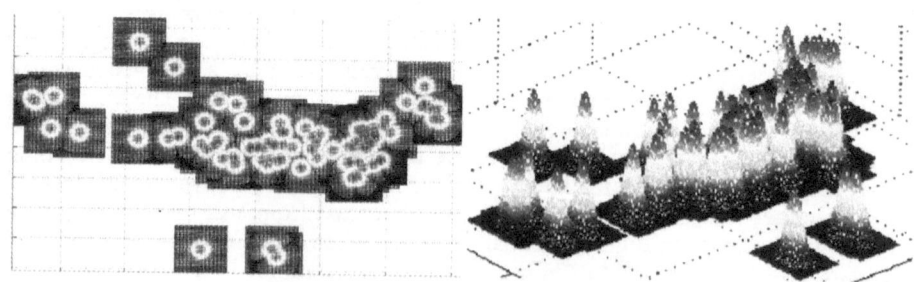

Figure 9.   2D and 3D feature maps

Figure 9.   2D and 3D feature maps

This further shows that as the vehicle moves, tractable features are continually building the map, where large objects form a multimodal distribution due to their continuity over several scan of the sensor without application of any data association algorithm. However, this does not limit application potential, any number of appropriate features could also be added, effectively increasing the robustness and reliability of the map for subsequent use.

8.1. Localization Results

In order to test the localization via feature map correlation, experiments were conducted in simulation so that the true path could be known with certainty. These trials produced the estimated paths for dead-reckoning only and with map correlation shown in Figure 10. The X and Y errors, shown in the plots below, yielded a mean X error (in meters) of 0.34 and 0.13, and a mean Y error of 0.21 and 0.03 for the dead reckoning and correlation localization paths, respectively. As can be seen, the vehicle s position is determined much more accurately using map correlation.

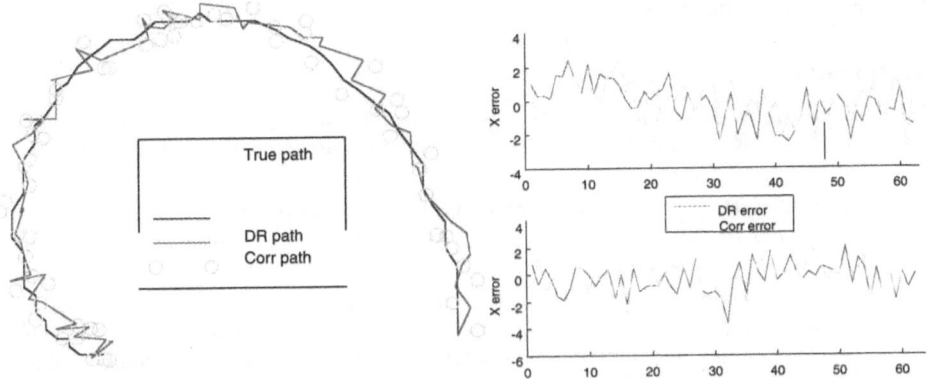

Figure 10.   Estimated paths using dead reckoning and map correlation; errors in X and Y.

9. Conclusion

This paper presents a new method for of combining information from two of the most common sensors used for obstacle detection and navigation of sub-sea vehicle, i.e. active sonar and underwater cameras, to construct a complete environmental map for navigation, using all possible sensor information. The main reason for this is that sonar and vision operate on different physical principles, and provide completely different type of information; point range data and passive two-dimensional images respectively. Also, completely different processing principles are used to manipulate and extract information from these two different sensor sources. A wide variety of computer vision algorithms exist for extracting color, texture, shape from an image (although most of them are not very reliable in the under water domain), whereas sonar data processing is mainly concerned with time of flight and amplitude information to determine object range and size. In natural sub-sea environments normally robust features, such as points and lines, and their associated feature extraction methods, turn out to be very fragile. This work demonstrates a new approach to combine multiple sensor information at a very low level to create a composite scene description - the sensor map, which is essentially the opposite to conventional data fusion methods based on feature extraction followed by fusion.

This paper also presents a method of representation for non-point features using probability theory, random variables, and Bayesian statistics to show that a feature map is essentially a conditional probability or maximum likely hood estimation of (observable and detectable) features. Bivariate Gaussian distributions have been used for the purpose of representation for their excellent transformation properties and compactness. Any other distribution could also be used for the purpose. A set of results from several offshore trials is also presented here.

Finally, sensor maps created during a vehicle run can then be compared against maps created and stored from previous missions, and improved localization can be achieved by maximizing correlation measures.

Acknowledgement

The authors would like to acknowledge the help of Mr. Stefan B. Williams during various stages of this work.

References

[1] R. Bahl, *Object classification using compact sector scanning sonars in turbid waters*, Proc. 2nd IARP Mobile Robots for Subsea Environments, Monterey, CA, vol. 23, pages 303-327, 1990.

[2] M. J. Chantler, D. M. Lane, D. Dai, and N. Williams, *Detection and Tracking of returns in sector scan sonar image sequences*, IEE Proc. Radar, Sonar Navigation, vol. 143, pp. 157-162, 1996.

[3], M.W.M.G. Dissanayake P. Newman H.F. Durrant-Whyte S. Clark and M. Csorba, *An Experimental and Theoretical Investigation into Simultaneous Localisation and Map Building*, Proceedings 6th International Symposium on Experimental Robotics, Sydney, NSW, Australia, pages 171 - 180, 1999.

[4] Alberto Elfes, *A Sonar-Based Real World Mapping and Navigation*, IEEE Journal of Robotics and Automation, volume 3, number 3, pages 249-265, 1987,

[5] Michael J. Kennish, *Practical Handbook of Marine Science*, CRC Press, Boca Raton, FL, 2nd edition, 1994.

[6] Lawrence E. Kinsler and Austin R. Frey, *Fundamentals of Acoustics*, John Wiley and Sons, Inc, second edition, 1950.

[7] I.S. Kweon & T. Kanade *High resolution terrain map from multiple sensor data*, IEEE Transactions on Pattern Analysis and Machine Intelligence, volume 14, number 2, pages 278 — 292, 1992.

[8] David M. Lane and J. P Stoner, *Automatic Interpretation of Sonar Imagery Using Qualitative Feature Matching*, IEEE Journal of Oceanic Engineering, vol. 19, no. 3, 1994, pp. 391-405.

[9], J.J. Leonard and H.F. Durrant-Whyte, *Simultaneous Map Building and Localization for an Autonomous Mobile Robot*, In Proc. IEEE Int. Workshop on Intelligent Robots and Systems, New York, USA, pages 1442 1447, 1991.

[10] S. Majumder S., Scheding and H. F. Durrant-Whyte, *Sensor fusion and map building for underwater navigation*, Proceedings of Australian Conference on Robotics and Automation, Melbourne, Australia, Aug 30-Sep 1, 2000.

[11] S. Thrun D. Fox and W. Burgard, *A probabilistic approach to concurrent mapping and localization for mobile robots*, Machine Learning, vol. 31, pp. 29-53, 1998.

[12] S.B. Williams, P. Newman, MWMG Dissanayake, J. Rosenblatt and H.F. Durrant-Whyte, *A decoupled, distributed AUV control architecture*, Proc. of 31st International Symposium on Robotics, Montreal, Canada, 2000.

Visually Realistic Mapping of a Planar Environment with Stereo

Luca Iocchi
University of Rome La Sapienza
Rome, Italy
iocchi@dis.uniroma1.it

Kurt Konolige
SRI International
Menlo Park, CA USA
konolige@ai.sri.com

Max Bajracharya
Massachusetts Institute of Technology
Cambridge, MA USA
maxb@mit.edu

Abstract: We present a hybrid technique for constructing geometrically accurate, visually realistic planar environments from stereo vision information. The technique is unique in estimating camera motion from two sources: range information from stereo, and visual alignment of images.

1. Mapping and Mobile Robots

Recent techniques in mapping using single-plane laser rangefinders on mobile robots have proven very successful in indoor environments [1,2,3,4,5]. These techniques match range scans to build up a floor model, or plan view. They make no assumptions about the geometry of the environment, and take advantage of the direct range measurements in reconstruction.

Image alignment techniques, on the other hand, attempt to simultaneously determine camera motion and 3D geometry from a sequence of images. They determine range only indirectly, as a byproduct of determining camera motion and matching images; there is a very large literature on this subject [6,7,8].

Both techniques have disadvantages. Range techniques are limited in their accuracy by the range measurements, which for mobile robots are typically much less precise than required for constructing a visually accurate model. Further, full 3D range sensors are expensive, power-hungry, and slow, and have yet to be deployed on mobile robots. On the other hand, image alignment, while it can yield visually precise results, suffers from several problems in a full 3D setting: high computational load, difficulty in matching, and ambiguity in determining camera motion. It is well-known that these problems are accentuated when dealing with just two views of an object, rather than a sequence of images [9].

In our work, we combine techniques from range mapping and image alignment to reconstruct visually-realistic, metrically precise maps from a mobile

D. Rus and S. Singh (Eds.): Experimental Robotics VII, LNCIS 271, pp. 521–532, 2001.

robot, using just a stereo sensor to provide range and image data. We are interested in two tasks:

- Reconstructing the planar geometry of the indoor environment, for robot navigation. The accuracy of this reconstruction need not be high, because current robot localization algorithms can deal with large uncertainties [10, 11].

- Providing a visually-realistic reconstruction of the environment for 3D virtual-reality viewing. In this case, although the geometry need not precisely reflect the real-world 3D geometry, images must be correctly texture-mapped and fused on the geometry.

As we will show, stereo range information from a short-baseline stereo rig is sufficient to accomplish (1), under suitable planarity assumptions. However, the camera motion estimated from range information is not accurate enough to visually fuse images into a convincing texture-mapped 3D reconstruction. Instead, we use correlation-based image alignment techniques to complement and fine-tune the geometrical matching process.

Although we believe the techniques presented here will generalize to more complex environments, in this study we rely on planar surfaces as the primary component of the environment model.

2. Stereo Range Data and the Planar Modeling Assumption

The input for our reconstruction technique comes from short-baseline (10 cm) stereo imagery, using 640 x 480 color images. We use wide-angle optics (4.8 mm lens) to capture a substantial field-of-view, including both sides of a corridor; but this comes at the expense of range precision. Figure 1 graphs the range precision of the stereo rig against distance; it should be noted that the range information is less accurate than this, because of various stereo-related effects such as smearing [12]. Figure 5 is an overhead view of the range returns along a corridor, showing that the geometric fidelity of the device is quite good, even at 10 meters.

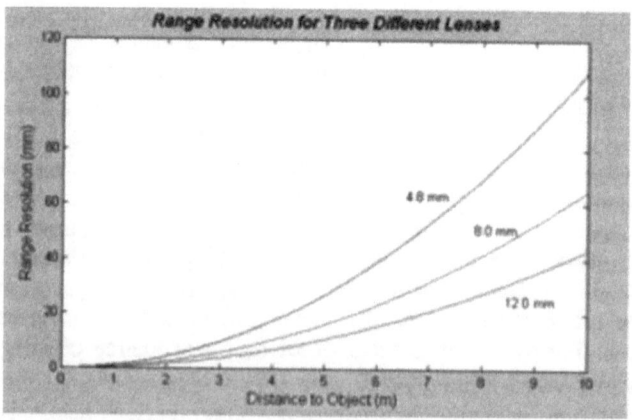

Figure 1. Stereo range resolution for several different lens focal lengths.

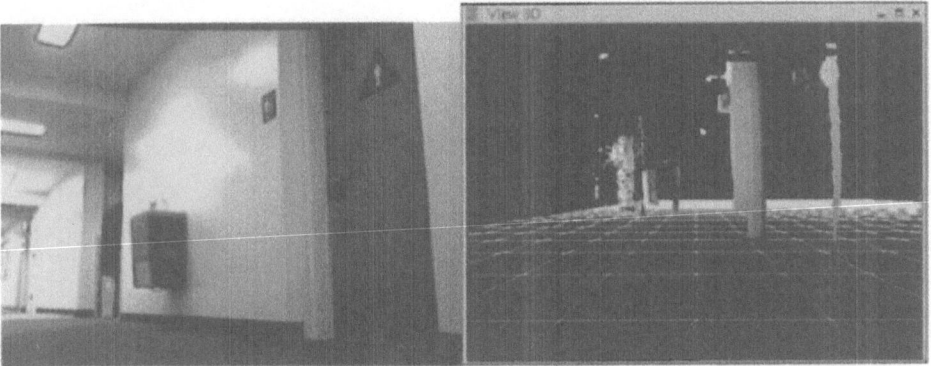

Figure 2. Color image (left) and reconstructed 3D mapping from range data (right). Note the many droupouts where there is no reliable stereo information.

To illustrate typical stereo data, Figure 2 shows a scene from the left stereo camera, and the reconstructed 3D data from stereo range. While the stereo found parts of the scene with good texture, it missed large areas that were uniform, e.g., the white walls. This problem of *dropouts* is the most serious obstacle to geometric stereo reconstruction.

Most indoor environments consist of large planar surfaces: walls, floors, ceilings, even furniture. These surfaces constitute the primary structure of the environment. By using planar surfaces, we can simplify some of the difficult problems in both image alignment and stereo range mapping.

- Range registration. Stereo sensors use triangulation to measure distance, and range error is related to the distance squared, which makes it impossible to get accurate range information at distances greater than a few meters. Range registration is the process of fusing range readings at successive robot positions, and it cannot be done reliably on the basis of stereo range alone. A planar surface assumption can correct for the lack of precision of stereo at distance.

- Stereo dropouts. One of the major shortcomings of stereo ranging is the lack of range information in non-textured areas. With the planar assumption, we recover these areas as part of a planar surface (wall, floor, etc.).

- Image registration. Using a planar assumption reduces the search space for image registration to a small number of parameters. The idea is similar to image mosaicing, which uses an affine assumption [13].

3. System Description

The basic task is to estimate robot motion on the ground plane, using information from stereo. Figure 3 shows the geometry involved: the robot's pose in a global 2D reference system (X,Z) is represented by three variables (x, z, θ), that correspond to the position and orientation of the left camera of the stereo rig. More specifically (x, y) represents the projection on the ground plane of the position of the optical center of the left camera, and θ is the orientation of the projection of the optical axes of the left camera. Therefore, for robot motion estima-

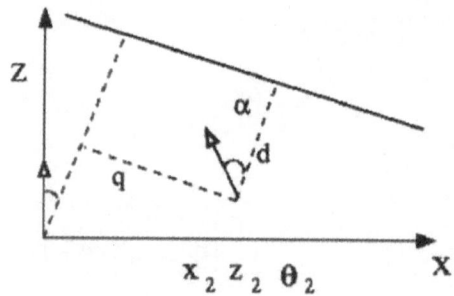

Figure 3. Robot motion geometry.

tion we add the constraint that the robot moves on a plane (the ground plane) and thus the three degrees of freedom are specified by (x,y,θ).

Information from stereo range is used to determine robot position and angle with respect to planar surfaces (α and d in Figure 3). In this way it is possible to partially correct possible errors from the robot's odometric system. However an accurate 3D reconstruction requires a higher precision in motion estimation and a fine measurement process is performed by using image alignment techniques.

Figure 4 is a synopsis of the 3D reconstruction system. As the robot acquires a new stereo pair, it is integrated into a growing map of the environment. First, from stereo range, a 3D Hough transform computes the major planar sur-

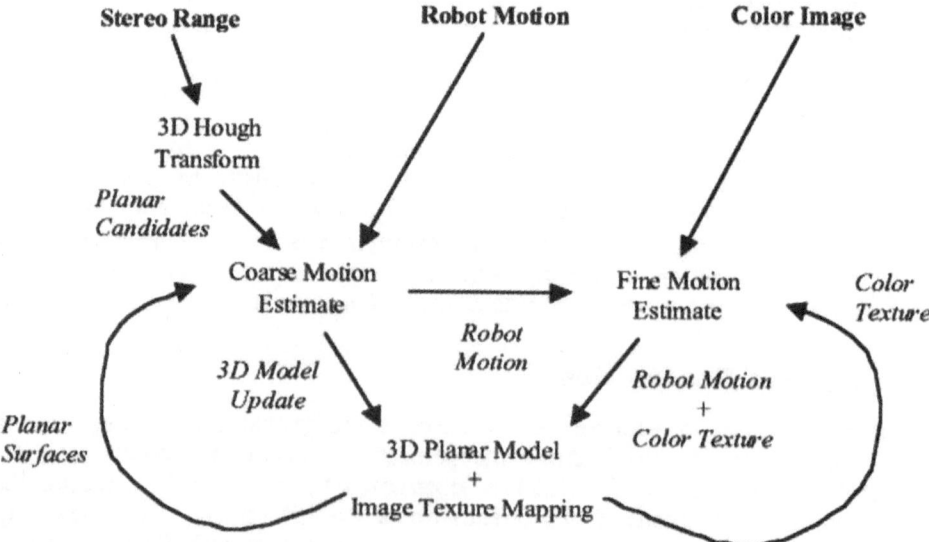

Figure 4. System description: range, robot motion, and image information are combined to create a textured 3D geometry.

faces in the image. These surfaces are fused with the current environment model, using information from robot motion encoders to give a coarse estimate for fusing. This estimate is not precise enough for accurate image alignment, so a second fine adjustment is made by correlating the new image with previous images mapped onto the geometry.

3.1. 3D Hough Transform

The Hough Transform allows for detecting the best fitting line/plane from a set of 2D/3D points, and it is very robust to noise due to occlusions and false positives [17,18]. The 3D HT is defined by the following transformation that is applied to every point (x,y,z) returned by the stereo device:

$$\rho = x\cos\theta + y\sin\theta\cos\varphi + z\sin\varphi$$

Every 3D point generates a curve in the Hough space (θ,φ,ρ) and every point in the Hough space corresponds to a plane. The main property of the HT is that given a set of 3D points all belonging to the same plane, the corresponding curves in the Hough space intersect each other in a single point of the Hough space that corresponds to that plane. Moreover, having defined a discretization of the Hough space in cells and computed for each cell the number of curves passing through it, the local maxima of this function correspond to the best fitting planes for a cluster of 3D points.

Plane extraction with the 3D HT has a computational complexity O(n^*m), where n is the number of 3D points returned by the stereo device and m is the size of a discretization of the dimensions (θ,φ) of the Hough space. This complexity typically permits real-time implementations (i.e. at most 100 ms cycle time) even with a large number of input 3D points (on the order of 100,000).

The accuracy of the method depends on the discretization of the Hough space and on the precision of the range sensor. In our setting, with images of size 640x480, the technique returns planes that have typical deviations of 3 to 5 degrees in α. At close range, the distance measured to walls d is very accurate, on the order of ±1 cm. Figure 5 shows a typical range result down a corridor, viewed from above. The corridor walls are clearly visible, with reasonable range precision out to 12 m (the readings in the middle of the corridor are from the ceiling).

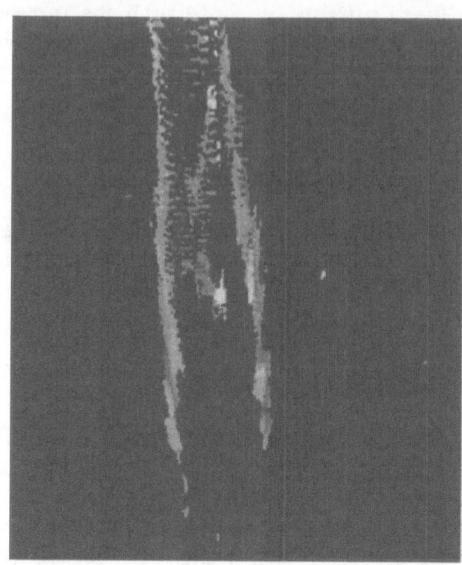

Figure 5. Stereo range returns along a corridor, viewed from above the corridor. Stereo cameras are at bottom of image; distance to the top is 12 m.

Given α and d, the wall section can be embedded within the correct 3D model plane. However, there is still ambiguity in pose within the model plane, and we use several matching techniques to recover the pose.

3.2. Map Building and Wall Reconstruction

The geometric map of the environment is represented by a set R of reference planes and by the corresponding texture extracted from the original images. Once planar surfaces are extracted from the stereo data (by means of the 3D HT described above), they are matched against the current 3D model in order to incrementally build the map of the environment. Since the robot has only moved a small amount (we usually choose around 1 m), odometry information is good enough to perform robust matching.

The extracted plane is matched against a set R of reference planes within the current map representation. If a match is found, the two features are merged together (possibly with a correction for reducing the position error of the robot as described in the next section), otherwise a new feature is added to the set R. Observe that under the assumption of small positioning error this step would not introduce false new features.

We make no assumption about the relative angles of the walls, e.g., the 90 degree assumption. However, the HT itself introduces a discretization of 5 degrees, which is enough to keep perpendicular walls exactly perpendicular. The registration is good enough so that, in small cycles, it is possible to rematch the original walls. Figure 6 shows the wall planes extracted and matched from 24 stereo pairs of the SRI offices. The robot completed a cycle about 20 m on a side, and the wall embedding was able to find the correct match at the end of the cycle to close the loop. In general, more sophisticated matching techniques will have to be employed in larger environments [1]. Note that the wall embed-

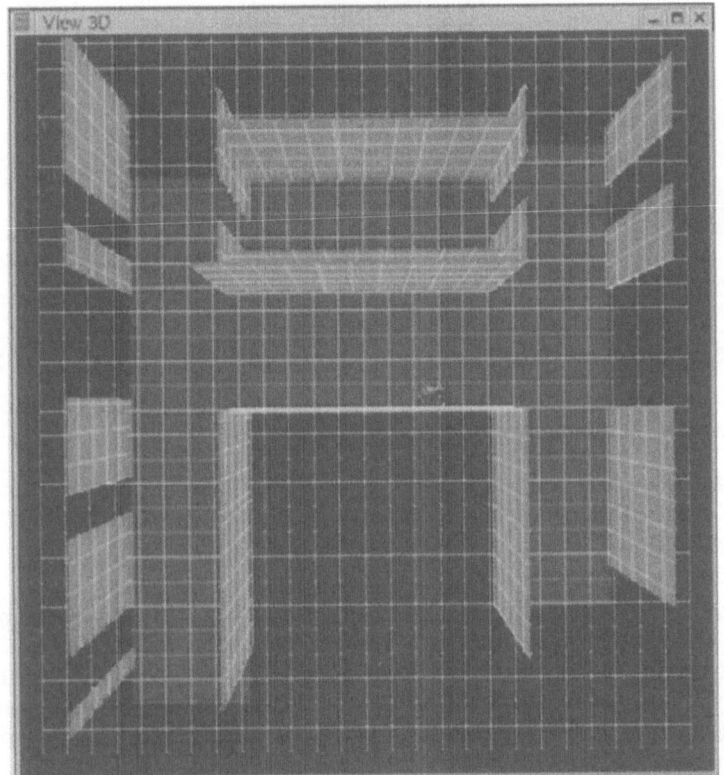

Figure 6. Geometric planes created from 24 stereo images of the SRI offices. Scale is approximately 20 m on a side.

ding process preserves fine structure: for example, the inner wall of the top area is distinct from the outer wall.

While odometry can provide a rough estimate of robot motion for the embedding process, it is not accurate enough to fuse wall textures from multiple images. Instead, we use two methods to determine robot motion along planar surfaces (the direction q in Figure 3). These methods are explained in the next section.

4. Image Alignment and Texture Fusion

The rough estimate of the plane position provided by robot motion and range information is not good enough to provide visually accurate rendering of the wall texture. For fine adjustment of the images we make use of two different techniques aiming at reducing the position error of the robot and thus image alignment. These two techniques differ in the use of the visual information acquired by the cameras.

- When the images acquired contain enough texture information, image correlation is used as a measure for the goodness of the alignment.

Figure 7. Wall texture reconstruction. Original image (top); transformed perpendicular view (left); final multi-image texture with holes (right).

* If image texture is not enough we try to detect structural elements in the environment (like door frames in a corridor) and to use these landmarks for the alignment.

   In either case, once we have determined the incremental camera pose estimate, we can reconstruct the image texture of the wall. We first extract the relevant image information, then transform it to a perpendicular view (Figure 7). Holes in the wall are discovered by finding objects behind the wall plane in the stereo range. Finally, multiple images along the wall can be fused to provide a complete wall texture.

4.1. Range Histogram Matching

One method of determining robot motion perpendicular to a planar surface is to match range information at the new pose against the previous one. The idea is to look at a horizontal band along the wall, and create a histogram of pixels that are in the plane of the wall (red) and not in the plane (blue). The two histograms of Figure 8 show sharp peaks around doorways, and can be easily matched. The variance of the peaks is around 5 cm, giving about a 5% average error for a 1 m movement.

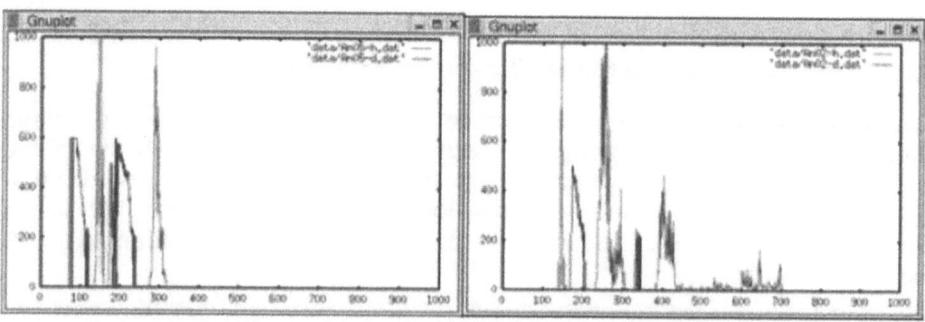

Figure 8. Histogram of range along a wall, from two different poses. Red peaks are in-wall returns, blue are off-wall returns.

4.2. Image Alignment by Correlation

Image alignment by correlation is peferentially used because of its higher precision. We search in the space of camera motions around the rough estimate, using image correlation as a goodness measure. Figure 9 shows the superposition of two texture maps taken from different robot positions. The left image uses a coarse estimate of camera motion, while the right is refined by search around this estimate. The fuzziness caused by misalignment is much reduced in the right superposition. Several points should be noted here about this process.

- Ambiguities exist in the alignment process, especially correlation between rotation and translation motions of the camera. Without range information from stereo, it is impossible to determine camera motion unambiguously, even if the images are aligned visually [14,15,16].
- The search is over two rotations and two translations of the camera, and can be computationally expensive. We have found techniques that make this search practical; these techniques isolate individual parameters or parameter pairs for optimization.

Unlike standard techniques for structure-from-motion, which rely on finding matching features between the two images, we simply use a hypothesize-and-test method, which is robust but can be computationally expensive. We are searching for robot movement constrained to planar motion -- a single component of rotation and two (orthogonal) components of translation. So overall, we are searching over four parameters: a rotation of the camera, a distance to the wall, a distance along the length of the wall, and a distance along the height of the wall (approximating the roll of the camera).

Doing the search along the two components of the wall is the least expensive; it only requires a translation of the two (rectified wall) images across each other. Changing the distance to the wall essentially requires the image to expand or shrink, requiring interpolation and, consequently, is more expensive.

Figure 9. Image alignment using correlation. Left image has initial overlay based on rough estimate; right image is refined estimate.

Changing the yaw angle, requires a computation of the wall points in 3D, translated back to 2D, and also requires interpolation.

The computation time to calculate the error between two images is proportional to the number of pixels which need to be compared. However, this number is reduced by using only points on the plane which have texture – only points that matched in the stereo matching process (points that didn't match in stereo are not likely to match when matching images) – and by using smaller image

The entire search procedure is done in a pyramid style, using 320x240 images to establish a rough estimate of parameters, and then a 640x480 image to refine the search. The search is done by gradient ascent (with look-ahead to prevent false local maxima). The search starts by fixing a rotation, and then finding the best distance to the wall. This involves a gradient search over this distance, and at each iteration, a search along the length and height of the wall. Then, in EM fashion, we use the new distance to the wall and adjust the angle, and repeat the process. The search is on the order of seconds (reaching up to a minute), running on a PII 400 MHz PC. Without doing the 640x480 refinement, it is considerably faster, but still on the order of seconds.

There is an ambiguity between rotation and translation: a rotation in the camera yaw is similar to a translation along the wall. Without the geometrical constraint furnished by stereo, the angular uncertainty would grow with every new pose, and the robot would quickly become lost. With the geometrical constraint, the only error that grows is translation along the wall plane. The maximum error for a single pose estimate is:

$$\Delta q = d(\tan(\alpha) - \tan(\alpha + \Delta \alpha),$$

where $\Delta \alpha$ is the maximum error in the wall angle determined from stereo. For a wall distance of 1m, at a 45 degree angle, the maximum error is 19 cm. Typical errors, of course, will be much less. We have not yet done any experiments to determine errors under real-world conditions.

5. Conclusion

We have built an experimental system that combines techniques from image processing and range fusion to create visually-realistic 3D environment models from a moving robot. The novelty of this approach lies in the combination of approaches, each of which exploits specific types of information to reduce ambiguity in the fusion process. The result is a 3D, texture-mapped planar model that can be used for virtual reality applications, as well as robot mapping and localization. The use of full 3D information makes the mapping more robust and of greater utility than techniques that use floor-plan scans only.

As proof of the viability of our system, we have constructed a model of the SRI offices over a space of about 30 m^2, using a total of 26 stereo pairs (see Figure 10).

Figure 10. A texture-mapped scene created of the SRI offices.

References

[1] Gutmann, J. S. and K. Konolige. Incremental Mapping of Large Cyclic Environments. *Proceedings of CIRA*, Monterey, CA (1999).

[2] F. Lu and E. Milios. Globally consistent range scan alignment for environment mapping. *Autonomous Robots*, 4:333–349, 1997.

[3] Thrun, S., W. Burgard, and D. Fox. A real-time algorithm for mobile robot mapping with applications to multi-robot and 3d mapping. In *Proceedings of ICRA*, San Francisco (2000).

[4] W. Burgard, D. Fox, H. Jans, C. Matenar, and S. Thrun. Sonar-based mapping of large-scale mobile robot environments using EM. In *Proc. of the International Conference on Machine Learning (ICML 99)*, 1999.

[5] P. Moutarlier and R. Chatila. Stochastic multisensory data fusion for mobile robot location and environment modelling. In *5th International Symposium on Robotics Research*, pages 85–94, 1989.

[6] Mandelbaum, R., G. Salgian and H.S. Sawhney. Correlation-based Estimation of Ego-Motion and Structure from Motion and Stereo. In *Proc. of the IEEE Intl. Conf. on Computer Vision*, Corfu (1999).

[7] C. Tomasi and T. Kanade. Shape and motion from image streams under orthography: A factorization approach. International Journal of Computer Vision, 9(2):137–154, November 1992.

[8] Z. Zhang and O. Faugeras. Estimation of displacements from two 3d frames obtained from stereo. *Trans. Pattern Analysis and Machine Intelligence*, 14(2):1141–1156, 1992.

[9] J. Oliensis. A critique of structure-from-motion algorithms.

[10] W. Burgard, D. Fox, D. Hennig, and T. Schmidt. Estimating the absolute position of a mobile robot using position probability grids. In *Proc. of the Fourteenth National Conference on Artificial Intelligence (AAAI 96)*, pages 896–901, 1996.

[11] A. C. Schultz and W. Adams. Continuous localization using evidence grids. Technical Report AIC-96-007, Naval Center for Applied Research in Artificial Intelligence, 1996.

[12] Konolige, K. Small Vision Systems: Hardware and Implementation. *Eighth International Symposium on Robotics Research*, Hayama, Japan (October 1997).

[13] Sawhney, H.S. and R. Kumar. True Multi-Image Alignment and its Application to Mosaicing and Lens Distortion Correction. *IEEE Transactions on Pattern Analysis and Machine Intelligence*, Vol 21, No. 3. (March, 1999).

[14] J. Alon and S. Sclaroff. Recursive Estimation of Motion and Planar Structure. *Proc. IEEE Conf. on Computer Vision and Pattern Recognition*, (2000).

[15] P. H. S. Torr, A. W. Fitzgibbon and A. Zisserman. The problem of degeneracy in structure and motion recovery from uncalibrated images. *IJCV* (2000).

[16] R. Szeliski and P. Torr. Geometrically constrained structure from motion: Points on planes. In European Workshop on 3D Structure from Multiple Images of Large Scale Environments, pages 171-186, Frieburg, Germany, June 1998.

[17] R. Duda, P. Hart. Use of the Hough Transformation to detect lines and curves in pictures. *Communications of ACM*, 15(1), 1972.

[18] L. Iocchi, D. Nardi. Hough Transform based localization for mobile robots. In N. Mastorakis (Ed.), Advances in Intelligent Systems and Computer Science, World Scientific Engineering Society, 1999.

Incorporation of Delayed Decision Making into Stochastic Mapping

John J. Leonard and Richard J. Rikoski
MIT Dept. of Ocean Engineering
Cambridge, MA 02139, U.S.A
jleonard@mit.edu and rikoski@mit.edu

Abstract: This paper presents a technique for incorporating delayed decision making into stochastic mapping algorithms for concurrent mapping and localization. The approach explicitly tracks the error correlations between current and previous vehicle states, enabling the initialization of map features using data from multiple time steps and improved data association decision-making. The method is illustrated using data from a ring of Polaroid sonar sensors from a B21 mobile robot, demonstrating the ability to perform CML with sparse and ambiguous data.

1. Introduction

The problem of concurrent mapping and localization (CML) for an autonomous mobile robot is stated as follows: starting from a initial position, a mobile robot travels through a sequence of positions and obtains a set of sensor measurements at each position. The goal is for the mobile robot to process the sensor data to produce an estimate of its position while concurrently building a map of the environment. This paper presents a new technique for performing CML in situations where individual measurements provide weak geometric constraints, making it necessary to initialize new map features using data from multiple vantage points. We apply the method in experiments using a B21 mobile robot and demonstrate accurate mapping of a simple geometric environment using data from a ring of Polaroid sonar sensors.

CML has been a central research topic in the robotics community, due to its theoretical challenges and critical importance for many mobile robot applications [1, 2, 3, 4]. In just the past few years, the research community has made tremendous strides towards the solution of this problem [5, 6, 7, 8], however there are still several critical open issues for future research. The current state-of-the-art in feature-based approaches to CML is characterized by nearest-neighbor techniques for data association and use of the extended Kalman filter (EKF) for state estimation [9, 3, 8, 10, 11]. From the titles of two seminal papers on CML by Smith, Self, and Cheeseman [9] and Moutarlier and Chatila [3], we refer to this class of feature-based methods for CML as "stochastic mapping". Stochastic mapping considers CML as a variable-dimension state estimation problem in which the size of the state space is increased or decreased as features are added or removed from the map. As the robot moves through its environment, it uses new sensor measurements to perform two basic operations: (1) adding new features to its state vector, and (2) updating concurrently its estimate of its own state and the locations of previously observed features in the

D. Rus and S. Singh (Eds.): Experimental Robotics VII, LNCIS 271, pp. 533–542, 2001.

environment.

Stochastic mapping differs from other applications of the EKF because the size of the state vector is modified adaptively as new features are added and removed from the map. Initially, no feature locations are known and the state vector is restricted to contain only the initial state of the robot.

Previously published work in stochastic mapping has effectively made two assumptions: (1) there is sufficient information in the set of measurements available from a single robot position to completely and consistently initialize a new feature into the map [9, 3, 8, 10, 11] and (2) a nearest-neighbor gating strategy is sufficient for determination of the correspondence between measurements and features. These two assumptions have limited the robustness of current CML implementations and have restricted the range of environments in which they can be applied.

In this paper, we describe a new extension to the state-of-the-art in CML by presenting a technique for delayed decision making that allows us to perform CML with sparse and noisy sonar data. The key innovations of the current paper are to add past vehicle positions to the state vector and to maintain explicitly estimates of the correlations between current and previous vehicle states. By incorporating past vehicle locations in the state vector, we are able to make improved data association and feature classification decisions and to initialize new map features by consistently combining data from multiple vantage points.

The motivation for the new approach is the following: if the sensor observations available from a single time step do not provide sufficient information to initialize the state estimate of a newly detected feature, then information from multiple vehicle positions must used. To maintain consistent error bounds, correlations between different vehicle locations must be taken into account by the CML algorithm. Further, decisions that are difficult based on the data from a single position (such as the disposition of an individual sonar return) can be made much easier when considered as delayed decisions, using data from multiple vehicle positions.

The structure of this paper is as follows: Section 2 reviews the current state-of-the-art for this type of algorithm, the following section describes our new approach in detail, and in Section 4 experimental results are presented.

2. Stochastic Mapping

Stochastic mapping is a feature-based concurrent mapping and localization algorithm that was first published by Smith, Self, and Cheeseman [9] and Moutarlier and Chatila [3]. The method assumes that there are n features in the environment, and that they are static. The true state at time k is designated by $\mathbf{x}[k] = [\mathbf{x}_r[k]^T \ \mathbf{x}_f[k]^T]^T$, where $\mathbf{x}_r[k]$ represents the location of the robot, and $\mathbf{x}_f[k]^T = [\mathbf{x}_{f_1}[k]^T \ \dots \ \mathbf{x}_{f_n}[k]^T]^T$ represent the locations of the environmental features. Let $\mathbf{z}[k]$ designate the sensor measurements obtained at time k, and Z^k designate the set of all measurements obtained from time 0 through time k.

Stochastic mapping algorithms for CML use the extended Kalman filter to compute recursively a state estimate $\hat{\mathbf{x}}[k|k] = [\hat{\mathbf{x}}_r[k|k]^T \ \hat{\mathbf{x}}_f[k]^T]^T$ at each discrete time step k, where $\hat{\mathbf{x}}_r[k|k]^T$ and $\hat{\mathbf{x}}_f[k]^T = [\hat{\mathbf{x}}_{f_1}[k]^T \ \dots \ \hat{\mathbf{x}}_{f_n}[k]^T]^T$ are the robot and feature state estimates, respectively. Based on assumptions about linearization and data association, this estimate is the approximate conditional mean of $p(\mathbf{x}[k]|Z^k)$:

$$\hat{\mathbf{x}}[k|k] \approx E[p(\mathbf{x}[k]|Z^k)] \tag{1}$$

Associated with this state vector is an estimated error covariance, $\mathbf{P}[k|k]$, which represents the errors in the robot and feature locations, and the cross-correlations between these states:

$$\mathbf{P}[k|k] = \begin{bmatrix} \mathbf{P}_{rr}[k|k] & \mathbf{P}_{rf}[k|k] \\ \mathbf{P}_{fr}[k|k] & \mathbf{P}_{ff}[k|k] \end{bmatrix} = \begin{bmatrix} \mathbf{P}_{rr}[k|k] & \mathbf{P}_{rf_1}[k|k] & \cdots & \mathbf{P}_{rf_n}[k|k] \\ \mathbf{P}_{f_1 r}[k|k] & \mathbf{P}_{f_1 f_1}[k|k] & \cdots & \mathbf{P}_{f_1 n}[k|k] \\ \vdots & \vdots & \ddots & \vdots \\ \mathbf{P}_{f_n r}[k|k] & \mathbf{P}_{f_n f_1}[k|k] & \cdots & \mathbf{P}_{f_n f_n}[k|k] \end{bmatrix}. \tag{2}$$

The method uses two models, a plant model and a measurement model. The plant model is used to make predictions of future vehicle positions based on a control input. In a two-dimensional implementation, a typical state model for the robot might be: $\mathbf{x}_r = [x_r \ y_r \ \phi \ v]^T$, representing the vehicle's east position, north position, heading, and speed, respectively. The simplest type of features are points, with feature f_i represented by $\mathbf{x}_{f_i} = [x_i \ y_i]^T$. The dynamic model for the motion of the robot is given by

$$\mathbf{x}[k+1] = \mathbf{f}(\mathbf{x}[k], \mathbf{u}[k]) + \mathbf{d}_\mathbf{x}(\mathbf{u}[k]), \tag{3}$$

where $\mathbf{f}(\cdot)$ is the plant model, $\mathbf{d}_\mathbf{x}(\mathbf{u}[k])$ is a white, Gaussian random process independent of $\mathbf{x}[0]$, with magnitude dependent on the control input $\mathbf{u}[k]$.

The measurement model $\mathbf{h}(\cdot)$ for the system is given by

$$\mathbf{z}[k] = \mathbf{h}(\mathbf{x}[k]) + \mathbf{d}_\mathbf{z}, \tag{4}$$

where $\mathbf{z}[k]$ is the vector of sensor measurements (e.g., range and bearing when using sonar). The observation model, $\mathbf{h}(\cdot)$, defines the nonlinear coordinate transformation from state to observation coordinates. The stochastic process $\mathbf{d}_\mathbf{z}$, is assumed to be white, Gaussian, and independent of $\hat{\mathbf{x}}[0]$ and $\mathbf{d}_\mathbf{x}$, and has covariance $\mathbf{R}$. Given these assumptions, an extended Kalman filter (EKF) is employed to estimate the state $\hat{\mathbf{x}}$ and covariance $\mathbf{P}$ given the measurements.

Data association is the process of determining the origin of sensor measurements. A decision must be made for each new measurement to determine if (1) it originates from one of the features currently in the map, (2) it originates from a new feature, or (3) it is spurious. In general, the data association problem is exponentially complex [12], and no general solution that can run in real-time has been published. Most published implementations of CML have used variations of "nearest-neighbor" gating techniques. For each feature in the vehicle state vector, predicted range and angle measurements are generated and are compared against the actual measurements using a weighted statistical distance in measurement space. For all measurements $\mathbf{z}_j[k]$ that can potentially be associated with feature $\hat{\mathbf{x}}_{f_i}[k]$, the innovation, $\boldsymbol{\nu}_{ij}[k]$, and the innovation covariance, $\mathbf{S}_{ij}[k]$, are constructed and the closest measurement within the gate defined by the Mahalanobis distance

$$\boldsymbol{\nu}_{ij}[k]^T \mathbf{S}_{ij}[k]^{-1} \boldsymbol{\nu}_{ij}[k] \leq \gamma, \tag{5}$$

is considered the most likely measurement of that feature [12].

Measurements that do not gate with any existing feature become candidates for the initialization of new features. Previously published methods have assumed that the state of the new feature, $\hat{\mathbf{x}}_{f_{n+1}}[k]$ can be computed using the measurement data available from a single vehicle position, using a feature initialization function $\mathbf{g}(\cdot)$:

$$\hat{\mathbf{x}}_{f_{n+1}}[k] = \mathbf{g}(\hat{\mathbf{x}}[k|k], \mathbf{z}_j[k]). \tag{6}$$

For example, for a sensor providing range and bearing measurements, $\mathbf{z}_j[k] = [r\ \theta]$, the feature initialization function for a point $\mathbf{g}(\cdot)$ takes the following form:

$$\hat{\mathbf{x}}_{f_{n+1}}[k] = \mathbf{g}(\hat{\mathbf{x}}[k|k], \mathbf{z}_j[k]) = \begin{bmatrix} x_r + r\cos(\phi + \theta) \\ y_r + r\sin(\phi + \theta) \end{bmatrix}. \tag{7}$$

The new feature is integrated into the map by expanding the state vector $\hat{\mathbf{x}}[k|k]$ and covariance $\mathbf{P}[k|k]$ as shown below:

$$\hat{\mathbf{x}}[k|k] \leftarrow \begin{bmatrix} \hat{\mathbf{x}}[k|k] \\ \hat{\mathbf{x}}_{f_{n+1}}[k] \end{bmatrix}, \tag{8}$$

$$\mathbf{P}[k|k] \leftarrow \begin{bmatrix} \mathbf{P}_{rr}[k|k] & \mathbf{P}_{rf}[k|k] & \mathbf{P}_{rf_{n+1}}[k|k] \\ \mathbf{P}_{fr}[k|k] & \mathbf{P}_{ff}[k|k] & \mathbf{P}_{ff_{n+1}}[k|k] \\ \mathbf{P}_{f_{n+1}r}[k|k] & \mathbf{P}_{f_{n+1}f}[k|k] & \mathbf{P}_{f_{n+1}f_{n+1}}[k|k] \end{bmatrix}, \tag{9}$$

where

$$\mathbf{P}_{f_{n+1}f_{n+1}}[k|k] = \mathbf{G_x}\mathbf{P}[k|k]\mathbf{G_x}^T + \mathbf{G_z}\mathbf{R}[k]\mathbf{G_z}^T, \tag{10}$$

$$\begin{bmatrix} \mathbf{P}_{f_{n+1}r}[k|k] & \mathbf{P}_{f_{n+1}f}[k|k] \end{bmatrix} = \begin{bmatrix} \mathbf{P}_{f_{n+1}r}[k|k] \\ \mathbf{P}_{f_{n+1}f}[k|k] \end{bmatrix}^T = \mathbf{G_x}\mathbf{P}[k|k], \tag{11}$$

$\mathbf{G_x}$ is the Jacobian of $\mathbf{g}$ with respect to the state vector and $\mathbf{G_z}$ is the Jacobian of $\mathbf{g}$ with respect to the measurement.

3. Incorporation of Delayed Decision Making

The key idea of our new approach is to expand the representation to add a number of previous vehicle locations to the state vector. We refer to these states as trajectory states. Each time the vehicle moves, the previous vehicle location is added to the state vector. We introduce the notation $\hat{\mathbf{x}}_{t_i}[k]$ to refer to the estimate of the state (position) of the robot at time i given all information up to time k. The complete trajectory of the robot for time step 0 through time step $k - 1$ is given by the vector

$\hat{\mathbf{x}}_t[k] = [\hat{\mathbf{x}}_{t_0}[k]^T \; \hat{\mathbf{x}}_{t_1}[k]^T \; \hat{\mathbf{x}}_{t_2}[k]^T \; \ldots \; \hat{\mathbf{x}}_{t_{k-1}}[k]]^T$. The complete state vector is:

$$\hat{\mathbf{x}}[k|k] = \begin{bmatrix} \hat{\mathbf{x}}_r[k|k] \\ \hat{\mathbf{x}}_t[k] \\ \hat{\mathbf{x}}_f[k] \end{bmatrix} = \begin{bmatrix} \hat{\mathbf{x}}_r[k|k] \\ \hat{\mathbf{x}}_{t_0}[k] \\ \hat{\mathbf{x}}_{t_1}[k] \\ \hat{\mathbf{x}}_{t_2}[k] \\ \vdots \\ \hat{\mathbf{x}}_{t_{k-1}}[k] \\ \hat{\mathbf{x}}_{f_1}[k] \\ \hat{\mathbf{x}}_{f_2}[k] \\ \hat{\mathbf{x}}_{f_3}[k] \\ \vdots \\ \hat{\mathbf{x}}_{f_{n-1}}[k] \\ \hat{\mathbf{x}}_{f_n}[k] \end{bmatrix}. \tag{12}$$

The associated covariance matrix is:

$$\mathbf{P}[k|k] = \begin{bmatrix} \mathbf{P}_{rr}[k|k] & \mathbf{P}_{rt}[k|k] & \mathbf{P}_{rf}[k|k] \\ \mathbf{P}_{tr}[k|k] & \mathbf{P}_{tt}[k|k] & \mathbf{P}_{tf}[k|k] \\ \mathbf{P}_{fr}[k|k] & \mathbf{P}_{ft}[k|k] & \mathbf{P}_{ff}[k|k] \end{bmatrix}, \tag{13}$$

or equivalently,

$$\mathbf{P}[k|k] = \begin{bmatrix} \mathbf{P}_{rr}[k|k] & \mathbf{P}_{rt_0}[k|k] & \cdots & \mathbf{P}_{rt_{k-1}}[k|k] & \mathbf{P}_{rf_1}[k|k] & \cdots & \mathbf{P}_{rf_n}[k|k] \\ \mathbf{P}_{t_0 r}[k|k] & \mathbf{P}_{t_0 t_0}[k|k] & \cdots & \mathbf{P}_{t_0 t_{k-1}}[k|k] & \mathbf{P}_{t_0 f_1}[k|k] & \cdots & \mathbf{P}_{t_0 f_n}[k|k] \\ \vdots & \vdots & \ddots & \vdots & \vdots & \ddots & \vdots \\ \mathbf{P}_{t_{k-1} r}[k|k] & \mathbf{P}_{t_{k-1} t_0}[k|k] & \cdots & \mathbf{P}_{t_{k-1} t_{k-1}}[k|k] & \mathbf{P}_{t_{k-1} f_1}[k|k] & \cdots & \mathbf{P}_{t_{k-1} f_n}[k|k] \\ \mathbf{P}_{f_1 r}[k|k] & \mathbf{P}_{f_1 t_0}[k|k] & \cdots & \mathbf{P}_{f_1 t_{k-1}}[k|k] & \mathbf{P}_{f_1 f_1}[k|k] & \cdots & \mathbf{P}_{f_1 f_n}[k|k] \\ \vdots & \vdots & \ddots & \vdots & \vdots & \ddots & \vdots \\ \mathbf{P}_{f_n r}[k|k] & \mathbf{P}_{f_n t_0}[k|k] & \cdots & \mathbf{P}_{f_n t_{k-1}}[k|k] & \mathbf{P}_{f_n f_1}[k|k] & \cdots & \mathbf{P}_{f_n f_n}[k|k] \end{bmatrix}. \tag{14}$$

New trajectory states are added to the state vector each time step by defining a new trajectory state $\hat{\mathbf{x}}_{t_k}[k] = \hat{\mathbf{x}}_r[k|k]$ and adding this to the state vector:

$$\hat{\mathbf{x}}[k|k] \leftarrow \begin{bmatrix} \hat{\mathbf{x}}_r[k|k] \\ \hat{\mathbf{x}}_{t_0}[k] \\ \hat{\mathbf{x}}_{t_1}[k] \\ \hat{\mathbf{x}}_{t_2}[k] \\ \vdots \\ \hat{\mathbf{x}}_{t_{k-1}}[k] \\ \hat{\mathbf{x}}_{t_k}[k] \\ \hat{\mathbf{x}}_f[k] \end{bmatrix}. \tag{15}$$

The state covariance is expanded as follows:

$$
\mathbf{P}[k|k] \leftarrow
\begin{bmatrix}
\mathbf{P}_{rr}[k|k] & \mathbf{P}_{rt_0}[k|k] & \cdots & \mathbf{P}_{rt_{k-1}}[k|k] & \mathbf{P}_{rt_k}[k|k] & \mathbf{P}_{rf}[k|k] \\
\mathbf{P}_{t_0 r}[k|k] & \mathbf{P}_{t_0 t_0}[k|k] & \cdots & \mathbf{P}_{t_0 t_{k-1}}[k|k] & \mathbf{P}_{t_0 t_k}[k|k] & \mathbf{P}_{t_0 f}[k|k] \\
\vdots & \vdots & \ddots & \vdots & \vdots & \vdots \\
\mathbf{P}_{t_{k-1} r}[k|k] & \mathbf{P}_{t_{k-1} t_0}[k|k] & \cdots & \mathbf{P}_{t_{k-1} t_{k-1}}[k|k] & \mathbf{P}_{t_{k-1} t_k}[k|k] & \mathbf{P}_{t_{k-1} f}[k|k] \\
\mathbf{P}_{t_k r}[k|k] & \mathbf{P}_{t_k t_0}[k|k] & \cdots & \mathbf{P}_{t_k t_{k-1}}[k|k] & \mathbf{P}_{t_k t_k}[k|k] & \mathbf{P}_{t_k f}[k|k] \\
\mathbf{P}_{fr}[k|k] & \mathbf{P}_{ft_0}[k|k] & \cdots & \mathbf{P}_{ft_{k-1}}[k|k] & \mathbf{P}_{ft_k}[k|k] & \mathbf{P}_{ff}[k|k]
\end{bmatrix},
\tag{16}
$$

where $\mathbf{P}_{t_k t_i}[k|k] = \mathbf{P}_{rt_i}[k|k]$, $\mathbf{P}_{t_k f}[k|k] = \mathbf{P}_{rf}[k|k]$, and $\mathbf{P}_{t_k t_k}[k|k] = \mathbf{P}_{rr}[k|k]$. The growth of the state vector in this manner increases the computational burden, however it is straightforward to delete old vehicle trajectory states and associated terms in the covariance, once all the measurements from a given time step have been either processed or discarded.

This process of adding past states is similar to a fixed-lag Kalman smoother [13]. In a fixed-lag smoother, states exceeding a certain age are automatically removed. In our approach, states are added and removed based on the data processing requirements of the stochastic mapping process. Unlike the fixed-lag smoother, states are not necessarily removed in the order in which they are added.

With the addition of prior vehicle states to the state vector, it now becomes possible to initialize new features using measurements from multiple time steps. For example, consider the initialization of a new feature using two measurements, $\mathbf{z}[k_1]$ and $\mathbf{z}[k_2]$, taken at time steps k_1 and k_2. The state of the new feature can be computed using a feature initialization function involving data from multiple time steps:

$$
\hat{\mathbf{x}}_{f_{n+1}} = g(\hat{\mathbf{x}}_{t_{k_1}}[k], \hat{\mathbf{x}}_{t_{k_2}}[k], [\mathbf{z}[k_1]^T \; \mathbf{z}[k_2]^T]^T).
\tag{17}
$$

For example, in two-dimensions if each measurement is a range-only sonar measurement, then the function $\mathbf{g}(\cdot)$ represents a solution for the intersection of two circles. The covariance for the new feature is initialized in a similar fashion as shown above in Equations 9 to 11, except that the Jacobian matrix $\mathbf{G_x}$ will contain additional non-zero terms corresponding to the trajectory states and the Jacobian matrix $\mathbf{G_z}$. The procedure is the same if the feature initialization function $\mathbf{g}(\cdot)$ is a function of measurements from more than two time steps. New feature initialization can also be performed using non-linear least squares [14] performed on many measurements, instead of using an explicit function $\mathbf{g}(\cdot)$.

To provide improved stability, the addition of new features to the state vector can be delayed to occur only when the initializing Jacobians indicate that the new feature estimate is well-conditioned. By examining the different possible initialization sets and choosing the Jacobian with the smallest values, the most stable initialization can be determined. In addition, one can incorporate an adaptive motion control step to direct the robot to move to a better vantage point that will yield a more stable initialization. By considering second-order derivatives, the robot can determine the optimal direction to move in order to obtain data that will yield the most stable initialization of a new feature.

Once a new feature is initialized, the map can be updated using all other previously obtained measurements that can be associated with the new feature. We call this procedure a "batch update". It allows the maximum amount of information to be extracted from all past measurements.

The ability to perform a batch update using many previous measurements provides a facility for making delayed data association decisions. If there is ambiguity about the correspondence between measurements and features, decisions can be postponed until additional information becomes available. Feature extraction is also simplified. The initialization of complex features in situations with high ambiguity can be greatly simplified by considering a batch of data obtained at multiple time steps.

4. Experimental Results

To demonstrate the approach, we present the application of the method to ring sonar data from a B21 mobile robot, shown in Figure 1. Figures 2 to 4 show the results of an experiment performed in a simple 6-foot by 8-foot room made of plywood. The data association and feature modeling techniques used in this experiment utilized a priori knowledge of the structure of the box, namely that each corner of the box was created by two walls, and that each wall was bounded at each end by a corner. More generic and powerful "anytime" data association techniques are in development and should be applicable to more complex scenarios. These results are reported here only as a means of illustrating how delayed decision making and batch reprocessing can be used for CML in a situation where individual measurements are sparse and provide weak constraints.

The data processing for this experiment proceeded as follows. First, the vehicle dead-reckoned around the room for 50 timesteps, collecting the data shown in Figure 2(a), but not initializing any features. At each time step, a new vehicle trajectory state was added to the system state vector, using Equations 15 and 16. The dead-reckoned vehicle trajectory is shown in Figure 2(b). After fifty time steps, a search was performed on all the sonar returns acquired to find a large subset of returns originating from a single point in the room. Because of the structure of the box, this point can safely be assumed to be one of the corners of the box, since they are visible from the largest number of sensing locations. From this subset of returns, the pair of returns obtained that provided the most stable initialization of a new point feature was selected, and used to initialize a new point feature. This feature corresponded to the upper left corner of the box, and serves as the "seed" feature for reconstruction of the box.

Next, a search was performed to find sonar returns that were tangent to lines drawn from the initial corner, corresponding to the two walls that comprise the corner. The initialization for each wall was based on a single measurement and the constraint that the wall passes through the corner. Subsequently, searches were performed to find a sonar returns consistent with new point features that lay close to the new walls. The chosen sonar returns were used, together with the state estimate for the respective walls, to initialize new point features corresponding to the upper right corner and lower left corner of the box. This process was continued around the box until the bottom right corner was initialized without any measurements as the intersection of two walls. An additional point feature, a crack on one of the walls, was discovered and mapped from a single measurement as a point constrained to lie on the wall. After the

Figure 1: B21 mobile robot in the plywood box.

initializations, nine constrained features (shown in Figure 3(a)) were mapped using nine range measurements (shown in Figure 3(b)).

Once these features were initialized, nearest-neighbor gating was performed between all of the remaining sonar measurements and the newly initialized map features. All the measurements that were uniquely matched to one of the nine features are shown in Figure 4(a). Finally, Figure 4(b) shows the result when all these measurements are applied in a single batch update, resulting in a dramatic reduction in the uncertainty ellipses for the estimated feature locations and in the complete trajectory of the vehicle.

5. Conclusion

This paper has described a technique for incorporating delayed decision making into stochastic mapping algorithms for CML. The approach enables the initialization of features using data from multiple time steps and improved data association decision-making. The method has been applied using data from a ring of Polaroid sonar sensors from a B21 robot, demonstrating the ability to perform CML with sparse and ambiguous data. The experiment shown above is quite simple, and utilized a priori knowledge of the environmental structure to solve the data association problem. However, the experiment provides one illustration of the benefits of adding past vehicle positions to the state vector, enabling stochastic mapping to be performed in situations where the state of a feature can only by partially observed from a single vehicle position. This approach should allow the development of robust CML implementations that use more complex objects as map features.

Work in progress is applying the technique in combination with decoupled stochastic mapping [10] to larger-scale and more complex experiments using both land and underwater robots.

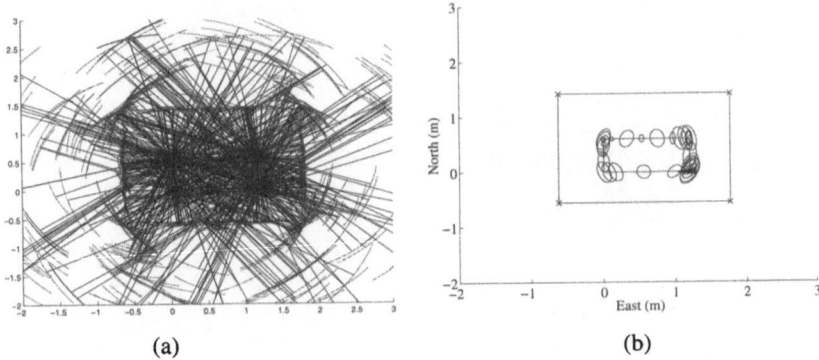

(a)   (b)

Figure 2: (a) Set of all measurements acquired over 50 time steps. Each sonar return is shown as a circular arc, with rays drawn from the center of the dead-reckoned robot position to the center of each arc. (b) Dead-reckoned vehicle trajectory, with 3-σ error ellipses.

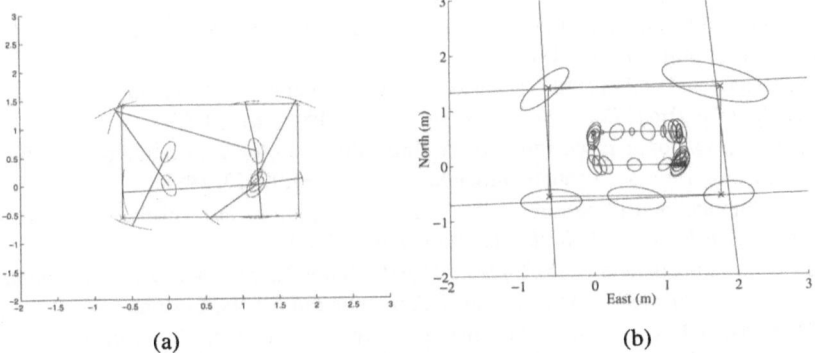

(a)   (b)

Figure 3: (a) Nine measurements used to initialize nine new features, starting with the corner in the upper left of the figure, and building in both directions around the room, closing the box in the lower right hand corner. (b) State estimates and 3-σ error ellipses for the nine initialized features.

Acknowledgements

This research has been funded in part by the Henry L. and Grace Doherty Assistant Professorship in Ocean Utilization, NSF Career Award BES-9733040, the MIT Sea Grant College Program under grant NA86RG0074 (project RCM-3), and the US Navy International Programs Office.

References

[1] R. Chatila and J.P. Laumond. Position referencing and consistent world modeling for mobile robots. In *IEEE International Conference on Robotics and Automation*, pages 138–145, 1985.

[2] W. K. Stewart. *Multisensor Modeling Underwater with Uncertain Information.* PhD thesis, Massachusetts Institute of Technology, 1988.

[3] P. Moutarlier and R. Chatila. An experimental system for incremental environment mod-

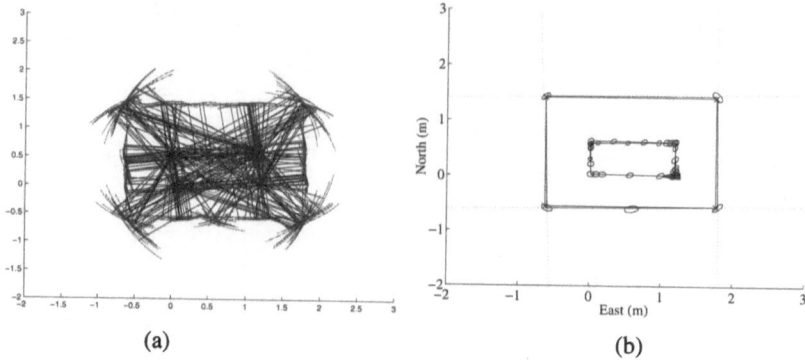

(a)   (b)

Figure 4: (a) Sonar measurements that uniquely gated with the nine initialized features, to be used in the batch update. (b) Feature location estimates, vehicle trajectory, and error ellipses after the batch update.

eling by an autonomous mobile robot. In *1st International Symposium on Experimental Robotics*, Montreal, June 1989.

[4] N. Ayache and O. Faugeras. Maintaining representations of the environment of a mobile robot. *IEEE Trans. Robotics and Automation*, 5(6):804–819, 1989.

[5] S. Thrun, D. Fox, and W. Burgard. A probabilistic approach to concurrent mapping and localization for mobile robots. *Machine Learning*, 31:29–53, 1998.

[6] J-S. Gutmann and K. Konolige. Incremental mapping of large cyclic environments. In *Proc. IEEE Int. Conf. Robotics and Automation*, 2000.

[7] J. A. Castellanos and J. D. Tardos. *Mobile Robot Localization and Map Building: A Multisensor Fusion Approach*. Kluwer Academic Publishers, Boston, 2000.

[8] M. W. M. G. Dissanayake, P. Newman, H. F. Durrant-Whyte, S. Clark, and M. Csorba. An experimental and theoretical investigation into simultaneous localization and map building. In *Sixth International Symposium on Experimental Robotics*, pages 265–274, March 1999.

[9] R. Smith, M. Self, and P. Cheeseman. Estimating uncertain spatial relationships in robotics. In I. Cox and G. Wilfong, editors, *Autonomous Robot Vehicles*, pages 167–193. Springer-Verlag, 1990.

[10] J. J. Leonard and H. J. S. Feder. A computationally efficient method for large-scale concurrent mapping and localization. In D Koditschek and J. Hollerbach, editors, *Robotics Research: The Ninth International Symposium*, pages 169–176, Snowbird, Utah, 2000. Springer Verlag.

[11] J. A. Castellanos, J. M. M. Montiel, J. Neira, and J. D. Tardos. The SPmap: A probabilistic framework for simultaneous localization and map building. *IEEE Trans. Robotics and Automation*, 15(5):948–952, 1999.

[12] Y. Bar-Shalom and T. E. Fortmann. *Tracking and Data Association*. Academic Press, 1988.

[13] B. D. O. Anderson and J. B. Moore. *Optimal filtering*. Englewood Cliffs, N.J.: Prentice-Hall, 1979.

[14] O. Faugeras. *Three-Dimensional Computer Vision: A Geometric Viewpoint*. MIT Press, 1993.

Tele-Autonomous Watercraft Navigation

Ray Jarvis
Intelligent Robotics Research Centre
Monash University
PO Box 35
VICTORIA 3800
Australia
Ray.Jarvis@eng.monash.edu.au

Abstract: This paper concerns the instrumenting of a small watercraft to support a hybrid navigation strategy which combines remote human supervisory guidance with reaction based obstacle avoidance. This style of control is called 'tele-autonomous'. Potential applications include search and rescue operations, coastal surveillance, water pollution source tracing and surface support for a submersible. Details are provided, the concept promoted and future plans sketched.

1. Introduction

It is now relatively easy to develop fully autonomous mobile robots for operations in well-structured factory environments [1, 2]. Tasks such as delivering components between workstations or carrying out surveillance duties can be accomplished autonomously and reliably in such domains.

In less structured environments such as in a home or in rugged outdoor situations [3, 4] a greater degree of intelligence based on sensor data acquisition and interpretation is required. The provision of autonomous system support becomes more expensive and, generally, reliability is poor.

Where the motivation has less to do with the reduction of human resources and more to do with safety and convenience, teleoperation, particularly in sensor-rich modes, has a lot to offer. Permitting humans to control complex navigation and manipulative tasks at a remote, hazardous site (as in mining, space exploration, undersea operations, in mine fields and nuclear plants) whilst, they, themselves, are in a safe and comfortable control centre offers many practical advantages.

Furthermore, such solutions can be delivered with reduced levels of artificial intelligence, since human judgement can be as fully engaged as the sensor feedback quality permits. With appropriate sensor feedback and sensitive and responsive feed-forward control, it is possible to extend 'teleoperation' into 'tele-existence' where the operator has the sensation of being at the remote site but without the danger or discomfort. Some forms of virtual reality can extend this type of activity towards creating computer-fabricated worlds within

D. Rus and S. Singh (Eds.): Experimental Robotics VII, LNCIS 271, pp. 543–550, 2001.

which people and machines interact to ultimately complete physical tasks with human intelligence and machine capabilities nicely matched.

This paper is concerned with capturing the essence of both autonomous functionality and sensor-rich teleoperation in a hybrid structure which flexibly provides a variable mix of automation and human intervention, in this case, with respect to watercraft navigation. The term 'tele-autonomous' has been invented to represent this hybrid approach. This mode is a type of supervisory control but where the degree of human involvement can range from peripheral to intense, depending on the mission goals and the time-varying environmental circumstances.

Possible applications include tracking the sources of pollution, search and rescue operations, surface support for submersibles and recreational activities.

We have chosen a recreational craft, a 'water-bike', as the platform for tele-autonomous experiments (See Figure 1). This craft is small (2 metres in length) and yet is capable of carrying a payload sufficient for appropriate instrumentation and power sources. The 'control centre' for tele-autonomous operations can be shore based (See Figure 2) or on-board another vessel.

Figure 1. Instrumented 'Water Bike'

2. Essential Requirements

The essential requirements for safe watercraft navigation are as follows:

1. The location of the craft should be known. This can be accomplished using Global Positioning Systems (GPS). The recent turning off of the 'selective availability' clock error by U.S.A. defence forces makes even stand-alone systems sufficiently accurate for these purposes, though both differential and phase modes can improve this considerably.

2. A naval map of the area, particularly in electronic form, allows path planing to be carried out.

3. Knowledge of surrounding vessels or hazards not shown in the naval maps

Figure 2. Shore Based Remote Control Centre (Home Base)

is important. This information can be provided by radar but only within accuracies of ± 25 metres up to tens of kilometres. Being able to relate the radar data to the map details is very important.

4. Steering and speed controls are needed. A number of steering servos systems which allow set bearings (supported by a flux gate compass) to be followed are readily available on the market. Accelerator control is not usually provided but can easily be achieved.

3. Auxiliary Instrumentation

The following extra instrumentation is used for our experiments:

1. A laser time-of-flight scanning rangefinder capable of measuring a one dimensional sweep of range data at 0.5 degree intervals over 180 degrees up to 50 metres with ± 3cm accuracy.

2. An optical gyroscope with relatively low drift to help with steering control.

3. Night vision video camera for low light operations.

4. Stereo colour cameras for stereo viewing by the operator.

5. A video cross bar switch to select various video sources for viewing from the control station.

6. Video transmitter(s).

7. Pitch/roll/bearing sensor.

8. Radio ethernet communications.

4. Implementation Details

Figure 3 shows the block schematic for the whole system. Two distinct communications subsystems are shown, one based on video transmitters, the other on radio ethernet, the latter with a return path from the control centre to the watercraft for steering and speed control.

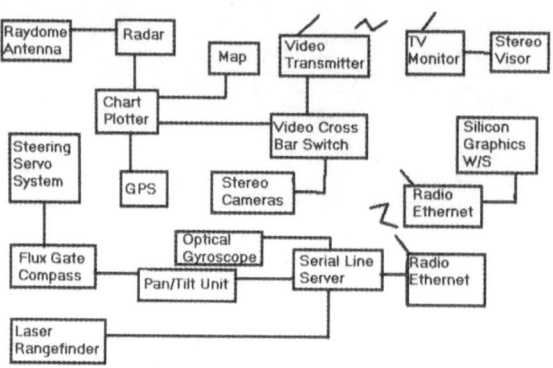

Figure 3. Tele-Autonomous System Block Schematic

A Raytheon RL 70CRC chart plotter system displays both GPS (Global Positioning System) derived position fixes on an electronic chart and radar scan data from a Raytheon system. In split screen mode both the mapping and radar data can be viewed simultaneously to scale and 'synchronised' in central location and orientation. A single colour video camera can provide both sets of data in a single image via a quality video radio transmitter to a remote (shore or boat based) home station.

A pair of stereo colour cameras (See 4) are genlocked so as to provide field sequential video through multiplexing electronics. A single video stream is radio transmitted from the camera pairs to a home base head mounted stereo visor which. demultiplexes the signals to the left and right screens. Alternatively, electronically switched glasses can be used to view the camera screen in stereo imagery on a standard TV receiver screen. It is planned to provide a night vision (0 lux) camera with its own infra-red illuminator for night operations. Switching video streams between the mapping/radar camera, the stereo camera pair and the night vision camera can be achieved using a video cross-bar switch is controlled via standard RS232 serial line to the serial line server (shown in Figure 3). Up to 8 input video channels can be handled so a number of other cameras in strategically useful positions can be easily added - including cameras which may be on a tethered submersible, for which the 'up top' vessel is the surface support base.

An Erwin Sick scanning rangefinder is used to detect relatively close obstacles floating in the water in front of the vessel. The rangefinder shown in Figure 4 is fitted with a mask to block out specular reflections from the water, which can frustrate the instrument. This rangefinder can provide ±3cm accuracy range up to 50 metres away at 1/2 degree intervals over a horizontal

Figure 4. Stereo Cameras

sweep of 180 degrees. The range data is channelled at 9600 baud via a RS232 serial link to the line server.

Figure 5. Erwin Sick rangefinder with Anti-specular Reflection 'Mask'

An optical gyroscope and a pitch/tilt sensor are also linked to the serial line server. The optical gyroscope can be used to aid steering control whilst the pitch/tilt sensor can indicate the altitude and rocking movements of the vessel to complement the interpretations made via the stereo cameras.

All of the instrumentation data scan provided via the serial line server (range, pitch/tilt, optical gyroscope) are transmitted via radio ethernet to a remote Silicon Graphics workstation at the home base in graphic from, allowing for easy interpretation by the operator. Control signals from the home base to the vessel are also transmitted via radio ethernet to the serial line server and from there to where they are acted upon via RS232 serial links, the serial line server being a multiple full duplex device.

An Raytheon sportspilot steering servo system directly operates the steering wheel, which in turn swivels the rear outboard petrol engine via cables.

The steering servo systems is designed to hold a set bearing by adjusting the steering wheel to maintain bearing data provided by a flux gate compass. The normal operation with a human pilot on board, is to point the vessel in the intended direction and then to engage the autopilot to maintain that bearing. If the pilot were to override this operation by taking hold of the steering wheel, the servo system simply takes over again to keep the new bearing once the wheel is released by the human pilot. Minor bearing adjustments can be made by pressing small buttons on the unit.

The way we have chosen to steer the craft involves a small amount of lateral thinking. Once the autopilot is engaged, steering bearing changes can be smoothly implemented by rotating the steering flux gate compass about a vertical axis in the opposite direction by an amount equal to the intended bearing change. The autopilot is 'tricked' into taking up a new bearing by trying to maintain what it takes to be original setting. A Directed Perception Inc. pan/tilt head is used for this purpose; it too is controlled via the serial line server by the operator at the home station. The pan axis is used to rotate the flux gate compass and the tilt axis to adjust the outboard motor accelerator via a light cable. Thus both steering and speed are controlled via this single pan/tilt head.

However, the pan axis does not permit a full ±180 degree rotation, there being a dead zone of ±30 degrees. If the vessel is required to make a round trip in straight line segments in a single loop, or complete several loops this could be a problem. A simple solution was found. It takes advantage of the possibility of 'taking over' the wheel to change the intended bearing. In our strategy the vessel is trying to maintain a pseudo single bearing but is being 'tricked' by rotating the compass. Stopping the wheel from turning whilst rotating the compass is like resetting the bearing. Very simply, a braking a solenoid is actuated when the pan angle is nearing the limit and the intention would be to move beyond this except for this limit. The solenoid activation is implemented using the tilt control in a position beyond the idling speed point of motor accelerator control to trip a microswitch. The solenoid pushes a brake pad against the steering wheel to hold it fixed, at which time the panning position of the pan/tilt head is moved back to where maximum rotation in either direction is possible. Then the wheel is released and the vessel is accelerated again in its original direction. Now, however, maximum steering range is re-established. Thus the maximum steering change potential can be reinstated when a limit is being approached and an effectively continuous bearing change accumulating beyond 360 degrees can be achieved. Absolute bearing data is provided on the chart display which uses a separate flux gate compass as a reference and the optical gyroscope can also be used to assist in making steering changes, although this should not be necessary with a bit of practice. Since the response to a bearing change command would be subject to the speed of the vessel, the gain of the servo system and other factors such as wind and currents, the optical gyroscope can indicate actual bearing changes over short terms, without the gyroscope drift being of significance. Whilst the map plus radar data can give a large scale view of navigational activity, the stereo camera view and the laser range finder scans provide localised range sensing which may prove valuable for manoeuvres close to obstacles.

An automatic obstacle avoidance mechanism can take over when the operator is perhaps occupied with other matters. Only this reactive autonomy has been considered so far; other aspects may be added later, depending on the task at hand. The Sick rangefinder can detect potential collisions with obstacle closer than 50 metres surrounding the vessel in the forward 180 degrees. The appropriate action is to slow the vessel, steer around obstacles and then to resume the original track. This strategy is currently being implemented. It is hoped that its success can be demonstrated when the paper is presented. Some aspects of the strategy relate to work with a semi-automatic wheelchair [5].

As indicated above, the tele-autonomous mode to be experimented with first is that of human intervention for global path planning with the support of mapping/GPS/radar data but to provide local obstacle avoidance autonomously by linking rangefinder data analysis with the steering and speed control once the overall plan is specified by the operator. The operator can intervene at will at any time. Thus, once a navigation mission has been planned and commenced, local obstacle avoidance can be carried out automatically and the operator will be alerted only when some situation too complex to handle autonomously arises.

At a later stage in our experiments it is intended that aspects such as route planning using map/GPS/radar data be automated once the operator has set the parameters of the overall mission. Once again, the operator will be free to take over at any time. Thus, parts of a mission may be almost completely carried out autonomously but other, more critical, stages can be under the direct and detailed control of the operator. Mixtures of automatic and human operator control can be used to suit particular missions and circumstances.

5. Conclusions

This paper has described preliminary work on a semi-autonomous watercraft, arguing the case for a mode of control called 'tele-autonomous', in which human high level supervision and low level activation modes are mixed according to the needs of the mission. Further work has yet to be done to demonstrate the full capability of such an approach and to properly gauge the best way in which human guidance and autonomous capabilities might be combined for various defined tasks, including search and rescue, submersible surface support and tracing water pollution sources.

References

[1] R. A. Jarvis and A. J. Lipton. GO-2 :- an autonomous mobile robot for a science museum. In *Proc. 4th International Conference on Control, Automation, Robotics and Vision, Westin Stamford, Singapore*, pages 260–266, 3-6 December, 1996.

[2] B. Hendry, R. A. Jarvis, and I. Bridger. An automated guided vehicle for industrial environments. In *Proc. ARA Conference on Robots for Australian Industries, Melbourne*, pages 121–131, 5-7 July, 1995.

[3] R. A. Jarvis. An autonomous heavy duty outdoor robotic tracked vehicle. In *Proc.*

International Conference on Intelligent Robots and Systems, Grenoble, France, pages 352–359, Sept. 8-12, 1997.

[4] C. Thorpe. Mixed traffic and automated highways. In *Proc. International Conference on Intelligent Robots and Systems, Grenoble, France*, pages 352–359, Sept. 8-12, 1997.

[5] R. A. Jarvis. A user adaptive semi-autonomous all-terrain robotic wheelchair. In *submitted to the 6th International Conference on Intelligent Autonomous Systems, Venice, Italy*, July 25-27, 2000.

An Underwater Vehicle Monitoring System and Its Sensors

S.K. Choi and O.T. Easterday

Autonomous Systems Laboratory
Department of Mechanical Engineering
University of Hawaii at Manoa
Honolulu, HI 96822 USA
schoi@hawaii.edu
easterd@eng.hawaii.edu

Abstract: This paper describes a virtual collaborative world simulator, DVECS (Distributed Virtual Environment Collaborative Simulator), for underwater robots and its underwater vehicle, SAUVIM (Semi-Autonomous Underwater Vehicle for Intervention Missions).   DVECS is used for testing unmanned underwater vehicles (UUVs) of both real and simulated worlds where interaction and cooperation of other real and simulated vehicles, obstacles, situations, conditions and disturbances in a hybrid, synthetic, virtual environment can be observed without physical intervention.   This virtual system can be used to determine: (1) the optimal performance and criteria for the cooperating vehicles and its relative application; (2) the determination of the advantages and disadvantages of collaborative application tasks between multiple UUVs; and (3) the optimal communication links between the cooperating vehicles and its remote control stations.   DVECS is used as a monitoring system.

1. Introduction

Even with the increased interest in the development of underwater robotic technology, the design, fabrication and analysis of autonomous underwater vehicles (AUVs) are still very complex and expensive.   The unpredictable and hazardous underwater environment is extremely unforgiving and remote.   With limitations in communication, an AUV must continuously operate in a fully autonomous or near autonomous modes.   These requirements immensely complicate the diagnosis and evaluation of an AUV s many subsystems.   In order to ensure reliability in these systems, it is imperative to obtain and maintain accurate software and hardware data. For these purposes, it is absolutely necessary to test and re-test these systems under severe or extreme conditions in a controlled laboratory environment before operational or sea-trial deployment.   In addition, many military, scientific and commercial tasks in open oceans often require multi-national participation, and it becomes a necessity to rehearse these operations before the actual operation; thus, establishing operational strategy and ensuring the success of the operation without releasing proprietary or secured materials.   This is where DVECS becomes the ultimate tool.

D. Rus and S. Singh (Eds.): Experimental Robotics VII, LNCIS 271, pp. 551–560, 2001.
© Springer-Verlag Berlin Heidelberg 2001

Several universities have conducted research in the graphic simulator arena. To mention a few, they are: (a) the Naval Postgraduate School and their NPS AUV Integrated Simulator for their NPS AUV [1]; (b) the University of Tokyo and their Multi-Vehicle Simulator for their Twin-Burger AUV [2]; and (c) the Autonomous Undersea Systems Institute and their Cooperative AUV Development Concept [3]. Both the NPS and UT systems were developed on the IRIX environment of the Silicon Graphics workstation, while the AUSI system runs on the Win32 environment on an Intel based system. All systems, however, are running the OpenGL graphic protocols, which is platform independent.

DVECS was developed with the sole objective of reducing the lead-time required for (1) the tedious aspects of pre-testing software and hardware before deployment and (2) the collaboration of various AUVs without having to consider the transportation of these vehicles. Much of DVECS is based and developed on the graphic test platform architecture for underwater vehicles by Yuh, Adivi and Choi [4], and the SGI GL based 3-dimensional graphics by Choi, Yuh and Takashige [5].

DVECS utilizes a similar software architecture of a combined hierarchical and heterarchical structure of the previous test platforms along with an OpenGL based simulation system and a variety of different wireless communications methods — radio frequency links, commercial cellular telephones, wireless Ethernet, wireless LAN and asynchronous transfer mode — for data transfer. Finally, DVECS incorporates a projection VR system that consists of a RGB high-resolution and high-refresh-rate projector and polarized eyewear with an emitter. Thus, this system creates an immersion effect.

2. DVECS

The design, testing and operation of AUVs and their control systems can benefit immensely from interactive, 3-dimensional (3D) computer simulations. In particular, developing and testing complex systems that involve multiple autonomous underwater robots operating in an uncontrolled environment is considerably safer and cost-effective in a controlled synthetic environment than a real environment, since the research vehicles are not placed at risk of loss or damage. Mission planning, monitoring and analysis can also benefit from an interactive, 3D virtual environment since its performance can be tested prior to actual sea-trials. For these reasons, the Distributed Virtual Environment Collaborative Simulator (DVECS) was developed to be used in hybrid synthetic simulations for testing real and virtual vehicles in a common environment and for mission collaboration, planning, monitoring and analysis of existing unmanned underwater vehicles (UUVs), as described in Figure 1.

DVECS architecture is designed to operate in a networked environment such that each component of the simulation can be run on a separate system, processor or virtual machine within a single computer; thus, distributing the computation load. This feature offers several advantages over a single system layout.

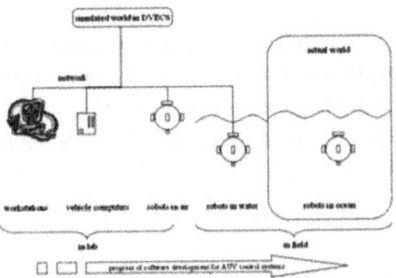

Figure 1: DVECS Development Environment

First, it is possible to modify or create new components with minimal restrictions on internal architecture. As long as each component adheres to specifically prescribed requirements for communication, language and operating system, the design and implementation of each component is irrelevant to the rest of the system.

Second, users can design and test their AUV simulations across the Internet using common servers. This allows computationally intensive 3D simulation of interaction between multiple objects in the virtual world to be simulated on one or more centralized high performance computers, while the processing requirements for the user s AUV simulations are no more than what their physical AUVs would normally require. Thus, this optimizes the computation time and allows users to accurately evaluate their vehicle s computation performance and requirements.

Finally, multiple simulated or physical entities can interact over a networked environment without requiring them to share code or knowledge of each other s capabilities so proprietary or secured algorithms can be tested in a common environment without making them public. This allows for collaboration from many different sectors of the underwater community that wish to evaluate their AUV in conjunction with pre-tested AUVs, as shown in Figure 2.

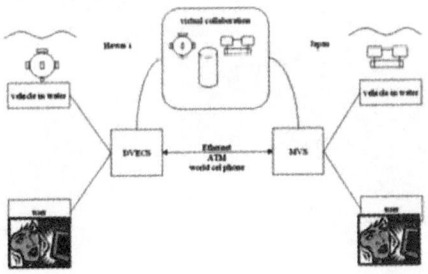

Figure 2: DVECS and MVS Collaboration

2.1 Vehicle Dynamics

DVECS, when in a monitoring mode, does not consider vehicle dynamics since transmitted data directly reflects the motions of the vehicle in the real-world environment. However, when in the simulation mode, the interacting system must consider the basic underwater vehicle dynamics. The following vector equation:

$$M\dot{V} + A(V)V + h = F$$

where $V \mathbin{\rfloor} R^6$ is the linear and angular velocities in the vehicle coordinates; $M \mathbin{\rfloor} R^{6x6}$ is the inertia matrix; $A \mathbin{\rfloor} R^{6x6}$ includes all the nonlinear dynamic terms with velocity terms; and $h \mathbin{\rfloor} R^6$ is a vector representing other forces and torques except $F \mathbin{\rfloor} R^6$, which represents the forces and torques generated by the thruster forces. [6]

2.2 Virtual User Interface

As mentioned, DVECS uses a multiple Silicon Graphics workstation setup that comprises of an Onyx, an Indy and an O^2, and interfaces with an Electrohome Virtual Reality Projection unit and Stereographics CrystalEyes eyewear and emitter system.

The DVECS software is a multi-layered C++ program modularized by its subsystems and utilizes the inheritance properties. It uses OpenGL graphics libraries to generate the background, vehicles and obstacles, and uses Open Inventor 3D toolkit protocols to create the 3-dimensional, virtual images.

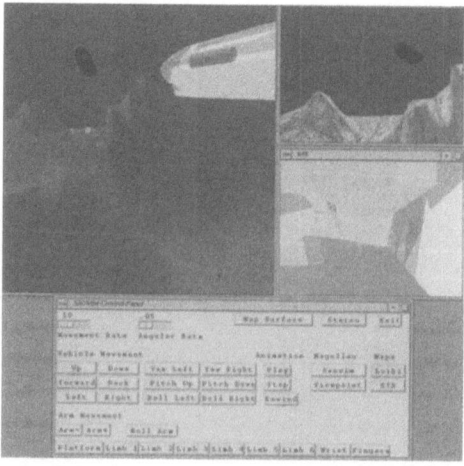

Figure 3: DVECS for AUV in an Undersea Environment Testing and Evaluation

Currently, DVECS consists of multiple windows that represent a 3-dimensional environment, a front camera view with video overlay capabilities, a manipulator camera view and a control panel. Figure 3 shows DVECS showing the Semi-Autonomous Vehicle for Intervention Missions (SAUVIM) at the Loihi Seamount test site off the Big Island of Hawaii.

3. SAUVIM

The Semi-Autonomous Undersea Vehicle for Intervention Missions (SAUVIM) is a class of AUV that extends the envelope of missions from the traditional wide area survey and fly-by missions to include close-in navigation and interactive tasks that are accomplished by means of its robotic manipulator and sensors. As seen in Figure 4, SAUVIM is a free-flooded deep-water AUV that has six primary pressure vessels, batteries, sensors and other ancillary systems mounted onto a 6061-aluminum frame.

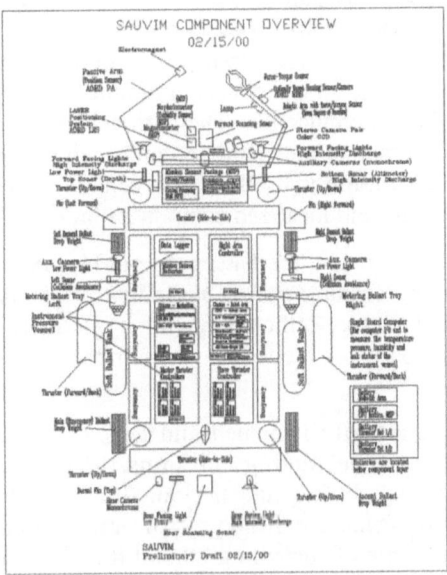

Figure 4: SAUVIM Components

The floatation foam is primary located on the topmost and lateral sides of the vehicle. Inside the frame on the upper of the two levels are aft lateral thruster, six primary pressure vessels, electrical junction boxes, forward lateral thruster, and primary sensor rack. On the lower level proceeding are battery banks, ballast carriage and robotic arm storage bay with the active and passive robotic arms slug underneath a retractable platform. Located outboard of the frame are the four vertical thrusters, three fins and decent ballast mounts. Out on each side of the vehicle are pylons, which are easily collapsible and contain a Tecnadyne Model-2010 brushless thruster that is capable of 130lbf output. The other six thrusters consist of two lateral and four vertical units that are a Tecnadyne Model-1020 model capable of 47lbf output. Figure 5 shows SAUVIM during its construction and assembly phases.

Figure 5: SAUVIM during Assembly

The vehicle has twin VME-bus computers housed in the middle set of pressure vessels. These are outfitted with Force Computing CPU boards based on the Motorola 68060 processor, navigation tasks will be dedicated primarily to the port-side bottle bus while arm control and coordination while be handled by the starboard-side VME CPU. The two computers will be linked together with a RS-232 serial line for time synching and backup communication along with a RJ-45 10-base-T twisted pair Ethernet connection. The various vehicle sensors (ranging sonar, scanning sonar, video cameras, etc.) and actuators (fins, thrusters, ballast release, etc.) are in a split configuration that allows the vehicle to self-recover from failure scenarios involving leakage and localized component failure.

To help insure fault tolerance, the batteries - six DeepSea Systems model SB48-18 and six SB24-38 - are feeding power buses that are isolated to each respective pressure vessel. These batteries, although having a relatively low energy density for a given volume were chosen for their ease of maintenance and potential turnaround time between missions.

The vehicle will be initially launched and tested in shallow water during the vehicle development and debugging missions. To facilitate easier systems maintenance and handling during this period, the floatation foam and pressure vessels employed on the vehicle will be pressure rated to only 600 feet of depth - upgrades to versions capable of sustaining 20,000 ft dives will be made; in so doing the dry mass of the SAUVIM will climb from 8,500 lbs to 13,500 lbs. The syntactic foam used on the full ocean depth vehicle will be standard glass micro-sphere and epoxy matrix foam, and the shallow water foam is rigid polyurethane foam. The SAUVIM will undergo shallow water testing with an abbreviated fairing. The front and aft areas will have fiberglass shields for flow development mounted to the frame. For the deep-water vehicle a much more complete carbon fiber/kevlar composite fairing shell will be fabricated for the vehicle to allow for drag minimization and some measure of collision damage absorption.

The main robotic manipulator is the Ansaldo/Maris Corporation Model 7080, which is an oil-filled 7-dof (degree-of-freedom), robotic manipulator that is capable of carrying 6kg payloads at full extension. The resolvers for commutation and positioning of the brushless motors within the arm are carried in the right-forward pressure vessel. The software and control architecture of the manipulator systems is being jointly developed with University of Genoa in Italy.

The vehicle and support systems are collapsible into one standard shipping container, facilitating deployment of the SAUVIM to worldwide locations. Command and control for the vehicle development missions will be done over a tether line that will feature serial and RJ-45 links to the Navigation computer CPU bottle. This link will swapped over to an acoustic modem towards that end of the vehicle systems testing in shallow water.

3.1 SAUVIM Navigation and Positioning Sensors

In order to successfully complete the far-field navigation to a worksite, close-in navigation around the site and manipulative tasks, SAUVIM carries an array of sensors. The first two classes of tasks are performed autonomously while the last class of task is performed in a terse command loop structure allowing for low-bandwidth command and control. SAUVIM will carry the following sensors for far-field navigation and collision avoidance: ranging sonar, scanning sonar, INS units, pressure sensors and a Doppler velocity log. For close-in navigation and intervention, task accomplishment will comprised of three custom-build Automatic Object Ranging and Dimensioning (AORD) sensors, which consist of a laser ranging array (LRA), passive arm (PA) and manipulator homing sensor (MHS). Consideration to carrying a stereo camera pair for ranging is also being considered, and other CCD cameras will be mounted for semi-autonomous event monitoring. Other supplemental sensors will be available to the SAUVIM Navigation and Arm Control computers via an RS-232 link to the oceanographic Mission Sensor Package (MSP), which will also be onboard.

The six ranging sonar, Perry Tritech PA200/20-S units, are bundled on RS-485 interfaces. Two bundles of three lead into each of the two VME computers. Mounted one to each of the six principle directions around the vehicle, these sonar, which feature a 200kHz pulse with a 20E conical beam-front and 100m effective range, are intended primarily for collision avoidance, and bottom and free-surface ranging. Based on pool tests, these units have a minimum range of 1.5m and repeatability within 5cm, and will be quite useful as altimeters and depth-meters for the vertically orientated units.

Two Imagenex 675kHz Model 881-000-105 radial scanning sonar will be used for bottom mapping and collision avoidance. One unit will be mounted on the forward end of the vehicle while the other will be mounted aft. The data return from these units is the sonic based analogue to a radar sweep. RS-485 based and highly programmable, these units will be used for collision avoidance as well as bottom characterization and mapping. The main challenge is to code the image and data garnishing software. These units feature and fan shaped beam of 1.7E horizontal by 30E vertical spread with a maximum range out to 100m with a 360E scan horizon capacity.

Within the forward pressure vessel is the primary vehicle primary inertial navigation sensor (INS), which is a Watson AHRS-BA303. Not a true INS, this unit is a RS-232-based gyro-unit that gives heading, pitch and roll angles and velocities. A secondary INS unit, which is the Precision Navigation Model TCM-2, will be housed within the Arm Control CPU pressure vessel. This unit features a 2-axis inclinometer and 3-axis fluxgate compass that will yield orientation much like the Watson unit.

Two pressure sensors will be aboard SAUVIM. These are both Data Instruments/Honeywell DS-2 units, which feature a 4-20mA current loop output. Each of these unit outputs will be feed to the arm control and navigation computers. These units have a range from 0-10,000psi absolute with a −1% FSO accuracy. These will be used for mid-water depth tracking and for bottom proximity tracking. Ballast release of the vehicle will be coordinated based on the output of these sensors primarily, and from the ranging sonar secondarily. To allow for dead reckoning via single-integration techniques, the addition of a Doppler Velocity Log (DVL) to the SAUVIM vehicle is being considered.

Most of the above sensors are intended for far-field navigation, vehicle positioning and collision avoidance although most will be used in the close-in navigation and intervention task phases as well. The next class of sensors is all primarily intended for use during the close-in navigation and intervention tasks. These are the three AORD sensors, which are being developed in-house.

The first is a laser ranging array (LRA) that in its second-generation configuration consisting of sixteen diode lasers in a grid array along with a CCD camera centered in the array. Data processing of the laser dot locations is done by a PC-104 based computer with an Imagenex frame grabber board. The effect of parallax migration of the laser dots across the array is exploited by this unit. The effective range in water was determined to be about 5-6 meters through fresh water. A four-laser unit was tested with about −5% relative positioning accuracy in a darkened room from 0.5 to about 5m. Testing and establishment of the accuracy of the underwater unit is now underway as well as making the software more robust to deal with turbid water conditions and unit calibration.

The next element of the AORD sensor array is the PA (Figure 6). This is a two segment robotic arm that is constructed from aluminum 6061 alloy. Filled with white #9 mineral oil, this unit has two three-axis gimbals at each of two end canisters and a single-axis hinge joint between the two arm segments. The larger, base canister is mounted to the retractable arm tray on the vehicle portside, opposite to the active arm. The tubing segments are approximately 0.5 and 0.75 meters in length each use neoprene bellows to allow joint flexibility and have wiring strung internally within the arm. The arm has seven JDK MicroDevices open-wiper potentiometer assemblies, which have a 10kOhm range. These are powered by a twin set of rails at 5VDC and all of the wiper voltages are feed back on individual lines and are routed to the arm control bottle VME computer Matrix A/D board.

Figure 6: Passive Arm Set-up

Typically, the Ansaldo arm will deploy the lower canister of the PA from a storage cradle position on the arm tray to the seabed bottom or the task area. Initial task sites will be highly structured and ferrous in nature. This will allow the electromagnet within the PA arm to attach to the task site. Deployment of the PA will allow the reference frame of the seabed and task site become known with respect to the SAUVIM coordinate system and will allow for active arm correct for vehicle drift due to currents and or second law action-reaction motions that the SAUVIM experiences.

The last element of the AORD suite is the MHS unit, which consists of a CCD camera and a SuperCircuits Model PC74WR with a 512x512 pixel resolution, mounted above the sixth joint of the Ansaldo arm. This unit is linked to a PC-104 (Real Time Devices CMH486DX100HR) computer also equipped with an Imagenex frame grabber board. The unit is programmed to identify circular bar codes and range to them to allow for final orientation and guidance of the manipulator. This unit gives the range and off-axis orientation to the bar code as well as an identifier number for the code. Effective range based on preliminary dry testing is from about 0.6m in to about 25cm. Relative accuracy is about –1-2cm in the near limit range.

The remaining sensors are supplementary sensors for the SAUVIM vehicle. These consist of video cameras and the MSP sensors. Video cameras are of both monochrome and color varieties and will be mounted onto SAUVIM. These consist of SuperCircuits PC75WR and PC74WR CCD camera units, respectively. Five monochrome cameras will be mounted, one to each side of the vehicle as well as the rear and two in the nose for general video feedback for any supervisor monitoring the vehicle progress topside. A color stereo pair of cameras will face into the arm workspace. The will be used for stereo ranging and visualization data gathering. These will be slaved to a PC-104 based architecture with one to two frame grabber boards.   The MSP sensors include a nephlometer, CTD sensor, 3-axis field magnetometer, and a pH and dissolved oxygen sensor. The outputs of these sensors can be polled from the MSP CPU over a RS-232 link if so needed during a SAUVIM mission and can be made available for navigation tasks.

4. Conclusion

This paper presents a brief description of the Distributed Virtual Environment Collaborative Simulator (DVECS) and the Semi-Autonomous Underwater Vehicle for Intervention Missions (SAUVIM) developed at the Autonomous Systems Laboratory of the University of Hawaii.   Various tests have shown that the combined system can greatly help reduce the development time of underwater vehicle hardware/software testing and verification. Further tests are scheduled to refine the monitoring system and vehicle to allow higher degree of robustness.

5. Acknowledgement

This project is partially supported by the National Science Foundation Grant Number INT-9603043 and partially supported by the Office of Naval Research Grant Numbers N00014-97-1-0961 and N00014-00-0629.

6. References

[1]   D.P. Brutzman, Y. Kanayama & M.J. Zyda, Integrated Simulation for Rapid Development of Autonomous Underwater Vehicles, Proc. of the IEEE Oceanic Engineering Society AUV92 Conference, Jun. 1992.

[2]   Y. Kuroda, K. Aramaki, T. Fujii & T. Ura, A Hybrid Environment for the Development of Underwater Mechatronic Systems, Proc. of the 1995 IEEE 21st International Conference on Industrial Electronics, Control and Instrumentation, Nov. 1995.

[3]   S.G. Chappell, R.J. Komerska, L. Peng & Y. Lu, Cooperative AUV Development Concept (CADCON) — An Environment for High-Level Multiple AUV Simulation, Proc. of the 11th International Symposium on Unmanned Untethered Submersible Technology, Aug. 1999.

[4]   J. Yuh, V. Adivi & S.K. Choi, Development of a 3D Graphic Test Platform for Underwater Robotic Vehicles, Proc. of the 2nd International Offshore and Polar Engineering Conference, Jun. 1992.

[5]   S.K. Choi, J. Yuh & G.Y. Takashige, Omni-Directional Intelligent Navigator, Underwater Robotic Vehicle: Design and Control, TSI Press, NM, 1995.

[6]   S.K. Choi & J. Yuh, Experimental Study of a Learning Control System with Bound Estimation for Underwater Robots, J. of Autonomous Robots, Mar. 1996.

Real Time Obstacle Detection for AGV Navigation Using Multi-baseline Stereo

Han Wang
School of Electrical & Electronic Engineering,
Nanyang Technological University, Singapore 639798
e-mail: hw@ntu.edu.sg

Jian Xu   Javier I Guzman
Gintic Institute of Manufacturing Technology, Singapore 638075

Ray A Jarvis
Department of Electrical & Computer Systems Engineering
Monash University, Melbourne, VIC 3168, Australia

Terence Goh   Chun Wah Chan
Defence Science Technology Agency, 1 Depot Road, Singapore 109679

Abstract: This paper reports the real time vision algorithm ELS, designed for obstacle detection and ground identification for AGV navigation using multi-baseline stereo. Our major contributions are (1) the Disparity Gradient Filter for noise reduction, (2) ground orientation detection using extended Hough Transform, and (3) the establishment of error model for obstacle detection. Experiments show that the algorithm works effectively in unstructured terrain.

1. Introduction

Vision based navigation for mobile robots has been implemented successfully in many places using different approaches [9, 1, 11]. Some are based on corners using structure from motion algorithms such as Droid by Harris [2], some are based on lines by Zhang [13], and some are based on neural network [4]. The CMU Navlab project has been a long success [3, 12] in unmanned guided vehicle navigation. Most of these systems required dedicated hardware such as the Systolic Array Processor (Warp) and DSP chips which are very costly.

Comparing with other vision approaches, multi-baseline stereo [5] has the advantage of offering dense range image at very high accuracy. For example, the Virtuoso by Webb *et al.* [6] and the Digiclops by PointGrey. They are commercially available using only a Pentium processor. Digiclops has been used in indoor mobile robot navigation using clustering techniques [7].

Our project is primarily designed for outdoor AGV navigation and the rich texture environment provides excellent features for the stereo matching algorithm. ELS takes disparity map as its input which is computed using the

D. Rus and S. Singh (Eds.): Experimental Robotics VII, LNCIS 271, pp. 561–568, 2001.

Sum of Squared-Difference (SSD) algorithm derived by Okutomi and Kanade [8]. Three cameras are used to form two baselines. Stereo images are taken simultaneously and no attempt is made for temporal integration. This gives us the advantage of detecting moving obstacles such as tree leaves and human figures.

The paper is organised following the sequence of the ELS algorithm. Firstly, the noise reduction method of disparity gradient limit is described. After the interpolation, the extended Hough Transform is introduced to find the ground orientation (slope). Finally, we give our conclusions, discuss existing problems and propose future works.

2. Filtering Using Disparity Gradient Limit

It is observed that large amount of noise exist and they are clustered due to the minimisation technique of SSD. Many low pass filters have been used such as the 1-D and 2-D median filter with little success. Further action is needed to clear them.

Assuming the world is continuous, a physical point must have its immediate neighbours or adjacent points. In another words, for any two neighbouring points, their difference in disparity must be sufficiently small. The disparity gradient is defined as

$$ DG = \sum_{i \in w} \frac{|D - D_i|}{d_i}, \qquad (1) $$

where w is a local window, D is the disparity of the pixel in concern, D_i is the neighbour's disparity and d_i is the *Euclidean* distance. A point is rejected if its disparity gradient is above a certain threshold (or DGL). DGL was first found in PMF [10] by Pollard *et al.* for disparity computing in a relaxation manner. We extend this idea in noise reduction. Figure 1 and 2 show the range points before and after the Disparity Gradient Limit filtering.

However, this method does not work well for occlusion because DGL is based on continuity of matter. Occlusion error has been largely eliminated by removing regions that have very high contrast.

3. Interpolation

SSD relies on texture. In some area, texture is not available (see Figure 3). Interpolation has to be carried out to fill up the gaps. We divide the region into 20x20cm grids. Inside each grid, the mean is used for interpolation. A blank grid is filled under the condition that it has more than one adjacent valid neighbours and the neighbouring grids are in opposite position. This is to prevent the algorithm from extrapolation. The result is shown in Figure 4.

4. Ground Orientation Detection

Extended Hough Transform is used to detect the ground orientation. The ground orientation is effectively a unit normal vector which is only two dimensional. For each 20x20 cm^2 grid, four facets produces four normals. Each normal is accounted for in the Hough Transform.

Figure 1. Range image of an office scene. The view point is set at the floor. Noise are visible under the floor (horizontal cluster in the middle of the image).

Figure 2. Results after the DGL filter

Surface normal is a unit vector of two degrees of freedom. Let

$$\mathbf{n} = (n_1, n_2, n_3)^T$$

represent the surface normal computed from one facet. It can be rearranged by dividing the vector with n_3, thus

$$\mathbf{n} = n_3(\frac{n_1}{n_3}, \frac{n_2}{n_3}, 1)^T$$

Setting $\frac{n_1}{n_3} = \alpha$ and $\frac{n_2}{n_3} = \beta$, the parametric space of (α, β) is defined and the extended Hough Transform can be applied.

Figure 5 shows the detected floor orientation with a drawing pin.

5. Error Model and Obstacle Detection

Obstacles are defined as object of 0.4m in height including negative obstacles such as pits and trenches. We define the error model for obstacle detection.

Figure 3. 3D reconstruction of ground floor with texture mapping. Spurious points are filtered out by the disparity gradient filter. In the region where low texture are detected, range information is not available (shown as black holes). The artificial lines are created by the graphics rendering.

Figure 4. The ground is divided into 20x20 cm^2 grids. The cylinder shows the maximum and minimum height of the grid with mean marked as a box. The blank regions are filled using linear interpolation techniques.

Given the image resolution w, baseline b and focal length f, for parallel

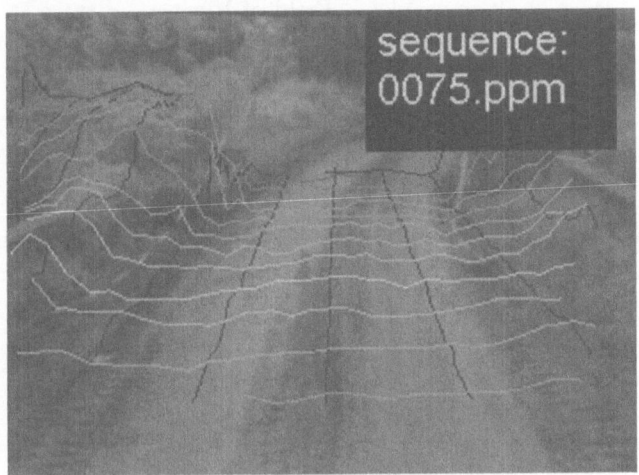

Figure 5. A 3D mesh has been placed on the original grey level image. The mesh shows 1x1 m^2 squares. The drawing pin shows the orientation of the ground.

stereo, the error in depth at distance Z is defined as,

$$e_z = \frac{Z^2}{bf - Z} \qquad (2)$$

where f is related with w by the horizontal field of view v (angle)

$$f = \frac{w}{2 \tan \frac{v}{2}}.$$

Assuming square pixels, the error in X (pointing right) and Y direction (pointing down) is given in

$$e_x = \frac{XZ}{fb - Z}, \qquad e_y = \frac{YZ}{fb - Z}. \qquad (3)$$

Let $\mathbf{e} = (e_x, e_y, e_z)^T$, the probability of an observed 3D point being in error is in Gaussian distribution,

$$P(\mathbf{e}) = e^{-\frac{1}{2}\mathbf{e}^T C^{-1}\mathbf{e}} \qquad (4)$$

where C is the observation covariance. $\mathbf{e}$ defines a volume encapsulates an observed 3D point. It is clear that the confidence of measurements decreases linearly as the distance in depth increases.

Table 1 shows the typical error of 0.4m object when the baseline $b = 10cm$ and $w = 320$ with $1/2$ sub-pixel accuracy.

	Field of View		
Range	30°	60°	90°
5	.0175	.0397	.0741
10	.0366	.0881	.1818
15	.0575	.1484	.3529
20	.0805	.2258	.6667

Table 1. Error of 0.4m object at different distance.(unit=metres)

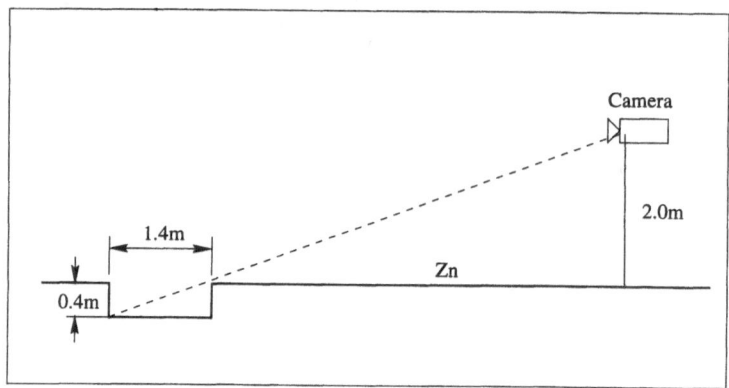

Figure 6. Negative obstacles can be identified at distance Z_n

5.1. Negative Obstacles

We need to identify negative obstacles as well. Figure 6 shows the relation of the distance Z_n from the camera to the pits of 1.4m wide. The camera is mounted at 2.0m above the ground and the identifiable size is 0.4m (obstacle size). Therefore, the distance can be found as

$$Z_n = \frac{2.0 \times 1.4}{0.4} = 7.0m$$

If the camera is mounted at higher location, the distance of detection is longer and it is more desirable, because more time is given to the navigation module. However, higher camera location will result in unstable imaging due to the rough terrain and longer distance means less confidence in data accuracy.

6. Performances

The system comprises of two Pentium-II 550MHz computers, one is dedicated to SSD and the other is for running ELS. It runs at 4Hz for 320x240x8 images and faster speed is expected with a better processor. It is able to detect obstacle of 0.4m at a distance close to 20 metres.

Figure 7. Obstacles are shown in red cross with 20cm grids and flat area is shown in 1m grids.

7. Conclusions and Future Work

DGL is effective in noise reduction and it outperforms traditional low-pass filter in this application. The error model of obstacle enables us to detect obstacle at very early stage. The use of extended Hough Transform makes reasonable estimation of the ground orientation. Further work is in progress to split the road into smaller region for early detection of slopes.

This is an on-going project. A speed of 10Hz is desirable for the robot to travel at 20Km/h. Some optimisation are required to further speedup the process. In addition, two sets of stereo cameras are in consideration in order to cover 90 degrees of horizontal field of view. Presently, the camera is fixed on the vehicle, when the vehicle reached the hill crest, the camera needs larger tilt-down angle. We are actively in search of techniques to solve this problem.

8. Acknowledgements

H Wang would like to thank Department of Electrical and Computer System Engineering, Monash University for providing him with the opportunity and research facilities during his study leave. Discussions with David Suter and Lindsay Kleeman have been very beneficial.

References

[1] S. Bohrer, T. Zielke, and V. Freiburg. An integrated obstacle detection framework for intelligent cruise control on motorway. In *Proceedings of Symposium of Intelligent Vehicles*, 1995.

[2] C. Harris. Determination of ego-motion from matched points. In *Proc. 3rd Alvey Vision Conference*, pages 233–236, 1987.

[3] M. Hebert and T. Kanade. 3-D vision for outdoor navigation. In *Image Under-*

standing Workshop, pages 593–601, Cambridge, MA, April 6-8, 1988. Morgan Kaufmann.

[4] T M Jochem, D A Pomerleau, and C E Thorpe. Vision guided lane transition. In *IEEE Symposium on Intelligent Vehicles*, Detroit, September 1995.

[5] T. kanade and T. Nakahara. Experimental results of multibaseline stereo. In *IEEE Special Workshop Passive Ranging*, Princeton, NJ, Oct. 1991.

[6] S B Kang, J A Webb, C L Zitnick, and T Kanade. A multibaseline stereo system with active illumination and real-time image acquisition. In *International Conference on Computer Vision*, Cambridge, MA, 1995.

[7] D Murray and J Little. Using real-time stereo vision for mobile robot navigation. *Autonomous Robots*, pages 161–171, 2000.

[8] M. Okutomi and T. Kanade. A multi-baseline stereo. *IEEE Transactions on Pattern Analysis and Machine Intelligence*, 15(4):353–363, 1993.

[9] L. Robert, M. Buffa, and M. Hebert. Weakly-calibrated stereo perception for rover navigation. In *Proceedings of International Conference on Computer Vision (ICCV95)*, 1995.

[10] Pollard SB, Mayhew JEW, and Frisby JP. PMF: A stereo correspondence algorithm using a disparity gradient limit. *Perception*, 14:449–470, 1985.

[11] S. Singh and B. Digney. Autonomous cross-country navigation using stereo vision. Technical Report CMU-RI-TR-99-03, Carnegie Mellon University, 1999.

[12] C. Thorpe, S. Shafer, and T. Kanade. Vision and navigation for the carnegie mellon navlab. In *Image Understanding Workshop*, pages 143–152, Los Angeles, CA, Feb. 23-25, 1987. Morgan Kaufmann.

[13] Z. Zhang. Estimating motion and structure from correspondence of line segments between two perspective images. *IEEE Transactions on Pattern Analysis and Machine Intelligence*, 17(2):1129–1139, 1995.

A new Generation of Light-weight Robot Arms and Multifingerd Hands

G. Hirzinger, J. Butterfaß, M. Fischer, M. Grebenstein, M. Hähnle,
H. Liu, I. Schaefer, N. Sporer, M. Schedl, R. Koeppe

DLR, German Aerospace Center
Institute of Robotics and Mechatronics
82234 Wessling, Germany
Gerd.Hirzinger@dlr.de

Abstract

The keynote lecture describes recent design and development efforts in DLR's robotics lab towards a new generation of ultra-light weight robots with articulated hands (Fig. 1). The design of fully sensorized joints with complete state feedback and the underlying mechanisms are outlined. The second joint torque-controlled light-weight arm generation is available now [1], as well as the second generation of a highly integrated 4 finger-hand with 13 actuators and more than 100 sensors [2]. Thus we hope that important steps towards a new generation of service and personal robots have been achieved, with space robotics becoming a major driver due to the need for advanced "robonaut" technologies.

References

[1] J. Butterfaß, M. Grebenstein, H. Liu, G. Hirzinger. DLR-Hand II: Next Generation of a Dextrous Robot Hand. IEEE International Conference on Robotics and Automation, Seoul, Korea, 2001.

[2] G. Hirzinger, A. Albu-Schäffer, M. Hähnle, I. Schaefer, and N. Sporer. A New Generation of Torque Controlled Light-weight Robots. IEEE International Conference on Robotics and Automation, Seoul, Korea, 2001.

D. Rus and S. Singh (Eds.): Experimental Robotics VII, LNCIS 271, pp. 569–570, 2001.
© Springer-Verlag Berlin Heidelberg 2001

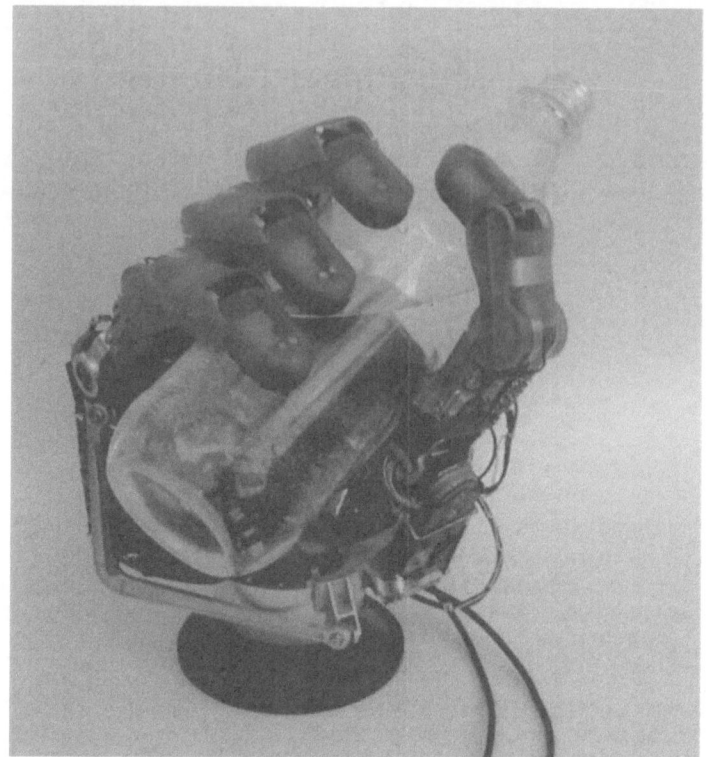

Figure 1: DLR Hand II and DLR Light-weight Robot II.

Lecture Notes in Control and Information Sciences

Edited by N. Thoma and M. Morari
1998–2001 Published Titles: